T. Matsumoto · M. Komuro
H. Kokubu · R. Tokunaga

Bifurcations

Sights, Sounds, and Mathematics

With 138 Illustrations
and 31 Color Plates in 78 Parts

Springer-Verlag
Tokyo Berlin Heidelberg
New York London Paris
Hong Kong Barcelona
Budapest

Takashi Matsumoto
Professor, Department of Electrical Engineering,
Waseda University, Ohkubo, 3-4-1, Shinjuku-ku,
Tokyo 169, Japan

Motomasa Komuro
Associate Professor, Department of Mathematics,
The Nishi-Tokyo University, Uenohara-machi, Kitatsurugun,
Yamanashi-Prefecture 409-01, Japan

Hiroshi Kokubu
Lecturer, Department of Mathematics, Kyoto University,
Kyoto 606-01, Japan

Ryuji Tokunaga
Lecturer, Institute of Information Science and Electronics,
University of Tsukuba, Tsukuba Science City 305, Japan

On the front cover: Two-parameter bifurcation diagram for the Double Scroll Circuit (see Section 1.5). © M. Komuro, R. Tokunaga, T. Matsumoto, L.O. Chua and A. Hotta, "A Global Bifurcation Analysis of the Double Scroll Circuit", Int. J. of Bifurcation and Chaos in Applied Sciences and Engineering, 1,1 (1991) 139-182.

ISBN-13:978-4-431-68245-5 e-ISBN-13:978-4-431-68243-1
DOI: 10.1007/978-4-431-68243-1

Printed on acid-free paper

Preface

Bifurcation originally meant "splitting into two parts." Namely, a system undergoes a bifurcation when there is a qualitative change in the behavior of the system. Bifurcation in the context of dynamical systems, where the time evolution of systems are involved, has been the subject of research for many scientists and engineers for the past hundred years simply because bifurcations are interesting.

A very good way of understanding bifurcations would be to see them first and study theories second. Another way would be to first comprehend the basic concepts and theories and then see what they look like. In any event, it is best to both observe experiments and understand the theories of bifurcations. This book attempts to provide a general audience with both avenues toward understanding bifurcations. Specifically,

(1) A variety of concrete experimental results obtained from electronic circuits are given in Chapter 1. All the circuits are very simple, which is crucial in any experiment. The circuits, however, should not be too simple, otherwise nothing interesting can happen. Albert Einstein once said "as simple as possible, but no more". One of the major reasons for the circuits discussed being simple is due to their piecewise-linear characteristics. Namely, the voltage-current relationships are composed of several line segments which are easy to build. Piecewise-linearity also simplifies rigorous analysis in a drastic manner.

(2) The piecewise-linearity of the circuits has far reaching consequences. It not only simplifies the analysis of bifurcations but also motivates one to develop a general theory of the bifurcations of piecewise-linear vector fields. Chapter 2 elaborates on this subject in an extensive manner, and describes the development of normal form theory. As a byproduct, an exact representation formula is derived for an arbitrary piecewise-linear mapping.

(3) Expositions of piecewise-linear bifurcations are certainly not enough because the deep results in bifurcation theory accumulated for the past hundred years are for smooth vector fields. Chapter 3 provides fundamental concepts and results in bifurcations as well as in dynamical systems for smooth vector

fields. The chapter also contains several of the most recent results on bifurcation theory.

For the past decade the number of publications on this subject matter has been literally exploding. This area itself appears to be undergoing a bifurcation! A major factor responsible for this is, apparently, growing interests in chaos. This book treats chaos within the framework of bifurcation phenomena, which we believe, is the way it should be treated.

If the reader has very little background in bifurcations, and wants to see what bifurcations/chaos look like, we would strongly recommend the reader to build one of the circuits discussed in Chapter 1 and see the trajectories on an oscilloscope. Only very few parts are necessary to build the circuits and the experiments are simple. In addition, all the circuits described in this book except for one, behave within audible frequencies so that one can listen to the sounds of bifurcations/chaos. It is a lot of fun. Then the reader can read Chapters 2 and 3 to acquire the theoretical background. We would also recommend the reader with good theoretical background build a circuit if he or she wants to see the reality of bifurcations/chaos. If the reader is an experimenter who finds difficulty in performing rigorous analysis of the system under study, we recommend a piecewise-linear model for the system. The general theory developed in Chapter 2 will then be greatly helpful. Some of the peicewise-liner analyses in Chapter 1 will also be helpful. Chapters 1 and 2 will of course be useful for the reader who wants to study bifurcations/chaos in general piecewise-linear systems. Even though this book is not written as a textbook on bifurcations, material given in Chapter 3 can be used as a handy introduction to bifurcations. Basic concepts and results are rigorously stated. An emphasis is put on ideas rather than on detailed technical proofs although some of the fundamental theorems accompany complete technical proofs. Many examples instead of exercises are given to facilitate the novice reader understand the subject.

When Springer-Verlag, Tokyo invited us to write this book, our first reaction was negative, simply because we felt that writing a book of this nature would demand too much work. Now we are happy that we changed our minds and completed the job. Our effort was worth making.

We would like to acknowledge all of the help we have received in shaping this book: S. Tanaka, K. Kobayashi, K. Tokumasu, K. Ayaki, T. Makise, K. Kuroda, A. Hotta, J. Noguchi, R. Fujimoto, H. Tomonaga, Y. Abe, S. Takahashi, K. Iori, A. Noro, all from Waseda University, N. Aoki from Tokyo Metropolitan University, Y. Takahashi from University of Tokyo, Y. Togawa from Science University of Tokyo. Thanks are also due to J. K. Hale from Georgia Institute of Technology, B. Deng from University of Nebraska, Lincoln, M. Kisaka from University of Osaka Prefecture, H. Ninomiya from Tokyo Institute of Technology and H. Oka from Ryukoku University for carefully reading the manuscript and giving us constructive criticisms. Our illustrators K. Noguchi and M. Tomoda took great pains to meet our demands and produced beautiful figures. Y. Kurimoto helped us prepare several figures. T. Shibata helped us during the final phase of this book in our struggle with LaTeX commands. Y. Hirahara, Y. Nitta and A. Deus of Springer-Verlag,

Tokyo spent a large amount of time and energy on this book. Last but certainly not least, we owe a great deal to M. Billings, who edited our English. Without her patience, this book could not have been born.

Takashi Matsumoto
Motomasa Komuro
Hiroshi Kokubu
Ryuji Tokunaga

Winter 1992

Table of Contents

Color Plates

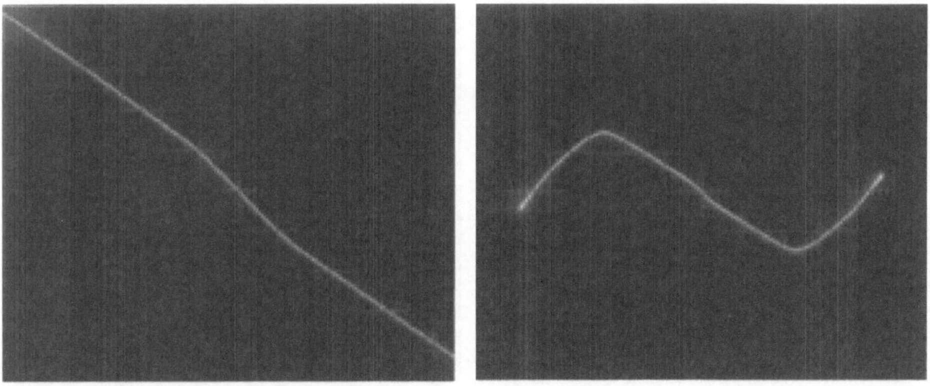

(a) (b)

Plate 1.1 (a) Measured nonlinear resistor characteristic of N. Horizontal scale :1V/division. Vertical scale: 1mA/division. (b) Characteristic of subcircuit N in Fig.2.1.2(a) for a large range. ©1986 Elsevier Science Publishers.

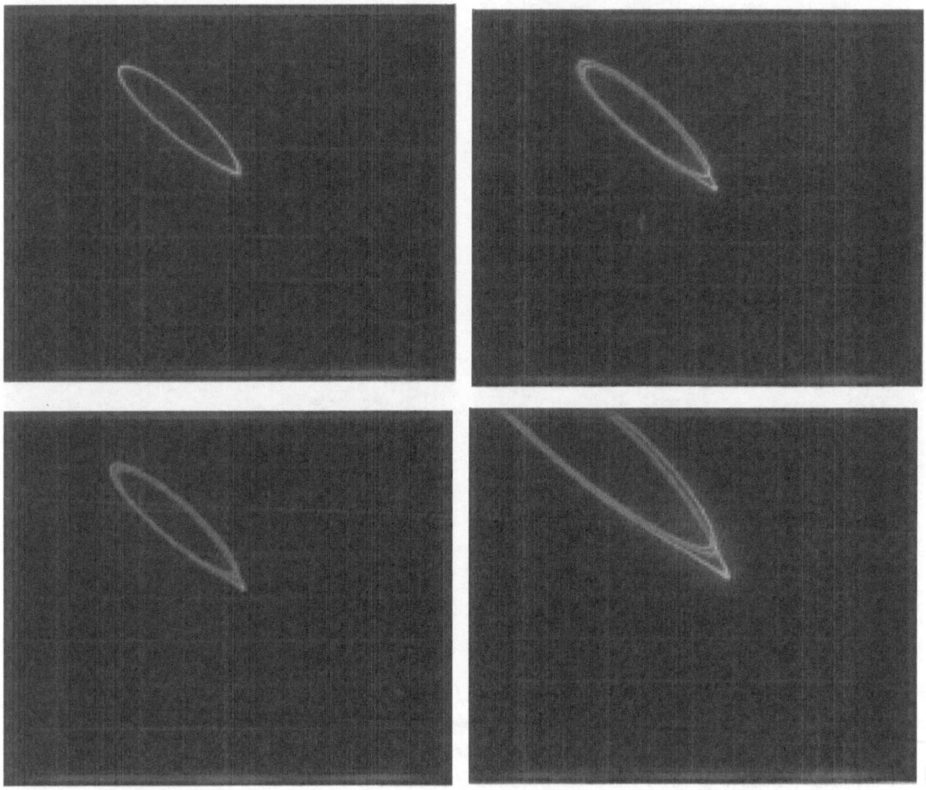

(a) (b)

(c) (d)

Plate 1.2 Attractors observed with the circuit of Fig. 2.1.2 projected onto the. (i_L, v_{C_1})-plane. Horizontal scale (except (e)): 2mA/ division. Vertical scale (except (e)): 2V/division. In (a)-(h) only one of two attractors is shown. (a) Periodic attractor shortly after Hopf bifurcation. (b) Period-2 attractor. (c) Period-4 attractor. (d) Enlarged version of (c). ©1986 Elsevier Science Publishers.

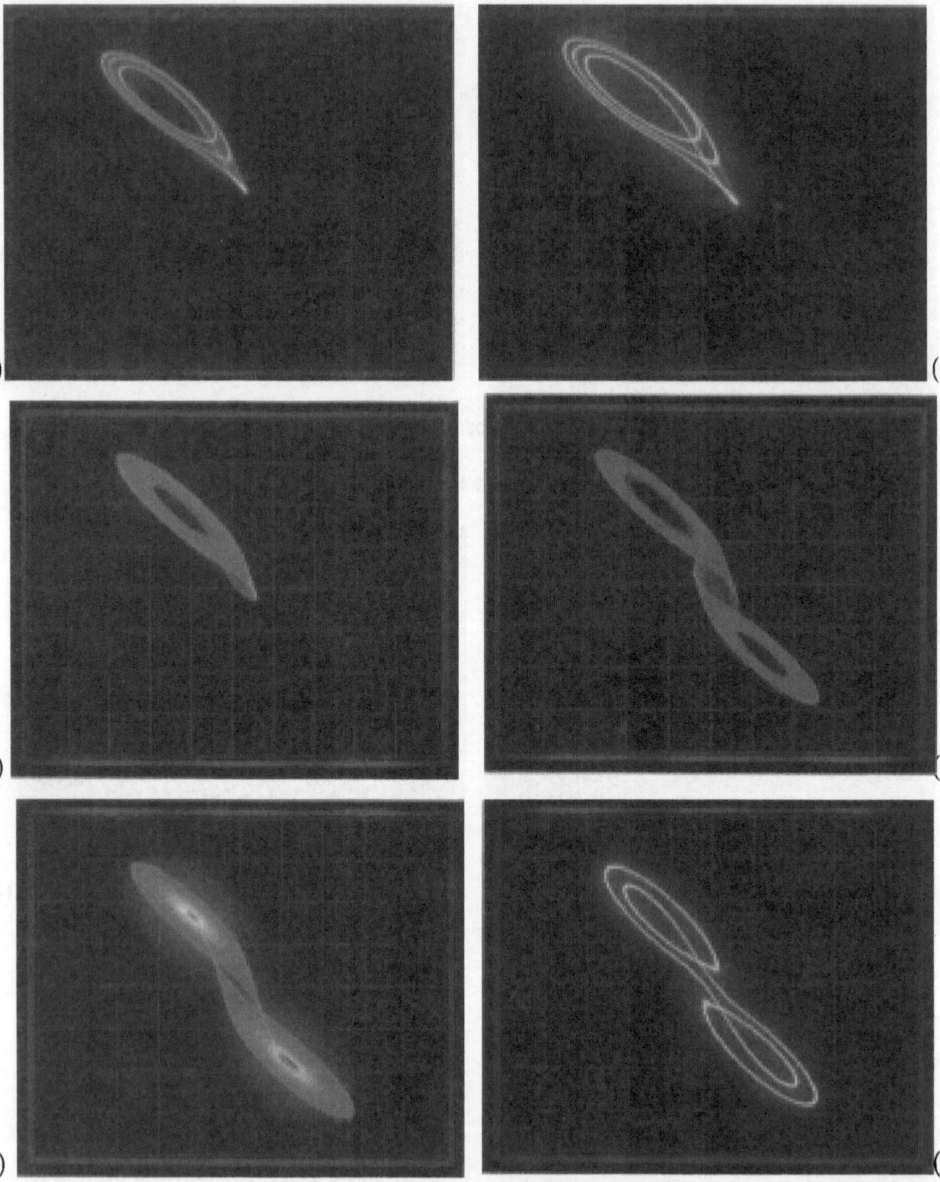

Plate 1.2 continued Horizontal scale: 1mA/division. Vertical scale:1V/division. (e) Rössler's spiral-type attractor. (f) Periodic window. (g) Trajectory near Rössler's screw-type attractor. (h) "Double- Scroll" attractor. (i) Trajectory near heteroclinicity. (j) Periodic Window ©1986 Elsevier Science Publishers.

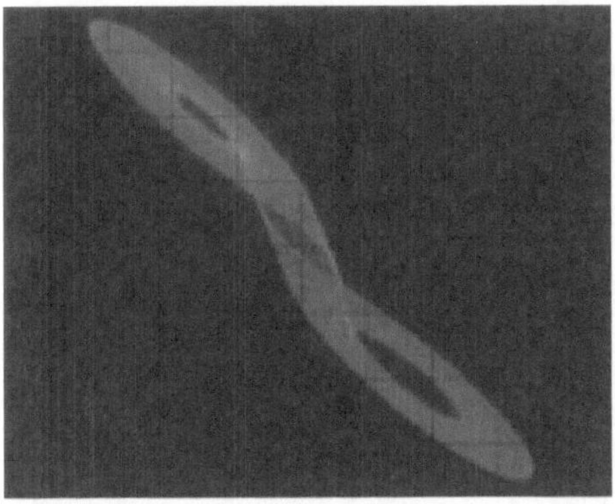
(a)

(b)

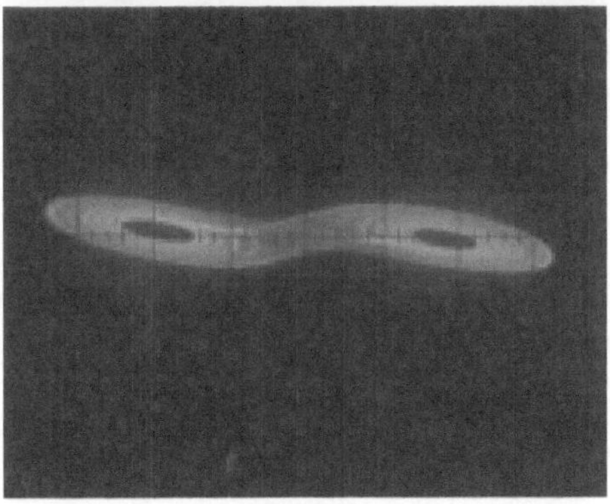
(c)

Plate 1.3 The Double Scroll. (a) Projection onto the (i_L, v_{C_1})-plane. (b) Projection onto the (i_L, v_{C_2})-plane. (c) Projection onto the (v_{C_1}, v_{C_2})-plane. voltage: $2V$/div. Current: $2mA$/div. ©1987 IEEE.

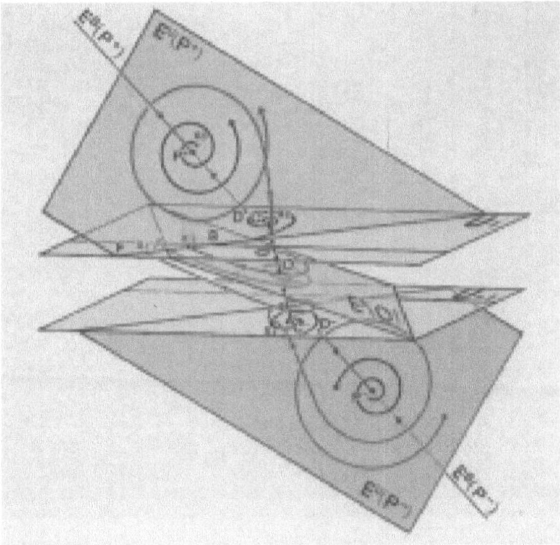

Plate 1.4 Typical trajectories of (DS 2). ©1987 IEEE.

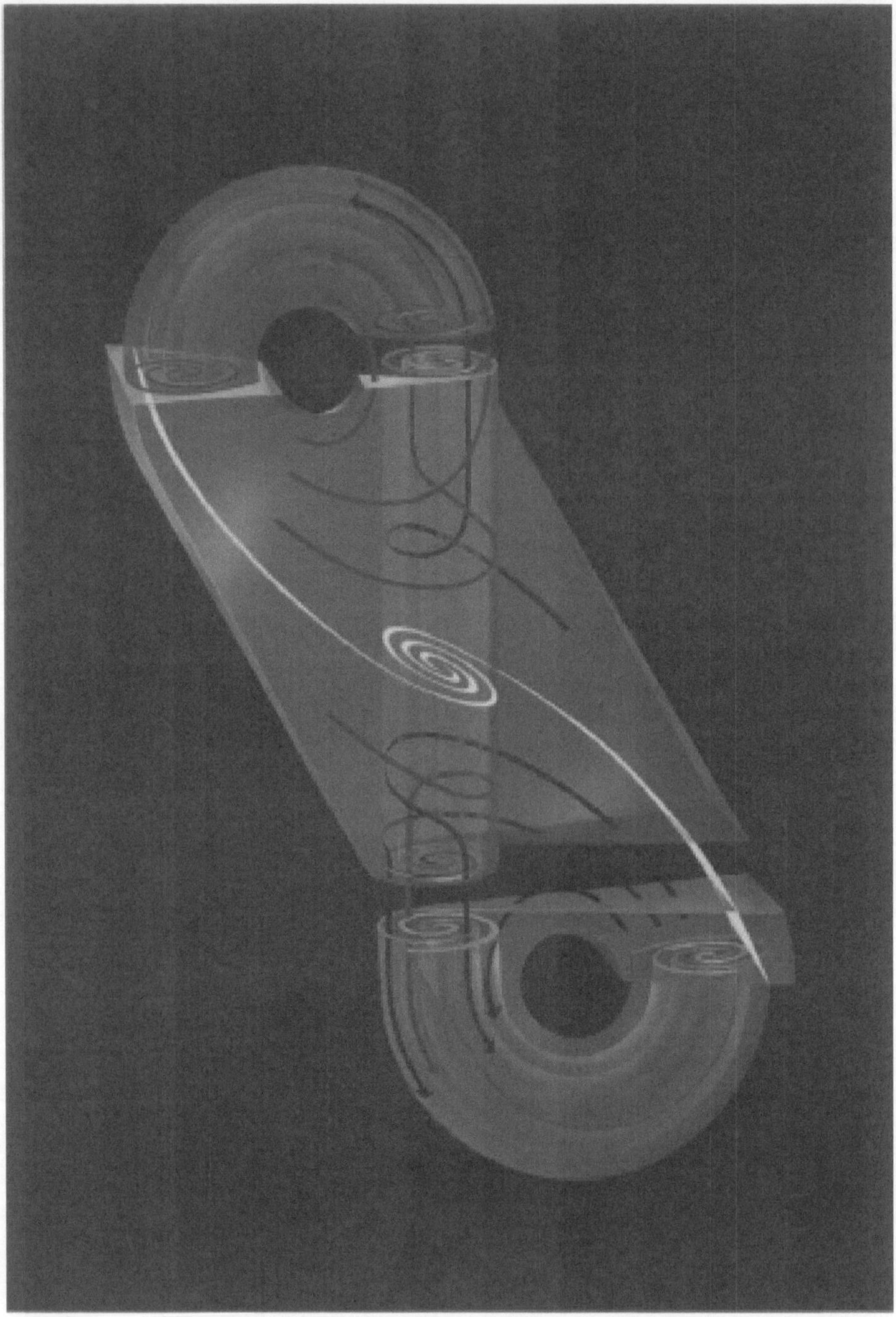

Plate 1.5 An abstract geometric model of the key features of the Double Scroll attractor. ©1987 IEEE.

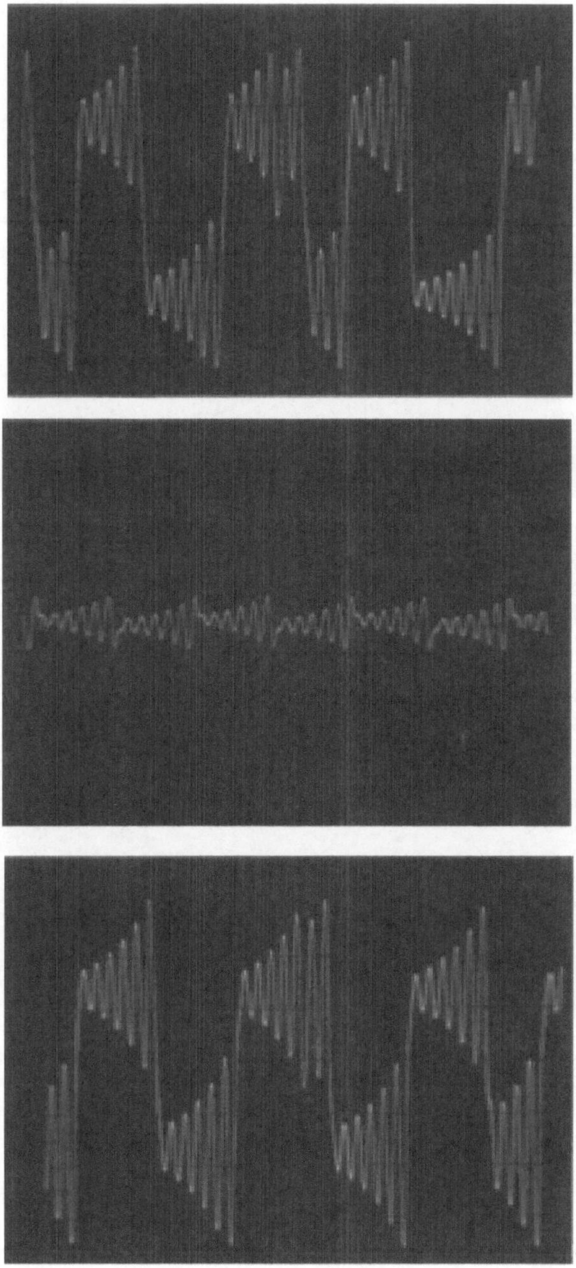

(a)

(b)

(c)

Plate 1.6 Time waveform corresponding to Plate1.2(h). (a) v_{C_1}. Horizontal scale: 1 ms/division. Vertical scale: 2 V/division. (b) v_{C_2}. Horizontal scale: 1 ms/division. Vertical scale: 2 V/division. (c) i_L. Horizontal scale: 1 ms/division. Vertical scale: 2 V/division. ©1986 Elsevier Science Publishers.

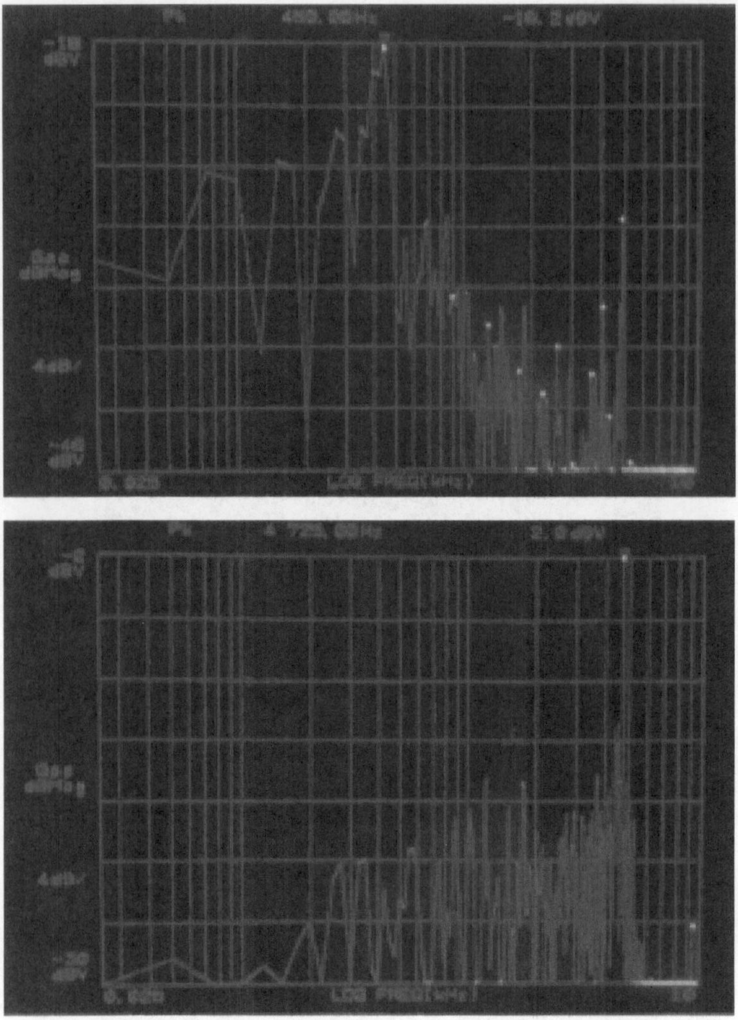

(a) (b)

Plate 1.7 Measured power spectra. (a) v_{C_1}. (b)v_{C_2}. ©1985 IEEE.

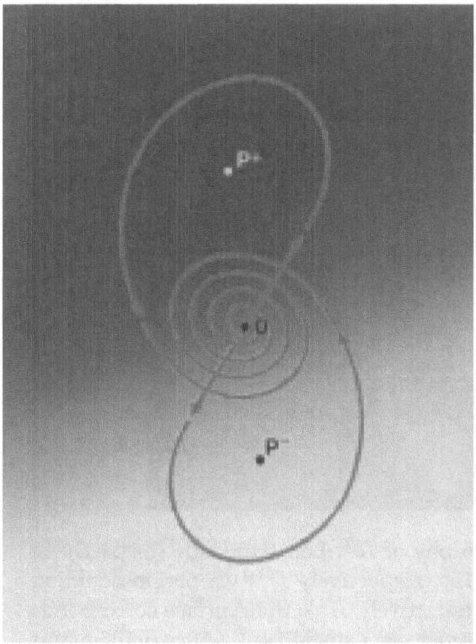

Plate 1.8 Homoclinic orbit of interest. ©1988 IEEE.

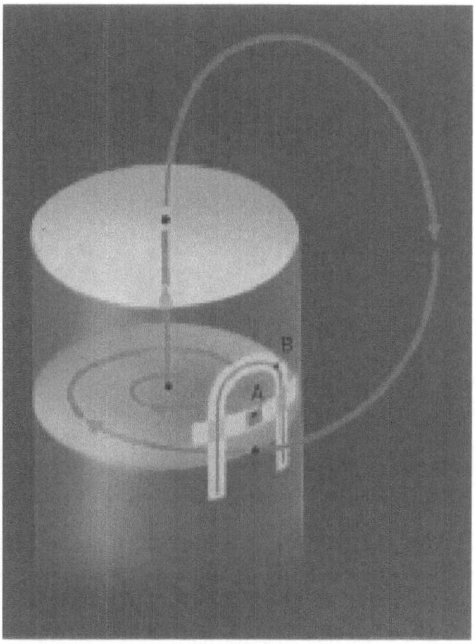

Plate 1.9 An illustration of how a horseshoe is formed. ©1988 IEEE.

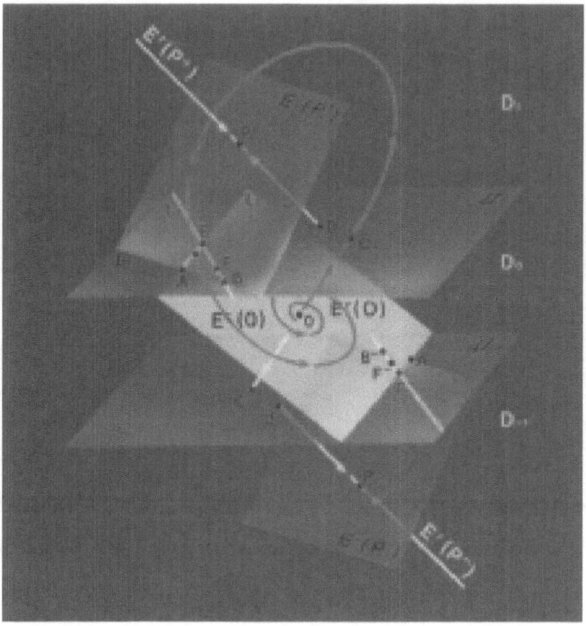

Plate 1.10 Homoclinic orbit together with various relevant sets in the state space. ©1988 IEEE.

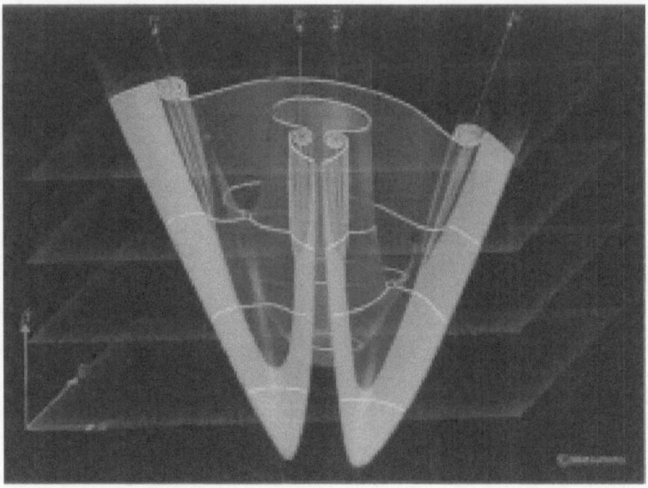

Plate 1.11 (α^*,β,ξ^*)-model bifurcation topography of Π^2. Left hand side (respectively right hand side) yellow surface corresponds to Π_0^2 (respectively Π_1^1) via period-doubling (respectively pitch-fork) bifurcation from Π_0^1 (respectively Π^1). Blue surface corresponds to Π^2. All homoclinic bifurcation sets are connected via a single family of periodic orbits Π^2. ©1991 World Scientific Publishing Company.

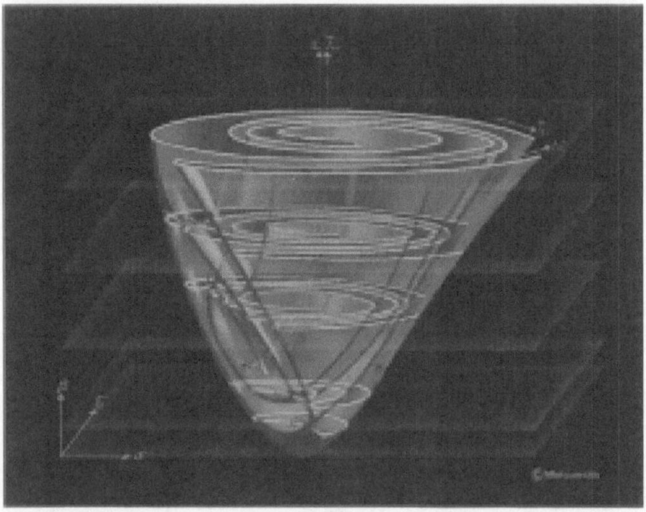

Plate 1.12 (α',β,ξ')-model bifurcation topography of π. ©1991 World Scientific Publishing Company.

Plate 1.13 (α',β,ξ')-model bifurcation topography of Π. ©1991 World Scientific Publishing Company.

(a)

Plate 1.14 (α', β)-bifurcation diagrams which are obtained by one-dimensional map (a) Global diagram. ©1991 World Scientific Publishing Company.

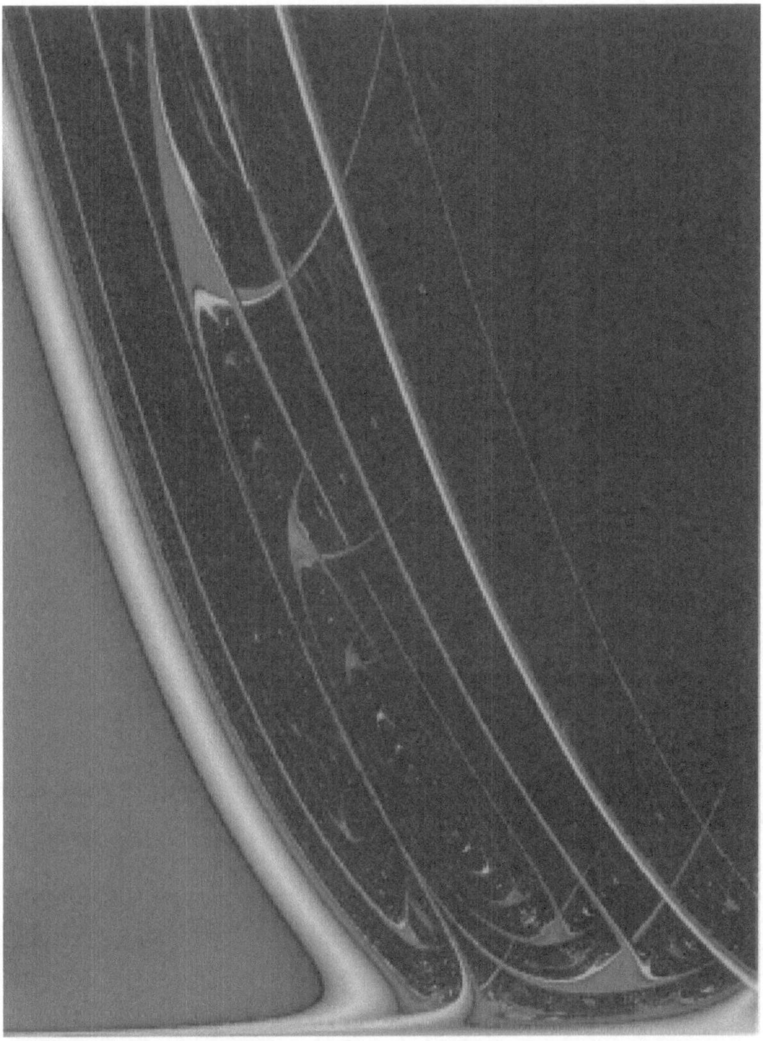

(b)

Plate 1.14 continued (b) Enlargement in a neighborhood of $\boldsymbol{r}$. ©1991 World Scientific Publishing Company.

(c)

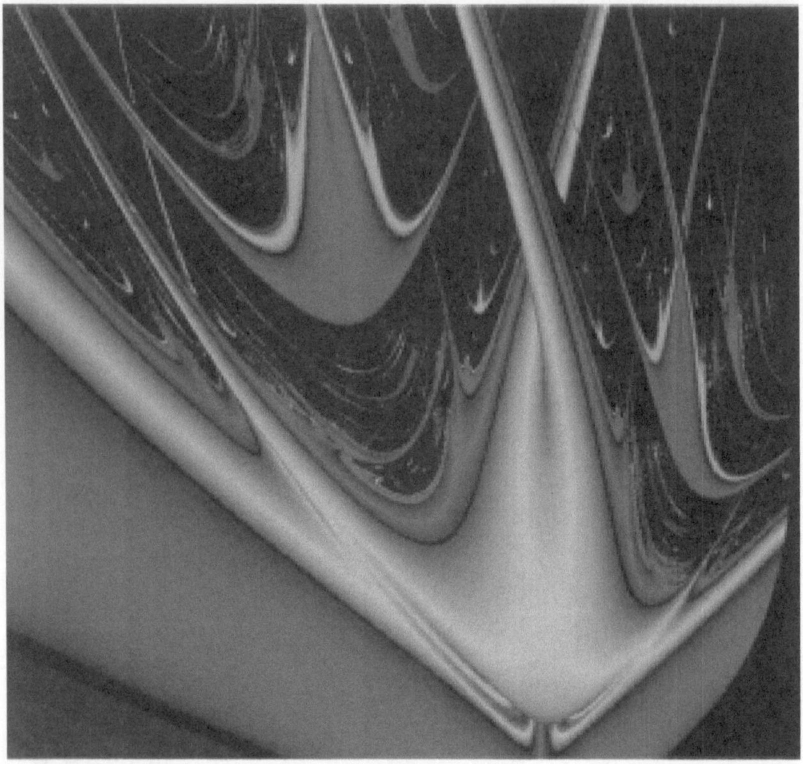

Plate 1.14 continued (c) Enlargement in a neighborhood of C^*. ©1991 World Scientific Publishing Company.

(a)

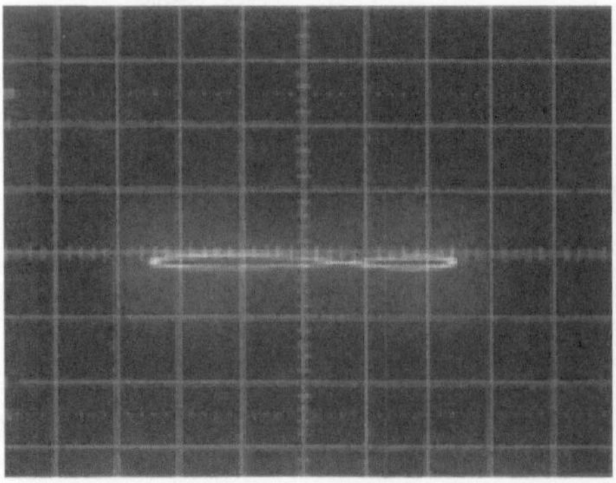

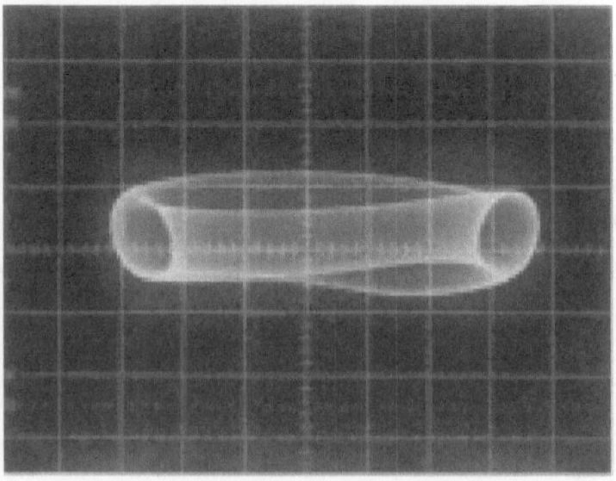

(b)

Plate 1.15 Attractors observed from the circuit of Fig.1.6.1 projected onto the (v_{C_1}, v_{C_2}) -plane. Horizontal scale (except (m)) : 0.5 V/division. Vertical scale (except (m)) : 0.5 V/division. In (a)-(l), only one of the two attractors is shown. (a) Period-1 orbit. (b) Two-Torus . ©1989 IEEE.

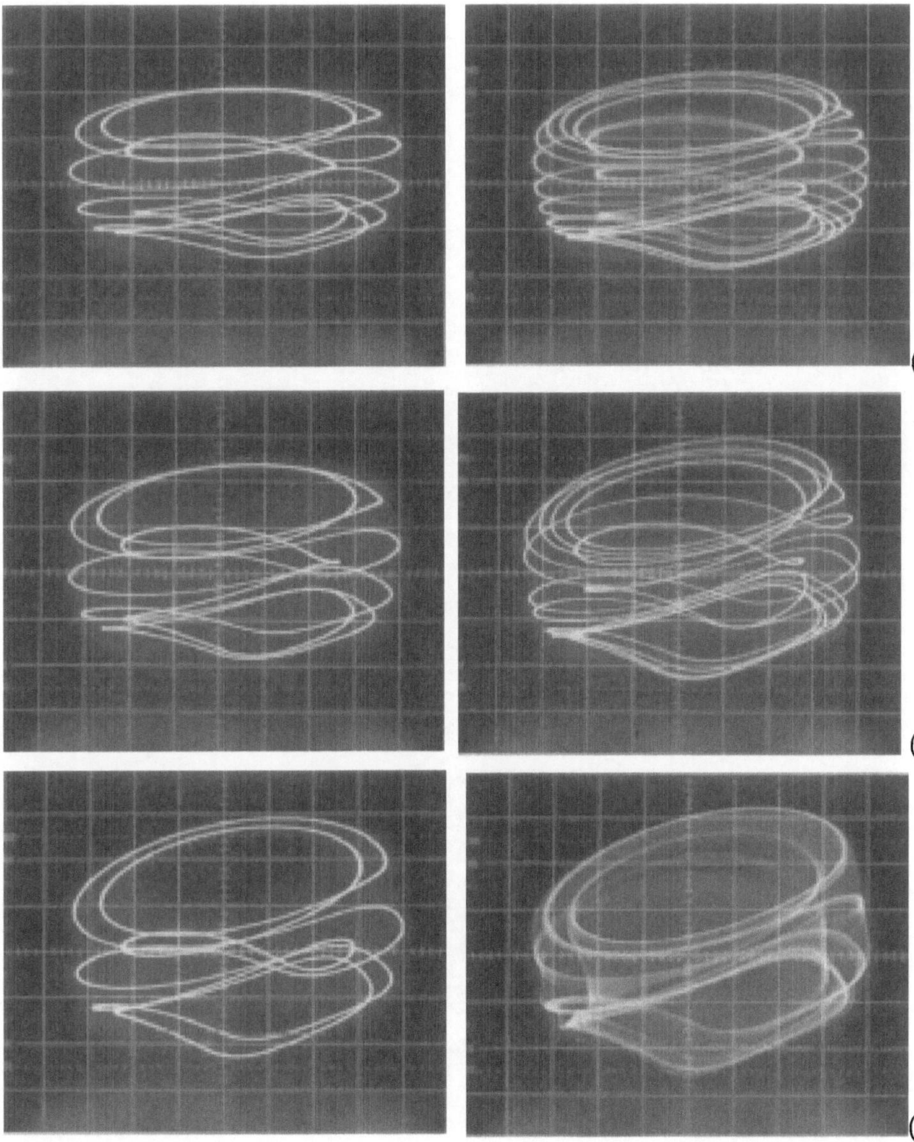

(c) (d) (e) (f) (g) (h)

Plate 1.15 continued (c) Period-8 orbit. (d) Period-15 orbit. (e) Period-7 orbit. (f) Period-13 orbit. (g) Period-6 orbit. (h) Chaotic attractor resulting from intermittency. ©1989 IEEE.

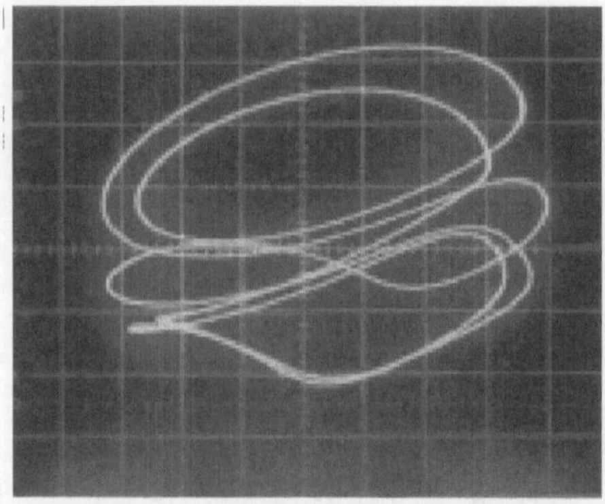

(i)

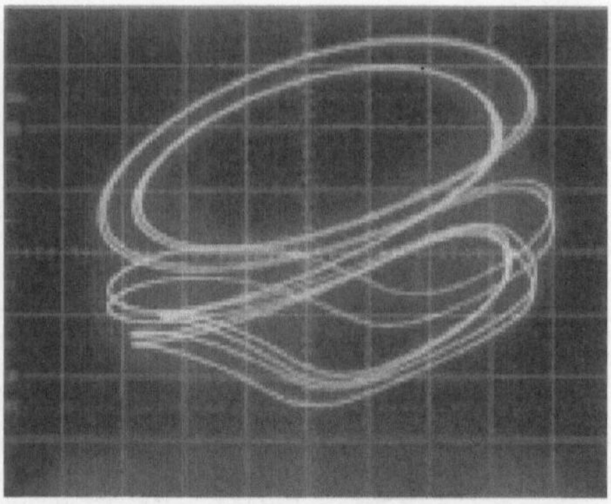

(j)

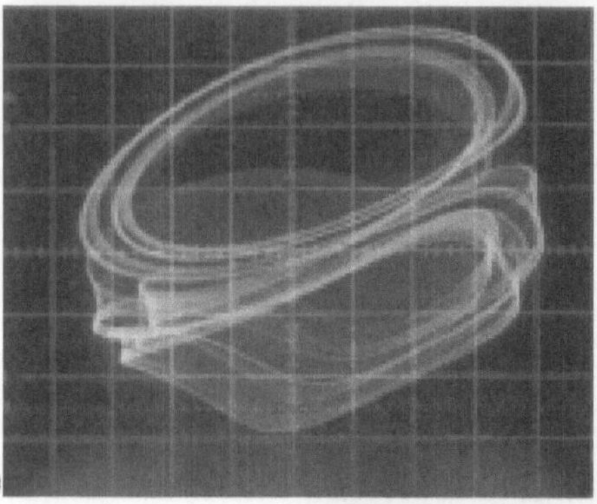

(k)

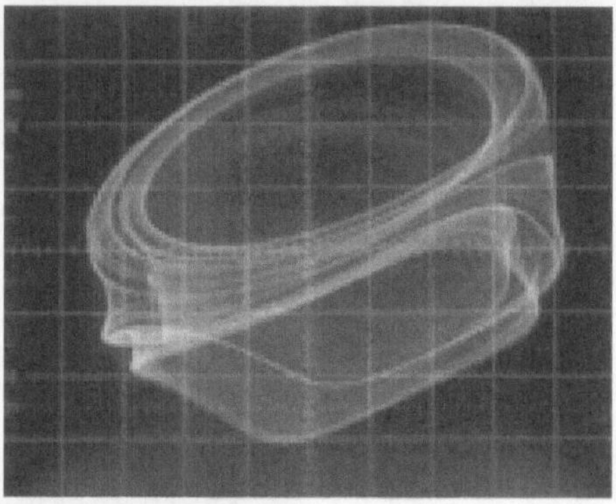

(l)

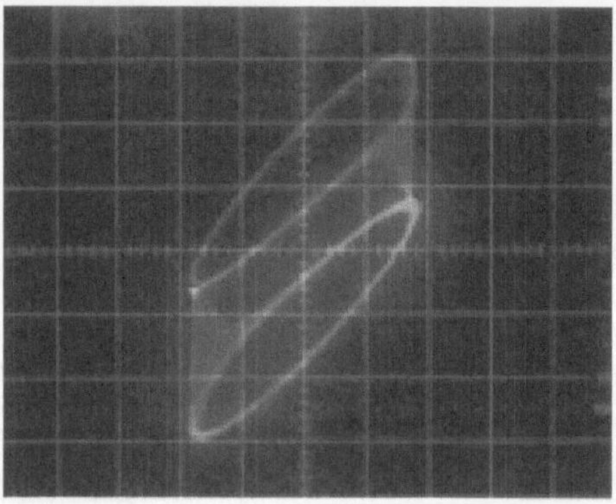

(m)

Plate 1.15 continued (i) Period-5 orbit. (j) Period-10 orbit. (k) Chaotic attractor resulting from period-doubling cascades. (l) Folded torus chaotic attractor. (m) Double Scroll-like attractor. Horizontal scale : 1.0 V/division. Vertical scale : 1.0 V/division.

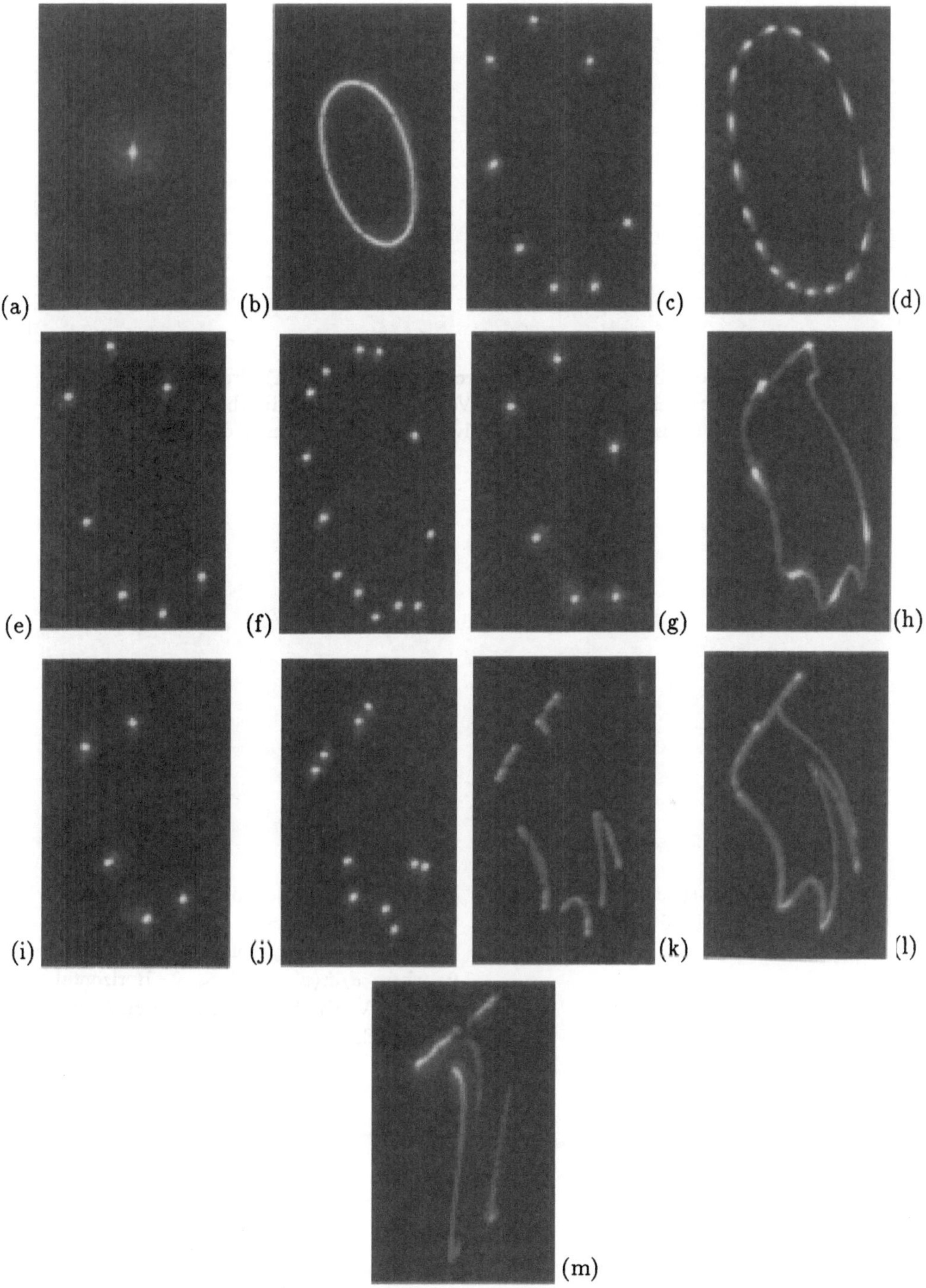

Plate 1.16 The cross section at $i_L = 0$, $v_{C_2} < 2$, of the corresponding trajectories from Plate.1.15. Vertical axis is v_{C_1} and the horizontal axis is v_{C_2}. ©1989 IEEE.

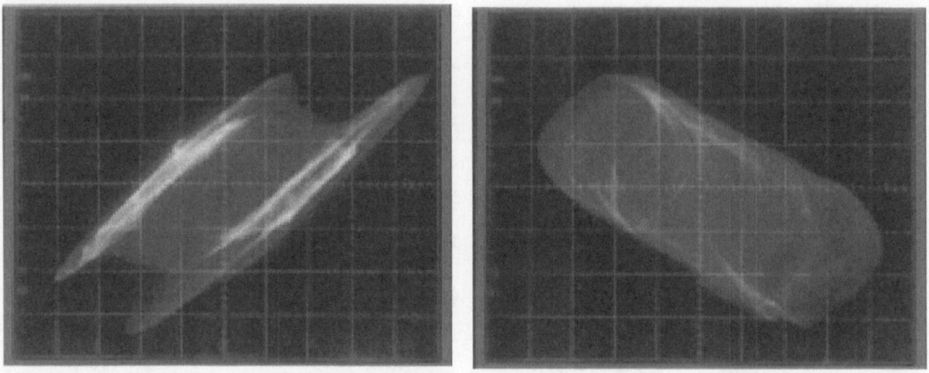

(a) (b

Plate 1.17 Attractor observed from the circuit of Fig.1.7.2. (a) Projection onto the (v_{C_1},v_{C_1}) plane. Horizontal scale: 0.5V/div. Vertical scale: 1 V/div. (b) Projection onto the (v_{C_1},i_{L_1}) plane. Horizontal scale: 0.5 V/div. Vertical scale: 1.8mA/div.

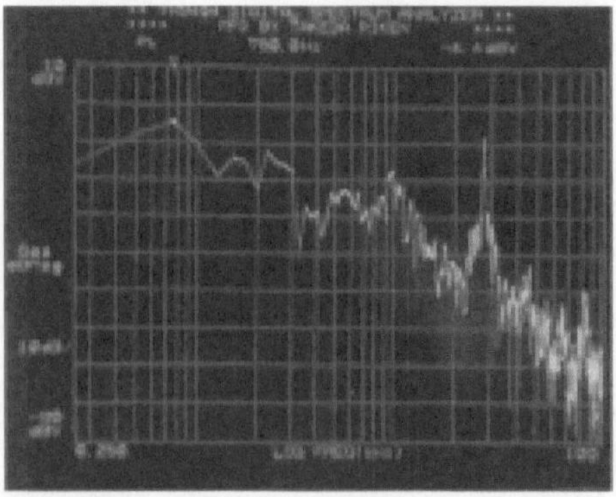

Plate 1.18 Power spectrum of the time waveform $v_{C_1}(t)$.

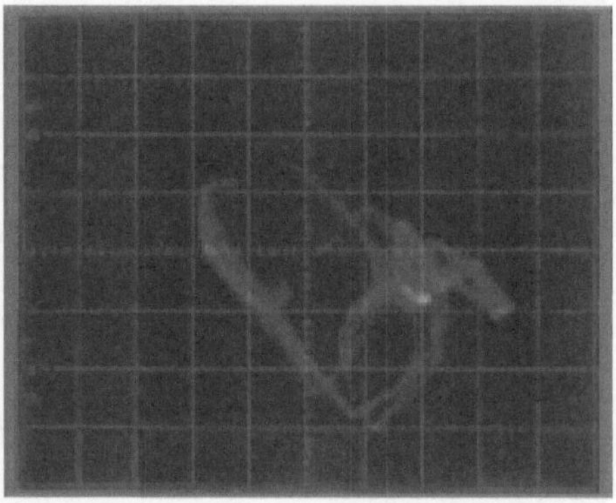

Plate 1.19 Projection onto the (v_{C_1}, i_{L_1})-plane of the cross section of the attractor on the hyperplane $v_{C_1} - v_{C_2} = 0$, where $d/dt(v_{C_1}, v_{C_2}) < 0$. Horizontal scale: 0.5V/div. Vertical scale: 1.8mA/div.

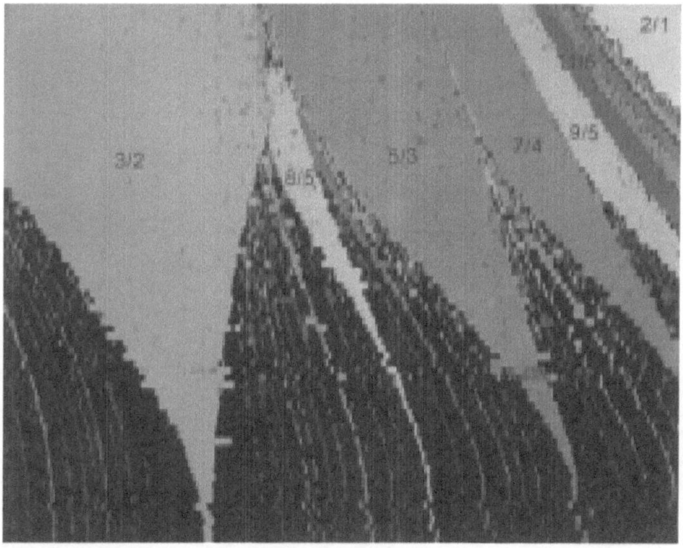

Plate 1.20 Two-parameter diagram of the neon bulb circuit in the $(f/f_0, E_0)$-space, where $1.40 \leq f/f_0 \leq 1.72, 0 \leq E_0 \leq 4V$, $R = 1M\Omega$, $C = 1nF$, battery $= 85V$, and $f_0 = 280$kHz is the oscillation frequency when $E_0 = 0$. See Plate 1.22 for color code.

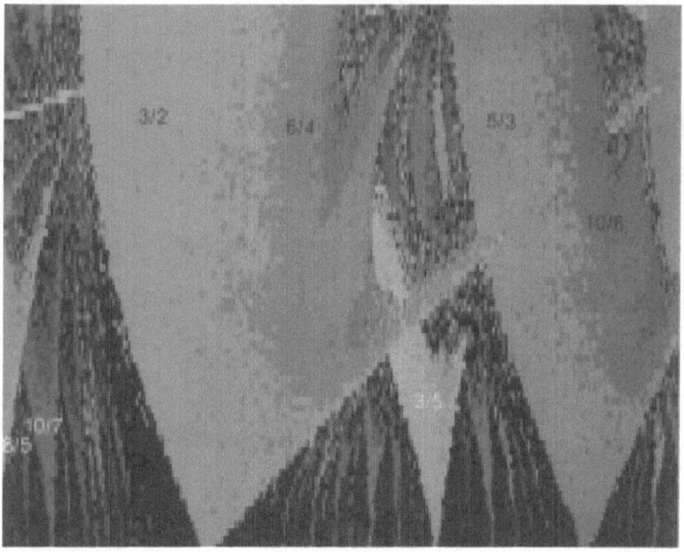

Plate 1.21 Bifurcation diagram of the neon bulb circuit in the $(E_0, f/f_0)$-space when the driving voltage source is "saw tooth". $1.40 \leq f/f_0 \leq 1.72$, $0 \leq E_0 \leq 4V$. See Plate 1.22 for color code.

Plate 1.22 Bifurcation diagram of the sine circle map in the (Ω, K)-space. Color code: period 1-2-4-8: white-pink-reddish pink-red, period 3-6-12: skyblue-lightgreen-green, period 5-10: light yellow-yellow, period 7-14: lightblue-blue. The numbers indicate rotation

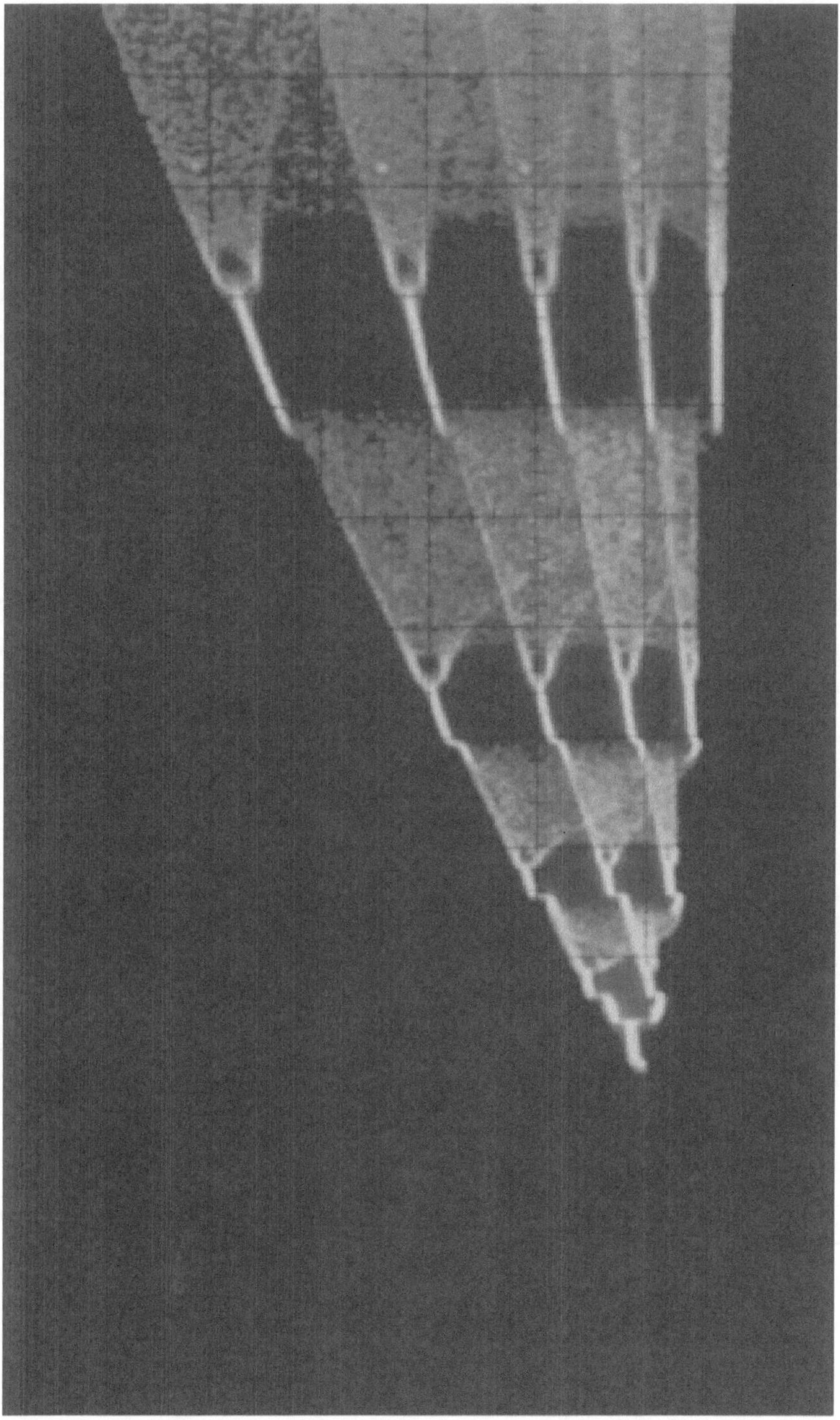

Plate 1.23 One-dimensional bifurcation diagram of the current i where the amplitude E is increased from 0 to 7.7V. ©1987 Elsevier Science Publishers.

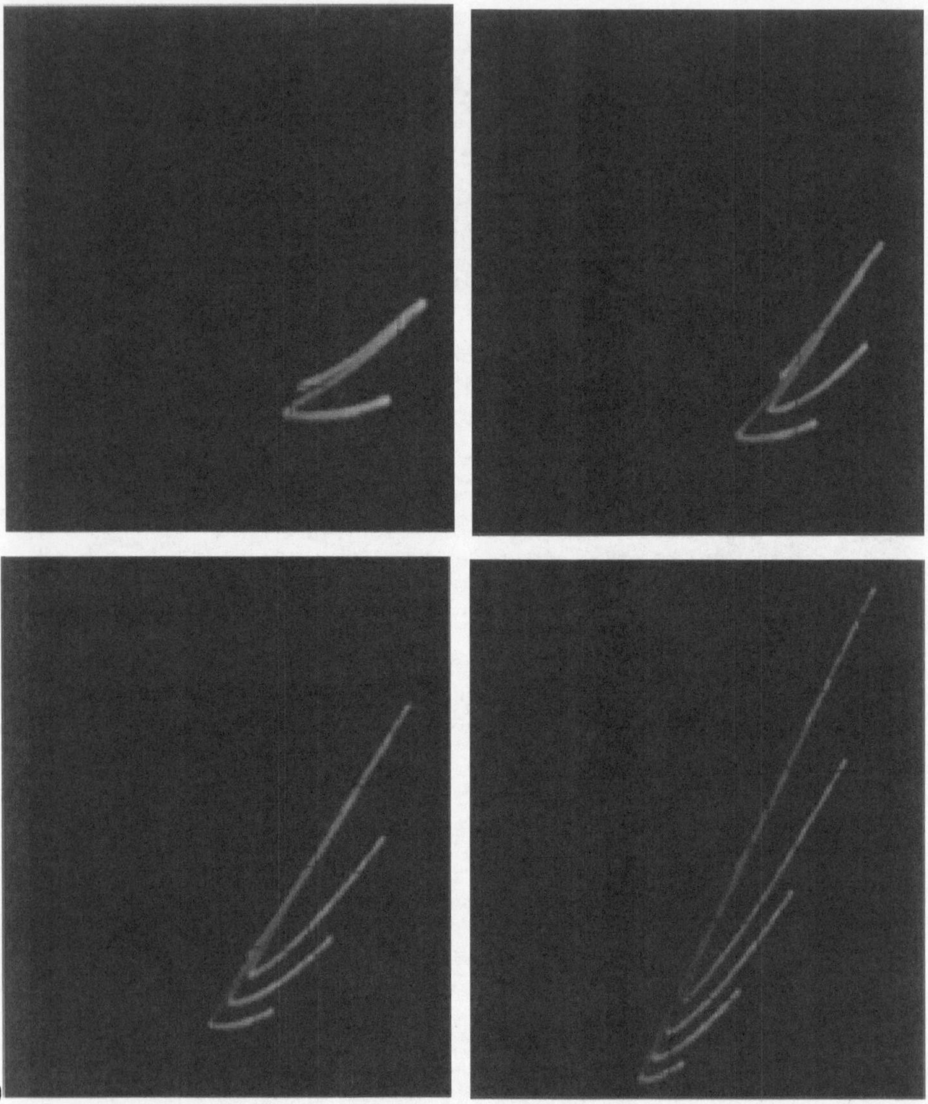

(a) (b) (c) (d)

Plate 1.24 One-dimensional bifurcation diagram of the current i where the amplitude E is increased from 0 to 7.7V. ©1987 Elsevier Science Publishers.

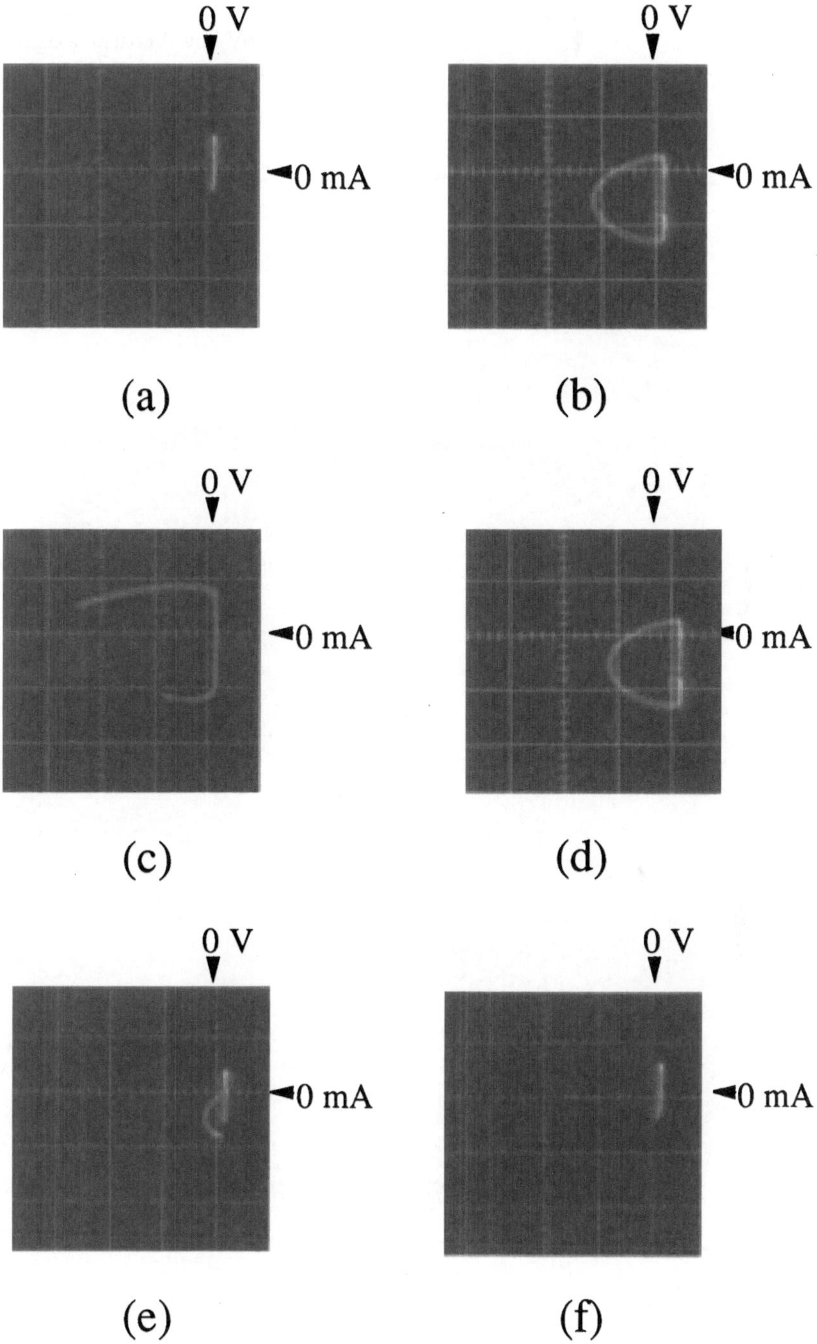

Plate 1.25 Cross sections at different phases in the voltage-current plane of the diode. Horizontal axis: 5.0V/div. Vertical axis 2.0mA/div.

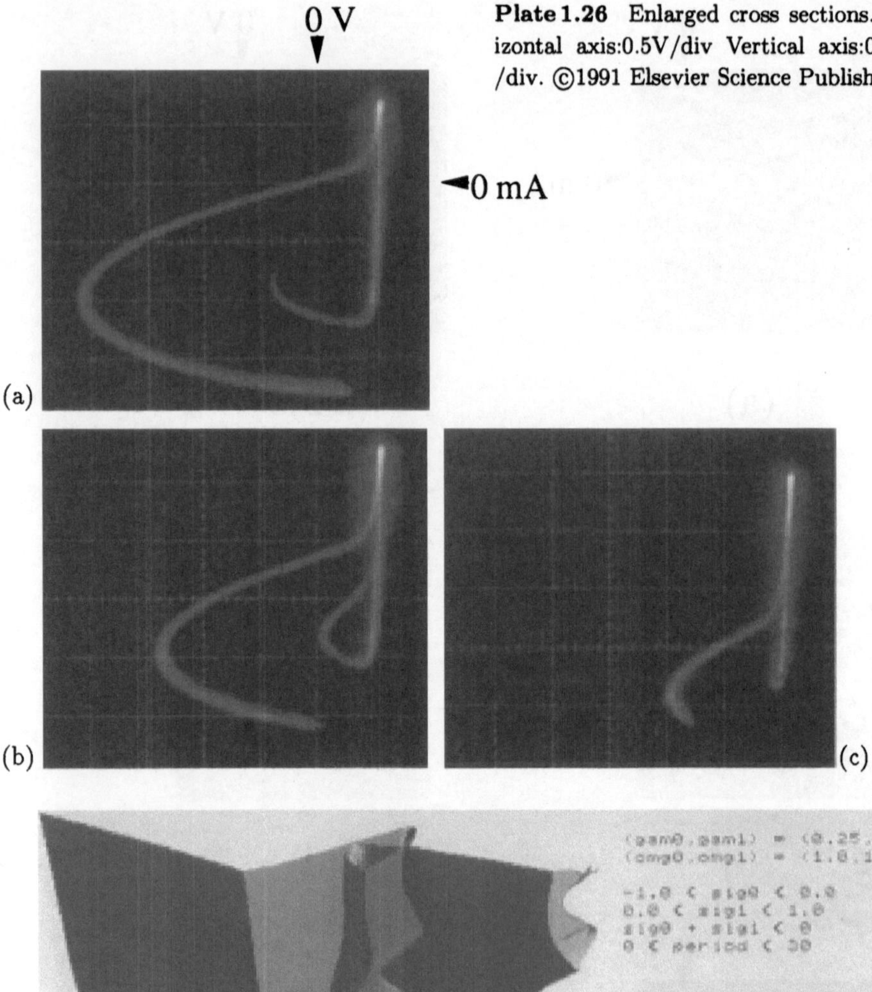

Plate 1.26 Enlarged cross sections. Horizontal axis:0.5V/div Vertical axis:0.5mA /div.

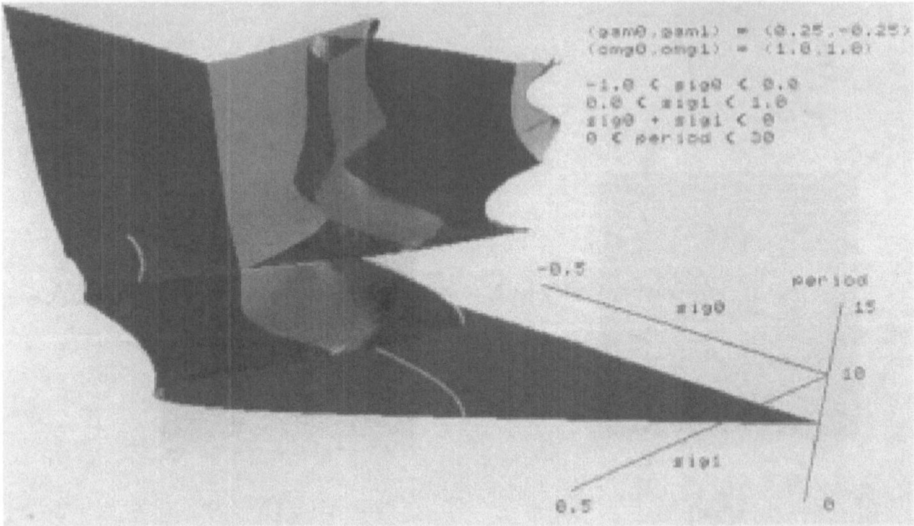

Plate 2.1 Periodic bifurcation topography. $(\gamma_0, \omega_0, \gamma_1, \omega_1) = (0.25, 1.0, -0.25, 1.0)$, $-1.0 \leq \gamma_0 \leq 0.0$, $0.0 \leq \gamma_1 \leq 1.0$, $\sigma_0 + \sigma_1 \leq 0$, $0 \leq$ *period* ≤ 30. Another half part of $\sigma_0 + \sigma_1 \geq 0$, is a mirror image with respect to a plane $\sigma_0 + \sigma_1 = 0$.

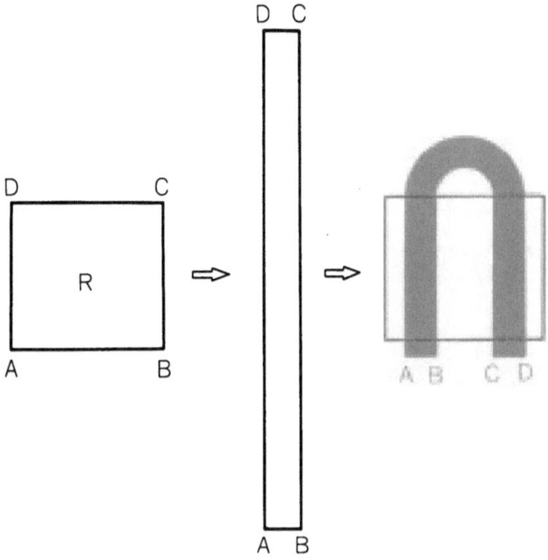

Plate 3.1 The process of forming a horseshoe.

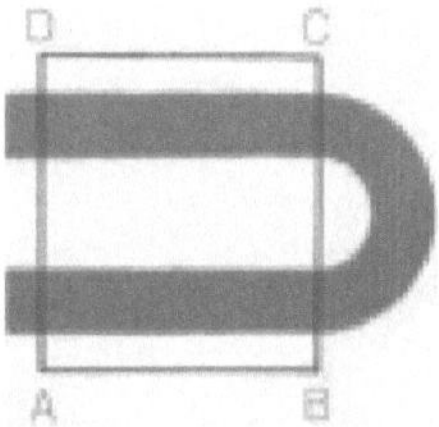

Plate 3.2 The inverse of the horseshoe map.

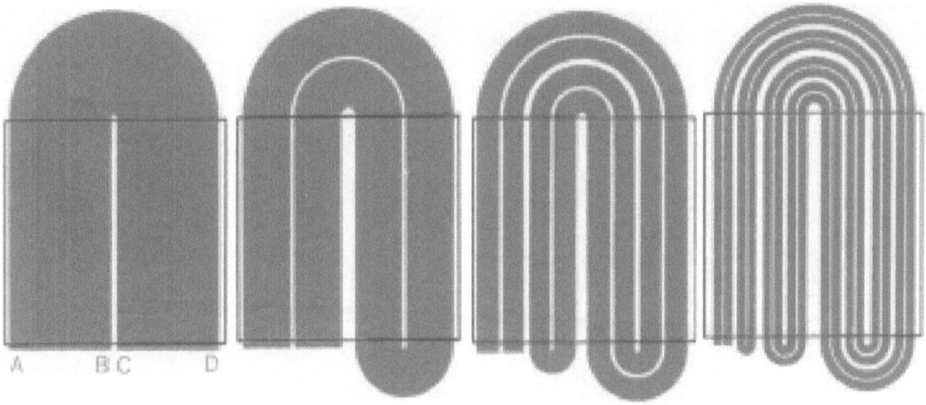

Plate 3.3 Several iterates $f^n(R)$ of the horseshoe map f. (a) $n = 1$. (b) $n = 2$. (c) $n = 3$. (d) $n = 4$.

Plate 3.4 The set $\bigcap_{k=-n}^{n} f^k(R)$ with $n = 4$.

1. Bifurcations Observed from Electronic Circuits

1.1 Introduction

The purpose of this chapter is to give concrete experimental as well as numerical results of bifurcation phenomena in electronic circuits together with reasonable theoretical justifications. All circuits described are so simple that high school students can build them. All of them except for one, in addition, behave within audible frequencies. We strongly recommend that the reader build at least one of them, and take a look at and listen to the bifurcations. It is a lot of fun.

A circuit is an interconnection of components. If there are n components, the interconnection imposes constraints:

Kirchhoff Voltage Law (KVL): The sum of the voltages along any closed loop is zero.

Kirchhoff Current Law (KCL): The sum of the currents at any node is zero.

It is easy to show that if there are m nodes, then only $m-1$ of KVL can be independent. Similarly, only $n-m+1$ of KCL can be independent.

There are three classes of components: (1)resistors (2)capacitors and (3)inductors. Let n_R, n_C and n_L be the number of resistors, capacitors, and inductors, respectively, so that $n = n_R + n_C + n_L$. Let $v_R \in \mathrm{IR}^{n_R}$,$v_C \in \mathrm{IR}^{n_C}$ and $v_L \in \mathrm{IR}^{n_L}$ be the vectors of resistor voltages, capacitor voltages, and inductor voltages, respectively. Similarly, let $i_R \in \mathrm{IR}^{n_R}$, $i_C \in \mathrm{IR}^{n_C}$ and $i_L \in \mathrm{IR}^{n_L}$ be the currents of resistors, capacitors, and inductors, respectively.

(1) Resistors are characterized by

$$f_R(v_R, i_R) = O, \quad f_R : \mathrm{IR}^{n_R} \times \mathrm{IR}^{n_R} \to \mathrm{IR}^{n_R} \,. \tag{1.1.1}$$

If a particular resistor, say R, is linear, then

$$v_R - Ri_R = 0,$$

which is Ohm's law and where R is the resistance. Formulation(1.1.1) includes a very large class of elements including transistors, vacuum tubes, etc. It also allows, for instance, a battery where

$$v_E - E = 0$$

and E is a fixed value.

(2) Capacitors are characterized by

$$f_C(v_C, q) = O, \quad f_C : \mathbb{R}^{nc} \times \mathbb{R}^{nc} \to \mathbb{R}^{nc}, \ \frac{dq}{dt} = i_C, \tag{1.1.2}$$

where $q \in \mathbb{R}^{nc}$ is the capacitor charge. If f_C is linear, then $q = Cv_C$ so that (1.1.2) is simplified to

$$C\frac{dv_C}{dt} = i_C \,. \tag{1.1.3}$$

(3) Inductors are characterized by

$$f_L(i_L, \phi) = O, \quad f_L : \mathbb{R}^{n_L} \times \mathbb{R}^{nc} \to \mathbb{R}^{n_L}, \ L\frac{d\phi}{dt} = v_L, \tag{1.1.4}$$

where $\phi \in \mathbb{R}^{nL}$ is the inductor flux. If f_L is linear then

$$L\frac{di_L}{dt} = v_L \,. \tag{1.1.5}$$

Given the limited amount of space, the circuits discussed in the book will have to be restricted to those studied by us, even though had it been possible, circuits studied by other people would have been included. Among those circuits we studied, the one described in Section 1.2, called the Double Scroll circuit, will be discussed thoroughly because (a) the circuit exhibits rich enough variety of bifurcations, and (b) we have rather complete knowledge of the circuit through our own work. Extensive literature on bifurcations in electronic circuits can be found in [Endo and Saito 1990]. See also [Matsumoto and Salam 1988].

1.2 The Double Scroll Circuit

1.2.1 Circuit and its Dynamics

The Double Scroll circuit is given by Fig. 1.2.1. It consists of two capacitors (C_1 and C_2), an inductor (L), a linear resistor (G) and only one nonlinear resistor (R) described by Fig. 1.2.1(b). In order to derive the dynamics of this circuit, let us first note that there are two independent KCL, e.g.,

$$\begin{aligned} i_{C_2} - i_L - i_G &= 0 \\ i_{C_1} - i_R - i_G &= 0 \,. \end{aligned}$$

On the other hand, there are three independent KVL, e.g.,

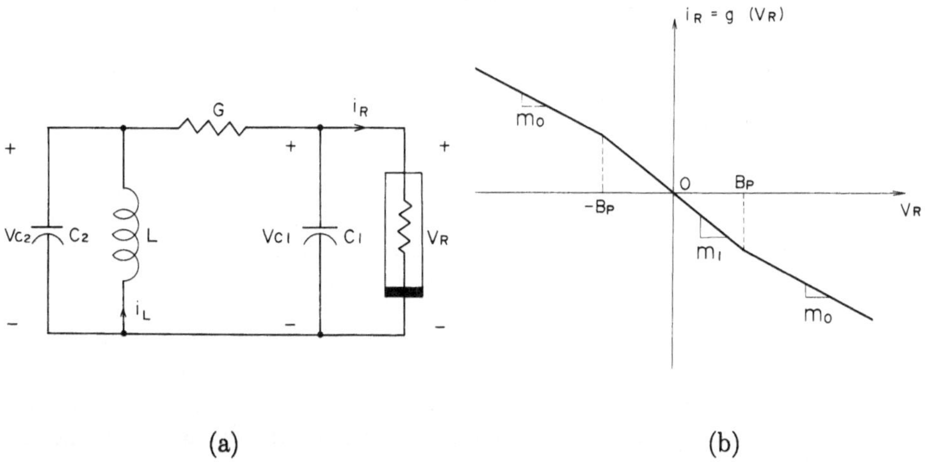

Fig. 1.2.1. The Double Scroll circuit. (a) Circuitry. (b) Nonlinear resistor characteristic. ©1985 IEEE.

$$
\begin{aligned}
v_{C_2} + v_L &= 0 \\
v_{C_1} + v_L + v_G &= 0 \\
v_{C_1} - v_R &= 0\,.
\end{aligned}
$$

Combining these equations with the element characteristics (see (1.1.1)-(1.1.5))

$$
\begin{gathered}
i_R = g(v_R), \qquad i_G = G v_G, \\
C_1 \frac{dv_{C_1}}{dt} = i_{C_1}, \quad C_2 \frac{dv_{C_2}}{dt} = i_{C_2}, \quad L\frac{di_L}{dt} = v_L,
\end{gathered}
\tag{1.2.1}
$$

one obtains:

$$
\boxed{
\begin{aligned}
C_1 \frac{dv_{C_1}}{dt} &= G(v_{C_2} - v_{C_1}) - g(v_{C_1}) \\
C_2 \frac{dv_{C_2}}{dt} &= G(v_{C_1} - v_{C_2}) + i_L \\
L\frac{di_L}{dt} &= -v_{C_2}\,.
\end{aligned}
}
\tag{DS 1}
$$

Several observations are in order:

(1) If Fig. 1.2.1(b) is *linear* i.e., if the graph is a single straight line, then nothing interesting can happen. The trajectory either converges to the origin or diverges to infinity. Even a sustained periodic orbit cannot occur!

(2) Note that the nonlinear resistor described by Fig. 1.2.1(b) is *active*, i.e., $v_R\, i_R = v_R\, g(v_R) \leq 0$. If it is *passive*, i.e., $v_R\, i_R \geq 0$, then again, nothing interesting can happen. This is due to the simple fact that a passive resistor always dissipates power so that the trajectory has no choice except for converging to a stable equilibrium point.

(3) Recall from (1.1.2) and (1.1.4) that the dimension of the state space is equal to the number of capacitors plus the number of inductors, which is three in the present circuit. If the dimension of the state space is two, then the most complicated behavior is a periodic orbit (see Chapter 3).

(4) If the inductor L is replaced with a capacitor, say C_3, then the dynamics becomes a *gradient system* [Matsumoto 1976] and hence the trajectory either converges to a stable equilibrium point or it diverges to infinity. Similarly, if C_1 and C_2 are replaced with inductors, then again the dynamics becomes a gradient system.

(5) There are simple and yet very important reasons for choosing a *piecewise-linear* resistor instead of other nonlinear resistors:

   (i) It is easy to implement. Recall that the nonlinear resistor is active, which automatically necessitates active circuit elements, e.g., transistors. If $g(\cdot)$ is, for example, a third order polynomial, it is extremely difficult, if not impossible, to implement;

   (ii) The piecewise-linearity simplifies rigorous analysis in a drastic manner. Namely, the state space can be decomposed into three regions in each of which the dynamics is *linear* so that any trajectory can be expressed as a composition of linear flows. One can write down, then, explicit bifurcation equations.

1.2.2 Implementation

We will observe in the next subsection, the C_1-bifurcations of the circuit given by Fig. 1.2.1 and described by (DS 1) where parameters other than C_1 are fixed as:

$$\begin{aligned} 1/C_2 &= 1.0, & 1/L &= 7, & G &= 0.7 \\ m_0 &= -0.5, & m_1 &= -0.8, & B_P &= 1\,. \end{aligned} \tag{1.2.2}$$

Appropriate rescaling must be done in order to implement the circuit with commercially available components. Among many choices of rescaling, one criterion worth taking into account is audibility: let the waveforms be within audible frequencies so that one can *listen* to the various bifurcations!

Let the circuit to be implemented be described by

$$\begin{aligned}\hat{C}_1\frac{d\hat{v}_{C_1}}{d\tau} &= \hat{G}(\hat{v}_{C_2}-\hat{v}_{C_1})-\hat{g}(\hat{v}_{C_1})\\ \hat{C}_2\frac{d\hat{v}_{C_2}}{d\tau} &= \hat{G}(\hat{v}_{C_1}-\hat{v}_{C_2})+\hat{g}(\hat{i}_L)\\ \hat{L}\frac{d\hat{i}_L}{d\tau} &= -\hat{v}_{C_2},\end{aligned} \tag{1.2.3}$$

where τ is a rescaled time. Rescaling here simply means a linear change of coordinate system and parameters:

$$\begin{aligned}&\tau=\alpha t, \quad &&\hat{v}_{C_1}=\beta v_{C_2}, \quad \hat{v}_{C_2}=\gamma v_{C_2}, \quad &&\hat{i}_L=\delta i_L\\ &\hat{C}_1=\varepsilon C_1, &&\hat{C}_2=\xi C_2, \quad \hat{L}=\eta L, &&\hat{G}=\theta G\\ &\hat{g}(v)=\kappa g(v)\,. &&&&\end{aligned} \tag{1.2.4}$$

In order to determine the nine parameters, first note that the break point B_P of Fig. 1.2.1(b) is 1 volt which is already well within a common value in transistor circuits . This suggests

$$\gamma=\beta=1\,. \tag{1.2.5}$$

If this is not a choice for γ and β, the last equation of (1.2.4) $\left(\hat{g}(v)=\kappa g(v)\right)$ should be replaced with a more general rescaling. The dynamics now reads

$$\begin{aligned}C_1\frac{dv_{C_1}}{dt} &= \frac{\alpha\kappa}{\beta\varepsilon}\left[\frac{\theta}{\kappa}G(v_{C_2}-v_{C_1})-g(v_{C_1})\right]\\ C_2\frac{dv_{C_2}}{dt} &= \frac{\alpha\theta}{\gamma\xi}G(v_{C_1}-v_{C_2})+\frac{\alpha\delta}{\gamma\xi}i_L\\ L\frac{di_L}{dt} &= -\frac{\alpha\gamma}{\delta\eta}v_{C_2},\end{aligned}$$

which suggest, among others,

$$\frac{\alpha\kappa}{\beta\varepsilon}=\frac{\theta}{\kappa}=\frac{\alpha\theta}{\gamma\xi}=\frac{\alpha\delta}{\gamma\xi}=\frac{\alpha\gamma}{\delta\eta}=1\,. \tag{1.2.6}$$

Note that $\theta=\kappa$ means that the two resistors should have the same rescaling, which is reasonable. Since $G=0.7$, a convenient choice of θ would be

$$\theta=10^{-3} \tag{1.2.7}$$

because, then

$$1/\hat{G}=\frac{1}{0.7\times10^3}\approx 1.42\ \mathrm{k}\Omega\ , \tag{1.2.8}$$

which is within easily available values.

Among the nine free parameters in (1.2.4), eight constraints have been imposed so far $\left((1.2.6)\text{-}(1.2.8)\right)$. Thus all the element values are a function of, say α:

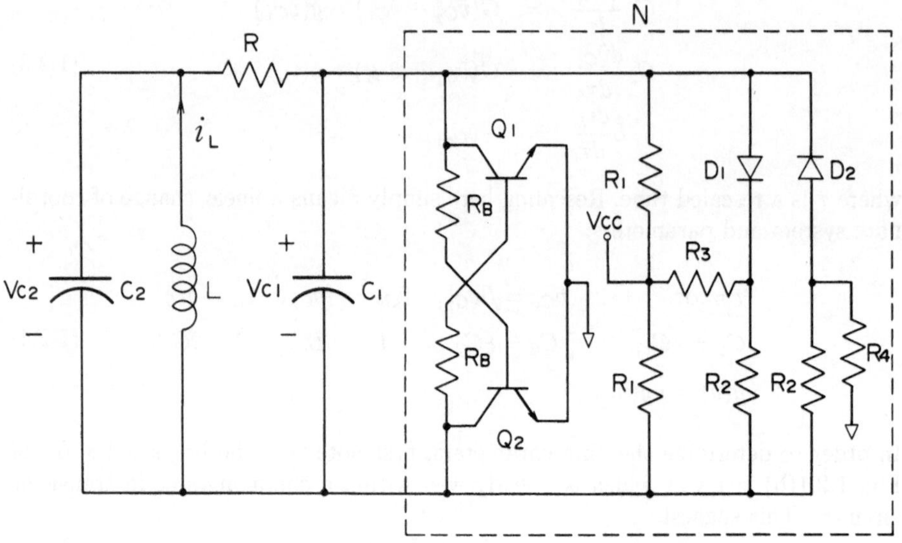

Fig. 1.2.2. Realization of the Double Scroll circuit. N realizes the nonlinear characteristic given in Fig. 1.2.1(b). $C_2 = 0.047\ \mu\text{F}$, $L = 6.8$ mH , $1/G = 1.21\ \text{k}\Omega$, $R_B = 56\ \text{k}\Omega$, $R_1 = 1\ \text{k}\Omega$, $R_2 = 3.3\ \text{k}\Omega$, $R_3 = 88\ \text{k}\Omega$, $R_4 = 39\ \text{k}\Omega$, $V_{CC} = 29$ V , Q_1 and Q_2 : 2SC1815, D_1 and D_2 : 1S1588. C_1: variable as a bifurcation parameter. See Subsection 1.2.2 for an explanation of these parameter values. ©1986 IEEE.

$$\hat{C}_1 = \varepsilon C_1 = 10^{-3}\alpha C_1, \quad \hat{C}_2 = 10^{-3}\alpha C_2,$$
$$\hat{L} = \frac{\alpha}{10^{-3}} L\,.$$

Since $C_2 = 1, L = 1/7$, a reasonable value of α would be in the order of 10^{-5} because then $\hat{C}_2$ would be in nF and $\hat{L}$ would be in mH. If, for instance,

$$\alpha = 4.5 \times 10^{-5},$$

then

$$\hat{C}_2 = 4.5 \text{ nF}\,. \tag{1.2.9}$$

Correspondingly,

$$\hat{L} = \frac{4.5 \times 10^{-5}}{10^{-3}} L \approx 6.4 \text{ mH}\,. \tag{1.2.10}$$

It follows from $\kappa = \theta = 10^{-3}$ that (Fig. 1.2.1(b))

$$\hat{m}_0 = 10^{-3} m_0, \ \hat{m}_1 = 10^{-3} m_1\,.$$

Fig. 1.2.2 shows the implementation of the circuit given in Fig.1.2.1. The subcircuit N enclosed by the broken line in Fig. 1.2.2 realizes the characteristic of Fig.

1.2.1(b). Plate 1.1(a) is a measured characteristic. Explanations of why N behaves as is shown in Plate 1.1(a) demands a hard core circuit analysis and it is slightly outside of the scope of this book, even though it is interesting. The reader is referred to [Matsumoto, Chua and Tokumasu 1986]. Since every element has a 15 % tolerance from the nominal value anyway, the final phase of the implementation is trial and error. In the particular experiment described below, the nominal values are:

$$\begin{array}{llll} C_2 = 0.047\ \mu\mathrm{F}\ , & L = 6.8\ \mathrm{mH}\ , & 1/G = 1.21\ \mathrm{k}\Omega\ , & R_B = 56\ \mathrm{k}\Omega \\ R_1 = 1\ \mathrm{k}\Omega\ , & R_2 = 3.3\ \mathrm{k}\Omega\ , & R_3 = 88\ \mathrm{k}\Omega\ , & R_4 = 39\ \mathrm{k}\Omega\ , \\ V_{CC} = 29\ \mathrm{V}\ , & Q_1 \text{ and } Q_2 : \text{2SC1815}, & D_1 \text{ and } D_2 : \text{1S1588}\ . & \end{array} \tag{1.2.11}$$

1.2.3 Experiments

Starting with 6.13 nF, we reduced the C_1 values manually to 4.5 nF. Plate 1.2 shows projections of the trajectories onto the (i_L, v_{C_1})-plane, in the order of decreasing C_1.

A Hopf Bifurcation For large values of C_1, nothing but a point was observed on the oscilloscope. This is a stable equilibrium point, or a point attractor, namely, all the trajectories settle down to it. After decreasing C_1 by an appropriate amount, a *periodic attractor* is born (Plate 1.2(a)), which signifies a *Hopf bifurcation.* It should be noted, however, that this is a piecewise-linear Hopf bifurcation instead of the Hopf bifurcation for smooth vector fields (Chapter 3) so that the periodic attractor born is not necessarilly small.

B Period-Doubling Bifurcations of the Periodic Orbit Further reductions in C_1 lead to Plate 1.2(b), a *period-doubling bifurcation* and then another period-doubling (Plate 1.2(c)). An enlarged picture is shown in Plate 1.2(d) where the period-4 nature of the trajectory is easier to see. Further period-doubling was difficult to observe.

C Chaotic Attractor (Rössler's Spiral-type) The next object observed is Plate 1.2(e), which does not appear to be periodic any more. It will be shown later that this attractor has the structure similar to that of Rössler's spiral-type attractor [Rössler 1979b]. Note, however, that the functional form of (DS 1) is entirely different from the Rössler equation (see Chapter 3). Since there is no unanimously accepted precise definition of chaos (see Chapter 3), terminology "chaotic" or "strange" attractor here is used to indicate nonperiodic attractor. In Section 1.4, however, discussions will be given under a precise definition of chaos.

D Saddle-Node Bifurcations of the Periodic Orbit and Periodic Window Upon further decreasing C_1, the chaotic attractor suddenly disappeared and a period-3 attractor was born (Plate 1.2(f)). A sudden birth (or death) of a periodic orbit signifies a saddle-node bifurcation. This period-3 attractor also exhibited a period-doubling bifurcation sequence and became another chaotic attractor (Plate 1.2(g)), which has a structure similar to Rössler's screw-type attractor. Therefore, the period-3 attractor is sandwiched between the chaotic attractors, and is called a *periodic window* (see Chapter 3).

E Interior Crisis (The Double Scroll) A further reduction in C_1 resulted in an interesting phenomenon: the attractor suddenly doubled its size (Plate 1.2(h)). Notice that the dynamics (DS 1) is symmetric with respect to the origin, i.e., it is invariant under $(v_{C_1}, v_{C_2}, i_L) \longrightarrow (-v_{C_1}, -v_{C_2}, -i_L)$. Therefore, in Plate 1.2(a)-(h), a "twin" attractor should coexist located symmetrically with respect to the origin. Indeed, by switching on and off the power supply (thereby changing the initial conditions), both attractors were observed. These twin (independent) attractors enlarge their size as C_1 decreases, and finally they collide with each other and emerge as a single attractor, which explains the observed size-doubling. This is called an *interior crisis* [Grebogi, Ott and Yorke 1982] and the attractor is called the Double Scroll because of its geometric shape (see Section 1.3).

F Near Heteroclinicity The two "holes" of the Double Scroll attractor became smaller and smaller, and the intensity of the orbit near the holes became significantly brighter than other parts (Plate 1.2(i)). This means that the orbit spends a very long period of time around the holes, which signifies that the circuit is on the verge of *heteroclinicity.* It should be noted, however, that heteroclinicity itself, by definition is not observable.

G Boundary Crisis After this, several periodic windows including Plate 1.2(j) were observed in between chaotic attractors and then finally the attractor suddenly disappeared. This signifies a *boundary crisis.* Namely, the attractor has been "quenched" upon colliding with saddle-type periodic orbit [Grebogi, Ott and Yorke 1982].

Remark 1.2.1. Note that so called analog computers constitute a very small subset in all electronic circuits in that in analog computer, its building blocks are integrators made up of capacitors and op. amps. The variables in a typical analog computer are merely node voltages of the integrator building block modules where the circuit current is completely irrelevant in the dynamics. On the other hand, our circuit consists of ordinary circuit elements, namely, resistors, capacitors, and an inductor. Both the current and voltage of each circuit element play a crucial role in the dynamics. Hence it would be misleading to confuse our circuit with an analog computer. Indeed, to call our circuit an "analog" computer would imply

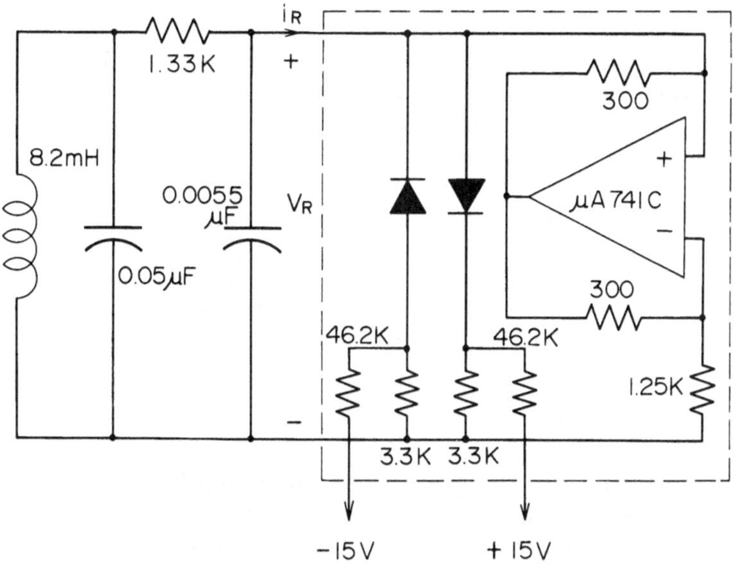

Fig. 1.2.3. Another realization of Fig. 1.2.1. ©1985 IEEE.

that all circuits, or for that matter all physical systems, are analog computers, which is absurd.

Remark 1.2.2. To the interested reader who wants to build the Double Scroll circuit, Fig. 1.2.3 is recommended because, first, the symmetry of the nonlinear resistor can be realized easily without worrying about the pair (Q_1, Q_2), and, second, the battery voltage is less than that of Fig. 1.2.2. Because of a great number of requests, we have produced many of the circuits given in Fig. 1.2.3. The interested reader can write to the authors for information.

H Sounds All circuit discussed in this book are so simple that high school students can build them and observe bifurcations on an oscilloscope. In addition, all the circuits except one, behave within audible frequencies so that one can listen to the sounds of bifurcation/chaos. Since we cannot convey the sounds of bifurcation/chaos in writing a book, we strongly recommend that the reader build at least one of the circuits described in this book, and take a look at and listen to its sounds. It is a lot of fun.

Since there are three quantities, v_{C_1}, v_{C_2} and i_L, in the circuit of Fig.1.2.1 one can connect two of them to oscilloscope terminals and connect the remaining terminal to a speaker. By tuning the C_1-values, one can watch and listen to bifurcation/chaos simultaneously. If a variable capacitor is not easily available, one can tune the linear conductance G and observe similar bibifurcation phenomena. When a Hopf bifurcation takes place out of a stable equilibrium point, one sees a circle popping out on an oscilloscope and hears a clean sinusoid-like sound. When a period-doubling occurs, one hears that another frequency is added to the original sound. With further period-doubling, one hears that other frequency components are added. When a periodic attractor bifurcates into a chaotic attractor, the sounds produced are extremely interesting. The Japanese newspaper Asahi Shinbun described these sounds sounding like "Avant Gard Jazz music." In fact, the contemporary composer H.Morimoto has composed several pieces of music inspired by the sounds from the Double Scroll circuit. It is rather interesting to compare "Avant Gard Jazz music" with the description "bag pipe" given by Van der Pol and Van der Mark for the neon bulb circuit discussed in Section 1.8.

1.2.4 Confirmations

This subsection confirms the experimentally observed bifurcations in the previous subsection via digital computer simulations of (DS 1). The subsection also provides simulated results of unstable orbits which cannot be observed by experiments. These unstable orbits often play important roles in the bifurcation structure (see Section 1.5).

An equilibrium point of (DS 1) satisfies

$$\begin{aligned} G(v_{C_2} - v_{C_1}) - g(v_{C_1}) &= 0 \\ G(v_{C_1} - v_{C_2}) + i_L &= 0 \\ v_{C_2} &= 0. \end{aligned}$$

Therefore, for fixed values of G, m_0, m_1, and B_P (see Fig. 1.2.1(b)), there are three equilibrium points:

$$\begin{aligned} P^+ &: \ v_{C_1} = k,\ v_{C_2} = 0,\ i_L = -Gk \\ O &: \ v_{C_1} = v_{C_2} = i_L = 0 \\ P^- &: \ v_{C_1} = -k,\ v_{C_2} = 0,\ i_L = Gk, \end{aligned}$$

where k and $-k$ are the positive and negative solutions of

$$Gv_{C_1} + g(v_{C_1}) = 0.$$

One can show that for any $C_1 > 0$, the origin O is always unstable. Other equilibrium points P^+ and P^- change their stability type as C_1 varies. Fig. 1.2.4(a) shows the location of the three equilibrium points. Note the equilibrium points are independent of C_1.

A Hopf Bifurcation Using the Routh formula [Routh 1905] one can show that for

$$1/C_1 < \frac{1}{2}(-3.5 + \sqrt{(3.5)^2 + 280}) \approx 6.8$$

P^+ and P^- are stable. At

$$1/C_1 = \frac{1}{2}(-3.5 + \sqrt{(3.5)^2 + 280})$$

a pair of eigenvalues crosses the imaginary axis and a (piecewise-linear) Hopf bifurcation occurs. Fig.1.2.4(b) shows two (coexisting) periodic attractors at

$$1/C_1 = 8.0\ .$$

Compare this with Plate 1.2(a) and note that any asymmetric periodic orbit must occur in a pair because (DS 1) is symmetric with respect to the origin.

B Period-Doubling Bifurcation As $1/C_1$ is increased slightly beyond 8.0, a period-doubling is initiated. Fig. 1.2.4(c) shows the period-2 attractors at (compare with Plate 1.2(b))

$$1/C_1 = 8.2,$$

while Fig. 1.2.4(d) (compare with Plate 1.2(c),(d)) shows the period-4 attractors at

$$1/C_1 = 8.44\ .$$

Further period-doubling was difficult to observe.

C Chaotic Attractor (Rössler's Spiral-type) At

$$1/C_1 = 8.5,$$

the attractor in Fig. 1.2.4 (e) no longer appears to be periodic (compare with Plate 1.2(e)). It has a structure similar to that of the Rössler spiral-type attractor. Namely, if one chooses an initial condition on the attractor, then the trajectory starts rotating outwards in a counterclockwise direction around the hole, thereby receding further and further away from center of the hole, where P^+ or P^- is located. After a certain random time interval, however, the trajectory returns to a point closer to P^+ or P^- and then repeats a similar process.

As was explained earlier, the symmetry of the dynamics implies that there is another spiral-type attractor located symmetrically with respect to the origin. This observation suggests that one attractor should still be present even if one replaces the three-segment function of Fig. 1.2.1(b) with the two-segment function in Fig. 1.2.4(f). This conjecture is confirmed in Fig. 1.2.4(g) where a spiral-type attractor is observed with a piecewise-linear resistor having only one break point. This has been also observed experimentally.

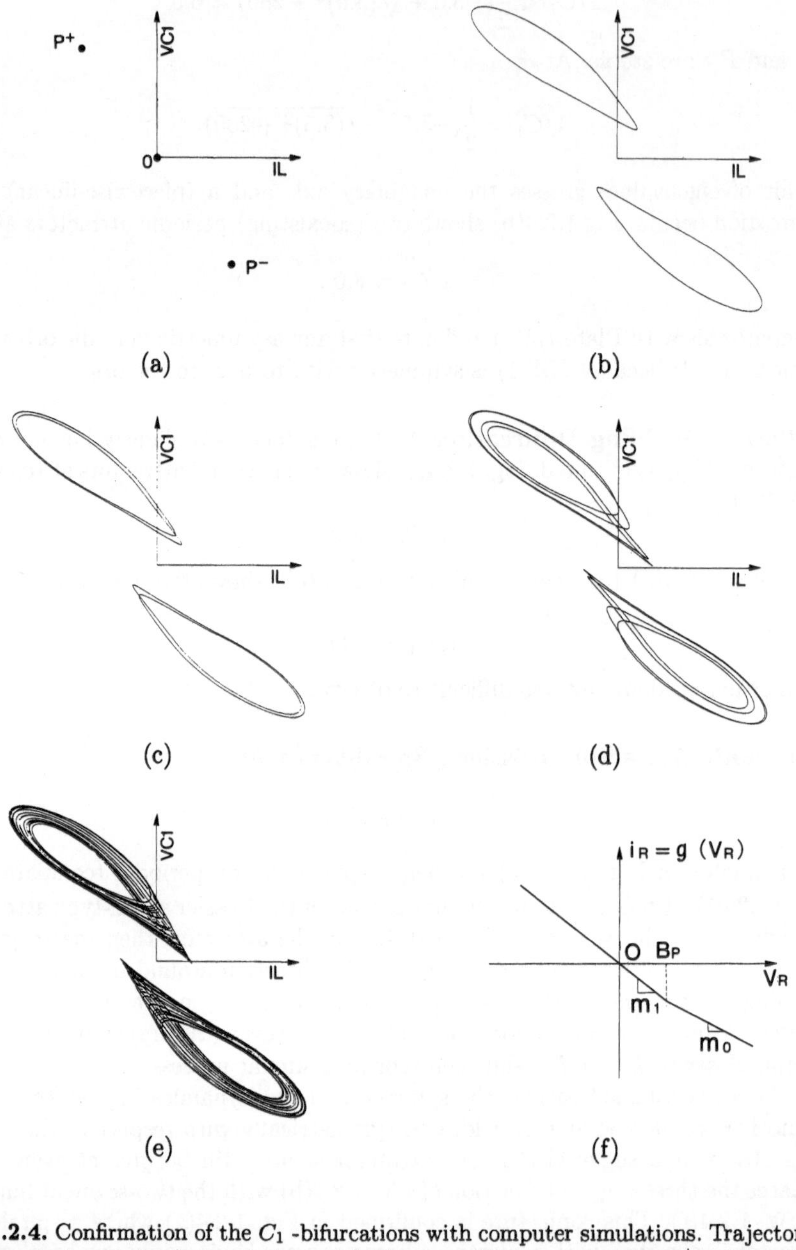

Fig. 1.2.4. Confirmation of the C_1 -bifurcations with computer simulations. Trajectories are projected onto the (i_L, v_{C_1})-plane. (a) Three equilibrium points: O,P^+andP^-. Here and after, the length of each arrow is 2.0. (b) Pair of periodic orbits at $1/C_1 = 8.0$. (c) Pair of period-2 orbits at $1/C_1 = 8.2$. (d) Pair of period-4 orbits at $1/C_1 = 8.44$. (e) Pair of Rössler's spiral-type attractor at $1/C_1 = 8.5$. (f) Modified resistor characteristic with a single break point. ©1986 Elsevier Science Publishers.

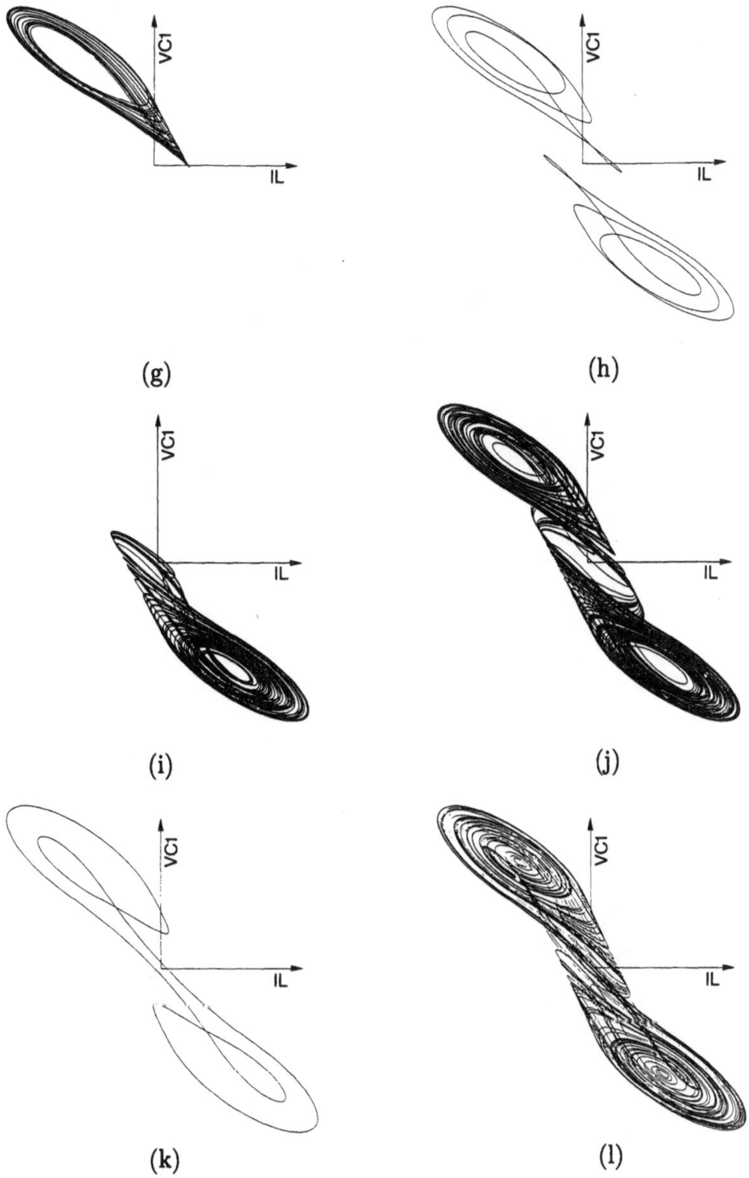

Fig. 1.2.4. continued (g) Spiral-type attractor obtained with the nonlinear resistor characteristic of (f). (h) Periodic window at $1/C_1 = 8.575$. (i) Rössler's screw-type attractor at $1/C_1 = 8.8066$. (j) The Double Scroll attractor at $1/C_1 = 9.0$. (k) Periodic window at $1/C_1 = 10.05$. (l) Near heteroclinicity at $1/C_1 = 9.78$. ©1986 Elsevier Science Publishers.

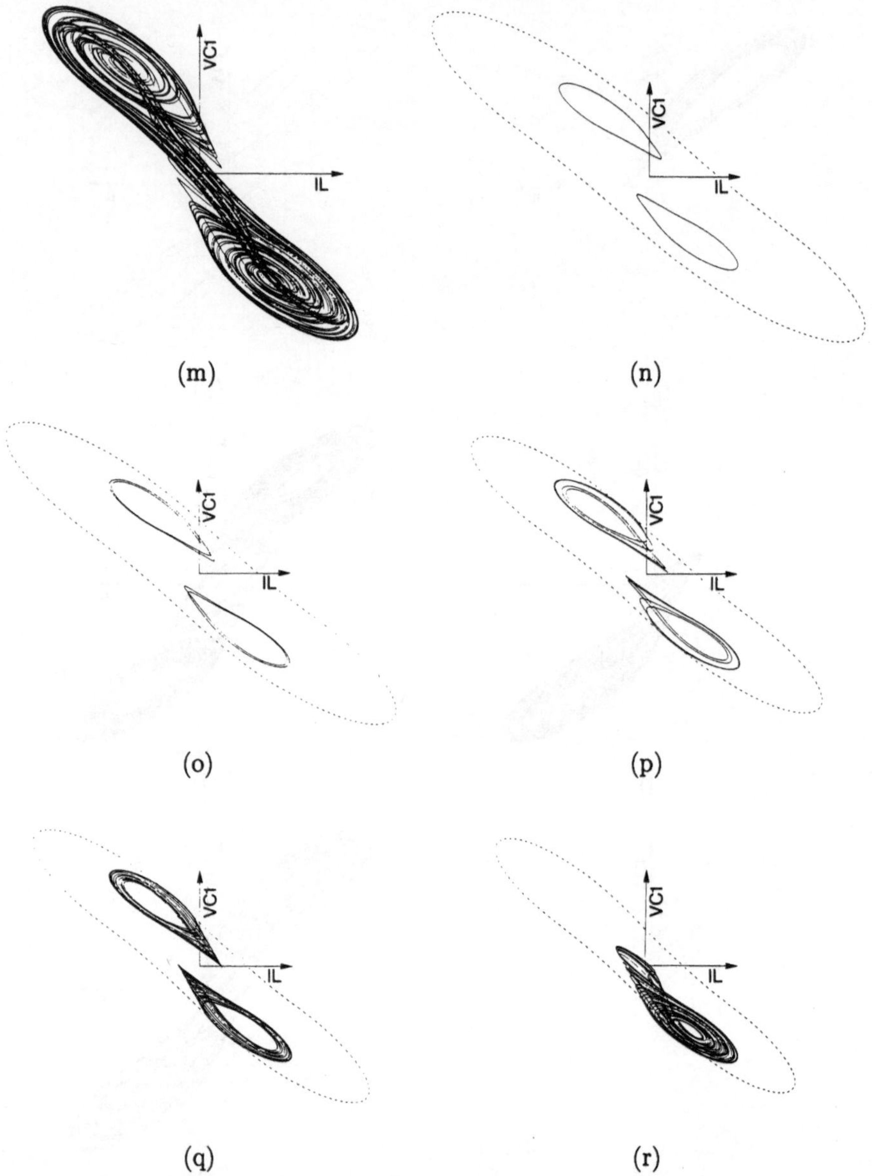

Fig. 1.2.4. continued (m) Near boundary crisis.(n) Periodic orbits of (b) together with the saddle-type periodic orbit ($1/C_1 = 8.0$). The length of the arrow along each axis is still 2.0 even though the scale of the figure is different from that of (b). (o) Period-2 orbits of (c) together with the saddle-type periodic orbit ($1/C_1 = 8.2$). (p) Period-4 orbits of (d) together with the saddle-type periodic orbit ($1/C_1 = 8.44$). (q) Spiral-type attractors of (e) together with the saddle-type periodic orbit ($1/C_1 = 8.5$). (r) Screw-type attractor of (i) together with the saddle-type periodic orbit ($1/C_1 = 8.8066$). ©1986 Elsevier Science Publishers.

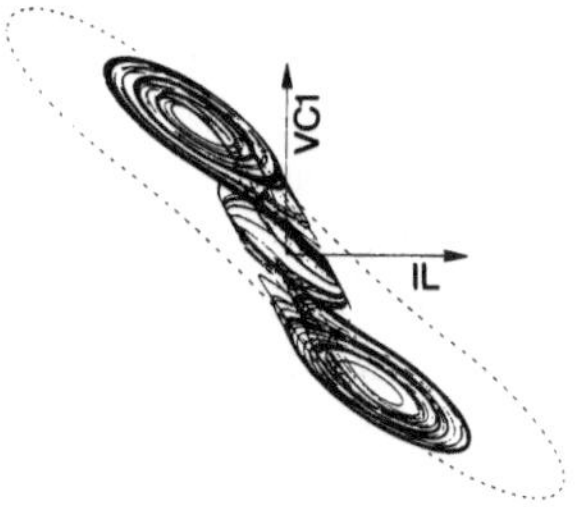

(s)

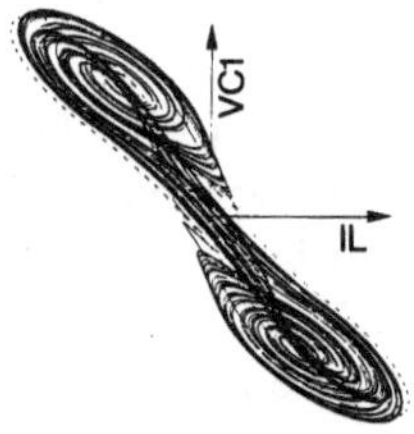

(t)

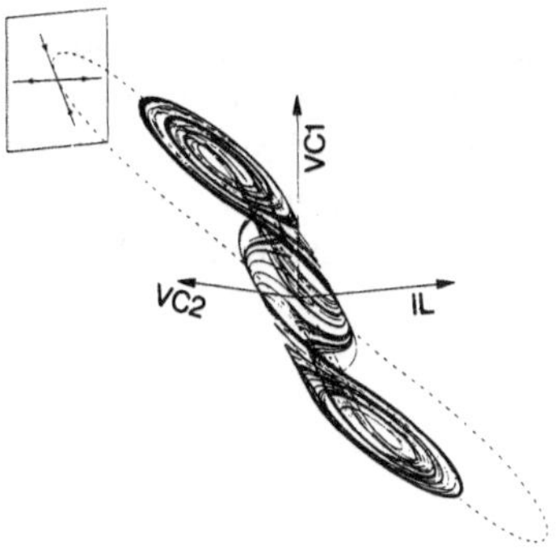

(u)

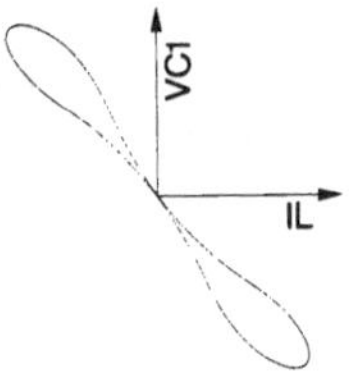

(v)

Fig. 1.2.4. continued (s) Double Scroll attractor of (j) together with the saddle-type periodic orbit ($1/C_1 = 9.0$). (t) Attractor observed shortly before the boundary crisis ($1/C_1 = 10.5$). It is about to collide with the saddle-type periodic orbit. (u) Cross section $v_{C_2} = 0$ for the Poincaré map of the saddle-type periodic orbit together with the stable and unstable eigen directions ($1/C_1 = 9.0$). (v) The saddle-type periodic orbit survives even after the death of the attractor ($1/C_1 = 13.0$). ©1986 Elsevier Science Publishers.

D Saddle-Node Bifurcation and Periodic Window As one continues tuning the bifurcation parameter C_1, one observes that the Rössler's spiral-type attractor persists up to

$$1/C_1 < 8.575 .$$

At

$$1/C_1 \approx 8.575,$$

a periodic window of Fig. 1.2.4(h) is observed (compare with Plate 1.2(f)). Further increase of $1/C_1$ results in a period-doubling cascade of this period-3 orbit and then again the chaotic attractor given in Fig. 1.2.4(i) which appears to have a structure similar to that of Rössler's screw-type attractor (compare with Plate 1.2(g)).

E Interior Crisis (The Double Scroll) As one increases $1/C_1$ further, the attractor abruptly enlarges itself and creates two holes located symmetrically with respect to the origin. The two attractors of Fig. 1.2.4(i), have collided with each other and quickly evolves into Fig. 1.2.4(j) (compare with Plate 1.2(h)), which corresponds to the parameter value

$$1/C_1 = 9.0 .$$

This is the Double Scroll. If one chooses an initial condition near the upper hole, then the trajectory starts rotating outwards in a counterclockwise direction around the hole. After a certain random time interval, the trajectory sometimes returns to a point closer to the upper hole. At some other times, however, it starts descending with respect to the v_{C_1}-axis in a spiral path and lands at a point near the lower hole. It then starts rotating outwards in a counterclockwise direction around the lower hole. Due to symmetry, the behavior after this descent is similar to the spiral excursion around the upper hole. This attractor persists over the parameter interval

$$8.81 < 1/C_1 < 10.05 .$$

However, at the parameter value

$$1/C_1 = 10.05,$$

the periodic window of Fig. 1.2.4(k) is observed (compare with Plate 1.2(j)). After this, several other strange-looking windows (not shown) are seen.

F Near Heteroclinicity The trajectory observed in Plate 1.2(i) corresponds to

$$1/C_1 \approx 9.78$$

and is given in Fig. 1.2.4(l). The trajectory almost hits P^+(or P^-), which is located inside the small hole, and spends an extremely long period of time near P^+(or P^-). Detailed analyses of heteroclinicity as well as homoclinicity in (DS 1) are given in Section 1.5. Chapters 2 and 3 also give general results on homoclinicity/heteroclinicity.

G Boundary Crisis Fig. 1.2.4(m) shows the attractor at

$$1/C_1 = 10.5 .$$

Suddenly, however, at

$$1/C_1 \approx 10.75,$$

the attractor *disappears*: (DS 1) diverges with any initial condition. This disappearing act provokes the interesting question as to how the attractor dies. A careful analysis suggests that this phenomenon is related to the simultaneous presence of a *saddle-type* periodic orbit encircling the attractor. To see this, let us go back to the parameter value $1/C_1 = 8.0$. With this value, a saddle-type periodic orbit has been found outside the periodic attractors. It is shown by the broken closed curve in Fig. 1.2.4(n). Since this orbit is not attracting, one cannot observe it on an oscilloscope. Nor, since it is saddle-type, can one observe it with a digital computer by integrating backward in time. Newton iteration was used to find it. As we increase C_1, the saddle-type periodic orbit shrinks gradually as shown in Fig. 1.2.4(o-s). At C_1=10.5, the attractor is located very close to the saddle-type periodic orbit (Fig. 1.2.4(t)). With a slight increase in $1/C_1$ beyond 10.5, the attractor appears to collide with the saddle-type periodic orbit. This collision provides a natural mechanism leading to the attractor's death. Note that if the attractor stayed away from the saddle-type periodic orbit, there would be no way for the trajectory in the attractor to escape. Fig. 1.2.4(u) shows the situation at $C_1 = 9.0$. The square in the upper left-hand corner is the Poincaré section (see Chapter 2) $v_{C_2} = 0$ and the arrows indicate the stable and unstable eigenspaces of the Poincaré map. This is looked at from an angle different from that of the other figures in order to show the relative positions of various sets. Now, if the attractor *collides* with the saddle-type periodic orbit, then it will provide an exit path for the trajectory to escape into the outer space. This is what happens at $1/C_1 \approx 10.75$. After this, no attractor is detected. The attractor seems to be quenched upon colliding with the saddle-type periodic orbit.

This corresponds to the boundary crisis [Grebogi, Ott and Yorke 1982] defined for one-dimensional and two-dimensional maps, i.e., an attractor suddenly disappears when it touches an unstable periodic point. This observation raises another interesting question: Does the saddle-type periodic orbit also die when it collides with the attractor? The answer is negative. It survives even after the attractor's death. Fig. 1.2.4(v) shows the saddle-type periodic orbit at

$$1/C_1 = 13.0 \, .$$

Although this orbit appears to be pinched at the center, it really is a simple (i.e., non-intersecting) closed curve in the three-dimensional state space. The saddle-type periodic orbit keeps shrinking in size while its period gets longer and longer as one increases $1/C_1$. Would this orbit persist for all values of $1/C_1$ or would it bifurcate into another object at a larger value of $1/C_1$? The answer will be given in Section 1.5.

1.2.5 Summary

Table 1.2.1 summarizes the sequence of bifurcations described above where the type of the eigenvalues at three equilibrium points are also given. (Recall that the nonlinearity in (DS 1) is a three-segment piecewise-linear function and that each linear region can have at most one equilibrium point. Hence the dynamics is strongly influenced by the eigenvalues at the equilibrium points.) Each equilibrium

Table 1.2.1. Summary of the C_1 bifurcations

$1/C_1$	Eigenvalues complex: $\sigma \pm \omega i$ real: γ O	$P^{\pm}$	Description of dynamics	Fig. 1.2.4
$0<1/C_1<6.8$	$\sigma_0<0,\gamma_0>0$	$\sigma_1<0,\gamma_1<0$	$P^{\pm}$ are point attractors	(a)
6.8		$\sigma_1=0,\gamma_1<0$	Hopf at $P^{\pm}$	
$6.8<1/C_1<8.2$		$\sigma_1>0,\gamma_1<0$	periodic around $P^{\pm}$	(b)
8.2			period-2	(c)
8.44			period-4	(d)
8.5			Spiral-type	(e)
8.575			period-3 window	(h)
$8.11<1/C_1<10.05$			Double Scroll	(j)
10.05			periodic window	(k)
9.78			near heteroclinicity	(l)
10.75			boundary crisis	(m)†
$10.75<1/C_1$			diverges from any initial condition except equilibrium points	

†Fig. 1.2.4(m) is obtained at $1/C_1 = 10.5$, which does not yet give rise to a crisis, though it is close to it.

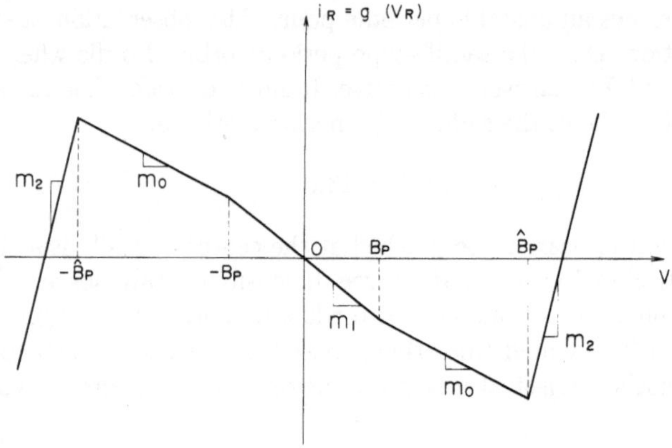

Fig. 1.2.5 A modified resistor characteristic. ©1985 IEEE.

point has a pair of complex-conjugate eigenvalues, $\sigma \pm \omega i$, and a real eigenvalue, γ.

Remark 1.2.3. If readers feel uncomfortable with Fig. 1.2.1(b) or Plate 1.2(a) because an actual circuit cannot be active everywhere, and a trajectory cannot

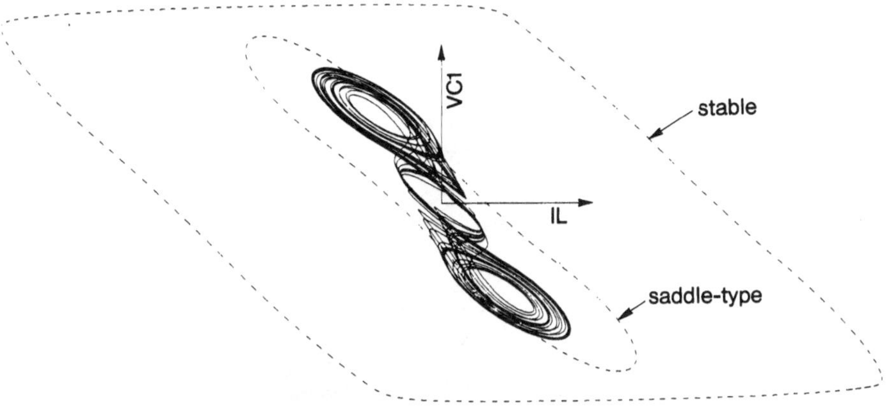

Fig. 1.2.6. The large periodic attractor together with the Double Scroll and the saddle-type periodic orbit. ©1985 IEEE.

diverge to infinity, they can simply replace Fig. 1.2.1(b), for instance, with Fig. 1.2.5. If $\hat{B}_P \geq 3$, the substituted characteristic curve has no effect on the attractor and on the saddle-type periodic orbit, because $|v_{C_1}(t)| < 3$ for all $t > 0$ on the attractor and on the saddle-type periodic orbit. The only difference is the additional appearance of a large (periodic attractor) as shown in Fig. 1.2.6 ($\hat{B}_P = 3,\ m_2 = 5$), where the trajectory does not diverge with any initial condition. As for the transistor circuit of Fig. 1.2.2, the v-i characteristic naturally becomes passive for large voltages. Plate 1.1(b) shows the v-i characteristic of the subcircuit N in Fig. 1.2.2(a) for a larger range.

Remark 1.2.4. The function $g(\cdot)$ of Fig. 1.2.1(b) does not have to be piecewise-linear to allow one to observe qualitatively the same attractor. Let us replace, for example, $g(\cdot)$ of Fig. 1.2.1(b) with the smooth cubic function

$$g(v_{C_1}) = a_0 v_{C_1}\left(\frac{a_1^2}{3}v_{C_1}^2 - 1\right), \tag{1.2.12}$$

where $a_0 = 0.8,\ a_1 = 0.1$. Then with $1/C_1 = 9,\ 1/C_2 = 0.5,\ 1/L = 7,\ G = 0.65$, the chaotic attractor of Fig. 1.2.7 is observed.

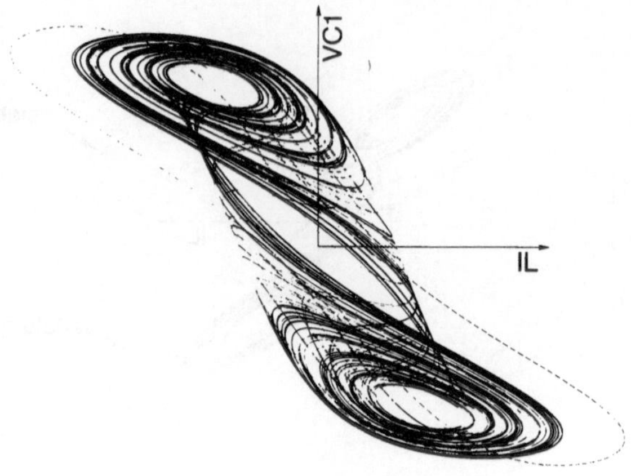

Fig. 1.2.7. The chaotic attractor and the saddle-type periodic orbit with the smooth resistor characteristic (1.2.12). ©1985 IEEE.

1.3 Structure of the Double Scroll

The purpose of this subsection is to study the structure of the Double Scroll attractor shown in Plate 1.2(i) and Fig. 1.2.4(j). Plate 1.3 shows the projections of the attractor which we will study.

1.3.1 Geometric Structure

First, note that the function $g(\cdot)$ of Fig. 1.2.1(b) is given by

$$\begin{aligned} g(v_R) &:= g(v_R; B_P, m_0, m_1) \\ &= \begin{cases} m_0 v_R + B_P(m_1 - m_0), & v_R \geq B_P \\ m_1 v_R, & |v_R| \leq B_P \\ m_0 v_R - B_P(m_1 - m_0), & v_R \leq -B_P, \end{cases} \end{aligned} \tag{1.3.1}$$

which satisfies

$$g(B_P v_R; B_P, m_0, m_1) = B_P g(v_R; 1, m_0, m_1). \tag{1.3.2}$$

Therefore, via the rescaling,

$$
\begin{aligned}
&x^1 := v_{C_1}/B_P, \quad x^2 := v_{C_2}/B_P, \quad x^3 := i_L/(B_P G),\\
&\tau := tG/C_2, \quad a := m_1/G, \quad b := m_0/G, \\
&\alpha := C_2/C_1, \quad \beta := C_2/(LG^2),
\end{aligned}
\tag{1.3.3}
$$

equation (DS 1) is transformed into

$$
\boxed{
\begin{aligned}
\frac{dx^1}{dt} &= \alpha(x^2 - h(x^1))\\
\frac{dx^2}{dt} &= x^1 - x^2 + x^3\\
\frac{dx^3}{dt} &= -\beta x^2
\end{aligned}
}
\tag{DS 2}
$$

$$
h(x^1) = \begin{cases} bx^1 + a - b, & x^1 \geq 1 \\ ax^1, & |x^1| \leq 1 \\ bx^1 - a + b, & x^1 \leq -1\,. \end{cases}
\tag{1.3.4}
$$

Here, we have abused our notation for time: it should have been "τ" instead of "t." There will be no confusion, however. Note that $h(x^1)$ includes both x^1 and $g(x^1)$. We begin with the following observations:

(1) Equation (DS 2) is symmetric with respect to the origin, i.e., the vector field is invariant under the transformation

$$
(x^1, x^2, x^3) \longrightarrow (-x^1, -x^2, -x^3).
$$

(2) Consider the equilibrium points

$$
\begin{cases} h(x^1) = 0 \\ x^2 = 0 \\ x^1 + x^3 = 0. \end{cases}
$$

It follows from the form of $h(\cdot)$ that (DS 2) has a unique equilibrium point in each of the following three subsets of $\mathbb{R}^3$:

$$
\begin{aligned}
D_1 &= \{(x^1, x^2, x^3)\,|x^1 \geq 1\}\\
D_0 &= \{(x^1, x^2, x^3)\,||x^1| \leq 1\}\\
D_{-1} &= \{(x^1, x^2, x^3)\,|x^1 \leq -1\}
\end{aligned}
$$

provided that $a, b \neq -1$. The equilibrium points are explicitly given by

$$\begin{aligned} P^+ &= (k, 0, -k) \in D_1 \\ O &= (0, 0, 0) \in D_0 \\ P^- &= (-k, 0, k) \in D_{-1}, \end{aligned}$$

where $k = (b-a)/(b+1)$.

(3) In each of D_1, D_0, and D_{-1}, (DS 2) is linear. In fact, letting

$$x = (x^1, x^2, x^3),\ q = (k, 0, -k)$$

and introducing the 3×3 real matrix

$$A(\alpha, \beta, c) = \begin{pmatrix} -\alpha c & \alpha & 0 \\ 1 & -1 & 1 \\ 0 & -\beta & 0 \end{pmatrix},$$

where A depends on α, β, and a parameter c, which is equal to a in D_0, and b in D_1 and D_{-1}, one can recast (DS 2) as:

$$\frac{dx}{dt} = \begin{cases} A(\alpha, \beta, b)(x - q), & x \in D_1 \\ A(\alpha, \beta, a), & x \in D_0 \\ A(\alpha, \beta, b)(x + q), & x \in D_{-1}. \end{cases}$$

The set of parameter values (α, β, a, b) corresponding to (1.3.3) are given by

$$(\alpha, \beta, a, b) = \left(9, \frac{100}{7}, -\frac{1}{7}, \frac{2}{7}\right).$$

Then the matrix

$$A_1 = A\left(9, \frac{100}{7}, \frac{2}{7}\right)$$

associated with the regions D_1 and D_{-1} has a real eigenvalue

$$\gamma \approx -3.94$$

and a pair of complex-conjugate eigenvalues

$$\sigma_1 \pm \omega_1 i \approx 0.19 \pm 3.05i.$$

Similarly, the matrix

$$A_0 = A\left(9, \frac{100}{7}, -\frac{1}{7}\right)$$

associated with the region D_0 has a real eigenvalue

$$\gamma_0 \approx 2.22$$

and a pair of complex-conjugate eigenvalues

$$\sigma_0 \pm \omega_0 i \approx -0.97 \pm 2.71i.$$

Let $E^r(P^\pm)$ be the eigenspace corresponding to the real eigenvalue γ_1 at $P^\pm$ and let $E^c(P^\pm)$ be the eigenspace corresponding to the complex eigenvalues $\sigma_1 \pm \omega_1 i$ at $P^\pm$. Similarly, let $E^r(O)$ and $E^c(O)$ be the eigenspaces corresponding to γ_0 and $\sigma_0 \pm \omega_0 i$, respectively. (Although in Chapter 3, E^c stands for the center eigenspace, there will be no confusion.) Then the eigenspaces are given explicitly by the following equations:

$$\begin{aligned}
E^r(P^\pm): &\quad \frac{x^1 \mp k}{\gamma_1^2+\gamma_1+\beta} = \frac{x^2}{\gamma_1} = \frac{x^3 \pm k}{-\beta} \\
E^c(P^\pm): &\quad (\gamma_1^2+\gamma_1+\beta)(x^1 \mp k) + \alpha\gamma_1 x^2 + \alpha(x^3 \pm k) = 0 \\
E^r(O): &\quad \frac{x^1}{\gamma_0^2+\gamma_0+\beta} = \frac{x^2}{\gamma_0} = \frac{x^3}{-\beta} \\
E^c(O): &\quad (\gamma_0^2+\gamma_0+\beta)x^1 + \alpha\gamma_0 x^2 + \alpha x^3 = 0\,.
\end{aligned}$$

The relative positions of the eigenspaces and related sets are described in Fig. 1.3.1, where

$$\begin{aligned}
&L_0 = E^c(O) \cap U_1, && C = E^r(O) \cap U_1 \\
&L_1 = E^c(P^+) \cap U_1, && D = E^r(P^+) \cap U_1 \\
&L_2 = \{x \in U_1 | f(x) \parallel U_1\}, && E = L_0 \cap L_2 \\
&A = L_0 \cap L_1, && F = \{x \in L_2 | f(x) \parallel L_2\} \\
&B = L_1 \cap L_2. &&
\end{aligned}$$

Here $f(x) \parallel L_2$ means that the vector field $f(x)$ defined by (DS 2) is parallel with L_2. It will be shown later that L_2 is a straight line segment.(see Lemma 1.4.5)

Since the dynamics is piecewise-linear, this picture (Fig. 1.3.1) already illustrates a great deal of important information. Let us describe the structure of the attractor. Within the parameter range of interest, $E^r(P^\pm)$ and $E^c(O)$ are stable while $E^c(P^\pm)$ and $E^r(O)$ are unstable. Therefore, in this subsection, we will use the following notation for the eigenspaces:

$$\begin{aligned}
E^s(P^\pm) &:= E^r(P^\pm) \quad & E^u(P^\pm) &:= E^c(P^\pm) \\
E^s(O) &:= E^c(O) \quad & E^u(O) &:= E^r(O).
\end{aligned}$$

Let φ^t be the flow generated by (DS 2) and pick an initial condition $x_0 \in E^u(P^+)$ in a neighborhood of P^+. Then, for $t > 0$, the trajectory $\varphi^t(x_0)$ starts wandering away from P^+ on $E^u(P^+)$. After winding around P^+ several times in a counterclockwise direction, it hits the plane U_1 at some time, say $t_1 : x_1 = \varphi^{t_1}(x_0)$.

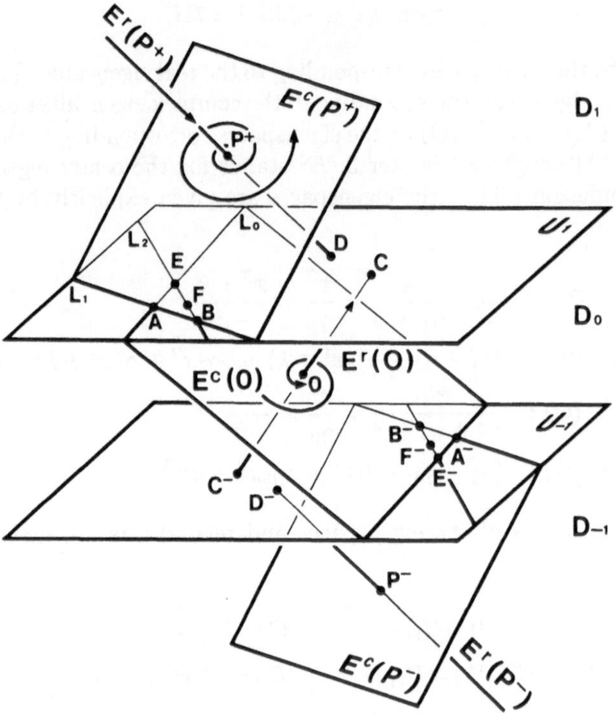

Fig. 1.3.1. Eigenspaces of the equilibrium points and related sets. ©1986 Elsevier Science Publishers.

The trajectory up to t_1 is a spiral because (DS 2) is linear in D_1 and $E^u(P^+)$ is invariant. Clearly, $x_1 \in L_0$. Note that the line L_2 is a straight line parallel to the x^3-axis because $\dot{x}^1$ is independent of x^3. Observe that L_2 separates the plane U_1 into two regions, one (to which A belongs) where $\dot{x}^1 < 0$ and another where $\dot{x}^1 > 0$. Since $\varphi^t(x_0)$ hits the plane U_1 downward (recall that the motion is counterclockwise) at $t = t_1$, one sees that x_1 belongs to the line segment $\overline{GB}$, where G is a point on L_0 to the left of and sufficiently far from A, i.e., $\dot{x}^1 < 0$ at x_1. The fate of $\varphi^t(x_1)$ depends crucially on which part of $\overline{GB}$ x_1 lies (see Plate 1.4).

Case 1: $x_1 = A$. Since the dynamics is linear in D_0, one can check analytically that $\varphi^t(A)$ never hits U_{-1} directly for the parameter values (1.3.3), i.e., the real part σ_0 of the complex-conjugate eigenvalues is negative and small compared to the imaginary part ω. Since $A \in E^s(O)$ and since $E^s(O)$ is invariant,

$\varphi^t(x_1)$ approaches the origin asymptotically as $t \to \infty$. The trajectory is a spiral with an infinite number of rotations for (DS 2) is linear in D_0 and $E^s(O)$ is invariant.

Case 2: x_1 lies in the interior of $\overline{AB}$. In this case $\varphi^t(x_1)$ has two components in the sense that its projection onto $E^s(O)$ approaches the origin asymptotically and its projection onto $\overline{OC} \subset E^u(O)$ wanders away from the origin. This means that $\varphi^t(x_1)$ moves up along a spiral with the central axis $\overline{OC}$ and then eventually hits U_1 again from below: $x_2 = \varphi^{t_2}(x_1)$. The number of rotations of $\varphi^t(x_1)$ around $\overline{OC}$ can get arbitrarily large without bounds if x_1 is very close to A. These processes naturally give rise to the map

$$\Psi : \overline{AB} \to U_1$$

defined by

$$\Psi(x_1) = x_2 .$$

The image $\Psi(\overline{AB})$ is a spiral with its center at C, which is tangent to L_0 at B. After hitting U_1, the trajectory $\varphi^t(x_2)$ has two components in the sense described above: one which stays in $E^u(P^+)$ and moves away from P^+ in a spiral manner and another in $E^s(P^+)$ which approaches P^+ asymptotically. Therefore, $\varphi^t(x_2)$ ascends in a spiral path with the central axis $\overline{DP^+}$ and flattens itself onto $E^u(P^+)$ from below.

Case 3: x_1 lies in the interior of $\overline{GA}$. $\varphi^t(x_1)$ has two components in the same sense as above. One component stays in $E^s(O)$ and asymptotically approaches O in a spiral manner. Another component stays in $E^u(P^+)$ and moves away from O on $\overline{OC^-}$. This means that $\varphi^t(x_1)$ descends along a spiral with the central axis $\overline{OC^-}$, hits U_{-1} at $x_2 = \varphi^{t_2}(x_1)$, and eventually enters region D_{-1}. The closer x_1 is to point A, the larger the number of rotations of $\varphi^t(x_1)$ around $\overline{OC^-}$. After entering into D_{-1}, the flow $\varphi^t(x_2)$ consists of two components: one which is in $E^u(P^-)$ and moves away from P^-, and another which stays in $E^s(P^-)$ and asymptotically approaches P^-. Therefore, $\varphi^t(x_2)$ descends spirally with the central axis $\overline{D^-P^-}$ and eventually flattens itself onto $E^u(P^-)$ from above.

In order to grasp the whole picture, pick a rectangle $abcd$ in D_1 in such a way that $\overline{ab}$ is on $E^u(P^+)$ and $\overline{bc}$ lies below $E^u(P^+)$, i.e., on the side to which D belongs. Fig. 1.3.2 shows how the rectangle $abcd$ changes its shape while flowing along φ^t. Suppose that the rectangle is thin enough and that it is chosen in such a way that the trajectories starting on the line segment $\overline{ef}$ hit L_1. Then, after hitting L_1, they approach the origin asymptotically in a spiral manner with infinitely many rotations. Trajectories starting in the rectangle $abfe$ stay within D_1 or return to D_1 eventually even if they once spend some time in D_0. Trajectories with initial states in the rectangle $cdef$ leave D_1, enter D_0, hit U_{-1} and enter D_{-1}. They turn around P^- and flatten themselves onto $E^u(P^-)$ from above.

Since (DS 2) is symmetric with respect to the origin, one sees that a similar argument applies to a rectangle $a^-b^-c^-d^-$ in region U_{-1}, located symmetrically

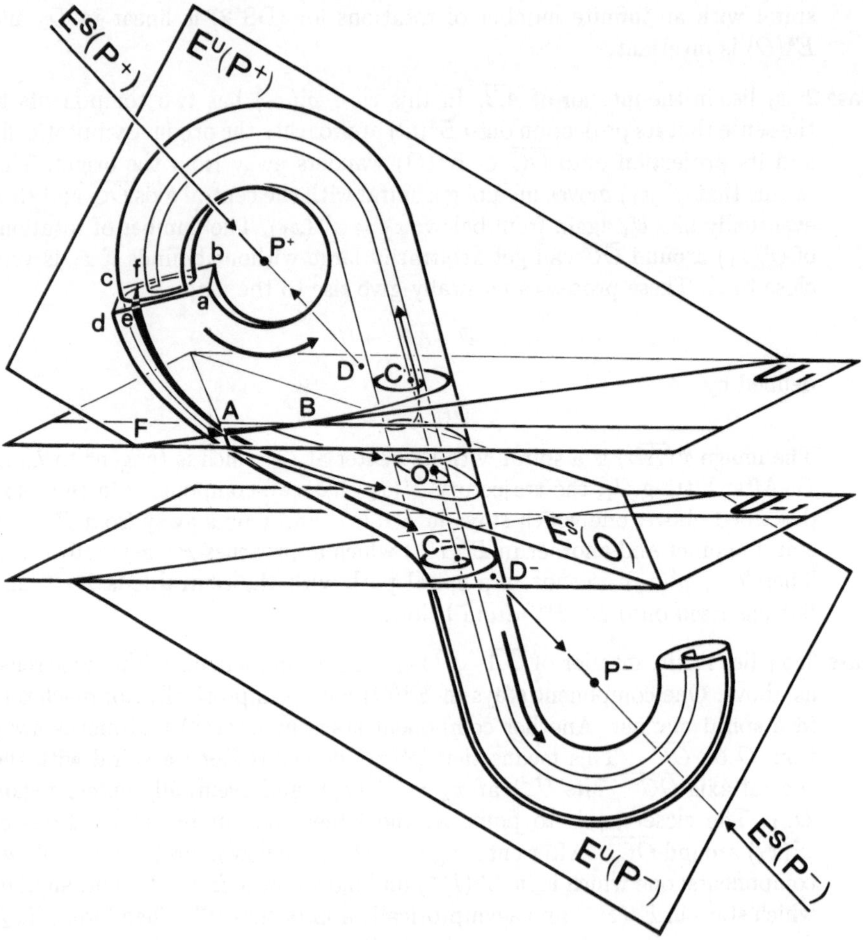

Fig. 1.3.2 Deformations of a rectangle which flows along the trajectories originating from points on the rectangle abcd. ©1985 IEEE.

with respect to origin. Assembling all the information, one obtains a whole picture (Fig. 1.3.3). Observe that the rectangle *abcd* is mapped into two spiral regions with infinitely many rotations: *abfe* is mapped into one spiral region and *cdef* into another spiral region. Note that $E^s(O)$ plays an important role in determining the fate of a trajectory after hitting U_1 or U_{-1}. It differentiates those trajectories which descend (respectively ascend) from those which remain in the upper part (respectively lower part).

A good way of describing the above attractor would be a "*double-scroll*" structure since two sheet-like objects are curled up together into spiral forms.

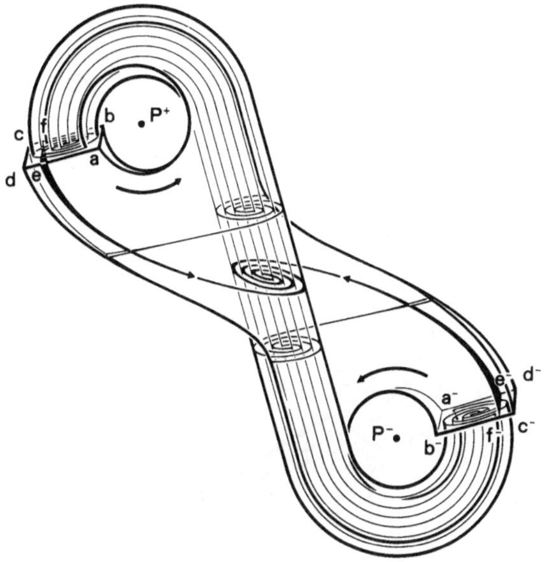

Fig. 1.3.3. Geometric structure of the attractor. ©1985 IEEE.

Locally, however, each sheet like object has a Cantor structure, i.e., infinitely many thin sheets. In order to see the structure more clearly let us look at the cross sections of the attractor. Fig. 1.3.4(b)-(g) shows cross sections of the attractor taken at

$$U(r) := \{(x^1, x^2, x^3) \mid x^1 = r\},\ r = 0, 0.5, 1.0, 1.25, 1.75, \text{and } 2.2\ .$$

Fig. 1.3.4(a) shows positions of the cross sections. On the cross section at $U(1.00)$ $\big($Fig. 1.3.4(d)$\big)$, various line segments and points are superimposed. One can clearly observe the double-scroll structure and how the scrolls flatten themselves gradually. The cross section at $U(1.75)$ $\big($Fig. 1.3.4(f)$\big)$ is particularly interesting. The sheet-like structure is clearly discernible: it is folded many times. Moreover, one can observe that the flattening of the left portion is sharper than that of the right portion so that the spirals still survive on the right portion while they flatten themselves on the left portion. This stems from the fact that trajectories rotate around P^+ in a counterclockwise direction and hence they flatten onto $E^u(P^+)$ as time continues. Note that theoretically the two scrolls are curled up infinitely many times even though numerical results show only several of them curled up. Plate 1.5 is an abstract picture showing only the key features of the attractor.

It is clear that the Double Scroll attractor has a structure quite different from the Lorenz [Lorenz 1963] and Rössler [Rössler 1976, 1979b] attractors. Recall that the Lorenz equation (see Chapter 3) at the popular parameter values $\sigma = 10, b = 8/3, r = 28$ has three equilibrium points: one at the origin, one in the half

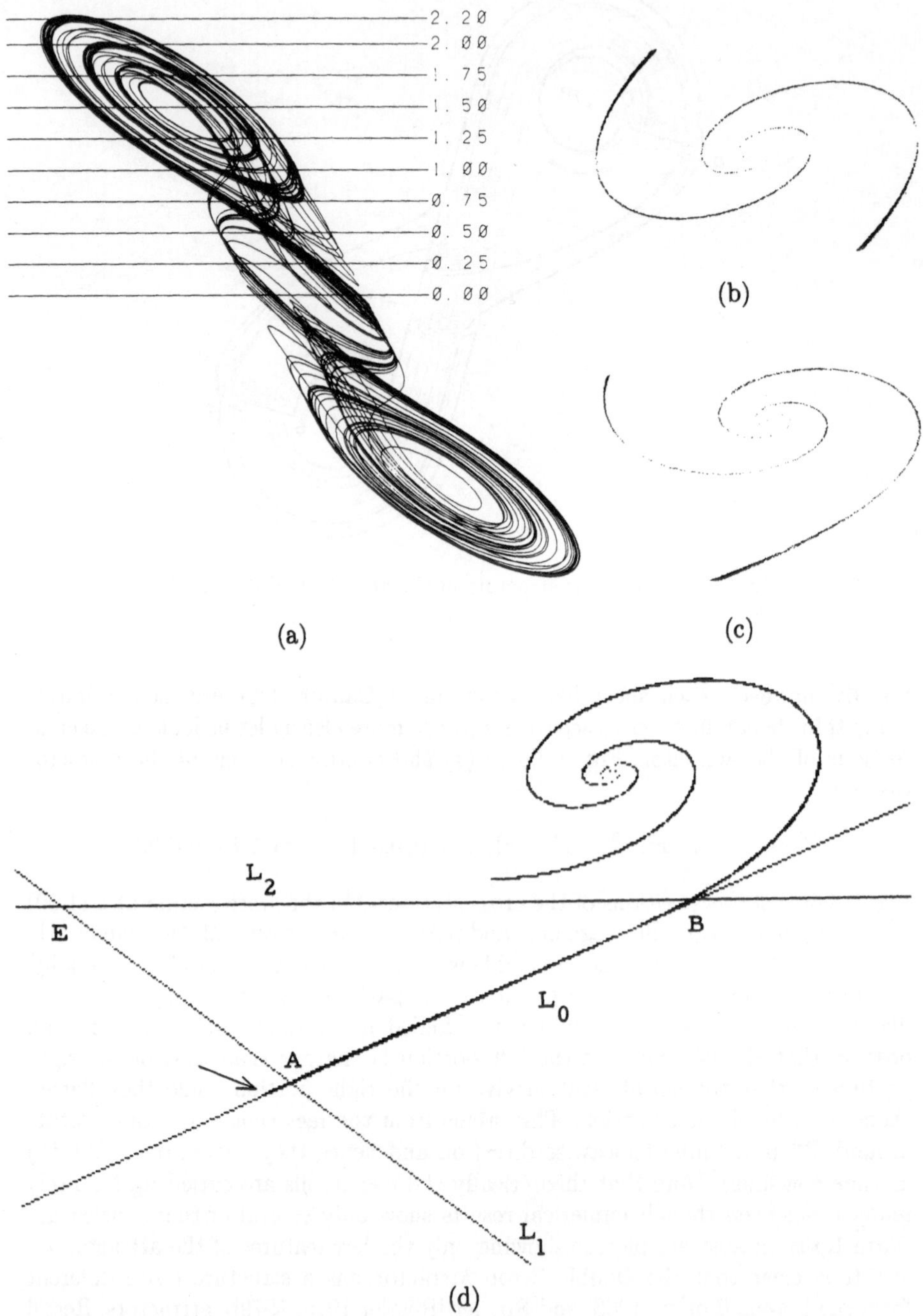

Fig. 1.3.4. Cross section of the attractor. (a) Locations of cross sections. (b) Cross section at $v_{C_1} = 0.00$. (c) Cross sections at $v_{C_1} = 0.50$ (d) Cross section at $v_{C_1} = 1.00$ with related sets. The arrow indicates the "tail" of the attractor. ©1985 IEEE.

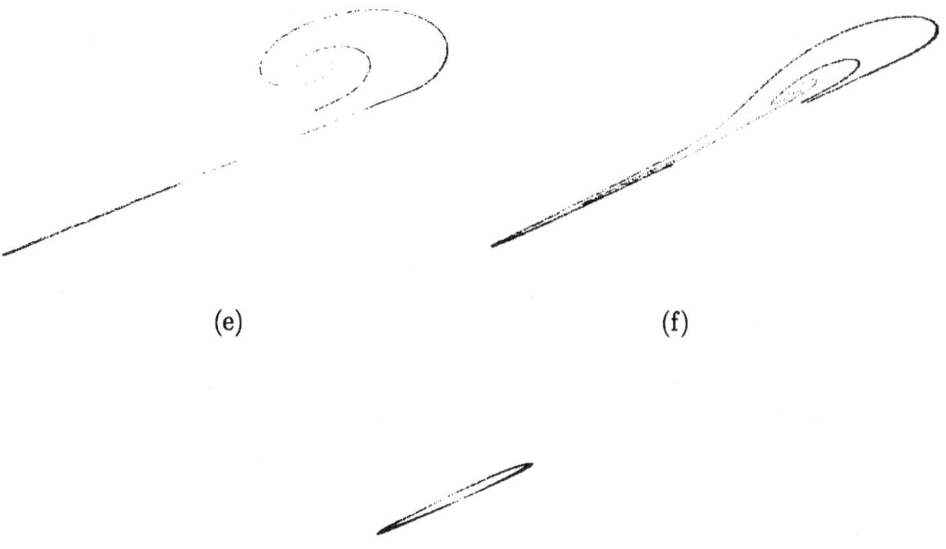

(e) (f) (g)

Fig. 1.3.4. continued (e) Cross section at $v_{C_1} = 1.25$. (f) Cross section at $v_{C_1} = 1.75$. (g) Cross section at $v_{C_1} = 2.20$. All of them have the same scaling. ©1985 IEEE.

space $x > 0$ and another in the half space $x < 0$. Note that the origin belongs to the Lorenz attractor and that the same is true in the Double Scroll. At the origin in the Lorenz attractor, however, all eigenvalues are real, whereas in our case, the origin has one positive real eigenvalue and a pair of complex-conjugate eigenvalues which is responsible for the curled up structure. Recall also that the Lorenz equation is symmetric with respect to the z-axis while (DS 2) is symmetric with respect to the origin. As for the Rössler equation, it has only two equilibrium points. Furthermore, the attractor does not contain any equilibrium point.

1.3.2 Lyapunov Exponents and Lyapunov Dimension

A Lyapunov Exponents First let us rewrite (DS 2) as

$$\frac{dx}{dt} = f(x),$$

where $x = (v_{C_1}, v_{C_2}, i_L)$ and let $\varphi^t(x_0)$ be the trajectory with an initial condition x_0. Lyapunov exponents are generalizations of Floquet exponents (defined only for periodic orbits) so that they make sense for the more general nonperiodic orbits. If Λ is a periodic orbit with period T and if $x_0 \in \Lambda$, then the eigenvalues of $D\varphi^T(x_0)$ (D stands for the derivative), denoted by $e^{\lambda_1}, e^{\lambda_2}$, and e^{λ_3}, are called the Floquet multipliers for Λ. The numbers λ_1, λ_2, and λ_3 are called the Floquet exponents. They give the rate of expansion and contraction of vectors in the tangent space $T_{x_0}\mathbb{R}^3$ along Λ. Since Λ is a closed curve, at least one of the three numbers, say e^{λ_1}, must be 1, and hence $\ln e^{\lambda_1} = 0$. If, in addition, $\ln e^{\lambda_2}, \ln e^{\lambda_3} < 0$, then Λ will be a periodic attractor. If $\ln e^{\lambda_2} < 0$ and $\ln e^{\lambda_3} > 0$, then Λ will be a saddle-type periodic orbit. Now let Λ be a non-periodic invariant set, e.g., a chaotic attractor. There is a technical difficulty in defining Floquet multipliers for Λ. Recall that for a closed orbit, the eigenvalues of $D\varphi^T(x_0)$ are well-defined since $D\varphi^T(x_0)$ maps $T_{x_0}\mathbb{R}^3$ onto itself. On the other hand, this definition is not valid for Λ if it is nonperiodic since $D\varphi^T(x_0)$ does not necessarily map $T_{x_0}\mathbb{R}^3$ onto itself for any t. The definition of Lyapunov exponents requires the invariance of the tangent subbundles. Suppose that for all $t > 0$, there are linear subspaces $E^1_{\varphi^t(x_0)} \supseteq E^2_{\varphi^t(x_0)} \supseteq E^3_{\varphi^t(x_0)}$ in $T_{\varphi^t(x_0)}\mathbb{R}^3$ and numbers $\mu_1(x_0) \geq \mu_2(x_0) \geq \mu_3(x_0)$ such that

$$D\varphi^t(x_0)E^k_{x_0} = E^k_{\varphi^t(x_0)}$$

$$\dim E^k_{\varphi^t(x_0)} = 4 - k$$

$$\mu_k(x_0) = \lim_{T\to\infty} \frac{1}{T} \ln \frac{|D\varphi^T(x_0)e|}{|e|},$$

$$\text{for all } e \in E^k_{x_0} - E^{k+1}_{x_0},\ k = 1, 2, 3.$$

Then the numbers $\mu_1(x_0), \mu_2(x_0)$, and $\mu_3(x_0)$ are called the *Lyapunov exponents* of x_0 in Λ if $x_0 \in \Lambda$. They give the average linearized expansion and contraction rates of nearby points along an orbit. Note that $E^1_{x_0} - E^2_{x_0}$ consists of vectors in $T_{x_0}\mathbb{R}^3$ which expand at the fastest rate, $E^2_{x_0} - E^3_{x_0}$ consists of vectors which expand at the second fastest rate, and the vectors in $E^3_{x_0}$ expand at the slowest rate. In many cases the vectors in $E^3_{x_0}$ are contracted if Λ is an attractor.

The conditions under which Lyapunov exponents exist are strong [Oseledec 1968] and are hard to check. Here we will only give our numerical results. They give, however, good quantitative information about the attractor described in the previous subsections. The computations are nontrivial since one does not know the invariant splittings $E^k_{\varphi^t(x_0)}, k = 1, 2, 3$. One can, however, compute $\mu_1(x_0)$, the largest exponent, numerically, provided that $\mu_1(x_0), \mu_2(x_0)$, and $\mu_3(x_0)$ are not too close to each other. In order to explain this, let $x_0 \in \Lambda$ and pick *any* $e \in T_{x_0}\mathbb{R}^3$. Then

$$\frac{1}{T} \ln \frac{|D\varphi^T(x_0)e|}{|e|} \tag{1.3.5}$$

would give $\mu_1(x_0)$ for T large, because the subspace with the fastest expansion rate would eventually dominate the others and the vector $D\varphi^T(x_0)e$ would fall in $E^1_{\varphi^T(x_0)} - E^2_{\varphi^T(x_0)}$ for any $e \in T_{x_0}\mathbb{R}^3, T$ large.

Computations of $\mu_2(x_0)$ and $\mu_3(x_0)$ need more care since $D\varphi^T(x_0)e$ is dominated by $E^1_{\varphi^T(x_0)} - E^2_{\varphi^T(x_0)}$ and one does not know how to compute $E^2_{\varphi^T(x_0)} - E^3_{\varphi^T(x_0)}$. In order to overcome this problem, we compute

$$\mu_1(x_0) + \mu_2(x_0) \tag{1.3.6}$$

instead of $\mu_2(x_0)$ alone. First note that the number (1.3.6) gives the average expansion or contraction rate of an area element of $E^1_{x_0} - E^3_{x_0}$. Let e_1, e_2 and vectors in $E^3_{x_0}$ be linearly independent. Then the exterior product $e_1 \wedge e_2$ (see [Fleming 1965]) is the parallelepiped generated by e_1 and e_2. Therefore,

$$\frac{1}{T}\ln\frac{|D\varphi^T(x_0)e_1 \wedge D\varphi^T(x_0)e_2|}{|e_1 \wedge e_2|} \tag{1.3.7}$$

would give (1.3.6) for T large. A numerical difficulty arises since $D\varphi^T(x_0)e_1$ and $D\varphi^T(x_0)e_2$ would eventually belong to or become very close to $E^1_{\varphi^t(x_0)} - E^2_{\varphi^t(x_0)}$ for the reason explained before. Hence, the angle between $D\varphi^T(x_0)e_1$ and $D\varphi^T(x_0)e_2$ gets smaller and smaller, and numerical inaccuracy will become serious. In order to overcome this difficulty, recall that the map:

$$(e_1, e_2) \to e_1 \wedge e_2$$

is bilinear , i.e., linear in each argument, and rewrite (1.3.7) as

$$\frac{1}{T}\ln\frac{|\left[D\varphi^T(x_0) \wedge D\varphi^T(x_0)\right](e_1 \wedge e_2)|}{|e_1 \wedge e_2|}, \tag{1.3.8}$$

where

$$D\varphi^T(x_0) \wedge D\varphi^T(x_0)$$

is the induced linear map. Since this is a 3×3 matrix and since

$$e_{12} := e_1 \wedge e_2 \tag{1.3.9}$$

is a 3-dimensional vector, we can compute (1.3.6) without the above difficulty. The initial vector e_{12} of (1.3.9) can be chosen arbitrarily for the same reason that e of (1.3.5) can be chosen arbitrarily, provided that $\mu_1(x_0) + \mu_2(x_0)$ dominates $\mu_1(x_0) + \mu_3(x_0)$ and $\mu_2(x_0) + \mu_3(x_0)$ by reasonable margins.

Finally, there is also a difficulty in computing $\mu_3(x_0)$ alone for the same reason as before. We compute, instead,

$$\mu_1(x_0) + \mu_2(x_0) + \mu_3(x_0), \tag{1.3.10}$$

which gives the average contraction or expansion rate of a volume element in $E^1_{x_0}$, assuming that $E^1_{x_0} = T_{x_0}\mathbb{R}^3$. An argument similar to the above shows that

$$\frac{1}{T}\ln\frac{|\left[D\varphi^T(x_0) \wedge D\varphi^T(x_0) \wedge D\varphi^T(x_0)\right](e_1 \wedge e_2 \wedge e_3)|}{|e_1 \wedge e_2 \wedge e_3|} \tag{1.3.11}$$

would eventually give (1.3.10), where

$$\text{span}\{e_1, e_2, e_3\} = E^1_{x_0} = T_{x_0}\mathbb{R}^3.$$

B Computations Based on the above algorithms, the Lyapunov exponent $\mu_1(x_0)$ was computed by solving the variational equation

$$\frac{dy}{dt} = Df(\varphi^t(x_0))y$$

with

$$y(0) = e, \ |e| = 1$$

and then computing

$$\frac{1}{T} \ln |y(T)|.$$

The result is

$$\mu_1(x_0) \approx 0.23, \tag{1.3.12}$$

where

$$\begin{cases} x_0 &= (-1.7713, 0.0527854, 1.74606) \\ e &= \left(\frac{1}{\sqrt{3}}, \frac{1}{\sqrt{3}}, \frac{1}{\sqrt{3}}\right) \\ T &= 3000. \end{cases}$$

Of course, one has to periodically renormalize $y(t)$ after a reasonable amount of time since $|y(t)|$ gets very large. More specifically, letting $T = n\tau$, one sees that

$$\begin{aligned} \frac{1}{T} \ln |y(T)| &= \frac{1}{n\tau} \ln |D\varphi^{n\tau}(y(0))| \\ &= \frac{1}{n\tau} \ln(|D\varphi^{\tau}(x((n-1)\tau)y((n-1)\tau)|/|y(0)|) \\ &= \frac{1}{n\tau} \sum_{k=0}^{n-1} \ln \frac{|D\varphi^{\tau}(x(k\tau))y(k\tau)|}{|y(k\tau)|}. \end{aligned}$$

If one renormalizes

$$|y(k\tau)| = 1$$

at each k, then

$$\frac{1}{T} \ln |y(T)| = \frac{1}{n\tau} \sum_{k=0}^{n-1} \ln |D\varphi^{\tau}(x(k\tau))y(k\tau)|.$$

With $T = 3000$, $\tau = 10$ and a Runge-Kutta step size 0.005, a fairly good convergence was achieved and (1.3.12) appeared to be insensitive to the initial tangent vector e and initial condition x_0.

In order to compute (1.3.6) let

$$y := D\varphi^{\tau}(x_0)e_1, \ z := D\varphi^{\tau}(x_0)e_2.$$

Then

$$\begin{aligned} \frac{d}{dt}(y \wedge z) &= \frac{dy}{dt} \wedge z + y \wedge \frac{dz}{dt} \\ &= [Df(\varphi^t(x_0))y] \wedge z + y \wedge [Df(\varphi^t(x_0))z] \\ &= [Df(\varphi^t(x_0)) \wedge I + I \wedge Df(\varphi^t(x_0))]\, y \wedge z, \end{aligned} \tag{1.3.13}$$

where I is the 3×3 identity matrix and

$$\begin{aligned} \left[Df(\varphi^t(x_0)) \wedge I\right](e_i \wedge e_j) &:= \left(Df(\varphi^t(x_0))e_i\right) \wedge e_j \\ \left[I \wedge Df(\varphi^t(x_0))\right](e_i \wedge e_j) &:= e_i \wedge \left(Df(\varphi^t(x_0))e_j\right) \end{aligned}$$

(An explicit formula is given in C below.) Therefore, solving the "two-dimensional" variational equation (1.3.13) with

$$|(y \wedge z)(0)| = |e_{12}| = 1,$$

one can compute

$$\frac{1}{T} \ln |(y \wedge z)(T)|.$$

Our computation gives

$$\mu_1(x_0) + \mu_2(x_0) \approx 0.23, \tag{1.3.14}$$

where x_0 and T are the same as before and

$$e_{12} = \left(\frac{1}{\sqrt{3}}, \frac{1}{\sqrt{3}}, \frac{1}{\sqrt{3}}\right).$$

Again (1.3.14) appeared to depend on neither x_0 nor e_{12}.

Finally, observing that

$$D\varphi^t(x_0) \wedge D\varphi^t(x_0) \wedge D\varphi^t(x_0) = \det D\varphi^t(x_0)$$

and

$$\frac{d}{dt} D\varphi^t(x_0) = \mathrm{tr} Df(\varphi^t(x_0)) \det D\varphi^t(x_0),$$

one can compute (1.3.11). Our computation with the same x_0 and T gives

$$\mu_1(x_0) + \mu_2(x_0) + \mu_3(x_0) \approx -1.55. \tag{1.3.15}$$

It follows from (1.3.12), (1.3.14), and (1.3.15) that

$$\begin{cases} \mu_1(x_0) \approx 0.23 \\ \mu_2(x_0) \approx 0 \\ \mu_3(x_0) \approx -1.78. \end{cases} \tag{1.3.16}$$

This shows that in the Double Scroll attractor certain line elements are expanded, certain area elements are preserved and certain volume elements are contracted.

This agrees with the sheet-like structure described in Subsection 1.3.1. The Lorenz attractor ($\sigma = 16, b = 4, r = 40$) has much sharper expansion and contraction rates than the Double Scroll attractor [Shimada and Nagashima 1979]:

$$\begin{cases} \mu_1(x_0) & \approx & 1.37 \\ \mu_2(x_0) & \approx & 0 \\ \mu_3(x_0) & \approx & -22.37. \end{cases} \tag{1.3.17}$$

Note also that in the Lorenz attractor, volume elements are contracted uniformly since its divergence $= -(\sigma + r + 1)$, a negative constant.

C Explicit Formula Here we will give an explicit formula for (1.3.13). Let $\{e_1, e_2, e_3\}$ be the standard basis for $\mathbb{R}^3$. Then $e_1 \wedge e_2 = e_{12}, e_2 \wedge e_3 = e_{23}, e_1 \wedge e_3 = e_{13}$ are the standard basis for $(\mathbb{R}^3_2)^*$, the set of all alternating bilinear functions on $\mathbb{R}^3 \times \mathbb{R}^3$ [Fleming 1965], where $\wedge$ denotes the exterior product. They satisfy,

$$e_i \wedge e_j = -e_j \wedge e_i, \quad e_i \wedge e_i = 0.$$

Since

$$Df(x) = \begin{pmatrix} -\alpha Dh(x^1) & \alpha & 0 \\ 1 & -1 & 1 \\ 0 & -\beta & 0 \end{pmatrix}$$

one can easily compute

$$\begin{aligned} Df(x) \wedge I &= \begin{pmatrix} -\alpha Dh(x^1) & 0 & 0 \\ 0 & -1 & 1 \\ 0 & \alpha & -\alpha Dh(x^1) \end{pmatrix} \\ I \wedge Df(x) &= \begin{pmatrix} -1 & 0 & 1 \\ 0 & 0 & 0 \\ -\beta & 0 & 0 \end{pmatrix}. \end{aligned}$$

Therefore,

$$Df(x) \wedge I + I \wedge Df(x) = \begin{pmatrix} -1 - \alpha Dh(x^1) & 0 & 1 \\ 0 & -1 & 1 \\ -\beta & \alpha & -\alpha Dh(x^1) \end{pmatrix}.$$

Remark 1.3.1. Theoretically, there is some difficulty in using this formula because $h(\cdot)$ is piecewise-linear and Dh has discontinuities at $x^1 = \pm 1$. Numerically however, there seem to be no problem if one chooses a small enough Runge-Kutta step size.

D Lyapunov Dimension The dimension of a chaotic attractor is one of the very few quantitative measures which are associated with chaotic attractors. Among the various different definitions of dimension of chaotic attractors [Young 1983], we compute the Lyapunov dimension since it naturally comes from Lyapunov exponents. We do not claim that this is the most appropriate one. Recall (1.3.16) and recall that our numerical results indicate that these numbers do not seem to depend on x_0. Assume that this is, in fact, the case. Then since $\mu_1 + \mu_2 > 0$ and since $\mu_1 + \mu_2 + \mu_3 < 0$, the Lyapunov dimension is given by

$$d_L = 2 + \frac{\mu_1 + \mu_2}{|\mu_3|} \approx 2.13. \qquad (1.3.18)$$

Let us compare this number with the Lorenz attractor. It follows from (1.3.17) that for the Lorenz attractor,

$$d_L = 2 + \frac{\mu_1 + \mu_2}{|\mu_3|} \approx 2.06. \qquad (1.3.19)$$

Both of them are fractals between 2 and 3 which agree with the sheet-like structure observed. Since (1.3.18) is greater than (1.3.19), one might say that our attractor is slightly "thicker" than the Lorenz attractor (with $\sigma = 16, b = 4, r = 40$).

E Time Waveforms and Power Spectra Plate 1.6 shows measured time waveforms from the real circuit while Plate 1.7 gives measured power spectra of v_{C_1} and v_{C_2}. The power spectrum of i_L is similar to that of v_{C_1}.

1.4 The Double Scroll Circuit is Chaotic in the Sense of Shil'nikov

1.4.1 Statement

This subsection shows that (DS 2) (see Subsection 1.3.1) is chaotic in the sense of Shil'nikov [Shil'nikov 1965] from the viewpoint that *chaos is a consequence of homoclinicity.* Consider (DS 2) and fix

$$a = -\frac{1}{7}, b = \frac{2}{7}, \alpha = 7, \qquad (1.4.1)$$

and let

$$J = [6.5, 10.5]. \qquad (1.4.2)$$

The goal of this subsection is to prove the following:

Proposition 1.4.1. *There is a value $\beta \in J$ such that (DS 2) with (1.4.1) is* **chaotic** *in the sense of Shil'nikov. Namely, there is a homoclinic orbit at the origin and the Poincaré return map contains a horseshoe in a neighborhood of the homoclinic orbit. Hence there is a positively and negatively invariant Cantor set Λ containing:*

(1) infinitely many saddle-type periodic orbits of arbitrarily long periods;

(2) uncountable number of bounded non-periodic orbits;

(3) a dense orbit.

Moreover, the horseshoe persists under perturbations.

Remark 1.4.2. Plate 1.8 shows the homoclinic orbit in question. Note that the symmetry of (DS 2) implies that homoclinic orbits appear in pairs. Plate 1.9 shows how a horseshoe is formed. In Plate 1.9, an appropriate coordinate change has been made in such a way that the stable eigenspace coincides with the (x^1, x^2)-plane and the unstable eigenspace coincides with the x^3-axis. Consider a cylinder as described in the figure and the narrow rectangle on it. If we let the rectangle flow along the solution trajectories, we see that the rectangle is compressed in the horizontal direction, stretched in the vertical direction, bent and then returned to the original cylinder, forming a horseshoe map. Note that a very thin object like A returns to an object B.

Remark 1.4.3. Shil'nikov's theorem we invoke here is its piecewise-linear version used in [Arneodo, Coullet and Tresser 1981, 1982]. Chapter 3 gives the Shil'nikov theorem for smooth vector fields.

Remark 1.4.4. Note that orbits in Λ will be extremely complicated because of (1)-(3) (see Chapter 3).

In order to accomplish the goal of this section, we need to show two facts:

Fact 1. There is a homoclinic orbit at the origin, i.e., an orbit which approaches the origin as $t \to +\infty$ and $t \to -\infty$;

Fact 2. The eigenvalues at the origin consist of a real $\gamma_0 > 0$ and a complex-conjugate pair $\sigma_0 \pm \omega_0 i, \sigma_0 < 0, \omega_0 \neq 0$, such that

$$|\sigma_0| < \gamma_0, \tag{1.4.3}$$

where the subscript "$_0$" indicates that the eigenvalues are associated with O.

Proofs of these facts are nontrivial. Plate 1.10 shows the homoclinic orbit together with relevant subsets in the state space. It is the same as Fig. 1.3.1 except that the homoclinic orbit of interest is added. Within the parameter range of (1.4.1) and (1.4.2), $E^r(P^\pm)$ and $E^c(O)$ are stable, while $E^r(O)$ and $E^c(P^\pm)$ are unstable. In order to characterize the homoclinicity in question, first note that the vector field on the left-hand side of L_2 is downward, while it is upward on the right-hand side of L_2. Recall that $E^r(O)$, the eigenspace at origin corresponding to the real eigenvalue, is unstable, and consider the trajectory $\varphi^t(C)$ starting at $C = E^r(O) \cap U_1$. If

$$\boxed{\varphi^{t_1}(C) \in \overline{AE}} \tag{1.4.4}$$

for some $t_1 > 0$, then $\varphi^t(C)$ tends to the origin asymptotically because $\overline{AE}$ is a part of $E^c(O)$, which is invariant under (DS 2). This clearly signifies homoclinicity.

Thanks to the piecewise-linearity of (DS 2), the homoclinicity condition (1.4.4) is extremely simple. Yet it is impossible to prove it by hand because one has to compute the *half-return time*, i.e., the time at which the trajectory hits U_1, which, in turn, is a zero of a transcendental equation involving sine, cosine, and exponential functions. Our purpose will be accomplished in two steps:

Step 1. Make an appropriate change of coordinate system so that (1.4.4) is more tractable;
Step 2. Perform a computer assisted proof.

1.4.2 The Class $\mathcal{L}$

Let $\mathcal{L}$ be the class of all three-region continuous piecewise-linear vector fields $f : \mathbb{R}^3 \to \mathbb{R}^3$ satisfying:

(1) $f(-x) = -f(x)$;

(2) f is proper (see Chapter 2);

(3) f has three equilibrium points, one at the origin O, one in the interior of D_1 (labeled P^+) and one in the interior D_{-1} (labeled P^-);

(4) The linear vector field in each region has a real eigenvalue (labeled γ_0 for D_0 and γ_1 for D_1 and D_{-1}), and a pair of compex-conjugate eigenvalues (labeled $\sigma_0 \pm \omega_0 i$ for D_0 and $\sigma_1 \pm \omega_1 i$ for D_1 and D_{-1}).

Lemma 1.4.5. *Let ξ be a linear vector field on $\mathbb{R}^3$ having a real eigenvalue γ and a pair of complex-conjugate eigenvalues $\sigma \pm \omega i$. Let U be any plane which is not parallel to the eigenspaces and which does not pass through the origin.*

(1)

$$L = \{x \in U |\ \xi(x) \parallel U\} \tag{1.4.5}$$

is a straight line.

(2) There is a coordinate system $x' = (x^{1'}, x^{2'}, x^{3'})$, such that ξ is transformed into real Jordan form

$$\xi(x') = \begin{pmatrix} \sigma & -\omega & 0 \\ \omega & \sigma & 0 \\ 0 & 0 & \gamma \end{pmatrix} \begin{pmatrix} x^{1'} \\ x^{2'} \\ x^{3'} \end{pmatrix} \tag{1.4.6}$$

and such that the equation for U in the new coordinate system is

$$U = \{(x^{1'}, x^{2'}, x^{3'}) |\ x^{1'} + x^{3'} = 1\}. \tag{1.4.7}$$

Proof. Choose vectors e_a, e_b, and e_c in $\mathbb{R}^3$ such that

(1) e_a is the real part of the complex eigenvector associated with $\sigma \pm \omega i$;

(2) e_b is the negative imaginary part of the complex eigenvector associated with $\sigma \pm \omega i$;

(3) e_c is the eigenvector associated with γ.

Let $Q = (e_a,\ e_b,\ e_c)$. Then $J = Q^{-1}MQ$ transforms an arbitrary 3×3 matrix with eigenvalues $\sigma \pm i\omega$ and γ into its real Jordan form. Hence, under this new coordinate system $x'' = Q^{-1}x$, ξ assumes the following real Jordan form:

$$\xi(x'') = \begin{pmatrix} \sigma & -\omega & 0 \\ \omega & \sigma & 0 \\ 0 & 0 & \gamma \end{pmatrix} \begin{pmatrix} x^{1''} \\ x^{2''} \\ x^{3''} \end{pmatrix}$$

where $x'' = \left(x^{1''},\ x^{2''},\ x^{3''}\right)$. Moreover, U is represented by

$$lx^{1''} + mx^{2''} + nx^{3''} = d,$$

where $l^2 + m^2 \neq 0$, $n \neq 0$, and $d \neq 0$ because U is not parallel to either eigenspace and does not pass through the origin. In the new x'' coordinate system, the three vectors e_a, e_b, and e_c are transformed into three orthonormal axes, the eigenspace spanned by e_a and e_b is transformed into the $(x^{1''}, x^{2''})$-plane, and the real eigenvector e_c is transformed into the $x^{3''}$-axis. The U-plane is of course transformed into another plane U'' which does not pass through the origin and is not parallel to the $(x^{1''}, x^{2''})$-plane. Our next goal is to rotate U'' so that it makes a 45° angle with the $(x^{1''}, x^{2''})$-plane, and intersects it at $x^{1''} = 1$. This can be achieved by choosing yet another coordinate system $x' = (x^{1'},\ x^{2'},\ x^{3'})$ such that the three orthonormal vectors, $e_a' = (1,0,0)$, $e_b' = (0,1,0)$, and $e_c' = (0,0,1)$, in the x'-coordinate system are transformed from e_1, e_2, and e_3 with the geometrical property which achieves the above transformation; namely, (i) make e_2 parallel to U''; (ii) make e_1 perpendicular to e_2, such that the tip of e_1 lies on U''; (iii) make e_1 and e_2 lie on the $(x^{1''}, x^{2''})$-plane; (iv) make $|e_2| = |e_1|$; (v) make $e_3 = (0,0,d_3)$, where d_3 is chosen so that the tip of e_3 lies on U''. The above requirements define e_1, e_2, and e_3 uniquely as

$$e_1 = \left(d/(l^2+m^2)\right)(l, m, 0)$$
$$e_2 = \left(d/(l^2+m^2)\right)(-m, l, 0)$$
$$e_3 = (d/n)(0, 0, 1)\ .$$

For the sake of simplicity, we are omitting the symbol "T" for a transpose of a vector. Namely, e_1, e_2 and e_3 are column vectors. Note that the new coordinate system x' is related to x'' by $x' = Q_1^{-1}x''$, where $Q_1 = (e_1,\ e_2,\ e_3)$. Since

$$\begin{aligned} Q_1 &= (e_1,\ e_2,\ e_3) \\ &= \begin{pmatrix} dl/(l^2+m^2) & -dm/(l^2+m^2) & 0 \\ dm/(l^2+m^2) & dl/(l^2+m^2) & 0 \\ 0 & 0 & d/n \end{pmatrix} \end{aligned}$$

and since

$$\begin{aligned} \begin{pmatrix} x^{1'} \\ x^{2'} \\ x^{3'} \end{pmatrix} &= Q_1^{-1}\begin{pmatrix} x^{1''} \\ x^{2''} \\ x^{3''} \end{pmatrix} \\ &= \begin{pmatrix} l/d(l^2+m^2) & m/d(l^2+m^2) & 0 \\ -m/d(l^2+m^2) & l/d(l^2+m^2) & 0 \\ 0 & 0 & n/d \end{pmatrix}\begin{pmatrix} x^{1''} \\ x^{2''} \\ x^{3''} \end{pmatrix}, \end{aligned}$$

we see that

$$\begin{aligned} \xi(x') &= Q_1^{-1}\begin{pmatrix} \sigma & -\omega & 0 \\ \omega & \sigma & 0 \\ 0 & 0 & \gamma \end{pmatrix} Q_1 \begin{pmatrix} x^{1''} \\ x^{2''} \\ x^{3''} \end{pmatrix} \\ &= \begin{pmatrix} \sigma & -\omega & 0 \\ \omega & \sigma & 0 \\ 0 & 0 & \gamma \end{pmatrix}\begin{pmatrix} x^{1''} \\ x^{2''} \\ x^{3''} \end{pmatrix}. \end{aligned}$$

Here we used the fact that $Q_1^{-1}JQ_1 = J$ because by choosing $|e_1| = |e_2|$ and $e_1 \perp e_2$, we see that the first two rows of Q_1 give rise to a planar rotation plus scale change. Therefore

$$U:\quad (l, m, n)\begin{pmatrix} x^{1''} \\ x^{2''} \\ x^{1''} \end{pmatrix} = d$$

is equivalent to

$$(l, m, n)Q \begin{pmatrix} x^{1'} \\ x^{2'} \\ x^{1'} \end{pmatrix} = d$$

which, in turn, is equivalent to

$$(d, 0, d) \begin{pmatrix} x^{1'} \\ x^{2'} \\ x^{1'} \end{pmatrix} = d$$

so that

$$x^{1'} + x^{3'} = 1.$$

□

For each $x \in L$, (1.4.5) implies that the inner product $\langle \xi(x), h \rangle = 0$, where $h := (1, 0, 1)$ is a normal vector to U in view of (1.4.7). Substituting $\xi(x')$ from (1.4.6) into the above inner product, and solving for $x^{2'}$, we see that L in (1.4.5) is a straight line defined by

$$L : x^{2'} = \tilde{\sigma} x^{1'} + \tilde{\gamma}(1 - x^{1'}), \; x^{3'} = 1 - x^{1'} \tag{1.4.8}$$

where $\tilde{\sigma} := \sigma/\omega$ and $\tilde{\gamma} := \gamma/\omega$.

Remark 1.4.6. The straight line L intersects the line $\{(x^{1'}, x^{2'}, x^{3'}) | x^{1'} = 1, x^{3'} = 0\}$ at $(x^{1'}, x^{2'}, x^{3'}) = (1, \tilde{\sigma}, 0)$.

For simplicity, we will often suppress the superscript + and write P instead of P^+. Note that the continuity of the vector field f implies that

$$\begin{aligned} &f(A) \parallel E^c(P), \quad f(A) \parallel E^c(O) \\ &f(B) \parallel L_1, \qquad f(E) \parallel L_0 \\ &f(C) \parallel E^r(O), \quad f(D) \parallel E^r(P). \end{aligned}$$

Since $f \in \mathcal{L}$ is a vector field on $\mathbb{R}^3$ and since f is a three-region system, there are, in general, nine parameters. The next objective here is to eliminate as many of these parameters as possible. Let M_0 (respectively M_1) be a 3×3 matrix representations of f on D_0 (respectively D_1 and D_{-1}) with respect to the original coordinates x. Let $\Psi_0 : D_0 \to \mathbb{R}^3$ and $\Psi_1 : D_1 \to \mathbb{R}^3$ denote appropriate affine maps which reduce M_0 and M_1 to the real Jordan form in (1.4.6) while simultaneously transforming the equation describing $U_{\pm 1}$ to the simplified form in (1.4.7):

(1)
$$\Psi_0(O) = O \tag{1.4.9}$$

$$\Psi_0(U_1) = V_0 := \{(x^{1'}, x^{2'}, x^{3'}) |\ x^{1'} + x^{3'} = 1\} \tag{1.4.10}$$

$$\Psi_0(U_{-1}) = V_0^- := \{(x^{1'}, x^{2'}, x^{3'}) |\ x^{1'} + x^{3'} = -1\} \tag{1.4.11}$$

$$\frac{1}{\omega_0} D\Psi_0(f(\Psi_0^{-1} x')) = f_0(x') := \begin{pmatrix} \tilde{\sigma}_0 & -1 & 0 \\ 1 & \tilde{\sigma}_0 & 0 \\ 0 & 0 & \tilde{\gamma}_0 \end{pmatrix} x' \tag{1.4.12}$$

where $\tilde{\sigma}_0 := \sigma_0/\omega_0$, $\tilde{\gamma}_0 := \gamma_0/\omega_0$, and $D\Psi_0$ denotes the derivative of Ψ_0.

(2)
$$\Psi_1(P) = O \tag{1.4.13}$$

$$\Psi_1(U_1) = V_1 := \{(x^{1''}, x^{2''}, x^{3''}) |\ x^{1''} + x^{3''} = 1\} \tag{1.4.14}$$

$$\frac{1}{\omega_1} D\Psi_1(f(\Psi_1^{-1} x'')) = f_1(x'') := \begin{pmatrix} \tilde{\sigma}_1 & -1 & 0 \\ 1 & \tilde{\sigma}_1 & 0 \\ 0 & 0 & \tilde{\gamma}_1 \end{pmatrix} x'' \tag{1.4.15}$$

where $\tilde{\sigma}_1 := \sigma_1/\omega_1$ and $\tilde{\gamma}_1 := \gamma_1/\omega_1$, and $D\Psi_1$ denotes the derivative of Ψ_1.

The set $\{f_0, V_0, \Psi_0\}$ (respectively $\{f_1, V_1, \Psi_1\}$) will be referred to as the D_0-unit (respectively D_1-unit). (see Fig. 1.4.1) The images of the points A, B, C, D, E, and F in Fig. 1.4.1 will be denoted by corresponding subscripts:

$$\begin{aligned} D_0: \quad & A_0 := \Psi(A), \quad B_0 := \Psi_0(B), \quad C_0 := \Psi(C), \\ & D_0 := \Psi(D), \quad E_0 := \Psi_0(E), \quad F_0 := \Psi(F), \\ D_1: \quad & A_1 := \Psi(A), \quad B_1 := \Psi_1(B), \quad C_1 := \Psi(C), \\ & D_1 := \Psi(D), \quad E_1 := \Psi_1(E), \quad F_1 := \Psi(F). \end{aligned}$$

Since A, B, C, D, E, and F are located on various intersection lines in Fig. 1.4.1, their images (under any affine map) must lie on the corresponding lines in the new coordinate systems. These lines are images of intersections between various eigenspaces ($E^c(O)$ or $E^r(O)$) with the plane U_1 in Fig. 1.4.1. In particular, it can be shown that

$$\Psi_0(E^c(O)) = \{(x^{1'}, x^{2'}, x^{3'}) |\ x^{3'} = 0\} \tag{1.4.16}$$

$$\Psi_0(E^r(O)) = \{(x^{1'}, x^{2'}, x^{3'}) |\ x^{1'} = x^{2'} = 0\} \tag{1.4.17}$$

$$\Psi_0(L_0) = \{(x^{1'}, x^{2'}, x^{3'}) |\ x^{1'} = 1, x^{3'} = 0\} \tag{1.4.18}$$

$$\Psi_0(L_2) = \{(x^{1'}, x^{2'}, x^{3'}) |\ x^{2'} = \tilde{\sigma}_0 x^{1'} + \tilde{\gamma}_0(1 - x^{1'}), \quad x^{3'} = 1 - x^{1'}\}. \tag{1.4.19}$$

Since $C = E^r(O) \cap U_1$, it follows from (1.4.17) and (1.4.10) that $C_0 = (0, 0, 1)$. Since $E = L_0 \cap L_2$, it follows that $F_0 \in \Psi_0(L_2)$ and $f_0(F_0) \parallel \Psi_0(L_2)$. Hence, the

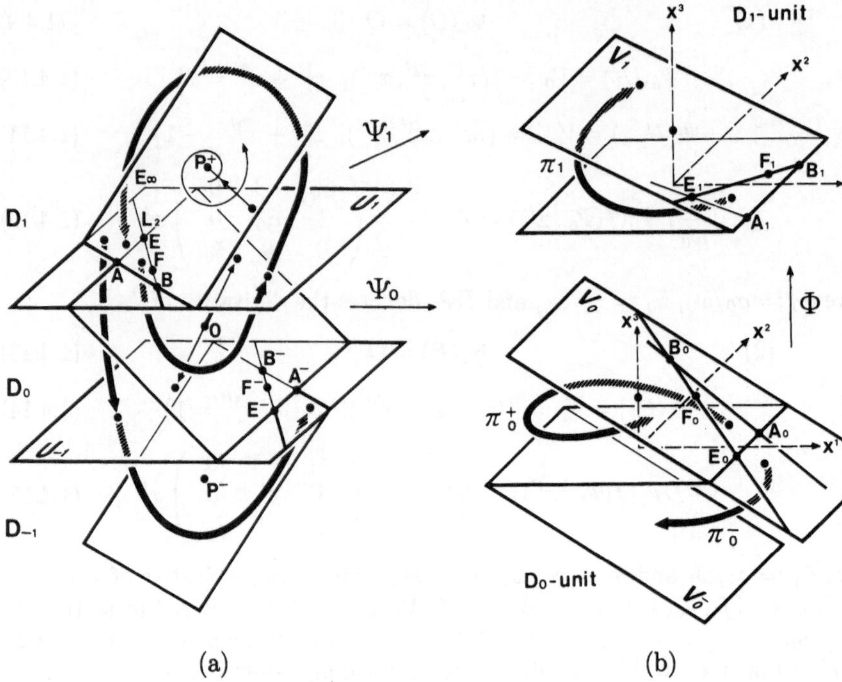

Fig. 1.4.1. Maps Ψ_0 and Ψ_1 together with relevant subsets. (a) Original state space with typical trajectories. (b) D_0-unit and D_1-unit and half-return maps. ©1986 IEEE.

coordinates of F_0 must satisfy

$$x^{2'} = \tilde{\sigma}_0 x^{1'} + \tilde{\gamma}_0(1 - x^{1'}),\ x^{3'} = 1 - x^{1'},$$
$$\frac{\tilde{\sigma}_0 x^{1'} - x^{2'}}{1} = \frac{x^{1'} + \tilde{\sigma}_0 x^{2'}}{\tilde{\sigma}_0 - \tilde{\gamma}_0} = \frac{\tilde{\gamma}_0 x^{3'}}{-1}. \quad (1.4.20)$$

Since A_0 lies on the line $\Psi_0(L_0)$, we can write $A_0 = (1, p_0, 0)$ for some $p_0 \in \mathbb{R}$. Since $B = L_1 \cap L_2$ and $f(B) \parallel L_1$, the coordinates of B_0 are determined by $B_0 \in \Psi_0(L_2)$ and $f_0(B_0) \parallel \overrightarrow{B_0A_0}$, where the arrow denotes the vector from B_0 to A_0. Since B_0, E_0, and F_0 all lie on the line $\Psi_0(L_2)$, it follows that

$$\overrightarrow{F_0B_0} = k_0\overrightarrow{E_0F_0}, \quad (1.4.21)$$

where $k_0 \in \mathbb{R}$. Similarly, we can derive the coordinates of A_1, B_1, C_1, D_1, E_1, and F_1 in the new coordinate system for the D_1-unit in Fig. 1.4.1 and obtain

$$\overrightarrow{E_1F_1} = k_1\overrightarrow{F_1B_1}, \quad (1.4.22)$$

where $k_1 \in \mathbb{R}$. Explicit coordinates for the image of all points in Fig. 1.4.1 are:

Points in the D_0-Unit ($\tilde{\sigma}_0 := \sigma_0/\omega_0, \tilde{\gamma}_0 := \gamma_0/\omega_0$)

$$A_0 = (1, p_0, 0) \tag{1.4.23}$$

where

$$p_0 := \tilde{\sigma}_0 + \frac{k_0}{\tilde{\gamma}_0}(\tilde{\sigma}_0^2 + 1), \; k_0 := \tilde{\gamma}_0(p_0 - \tilde{\sigma}_0)/(\tilde{\sigma}_0^2 + 1) \tag{1.4.24}$$

$$\begin{aligned} B_0 = &\left(\tilde{\gamma}_0(\tilde{\gamma}_0 - \tilde{\sigma}_0 - p_0)/Q_0, \tilde{\gamma}_0\left[1 - p_0(\tilde{\sigma}_0 - \tilde{\gamma}_0)\right]/Q_0,\right. \\ &\left. 1 - \tilde{\gamma}_0(\tilde{\gamma}_0 - \tilde{\sigma}_0 - p_0)/Q_0\right) \end{aligned} \tag{1.4.25}$$

$$Q_0 := (\tilde{\sigma}_0 - \tilde{\gamma}_0)^2 + 1$$

$$C_0 = (0, 1, 1) \tag{1.4.26}$$

$$E_0 = (1, \tilde{\sigma}_0, 0) \tag{1.4.27}$$

$$\begin{aligned} F_0 = &\left(\tilde{\gamma}_0(\tilde{\gamma}_0 - 2\tilde{\sigma}_0)/Q_0, \tilde{\gamma}_0\left[1 - \tilde{\sigma}_0(\tilde{\sigma}_0 - \tilde{\gamma}_0)\right]/Q_0,\right. \\ &\left. (\tilde{\sigma}_0^2 + 1)/Q_0\right). \end{aligned} \tag{1.4.28}$$

Points in the D_1-Unit ($\tilde{\sigma}_1 := \sigma_1/\omega_1, \tilde{\gamma}_1 := \gamma_1/\omega_1$)

$$A_1 = (1, p_1, 0) \tag{1.4.29}$$

where

$$p_1 := \tilde{\sigma}_1 + k_0(\tilde{\sigma}_1^2 + 1)/\tilde{\gamma}_1, \; k_1 := \tilde{\gamma}_1(p_1 - \tilde{\sigma}_1)/(\tilde{\sigma}_1^2 + 1) \tag{1.4.30}$$

$$B_1 = (1, \tilde{\sigma}_1, 0) \tag{1.4.31}$$

$$D_1 = (0, 0, 1) \tag{1.4.32}$$

$$\begin{aligned} E_1 = &\left(\tilde{\gamma}_1(\tilde{\gamma}_1 - \tilde{\sigma}_1 - p_1)/Q_1, \tilde{\gamma}_1\left[1 - p_1(\tilde{\sigma}_1 - \tilde{\gamma}_1)\right]/Q_1,\right. \\ &\left. 1 - \tilde{\gamma}_1(\tilde{\gamma}_1 - \tilde{\sigma}_1 - p_0)/Q_1\right) \end{aligned} \tag{1.4.33}$$

$$Q_1 := (\tilde{\sigma}_1 - \tilde{\gamma}_1)^2 + 1 \tag{1.4.34}$$

$$\begin{aligned} F_1 = &\left(\tilde{\gamma}_1(\tilde{\gamma}_1 - 2\tilde{\sigma}_1)/Q_1, \tilde{\gamma}_1\left[1 - \tilde{\sigma}_1(\tilde{\sigma}_1 - \tilde{\gamma}_1)\right]/Q_1,\right. \\ &\left. (\tilde{\sigma}_1^2 + 1)/Q_1\right). \end{aligned} \tag{1.4.35}$$

Note that k_0 cannot be obtained directly from (1.4.24) since it depends on p_0, which in turn depends on k_0. A similar situation applies to k_1 in (1.4.30). However, they can be obtained from the relationship

$$k_0 = 1/k_1 = k := -\gamma_0/\gamma_1, \tag{1.4.36}$$

which will be derived in Lemma 1.4.11 below. The relationship

$$k_0 k_1 = 1 \tag{1.4.37}$$

follows from the ratio between the lengths of the following vectors (see Fig. 1.4.1):

$$\frac{|\overrightarrow{F_0B_0}|}{|\overrightarrow{E_0F_0}|} = \frac{|\overrightarrow{F_1B_1}|}{|\overrightarrow{E_1F_1}|}.$$

1.4.3 Equivalence and Conjugacy Classes of $\mathcal{L}$

Definition 1.4.7.

(1) Two vector fields f and f' on $\mathbb{R}^n$ are *linearly equivalent* if there is a linear $G : \mathbb{R}^n \to \mathbb{R}^n$ and a real number $\nu > 0$ such that

$$G \circ f = \nu(f' \circ G), \tag{1.4.38}$$

where $\circ$ denotes a composition.

(2) Linearly equivalent vector fields f and f' are *linearly conjugate* if $\nu = 1$ in (1.4.38).

Note that linear equivalence is a special case of a general equivalence discussed in Chapter 3. Similarly, linear conjugacy is a special case of a general conjugacy where G is a homeomorphism. Note also that it is difficult to prove (general) equivalence or conjugacy of two vector fields. It turns out, however, in terms of the linear equivalence (respectively the linear conjugacy), and for the particular class $\mathcal{L}$, one can explicitly classify equivalence (respectively conjugacy) classes (Proposition 1.4.8 and 1.4.9).

Proposition 1.4.8.

(1) For each

$$\{\sigma_0, \omega_0, \gamma_0, \sigma_1, \omega_1, \gamma_1\} \tag{1.4.39}$$

there is an $f \in \mathcal{L}$ having these eigenvalue parameters if and only if

$$\omega_0 > 0, \omega_1 > 0, \text{ and } \gamma_0\gamma_1 < 0. \tag{1.4.40}$$

(2) Two vector fields f and $f' \in \mathcal{L}$ are linearly conjugate to each other if and only if they have identical eigenvalues, i.e.,

$$\begin{matrix} \sigma_0 = \sigma_0' & \omega_0 = \omega_0' & \gamma_0 = \gamma_0' \\ \sigma_1 = \sigma_1' & \omega_1 = \omega_1' & \gamma_1 = \gamma_1'. \end{matrix} \tag{1.4.41}$$

Proposition 1.4.9. *Let*

$$\begin{aligned}&\tilde{\sigma}_0 := \frac{\sigma_0}{\omega_0}, \quad \tilde{\gamma}_0 := \frac{\gamma_0}{\omega_0}, \quad \tilde{\sigma}_1 := \frac{\sigma_1}{\omega_1}, \quad \tilde{\gamma}_1 := \frac{\gamma_1}{\omega_1}, \\ &k := -\frac{\gamma_0}{\gamma_1},\end{aligned} \tag{1.4.42}$$

which will be called the normalized eigenvalue parameters.

(1) For each

$$\{\tilde{\sigma}_0, \tilde{\gamma}_0, \tilde{\sigma}_1, \tilde{\gamma}_1, k\}$$

there is an $f \in \mathcal{L}$ having these normalized eigenvalue if and only if

$$k > 0 \text{ and } \tilde{\sigma}_0\tilde{\gamma}_1 < 0 . \tag{1.4.43}$$

(2) Two vector fields f and $f' \in \mathcal{L}$ are linearly equivalent if and only if they have identical normalized eigenvalue parameters. Moreover, the scale ν in (1.4.38) is given by

$$\nu = \frac{\omega_0}{\omega_0'} = \frac{\omega_1}{\omega_1'} . \tag{1.4.44}$$

Note that the eigenvalues of two distinct vector fields having identical normalized eigenvalue parameters are generally not identical because one more parameter must be specified in order to identify the eigenvalues uniquely. It follows from Proposition 1.4.8 that two vector fields having identical normalized eigenvalue parameters are generally not linearly conjugate to each other. Equation (1.4.44) implies that the additional condition $\omega_0 = \omega_0'$ is needed for linear conjugacy.

Lemma 1.4.10. *Propositions 1.4.8 and 1.4.9 are equivalent.*

Proof. ($\Longrightarrow$) If Proposition 1.4.8 holds, then (1.4.40) implies that $\gamma_0\gamma_1/\omega_0\omega_1 < 0$ and hence (1.4.43) is satisfied. Next, given any $\{\tilde{\sigma}_0, \tilde{\gamma}_0, \tilde{\sigma}_1, \tilde{\gamma}_1, k\}$ satisfying (1.4.43) define

$$\{\sigma_0, \omega_0, \gamma_0, \sigma_1, \omega_1, \gamma_1\} := \{\tilde{\sigma}_0, 1, \tilde{\gamma}_0 - \tilde{\sigma}_1\tilde{\gamma}_0/\tilde{\gamma}_1 k, -\tilde{\gamma}_0/\tilde{\gamma}_0 k, -\tilde{\gamma}_0/k\} . \tag{1.4.45}$$

Since $\omega_1 := -\tilde{\gamma}_0/\tilde{\gamma}_1 k > 0$ and since $\gamma_0\gamma_1 = -\tilde{\gamma}_0^2/k < 0$, (1.4.45) satisfies (1.4.40). Therefore, Proposition 1.4.8 implies the existence of an $f \in \mathcal{L}$ associated with (1.4.45). This proves (1) of Proposition 1.4.9. To prove (2) of Proposition 1.4.9, suppose that f and f' are linearly equivalent and hence $G \circ f = (\omega_0/\omega_0')f' \circ G$ for some G. Then the two vector fields f and $(\omega_0/\omega_0')f'$ are linearly conjugate and must have identical eigenvalue parameters $\{\sigma_0, \omega_0, \gamma_0, \sigma_1, \omega_1, \gamma_1\}$. It follows that the eigenvalue parameters of f' are given by

$$\left\{\frac{\sigma_0\omega_0'}{\omega_0}, \omega_0', \frac{\gamma_0\omega_0'}{\omega_0}, \frac{\sigma_1\omega_0'}{\omega_0}, \frac{\omega_1\omega_0'}{\omega_0}, \frac{\gamma_1\omega_0'}{\omega_0}\right\} .$$

Using (1.4.42), we obtain the following normalized eigenvalue parameters of f':

$$\left\{\frac{\sigma_0}{\omega_0}, \frac{\gamma_0}{\omega_0}, \frac{\sigma_1}{\omega_1}, \frac{\gamma_1}{\omega_1}, -\frac{\gamma_0}{\gamma_1}\right\} = \{\tilde{\sigma}_0, \tilde{\gamma}_0, \tilde{\sigma}_1, \tilde{\gamma}_1, k\},$$

which are identical to those of f. Next suppose that

$$\left\{\frac{\sigma_0}{\omega_0}, \frac{\gamma_0}{\omega_0}, \frac{\sigma_1}{\omega_1}, \frac{\gamma_1}{\omega_1}, -\frac{\gamma_0}{\gamma_1}\right\} = \left\{\frac{\sigma_0'}{\omega_0'}, \frac{\gamma_0'}{\omega_0'}, \frac{\sigma_1'}{\omega_1'}, \frac{\gamma_1'}{\omega_1'}, -\frac{\gamma_0'}{\gamma_1'}\right\}.$$

Then

$$\{\sigma_0, \omega_0, \gamma_0, \sigma_1, \omega_1, \gamma_1\} = \left(\frac{\omega_0}{\omega_0'}\right)\{\sigma_0', \omega_0', \gamma_0', \sigma_1', \omega_1', \gamma_1'\}$$

and, hence, f and $(\omega_0/\omega_0')f'$ are linearly conjugate to each other.

($\Longleftarrow$) If Proposition 1.4.9 holds, then given an $f \in \mathcal{L}$, we have $k = -\gamma_0/\gamma_1 > 0$ in view of (1.4.43), and hence $\gamma_0\gamma_1 < 0$. Moreover, $\omega_0 > 0$ and $\omega_1 > 0$ by definition. Therefore (1.4.40) is satisfied. Next, let (1.4.40) be satisfied. Then the normalized eigenvalue parameters.

$$\left\{\frac{\sigma_0}{\omega_0}, \frac{\gamma_0}{\omega_0}, \frac{\sigma_1}{\omega_1}, \frac{\gamma_1}{\omega_1}, -\frac{\gamma_0}{\gamma_1}\right\}$$

clearly satisfy (1.4.43). It follows from Proposition 1.4.8 that there is an $f' \in \mathcal{L}$ having these normalized eigenvalue parameters and $(\omega_0/\omega_0')f' \in \mathcal{L}$ is linearly conjugate to f. Hence $(\omega_0/\omega_0')f'$ and f have identical eigenvalue parameters; namely $\{\sigma_0, \omega_0, \gamma_0, \sigma_1, \omega_1, \gamma_1\}$. Therefore f is the vector field sought.

To prove (2) of Proposition 1.4.8, let $G \circ f = f' \circ G$ for some G. Then f and f' have identical normalized eigenvalue parameters

$$\left\{\frac{\sigma_0}{\omega_0}, \frac{\gamma_0}{\omega_0}, \frac{\sigma_1}{\omega_1}, \frac{\gamma_1}{\omega_1}, -\frac{\gamma_0}{\gamma_1}\right\} = \left\{\frac{\sigma_0'}{\omega_0'}, \frac{\gamma_0'}{\omega_0'}, \frac{\sigma_1'}{\omega_1'}, \frac{\gamma_1'}{\omega_1'}, -\frac{\gamma_0'}{\gamma_1'}\right\}$$

and $\nu = \omega_0/\omega_0' = 1$. Therefore f and f' have identical eigenvalue parameters. Conversely, if f and f' have identical eigenvalue parameters, then they have identical normalized eigenvalue parameters and $\nu = \omega_0/\omega_0' = 1$. It follows from Proposition 1.4.9 (2) that $G \circ f = f' \circ G$ and hence f is linearly conjugate to f'. □

Now we will prove Proposition 1.4.9, for which we need the following:

Lemma 1.4.11. *Let $\mu := (\tilde{\sigma}_0, \tilde{\gamma}_0, \tilde{\sigma}_1, \tilde{\gamma}_1, k)$ and let $f[\mu]$ denote the family of all vector fields in $\mathcal{L}$ having the same normalized eigenvalue parameters μ. Let $\overrightarrow{OP}, \overrightarrow{OA}, \overrightarrow{OB}$, and $\overrightarrow{OE}$ denote the vectors from the origin O in Plate 1.10 to P, A, B, and E, respectively.*

(1) All polyhedra whose vertices consist of the origin and the four points belonging to the family $f[\mu]$ are similar in the sense that

$$\overrightarrow{OP} = l\overrightarrow{OA} + m\overrightarrow{OB} + n\overrightarrow{OE}, \tag{1.4.46}$$

where $l = l(\mu), m = m(\mu)$, and $n = n(\mu)$ are real numbers which depend only on μ and, hence, are identical for all vector fields in $f[\mu]$.

(2) The numbers k_0, k_1, and k defined in (1.4.24), (1.4.30) and (1.4.42) are related by

$$k = k_0 = \frac{1}{k_1} . \tag{1.4.47}$$

(3) There is an $f \in f[\mu]$ if and only if

$$\tilde{\gamma}_0 \tilde{\gamma}_1 < 0 \textit{ and } k > 0 . \tag{1.4.48}$$

Proof is essentially matrix computations on $\mathbb{R}^3$. We will give only a rough sketch of this proof. First, one can show that $\overrightarrow{OA}, \overrightarrow{OB}$ and $\overrightarrow{OE}$ are linearly independent. Second, since Ψ_i maps $\{A, B, E\}$ into $\{A_i, B_i, E_i\}$, $i = 0, 1$, one can compute explicit formulas for

$$W_i := [A_i, B_i, E_i]^{-1} J_i [A_i, B_i, E_i]$$

where

$$J_i x := \frac{1}{\omega_i} D\Psi_i \left(f \left(\Psi_i^{-1}(x) \right) \right) , \; i = 0, 1 ,$$

from which one can compute relationships between normalized eigenvalue parameters. See [Chua, Komuro and Matsumoto 1986] for details.

Proof of Proposition 1.4.9 Statement (1) of Proposition 1.4.9 is equivalent to (3) of Lemma 1.4.11. It remains to prove (2) of Proposition 1.4.9.

Necessity. Suppose that there is a nonsingular G and a $\nu > 0$ with $G \circ f = \nu f' \circ G$. Then the eigenvalues of f and f' must satisfy $\sigma_i \pm \omega_i \sqrt{-1} = \nu \sigma_i' \pm \nu \omega_i' \sqrt{-1}$ and $\gamma_i = \nu \gamma_i'$ $(i = 0, 1)$. It follows from (1.4.42) that their respective normalized eigenvalue parameters are identical.

Sufficiency. Let $\sigma_i \pm \omega_i \sqrt{-1}$ $(\omega_i > 0)$ and $\gamma_i \neq 0$ $(i = 0, 1)$ denote the eigenvalues of $f \in f[\mu]$. Let $P = \overrightarrow{OP}, A = \overrightarrow{OA}, B = \overrightarrow{OB}, E = \overrightarrow{OE}$ for f and let P', A', B', E' be similarly defined for f'. Then there are 3×3 matrices M_i and M_i' $(i = 0, 1)$ such that

$$f(x) = \begin{cases} M_1(x - P), & x \in D_1 \\ M_0 x, & x \in D_0 \\ M_1(x + P), & x \in D_{-1} \end{cases}$$

and

$$f'(x) = \begin{cases} M_1'(x - P'), & x \in D_1' \\ M_0' x, & x \in D_0' \\ M_1'(x + P'), & x \in D_{-1}' \end{cases} \tag{1.4.49}$$

where D_i and D_i' $(i = 0, \pm 1)$ are the affine regions of f and f', respectively. It follows from the continuity of f and f' that

$$\begin{aligned} M_0(A, B, E) &= M_1(A-P, B-P, E-P) && (1.4.50)\\ M_0'(A', B', E') &= M_1'(A'-P', B'-P', E'-P'), && (1.4.51)\end{aligned}$$

where $(\cdot)$ denotes a 3×3 matrix made up of the various column vectors defined above.

Now recall (1.4.12) and (1.4.15), which are obtained by Ψ_0 and Ψ_1, respectively. It follows from (1.4.9) and (1.4.13) that Ψ_0 and Ψ_1 can be represented by

$$\Psi_0(x) = \Phi_0 x \qquad (1.4.52)$$

and

$$\Psi_1(x) = \Phi_1(x - P) \qquad (1.4.53)$$

respectively, where Φ_0 and Φ_1 are 3×3 matrices to be determined. Since Ψ_0 maps $\{A, B, E\}$ onto $\{A_0, B_0, E_0\}$, we have

$$\Phi_0(A, B, E) = (A_0, B_0, E_0)$$

and hence

$$\Phi_0 = (A_0, B_0, E_0)(A, B, E)^{-1}. \qquad (1.4.54)$$

Similarly, since Ψ_1 maps $\{A, B, E\}$ onto $\{A_1, B_1, E_1\}$, we have

$$\Phi_1(A-P, B-P, E-P) = (A_1, B_1, E_1)$$

and hence

$$\Phi_1 = (A_1, B_1, E_1)(A-P, B-P, E-P)^{-1}. \qquad (1.4.55)$$

It follows from (1.4.12) and (1.4.15) that

$$\frac{1}{\omega_i}(\Phi_i M_i {\Phi_i}^{-1}) = J_i, \ (i = 0, 1), \qquad (1.4.56)$$

where

$$J_i = \begin{pmatrix} \tilde{\sigma}_i & -1 & 0 \\ 1 & \tilde{\sigma}_i & 0 \\ 0 & 0 & \tilde{\gamma}_i \end{pmatrix}. \qquad (1.4.57)$$

Now, by hypothesis f and f' have identical normalized eigenvalue parameters. Hence, the Jordan forms J_0 and J_0' (corresponding to M_0 and M_0' respectively) are identical. Substituting (1.4.54) into (1.4.56), we obtain

$$\begin{aligned} J_0 &= \frac{1}{\omega_0}(A_0, B_0, E_0)(A, B, E)^{-1} M_0(A, B, E)(A_0, B_0, E_0)^{-1} \\ &= \frac{1}{\omega_0'}(A_0, B_0, E_0)(A', B', E')^{-1} {M_0}'(A', B', E')(A_0, B_0, E_0)^{-1} = J_0'. \end{aligned} \qquad (1.4.58)$$

Let us define a linear map $G : \mathbb{R}^3 \to \mathbb{R}^3$ and a real number $\nu > 0$ by

$$G := (A', B', E')(A, B, E)^{-1}, \; \nu := \omega_0 / \omega_0' \, . \tag{1.4.59}$$

Premultiplying both sides of (1.4.58) by $\omega_0(A', B', E')(A_0, B_0, E_0)^{-1}$ and postmultiplying both sides by $(A_0, B_0, E_0)(A', B', E')^{-1}$, we obtain

$$(A', B', E')(A, B, E)^{-1} M_0 (A, B, E)(A', B', E')^{-1} = \left(\frac{\omega_0}{\omega_0'}\right) M_0'. \tag{1.4.60}$$

Substitution of (1.4.59) into (1.4.60) yields

$$\nu M_0' = G M_0 G^{-1}. \tag{1.4.61}$$

Equation (1.4.49) implies

$$\nu f'(x)|_{D_0} = G\left(f(G^{-1}x)|_{D_0}\right), \; x \in D_0'. \tag{1.4.62}$$

Now rewrite (1.4.46) in the following vector form:

$$P = (A, B, E)(l, m, n)^T, \; P' = (A', B', E')(l, m, n)^T. \tag{1.4.63}$$

Then

$$GP = G(A, B, E)(l, m, n0^T = (A', B', E')(l, m, n)^T = P'. \tag{1.4.64}$$

Now solving (1.4.51) for M_1' and (1.4.50) for M_1 and using (1.4.61) and (1.4.59) repeatedly, we obtain

$$\begin{aligned} \nu M_1' &= M_0'(A', B', E')(A' - P', B' - P', E' - P')^{-1} \\ &= G M_0 G_{-1} (A', B', E') \left(G(A - P), G(B - P), G(E - P)\right)^{-1} \\ &= G M_0 (A, B, E)(A - P, B - P, E - P)^{-1} G^{-1} \\ &= G M_1 G^{-1}. \end{aligned} \tag{1.4.65}$$

Now for any $x \in D_{\pm 1}'$, (1.4.49) implies

$$\begin{aligned} \nu f'(x)|_{D_{\pm 1}'} &= \nu M_1'(x \mp P') \\ &= G M_1 G^{-1}(x \mp P') && \text{(in view of (1.4.65))} \\ &= G M_1 (G^{-1}x \mp P) && \text{(in view of (1.4.64))} \\ &= G f(G^{-1}x)|_{D_{\pm 1}'} && \text{(in view of (1.4.49))}. \end{aligned} \tag{1.4.66}$$

Equations (1.4.62) and (1.4.66) together imply

$$\nu f'(x) = G f(G^{-1}x) \tag{1.4.67}$$

for all $x \in D'_0 \cup D'_{\pm 1}$. Hence, (1.4.38) holds and f and f' are linearly equivalent. □

Proposition 1.4.8 is equivalent to Proposition 1.4.9, and it allows us to partition all vector fields in $\mathcal{L}$ into linearly conjugate equivalence classes, each one parametrized by the eigenvalues $\sigma_0 \pm \omega_0 i, \gamma_0, \sigma_1 \pm \omega_1 i$, and γ_1. Since all vector fields in $\mathcal{L}$ having the same eigenvalues have identical qualitative behavior, it suffices to investigate only one member in each class. In fact, one can derive an explicit normal form in terms of the eigenvalues for each class (see Chapter 2).

1.4.4 Subset $\mathcal{L}_{\mathcal{DS}}$

Definition 1.4.7 implies that in so far as qualitative behavior is concerned, we only need to study one member of each linearly equivalent class. Proposition 1.4.9 implies that we can, without loss of generality, choose a convenient vector field $f \in \mathcal{L}$ having a given set of normalized eigenvalue parameters $\{\tilde{\sigma}_0, \tilde{\gamma}_0, \tilde{\sigma}_1, \tilde{\gamma}_1, k\}$ as defined in (1.4.42), where $\tilde{\gamma}_0\tilde{\gamma}_1 < 0$ and $k > 0$. Note that a piecewise-linear vector field with $\{\tilde{\sigma}_0, \tilde{\gamma}_0, \tilde{\sigma}_1, \tilde{\gamma}_1, k\}$ may be discontinuous at the boundary planes U_1 and U_{-1} and, hence, is not a member of $\mathcal{L}$ even though it satisfies (1)-(4) of Definition 1.4.7. Proposition 1.4.9 therefore provides the additional necessary and sufficient condition (1.4.40) for the existence of a continuous vector field with the given parameters. Namely, this eigenvalue condition asserts that the real eigenvalue associated with the equilibrium point P^+ (respectively P^-) must be opposite in sign to that at O. Hence, trajectories along the real eigenvector at P^+ (respectively P^-) and those at O must have opposite stability properties.

Remark 1.4.12. This eigenvalue condition (1.4.40) is not necessary for continuity of the vector field if we allow the piecewise-linear system to have only one equilibrium point instead of three.

Since our main motivation in this subsection is to characterize the Double Scroll where $\gamma_0 > 0$, we will restrict our analysis to the following subset $\mathcal{L}_{\mathcal{DS}} \subset \mathcal{L}$:

$$\mathcal{L}_{\mathcal{DS}} := \{f(\tilde{\sigma}_0, \tilde{\gamma}_0, \tilde{\sigma}_1, \tilde{\gamma}_1, k) \mid \tilde{\sigma}_0 < 0, \tilde{\gamma}_0 < 0, \tilde{\sigma}_1 < 0, \tilde{\gamma}_1 < 0, k > 0\}, \quad (1.4.68)$$

where $\{\tilde{\sigma}_0, \tilde{\gamma}_0, \tilde{\sigma}_1, \tilde{\gamma}_1, k\}$ are the normalized eigenvalue parameters. Therefore, the eigenvalue pattern of any member of $\mathcal{L}_{\mathcal{DS}}$ at the equilibrium point P^+ (respectively P^-) must be a mirror image (except for scales) of that at the origin O. The eigenspaces of a typical vector field $f \in \mathcal{L}$ are shown in Fig. 1.4.1(a), along with two typical trajectories. The upper trajectory Γ_1 in Fig. 1.4.1(a) originates from some point on U_1, moves downward, turns around (before reaching U_{-1}), and returns to U_1 after a finite amount of time. It continues to move upward before turning around and returns once again to U_1. This typical trajectory can never penetrate the upper plane because this plane is an eigenspace and is therefore invariant. This defines a return map (Poincaré map) from a subset $S \subset U_1$ into S. We can decompose this Poincaré map into two components: a "half-return map", which maps the initial point on U_1 to the first return point on U_1, and a "second half-return map," which maps the first return point to the second return point on U_1.

The lower trajectory Γ_2 in Fig. 1.4.1(a) also originates from U_1, moves downward, penetrates U_{-1}, and after some finite amount of time, turns around, and returns to U_{-1} a second time. By $f(-x) = -f(x)$, however, we can identify each return point x in U_{-1} by its reflected image $-x$ in U_1. Similarly, the portion of Γ_2 below U_{-1} can be identified with a corresponding version of Γ_1 above U_1. Through this identification, both typical types of trajectories Γ_1 and Γ_2 actually define the same Poincaré map, which in turn is simply the composition of two half-return maps.

Unfortunately, the half-return maps in Fig. 1.4.1(a) cannot in general be computed by an explicit formula or algorithm because the coordinates of the return points can only be found by solving a pair of transcendental equations. Since these half-return maps will be used in a crucial way to prove that the Double Scroll circuit is chaotic in the sense of Shil'nikov, we must find a new coordinate system so that half-return maps can easily be computed and their errors rigorously estimated. Our approach for deriving this new coordinate system is to work with the greatly simplified but equivalent real Jordan forms of the regions D_0 and D_1 in Fig. 1.4.1(a), namely, the D_0-unit and the D_1-unit in Fig. 1.4.1(b) described earlier.

A Half-Return Map π_0 Consider first the D_0-unit at the bottom of Fig. 1.4.1(b), representing the image of D_0 in Fig. 1.4.1(a) under the affine map Ψ_0 (recall (1.4.9)-(1.4.12)). Points A, B, and E in D_0 map into A_0, B_0, and E_0, respectively. Since L_2 maps into the straight line L_{2_0} passing through B_0 and E_0, it follows from the definition of L_2 and the qualitative nature of trajectories in D_0 that the vector field $f_0(x)$ has a downward[1] component for all x to the right of L_{2_0}, and an upward component to the left. Hence, any trajectory originating inside the triangular region

$$\triangle A_0B_0E_0 := \{x \in V_0 \mid x \text{ is bounded within triangle } A_0B_0E_0\} \tag{1.4.69}$$

must move down initially. But because the x^3-axis in the D_0-unit is the image of an unstable eigenvector, this trajectory must move toward V_0 as depicted by the upper trajectory in the D_0-unit. This trajectory defines the map

$$\pi_0^+ : \triangle A_0B_0C_0 \to V_0 \tag{1.4.70}$$

via

$$\pi_0^+(x) = \varphi_0^T(x), \tag{1.4.71}$$

where $\varphi_0^T(x)$ denotes the trajectory (in the D_0-unit) from x to the first return point where the trajectory first hits V_0 at some time $T > 0$, where

[1]Throughout this subsection, "downward component" or "moving down" (respectively "upward component" or "moving up") means the vector field enters the boundary plane V_0 from above (respectively leaves V_0 from below).

$$T = T(x) := \inf\{t > 0 \mid \varphi_0^t(x) \in V_0\}. \tag{1.4.72}$$

Here we assume that $\varphi^t(x)$ does not hit U_{-1} before time T.

Consider next a typical trajectory originating from a point in the angular region

$$\angle A_0B_0E_0 := \{x \in V_0 \mid x \text{ lies within the wedge-like extension of } \triangle A_0B_0E_0\} \tag{1.4.73}$$

to the right of $\overline{A_0E_0}$ in the D_0-unit as depicted in Fig. 1.4.1(b). This trajectory must move downward (because it originates to the right of L_{2_0}) and eventually hits V_0^-. This trajectory corresponds to the portion of Γ_2 within D_0 in Fig. 1.4.1(a) and defines the map

$$\pi_0^- : \angle A_0B_0E_0 \backslash \triangle A_0B_0E_0 \to V_0^- \tag{1.4.74}$$

via

$$\pi_0^-(x) = \varphi_0^T(x), \tag{1.4.75}$$

where

$$T = T(x) := \inf\{t > 0 \mid \varphi_0^t(x) \in V_0^-\} \tag{1.4.76}$$

is the time this trajectory first penetrates V_0^-. By identifying this return point in V_0^- with its reflected image in V_0, we can define the following half-return map

$$\pi_0 : \angle A_0B_0E_0 \to V_0 \tag{1.4.77}$$

by

$$\pi_0(x) = \begin{cases} \pi_0^+(x), & x \in \triangle A_0B_0E_0 \\ -\pi_0^-(x), & x \in \angle A_0B_0E_0 \backslash \triangle A_0B_0E_0. \end{cases} \tag{1.4.78}$$

In order to derive a formulas for $\pi_0^+(x)$ and $\pi_0^-(x)$, look at the triangular region $\triangle A_0B_0E_0$ on V_0 and the angular region $\angle A_0B_0E_0$ on V_0 as shown in Fig. 1.4.2. Since the x^3-coordinate of each point (x^1, x^2, x^3) on V_0 is simply $x^3 = 1 - x^1$, it suffices to specify each point on V_0 by its (x^1, x^2)-coordinate. The next step is to define a local coordinate system (u, v) on V_0 so that each point $x_0 = (x^1, x^2)^T \in \angle A_0B_0E_0$ is uniquely specified in terms of (u, v) such that $\pi_0^+(x)$ and $\pi_0^-(x)$ can be expressed in terms of u and v. We will define local (u, v)-coordinate as a weighted sum of the four corner points A_0, B_0, E_0, and F_0, whose (x^1, x^2)-coordinate have already been found in (1.4.23), (1.4.25), (1.4.27), and (1.4.28), in terms of the normalized eigenvalue parameters, namely,

$$x_0(u, v) = u\,[vA_0 + (1 - v)E_0] + (1 - u)\,[vB_0 + (1 - v)F_0] \tag{1.4.79}$$

where $0 \le u < \infty$ and $0 \le v \le 1$. Here we have abused our notation by denoting the (x^1, x^2)-coordinate of the four corner points by A_0, B_0, E_0, and F_0, respectively. Note that $x_0(1, 1) = A_0$, $x_0(1, 0) = E_0$, $x_0(0, 1) = B_0$ and $x_0(0, 0) = F_0$. Note also that all points along the line segments $\overline{E_0A_0}$ and $\overline{F_0B_0}$ have a u-coordinate equal

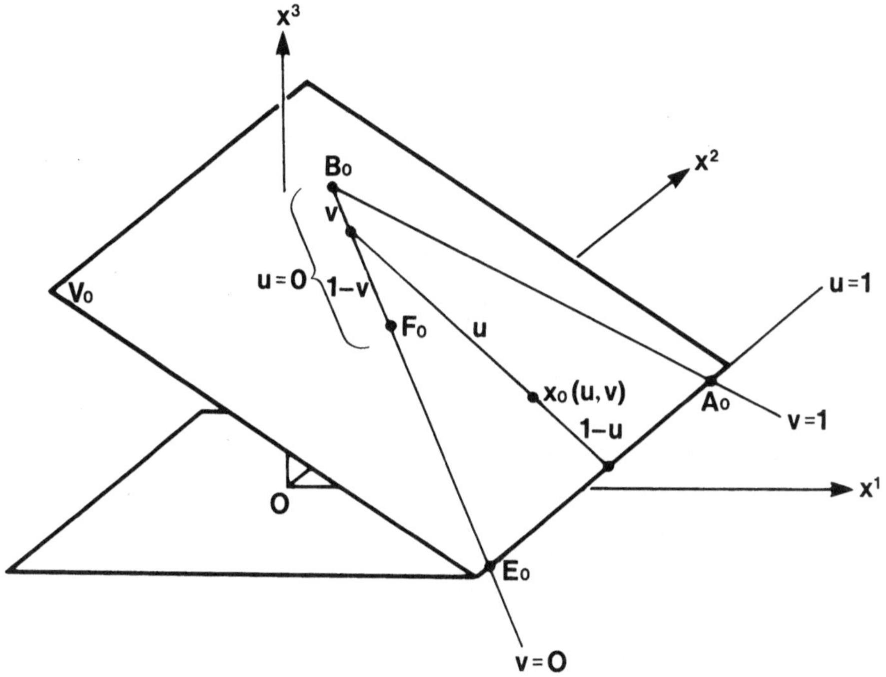

Fig. 1.4.2. The (u, v)-coordinate system. ©1986 IEEE.

to 1 and 0, respectively. Similarly, all points along the line segments $\overline{B_0A_0}$ and $\overline{F_0E_0}$ have a v-coordinate equal to 1 and 0, respectively. A typical point (u_0, v_0) can be identified as the intersection between the $u = u_0$ coordinate line and the $v = v_0$ coordinate line. All points inside the triangular region $\triangle A_0B_0E_0$ have $0 < u < 1$, and all points inside the angular region $\angle A_0B_0E_0$ outside $\triangle A_0B_0E_0$ have $1 < u < \infty$. Hence, in terms of the (u, v)-coordinate system, (1.4.69) and (1.4.73) have the following equivalent form:

$$\triangle A_0B_0E_0 = \{x_0(u,v) \mid (u,v) \in [0,1] \times [0,1]\} \tag{1.4.80}$$

$$\angle A_0B_0E_0 = \{x_0(u,v) \mid (u,v) \in [0,\infty] \times [0,1]\}\,. \tag{1.4.81}$$

One of the main reasons for choosing the (u, v)-coordinate system is that the sets of interest on V_0 (respectively V_1) are of the form v =constant (respectively u =constant). Recalling (1.4.79), let

$$A_{0v} = x_0(1,v), \;\; B_{0v} = x_0(0,v) \tag{1.4.82}$$

and consider the line segment

$$\{x_0(u,v) \mid u \in [0,1]\} = \{uA_{0v} + (1-u)B_{0v} \mid u \in [0,1]\}, \tag{1.4.83}$$

where v is fixed.

Proposition 1.4.13. *Let*

$$\eta^{\pm}(v,t) := \frac{\langle \varphi_0^t(B_{0v}), h\rangle \mp 1}{\langle \varphi_0^t(B_{0v} - A_{0v}), h\rangle} \tag{1.4.84}$$

$$h := (1,0,1)^T,$$

$$\tau := \inf\left\{t \geq 0 \mid \eta^{\pm}(v,t) = u\right\}, \tag{1.4.85}$$

where $\langle \cdot\,,\cdot\rangle$ denotes the usual inner product. Then

$$\begin{aligned} &\pi_0^{\pm}\left(\{x_0(u,v) \mid u \in [0,1]\}\right) = \\ &\left\{ e^{\sigma_0\tau} \begin{bmatrix} \cos\tau & -\sin\tau \\ \sin\tau & \cos\tau \end{bmatrix} x_0(\eta^{\pm}(v,\tau),v) \mid \tau \in I^{\pm}(v) \right\} \end{aligned} \tag{1.4.86}$$

where $I^{\pm}(v)$ is the set of τ for which (1.4.85) holds for $u \in [0,1]$.

Proof. The dynamics in the D_0-unit is (1.4.12). The trajectory $\varphi_0^t(x_0)$ from a point $x_0 = (x_0^1, x_0^2, x_0^3)$ is given by

$$\varphi_0^t(x_0) = \begin{pmatrix} e^{\sigma_0 t}\cos t & -e^{\sigma_0 t}\sin t & 0 \\ e^{\sigma_0 t}\sin t & e^{\sigma_0 t}\cos t & 0 \\ 0 & 0 & e^{\gamma_0 t} \end{pmatrix} \begin{pmatrix} x_0^1 \\ x_0^2 \\ x_0^3 \end{pmatrix}. \tag{1.4.87}$$

Since $A_{0v} \to \varphi_0^t(A_{0v})$, $B_{0v} \to \varphi_0^t(B_{0v})$ and since for a *fixed* t, $\varphi_0^t(\,\cdot\,)$ in (1.4.87) is *linear*, the straight line segment $\overline{A_{0v}B_{0v}}$, joining A_{ov} and B_{0v} in Fig. 1.4.2 maps into a straight line segment $\overline{\varphi_0^t(A_{0v})\varphi_0^t(B_{0v})}$, joining $\varphi_0^t(A_{0v})$ and $\varphi_0^t(B_{0v})$. Now if we let $\hat{x}_0 := \varphi_0^t(x_0)$, then $\hat{x}_0$ must divide the length of the vector $\overrightarrow{\varphi_0^t(A_{0v})\varphi_0^t(B_{0v})}$ into the same proportion as x_0 (see Fig. 1.4.2) divides the vector $\overrightarrow{A_{0v}B_{0v}}$ into lengths u and $1-u$, respectively. In particular

$$u = \frac{u}{u+(1-u)} = \frac{|\overrightarrow{\hat{x}_0\varphi_0^t(B_{0v})}|}{|\overrightarrow{\varphi_0^t(A_{0v})\varphi_0^t(B_{0v})}|}$$

$$= \frac{\langle \overrightarrow{\hat{x}_0\varphi_0^t(B_{0v})}, h\rangle}{\langle \overrightarrow{\varphi_0^t(A_{0v})\varphi_0^t(B_{0v})}, h\rangle} \tag{1.4.88}$$

$$= \frac{\langle \varphi_0^t(B_{0v}), h\rangle - \langle \hat{x}_0, h\rangle}{\langle \varphi_0^t(B_{0v} - A_{0v}), h\rangle}. \tag{1.4.89}$$

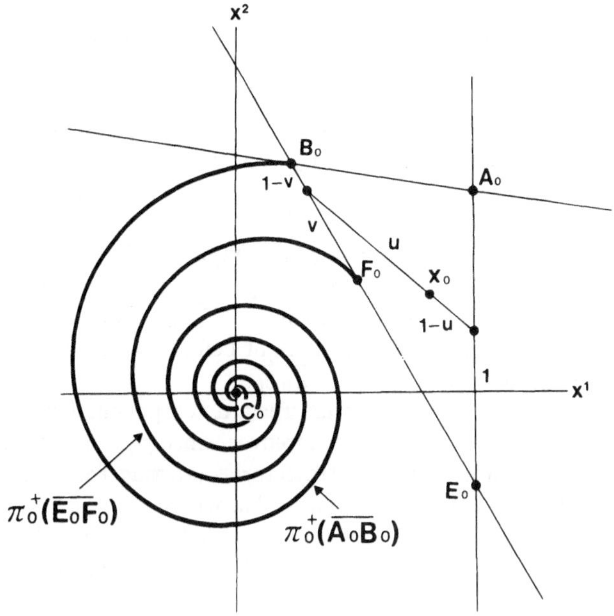

Fig. 1.4.3. Typical π_0^+-images. ©1986 Elsevier Science Publishers.

Note that (1.4.89) is simply the ratio between the projections along the normal vector h of the vectors in the numerator and the denominator in (1.4.88), respectively. But

$$\langle \hat{x}_0, h \rangle = \langle (\hat{x}_0^1, \hat{x}_0^2, \hat{x}_0^3), (1,0,1) \rangle = \hat{x}_0^1 + \hat{x}_0^3 = 1 \text{ (respectively -1)} \tag{1.4.90}$$

since $\hat{x}_0$ lies on V_0 (respectively V_0^-). Substituting $\langle \hat{x}_0, h \rangle = \pm 1$ into (1.4.89), we obtain

$$u = \frac{\langle \varphi_0^t(B_{0v}), h \rangle - 1}{\langle \varphi_0^t(B_{0v} - A_{0v}), h \rangle} \,. \tag{1.4.91}$$

Letting τ be defined by (1.4.85) and using (1.4.89) and (1.4.91), we obtain (1.4.86). □

Remark 1.4.14. Since $\tilde{\sigma}_0 < 0$, (1.4.86) tells us that the image $\pi_0^\pm(\{x_0(u,v) \mid u \in [0,1]\})$ is a union of shrinking spirals or subsets of shrinking spirals. A typical situation is depicted in Fig. 1.4.3.

Remark 1.4.15. Since any initial point $x_0(1,v)$ lies on the stable eigenspace $\Psi_0(E^c(O))$, $\varphi_0^t(x_0(1,v))$ may not return to V_0 but instead may converge to the origin O as $t \to \infty$. In this case, however, it is convenient to define $\pi_0^+(x_0(1,v)) :=$

$C_0 = \Psi_0(C)$ since we have earlier identified C_0 and O as the same point. It follows from this definition that $\eta^+(v,t) \to 1$ as $t \to \infty$.

Remark 1.4.16. It can be shown that the vector field $f_0(E_0)$ is directed from E_0 to A_0, $f_0(B_0)$ is directed from A_0 to B_0, and $f_0(F_0)$ is directed from F_0 to B_0, as shown in Fig. 1.4.2. It follows from the continuity of $f(x)$ that the vectors along the line segment $\overline{B_0F_0}$ are as depicted in Fig. 1.4.2.

Since the vector field $f(x)$ has a downward component for all x to the right of the line segment $\overline{E_0F_0}$ in Fig. 1.4.2, and since $f(x)$ is directed to the right for all $x \in \overline{E_0F_0}$, it follows that trajectories starting on $\overline{E_0F_0}$ or slightly to the right of $\overline{E_0F_0}$ will first move downward towards the right before returning to V_0. Hence, $\pi_0^+(x)$ is continuous even along the point on $\overline{E_0F_0}$.

In contrast, the vector field $f(x)$ has an upward component for all x to the left of the line segment $\overline{F_0B_0}$ in Fig. 1.4.2. Moreover, since $f(x)$ is directed to the left for all $x \in \overline{F_0B_0}$, it follows that the trajectories starting from points along $\overline{F_0B_0}$ will first move upward before returning to V_0, whereas trajectories starting from points arbitrarily close to $\overline{F_0B_0}$ (but on the right-hand side) will first move downward and return to V_0 after a much shorter time. Consequently, $\pi_0^+(x)$ is discontinuous along $\overline{F_0B_0}$. For convenience, we will define

$$\pi_0^+(x) = x \text{ for all } x \in \overline{F_0B_0}.$$

In other words, we define each point $x \in \overline{F_0B_0}$ as a fixed point of $\pi_0^+(x)$ and, hence,

$$\eta^+(v,t) := 0 \text{ at } t = 0.$$

Remark 1.4.17. Between $t = 0$ and $t = \infty$, $\eta^+(v,t)$ is a continuous but not necessarily monotonic function of t.

Remark 1.4.18. In general, τ is a discontinuous function of u and, hence, of the initial point x_0. This shows that it is, in general, impossible to express τ as a continuous function of x_0.

B Half-Return Map π_1 Consider next the D_1-unit at the top of Fig. 1.4.1(b), representing the image of D_1 in Fig. 1.4.1(a) under Ψ_1 (recall (1.4.13)-(1.4.15)). Three points A, B, and E in D_1 map into A_1, B_1, and E_1, respectively. Here we abuse our notation by using the same symbol D_1 to denote the top region in Fig. 1.4.1(a) and a point on the x^3-axis in the D_1 unit in Fig. 1.4.1(b). We will use the same notations as above with the exception that each subscript "$_1$" corresponds to the D_1-unit. Hence, we define again a local coordinate system (u,v) such that the line segments $\overline{E_1F_1}$ and $\overline{A_1B_1}$ in V_1 in Fig. 1.4.2 correspond to the $v = 0$ and $v = 1$ coordinate lines, respectively. Likewise, the lines segments $\overline{F_1B_1}$ and $\overline{E_1A_1}$ correspond to the $u = 0$ and $u = 1$ coordinate line, respectively. Any point x_1 inside the angular region bounded by (the extension of) $\overline{B_1A_1}$ and (the extension of) $\overline{B_1E_1}$ is uniquely identified by

$$x_1(u,v) = u\,[vA_1 + (1-v)E_1] + (1-u)\,[vB_1 + (1-v)F_1]$$
$$\text{for } 0 \le u < \infty \text{ and } 0 \le v \le 1.$$

Under this local coordinate system, we can define the triangular region $\triangle A_1B_1E_1$ and the angular region $\angle A_1B_1E_1$ as:

$$\begin{aligned} \triangle A_1B_1E_1 &:= \{x_1(u,v) \mid (u,v) \in [0,1] \times [0,1]\} \\ \angle A_1B_1E_1 &:= \{x_1(u,v) \mid (u,v) \in [0,\infty] \times [0,1]\}. \end{aligned}$$

Finally, we define the half-return map

$$\pi_1(x) : \angle A_1B_1E_1 \to V_1$$

via

$$\pi_1(x) = \varphi_1^{-T}(x),$$

where $\varphi_1^{-T}(x)$ denotes the flow (in the D_1-unit) from x to the first return point where the trajectory first strikes V_1 at some reverse time $-T < 0$, where

$$T = T(x) := \inf\{t > 0 \mid \varphi_1^{-t}(x) \in V_1\}.$$

The next proposition can be proved in a manner similar to that of Proposition 1.4.13.

Proposition 1.4.19. *Let*

$$A_{1u} = x_1(u,1), \; E_{1u} = x_1(u,0) \tag{1.4.92}$$

and consider line segment

$$\{x_1(u,v) \mid v \in [0,1]\} = \{vA_{1u} + (1-v)E_{1u} \mid v \in [0,1]\},$$

where u is fixed. Then

$$\pi_1(\{x_1(u,v) \mid v \in [0,1]\}) = \left\{ e^{\sigma_1\tau} \begin{pmatrix} \cos\tau & \sin\tau \\ -\sin\tau & \cos\tau \end{pmatrix} x_1(u,\xi(u,\tau)) \,\middle|\, \tau \in I(u) \right\},$$

where

$$\begin{aligned} \xi(u,t) &:= \frac{\langle \varphi_1^{-t}(B_{1u}), h\rangle - 1}{\langle \varphi_1^{-t}(E_{1u} - A_{1u}), h\rangle} \\ \tau &:= \inf\{t \ge 0 \mid \xi(u,t) = v\} \end{aligned} \tag{1.4.93}$$

and $I(u)$ is the set of t for which (1.4.93) holds for $v \in [0,1]$.

A typical situation is shown in Fig. 1.4.4. The image of $\overline{F_1B_1}$ under π_1 is *spiral*$[F_1W_1D_1]$ in Fig. 1.4.4. The image of the line segment $\overline{E_1A_1}$ is shown as part

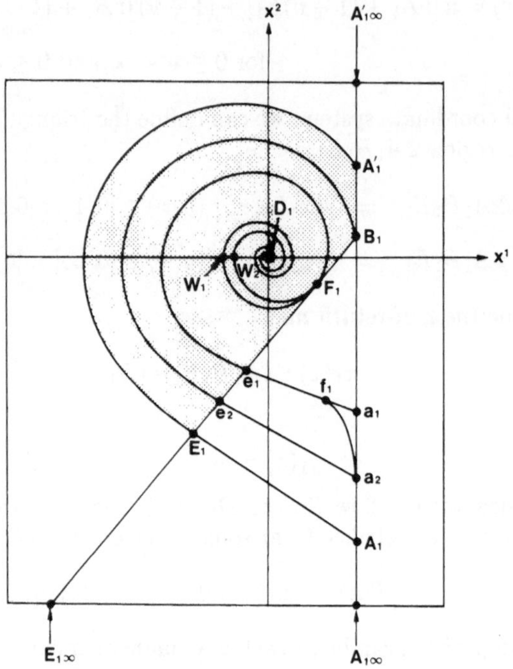

Fig. 1.4.4. Typical π_1-images. ©1986 IEEE.

of a large *spiral*$[E_1A_1']$. The continuation of this spiral to the right of A_1' is the image of the extension of $\overline{E_1A_1}$ beyond A_1.

Now let f_1 be the inverse image of F_1 in Fig. 1.4.4, i.e., $\pi_1(f_1) = F_1$. Similarly, let the inverse image of $\overline{F_1B_1}$ be denoted by *curve*$[f_1a_2]$. Since the region bounded by *closed curve*$[e_1e_2a_2f_1e_1]$ is found to map into the region bounded by *arc*$[e_1F_1]$, line $\overline{F_1B_1}$, *arc*$[B_1e_2]$, and line $\overline{e_2e_1}$, whereas the neighboring region bounded by the *closed curve*$[f_1a_1a_2f_1]$ is mapped into the region bounded by *closed curve*$[F_1W_1D_1W_2F_1]$, it follows that $\pi_1(x)$ is discontinuous along the *curve* $[f_1a_2]$, in addition to already being discontinuous along $\overline{E_1F_1}$. These additional discontinuity points occur when we choose our parameters close to those which gave us the Double Scroll. They may not, however, occur inside $\triangle A_1B_1E_1$ when other parameters are chosen.

Let us summarize the behavior of π_1 in Fig. 1.4.4:

(1)

$$\pi_1(\triangle A_1B_1E_1) = \text{ a fan-like closed region } 2\, A_1'B_1E_1 \text{ (shown half-toned).}$$

(2)

$$\pi_1(\overline{B_1a_2}) = D_1.$$

Here $\pi_1(\overline{B_1 a_2})$ actually maps into the origin in the unstable eigenspace $\Psi(E^c(P))$, which becomes a stable equilibrium point under the reverse flow φ_1^{-t}. It is convenient to identify the origin with $D_1 = \Psi_1(P^+)$ in V_1.

(3) Since π_1 is discontinuous along $\overline{E_1 F_1}$, we will define (as in π_0)

$$\pi_1(x) := x \text{ for all } x \in \overline{E_1 F_1}.$$

In particular

$$\pi_1(f_1) = \pi_1(F_1) = F_1.$$

With this definition, π_1 becomes continuous at $\overline{E_1 F_1}$.

(4) π_1 is one-to-one at all points inside the triangular region $\triangle A_1 B_1 E_1$ and its boundaries except the points along the line segment $[\overline{B_1 a_2})$ and the isolated point f_1, i.e., on $\triangle A_1' B_1 E_1 \backslash ([\overline{B_1 a_2}) \cup \{f_1\})$.

(5) π_1^{-1} is well-defined at all points in the fan-like region $2\, A_1' B_1 E_1$, except for the two isolated points F_1 and D_1.

(6) *Spiral* $[F_1 W_1 D_1]$ is the set of discontinuous points of π_1^{-1}. The function π_1^{-1} is discontinuous at these points because $\pi_1^{-1}(x) \to$ *curve*$[f_1 a_2]$ from the right as $x \to W_1$ from the right, whereas $\pi_1^{-1}(x) \to \overline{F_1 B_1}$ from the right as $x \to W_1$ from the left. This follows because the return map π_1 is discontinuous along the *curve*$[f_1 a_1]$, and because $\pi_1(\textit{curve}[f_1 a_1]) = \overline{F_1 B_1} = \pi_1^{-1}(\textit{curve}[F_1 W_1 D_1])$.

Using the above properties, we can now define the *inverse half-return map* π_1^{-1} as

$$\pi_1^{-1} : 2\, A_{1\infty}' B_1 E_{1\infty} \to \angle A_1 B_1 E_1, \tag{1.4.94}$$

where

$$2\, A_{1\infty}' B_1 E_{1\infty} := \left\{ (x^1, x^2, x^3) \in V_1 \mid x^2 \geq \sigma_1 x^1 + \gamma_1 (1 - x^1),\ x^1 \leq 1 \right\} \tag{1.4.95}$$

is the region above $\overline{B_1 E_1}$ and to the left of $\overline{A_1 A_1'}$ in Fig. 1.4.4, and where

$$\begin{aligned} \pi_1^{-1}(D_1) &:= B_1 \\ \pi_1^{-1}(F_1) &:= f_1. \end{aligned}$$

Note that π_1^{-1} is discontinuous along *curve*$[F_1 W_1 D_1]$.

C The Map Φ Since the D_0-unit and the D_1-unit have different coordinate systems, we naturally need a coordinate patch defined by

$$\Phi := (\Psi_1|_{U_1}) \circ (\Psi_0|_{U_1})^{-1}, \tag{1.4.96}$$

where $\Psi_1|_{U_1}$ and $\Psi_0|_{U_1}$ denote the restriction of Ψ_1 and Ψ_0 to U_1. Again, since $x_i^3 = 1 - x_i^1$, it suffices to find the explicit formula relating $(x_0^1, x_0^2) \in D_0$ to $(x_1^1, x_1^2) \in D_1$. Since $A_0 = (1, p_0, 0) \to A_1 = (1, p_1, 0)$, we have

$$\begin{pmatrix} x_1^1 \\ x_1^2 \end{pmatrix} = \Phi \begin{pmatrix} x_0^1 \\ x_0^2 \end{pmatrix} = L \begin{pmatrix} x_0^1 - 1 \\ x_0^2 - p_0 \end{pmatrix} + \begin{pmatrix} 1 \\ p_1 \end{pmatrix}$$

so that

$$\begin{pmatrix} x_1^1 - 1 \\ x_1^2 - p_1 \end{pmatrix} = L \begin{pmatrix} x_0^1 - 1 \\ x_0^2 - p_0 \end{pmatrix} \quad \text{for any } (x_0^1, x_0^2) \in D_0. \tag{1.4.97}$$

Since $B_0 := (B_{0x^1}, B_{0x^2}) \to B_1 := (B_{1x^1}, B_{1x^2})$ and $E_0 := (E_{0x^1}, E_{0x^2}) \to E_1 := (E_{1x^1}, E_{1x^2})$, it follows from the action of L in (1.4.97) that

$$\begin{aligned} \begin{pmatrix} B_{1x^1} - A_{1x^1} \\ B_{1x^2} - A_{1x^2} \end{pmatrix} &= L \begin{pmatrix} B_{0x^1} - A_{0x^1} \\ B_{0x^2} - A_{0x^2} \end{pmatrix} \\ \begin{pmatrix} E_{1x^1} - A_{1x^1} \\ E_{1x^2} - A_{1x^2} \end{pmatrix} &= L \begin{pmatrix} E_{0x^1} - A_{0x^1} \\ E_{0x^2} - A_{0x^2} \end{pmatrix}, \end{aligned} \tag{1.4.98}$$

which imply

$$L = \begin{pmatrix} B_{1x^1} - A_{1x^1} & E_{1x^1} - A_{1x^1} \\ B_{1x^2} - A_{1x^2} & E_{1x^2} - A_{1x^2} \end{pmatrix} \begin{pmatrix} B_{0x^1} - A_{0x^1} & E_{0x^1} - A_{0x^1} \\ B_{1x^2} - A_{1x^2} & E_{0x^2} - A_{0x^2} \end{pmatrix}^{-1}.$$

Substituting (1.4.23), (1.4.25), (1.4.27), (1.4.29), (1.4.31), and (1.4.33) for the respective components of A_i, B_i, E_i into (1.4.98), we obtain the following formula for L:

$$\begin{aligned} L &= \frac{(\tilde{\sigma}_1^2 + 1)k_1)}{(\tilde{\sigma}_0^2 + 1)(k_0 + 1)Q_1\tilde{\gamma}_1} \begin{pmatrix} L_{11} & L_{12} \\ L_{21} & L_{22} \end{pmatrix} \\ L_{11} &= -\tilde{\gamma}_1(k_0 + 1)\left(Q_0 + \tilde{\gamma}_0(\tilde{\sigma}_0 - \tilde{\gamma}_0)(k_1 + 1)\right) \\ L_{12} &= \tilde{\gamma}_0\tilde{\gamma}_1(k_0 + 1)(k_1 + 1) \\ L_{21} &= -\tilde{\gamma}_0(k_1 + 1)(\tilde{\sigma}_0 - \tilde{\gamma}_0)\left(\tilde{\sigma}_1(\tilde{\sigma}_1 - \tilde{\gamma}_1) + 1\right) \\ &\quad -\tilde{\gamma}_1(k_0 + 1)(\tilde{\sigma}_1 - \tilde{\gamma}_1)\left(\tilde{\sigma}_0(\tilde{\sigma}_0 - \tilde{\gamma}_0) + 1\right) \\ L_{22} &= \tilde{\gamma}_0(k_1 + 1)\left(Q_1 + \tilde{\gamma}_1(\tilde{\sigma}_1 - \tilde{\gamma}_1)(k_0 + 1)\right), \end{aligned}$$

where $Q_i := (\tilde{\sigma}_i - \tilde{\gamma}_i)^2 + 1$, $k_0 := k$, and $k_1 := 1/k$.

Note that L is expressed directly in terms of the normalized eigenvalue parameters $\{\tilde{\sigma}_0, \tilde{\gamma}_0, \tilde{\sigma}_1, \tilde{\gamma}_1, k\}$.

D Poincaré Map π We will now use the half-return maps π_0 and π_1 and the map Φ to define a Poincaré map

$$\pi : V_1' \to V_1', \tag{1.4.99}$$

where

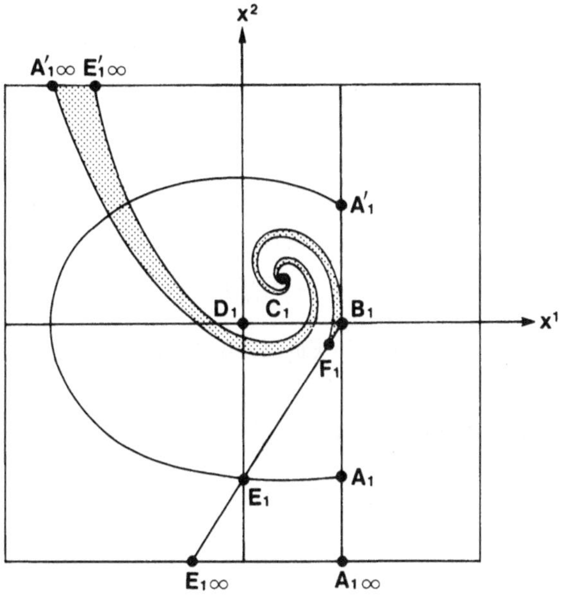

Fig. 1.4.5. Various images under π_2 or $\Phi\pi_0$ at a typical set of normalized eigenvalue parameters. ©1986 IEEE.

$$V_1' := \{(x^1, x^2) \in V_1 \mid x^1 \leq 1\}$$

by

$$\begin{aligned}\pi(x) &= \pi_1^{-1}\Phi\pi_0\Phi^{-1}(x), \quad \text{if } x \in \angle A_1B_1E_1 \\ &= \Phi\pi_0\Phi^{-1}\pi_1^{-1}(x), \quad \text{if } x \in V_1'\backslash\angle A_1B_1E_1 .\end{aligned}$$

Note that $\pi(\angle A_1B_1E_1) \subset \angle A_1B_1E_1$ and π_1^{-1} is well-defined for all $x \in V_1'\backslash\angle A_1B_1E_1$ in view of (1.4.94) and (1.4.95). Let

$$\pi_2 = \Phi\pi_0\Phi^{-1} .$$

Then

$$\pi(x) = \begin{cases} \pi_1^{-1}\pi_2(x), & x \in \angle A_1B_1E_1 \\ \pi_2\pi_1^{-1}(x), & x \in V_1'\backslash\angle A_1B_1E_1 . \end{cases}$$

Instead of studying the full return map π directly, we will look at π_1 and π_2 separately. Namely, we will check how a subset of V_1 returns to V_1 in (half) positive time (π_1) and how another subset of V_1 returns to V_1 in (half) negative time (π_2). Let $A_{i\infty}$(respectively $E_{i\infty}$) be an extension of A_i(respectively E_i) along $\overline{A_iB_i}$(respectively $\overline{E_iB_i}$),$i = 0, 1$. The following images will be important:

$$\begin{aligned}
curve[B_1C_1] &:= \pi_2(\overline{A_1B_1}) = \Phi\pi_0(\overline{A_0B_0}) \\
curve[F_1C_1] &:= \pi_2(\overline{F_1E_1}) = \Phi\pi_0(\overline{F_0E_0}) \\
curve[C_1A'_{1\infty}] &:= \pi_2(\overline{A_1A_{1\infty}}) = \Phi\pi_0(\overline{A_0A_{0\infty}}) \\
curve[C_1E'_{1\infty}] &:= \pi_2(\overline{E_1E_{1\infty}}) = \Phi\pi_0(\overline{E_0E_{0\infty}}) \\
curve[E_1A'_1] &:= \pi_1(\overline{E_1A_1}).
\end{aligned}$$

A typical situation is shown in Fig. 1.4.5. Note that $C_1 := \Phi(C_0) = \Psi(C)$.

1.4.5 Completion of the Proof

Our goal is to prove that there is a $\beta \in J$ such that

(1) $C_1 \in \pi_1(\overline{E_1A_1})$.

(2) $|\alpha_0| < \gamma_0$.

Recall (1.4.1), (1.4.2), (1.4.12) and (1.4.15). In the following arguments, we will need various numerical bounds on $\sigma_i, \omega_i, \gamma_i, \tilde{\sigma}_i, \tilde{\gamma}_i$ and $k_i/\tilde{\gamma}_i$, $i = 0, 1$, and many arithmetic operations thereafter. Since (DS 1) or (DS 2) is a third order system, the eigenvalues are zeros of cubic equations. Recall that the Cardano formula for solving $x^3 + ax^2 + bx + c = 0$ requires first the transformation $p = b - a^2/3$, $q = 2a^3/27 - ab/3 + c$, and then is followed by the computation of the two quantities $m, n = \left[-q/2 \pm (q^2/4 + p^3/27)^{1/2}\right]^{1/3}$, which involve both square roots and cubic roots. Finally, the three zeros are given by $m + n, \omega m + \omega^3 n, \omega^2 m + \omega n$, where $\omega = (-1 + \sqrt{3}i)/2$. In addition to the eigenvalue computations, there are a great number of arithmetic operations on them. In principle, it is possible to arrive at numerical values together with exact error bounds by hand. In practice, however, this is formidable. Obviously, usage of a higher order precision format of a computer does not solve this problem. One way of surmounting this difficulty is to give *verifiable* error bounds to all the values involved and prove them with the aid of a computer. In order to facilitate this, one must take into account every possible error. Specifically, errors are induced by

(1) eigenvalue computations;

(2) arithmetic operations, i.e., $+, -\times, \div$;

(3) conversion of a real number to and from a machine-represented number.

In order to accomplish this, all operations must be transparent with no intervening black box. Namely, every operation of the computer must be under our control so that every possible error can be taken into account. Since the complete details of this particular subject are slightly outside of the scope of the book, we will give below a brief description of the work reported in [Matsumoto, Chua and Ayaki 1988] and refer to it for specific numerical bounds. The goal is to reduce all the computations involved to the following four logical operations:

AND, OR, NOT, XOR

and to perform an interval analysis [Moore 1979], which is a method of computing intervals containing the true values. Since the algorithm makes everything transparent all the way down to the logical operations, one could say that a special purpose computer has been built (by emulation). The key function of the machine is its capability to perform a floating point interval analysis.

Let $X = [X_{\min}, X_{\max}]$ and $Y = [Y_{\min}, Y_{\max}]$ be intervals of $\mathbb{R}$ where $X_{\min}, X_{\max}, Y_{\min}$ and $Y_{\max}$ are machine-representable. Such intervals are called machine-representable intervals. Arithmetic operations on X and Y need care. For instance

$$\{x + y \mid x \in X, y \in Y\},$$

where $+$ is the addition on $\mathbb{R}$, may not be machine-representable. A careful analysis should be done in such a way that the machine computes a machine-representable interval Z such that

$$X \boxed{+} Y := Z \supseteq \{x + y \mid x \in X, y \in Y\}.$$

Operations, $\boxed{-}$, $\boxed{\times}$ and $\boxed{\div}$ are defined similarly.

For the eigenvalue computation, the interval bisection is used, which necessitates only arithmetic operations. The actual algorithm and coding are far from trivial and, in the interest of simplicity, are omitted here. The algorithm completely emulated the VS-FORTRAN on IBM 3090 simply because it is reliable in addition to being easily accessible. The algorithm, therefore, operates on hexadecimal numbers ($0 \sim F$), and the outputs (the bounds) are given in [Matsumoto, Chua and Ayaki 1988].

Lemma 1.4.20.

(*1*) $\sigma_0, \omega_0, \omega_1, \gamma_1, \tilde{\sigma}_0, \tilde{\gamma}_1$ *and* $k_0/\tilde{\gamma}_0$ *are monotonically increasing in* β *on* J.

(*2*) $\gamma_0, \sigma_1, \tilde{\sigma}_1, \tilde{\gamma}_0$ *and* $k_1/\tilde{\gamma}_1$ *are monotonically decreasing in* β *on* J.

Proof. Recall (1.3.1) – (1.3.4) and note that m_i, $i = 0, 1$, is the slope of the piecewise-linear resistor. The real eigenvalue γ_i corresponding to $m = m_i$, $(i = 1, 0)$, is a real zero of the characteristic polynomial

$$\lambda^3 + (\alpha m + 1)\lambda^2 + (\alpha m - \alpha + \beta)\lambda + \alpha\beta m = 0, \tag{1.4.100}$$

which gives

$$\beta = \alpha - \lambda(\lambda+1) - \frac{\alpha^2 m}{\lambda + \alpha m}. \tag{1.4.101}$$

If $\alpha > 0$ and $\alpha m > 1$, then $\beta : (-\infty, -\alpha m) \to \mathbb{R}$ is an increasing bijection while if $\alpha > 0$ and $\alpha m < 0$, then $\beta : (-\alpha m, \infty) \to \mathbb{R}$ is a decreasing bijection. Hence, for $\alpha = 7$ and $m_0 = -1/7$ (respectively $m_1 = 2/7$), γ_0 (respectively γ_1) decreases (respectively increases). Note that coefficients and zeros of (1.4.100) satisfy

$$\begin{aligned} 2\sigma_i + \gamma_i &= -(\alpha m_i + 1) \\ \sigma_i^2 + \omega_i^2 + 2\sigma_i\gamma_i &= \alpha(m_i - 1) + \beta \\ \gamma_i(\sigma_i^2 + \omega_i^2) &= -\alpha\beta m_i, \end{aligned}$$

from which it follows that

$$\begin{aligned} \sigma_i &= -\frac{1}{2}(\alpha m_i + 1 + \gamma_i) \\ \omega_i^2 &= -\frac{1}{4}(\alpha m_i - 1 - \gamma_i)^2 - \frac{\alpha^2 m_i}{\gamma_i + \alpha m_i}. \end{aligned} \tag{1.4.102}$$

These formulas prove monotonicity of σ_i and ω_i. Monotonicity of $\tilde{\sigma}_0$, $\tilde{\gamma}_0$ and $\tilde{\gamma}_1$ follows from $\tilde{\sigma}_0 < 0$, $\tilde{\gamma}_0 > 0$ and $\omega_0 > 0$, and $\tilde{\sigma}_0 = \sigma_0/\omega_0$, $\tilde{\gamma}_0 = \gamma_0/\omega_0$. Finally, monotonicity of $k_0/\tilde{\gamma}_0$ and $k_1/\tilde{\gamma}_1$ follow from

$$\begin{aligned} \frac{k_0}{\tilde{\gamma}_0} &= -\frac{\gamma_0/\gamma_1}{\gamma_0/\omega_0} = -\frac{\omega_0}{\gamma_1} \\ \frac{k_1}{\tilde{\gamma}_1} &= -\frac{\gamma_1/\gamma_0}{\gamma_1/\omega_1} = -\frac{\omega_1}{\gamma_0}. \end{aligned}$$

□

Recall (1.4.68), (1.4.1) and (1.4.2). The following is a direct consequence of Lemma 1.4.20

Proposition 1.4.21. *For all $\beta \in J$ with (1.4.1), the vector field (DS 2) belongs to $\mathcal{L}_{DS}$.*

This proposition tells us that Proposition 1.4.1 can be proved in terms of the convenient coordinate systems we have derived. Now recall (1.4.82) and (1.4.92) and note that

$$A_0 = x_0(1,1) \quad B_0 = x_0(0,1)$$

$$A_1 = x_1(1,1) \quad E_1 = x_1(1,0)\,.$$

It follows from Proposition 1.4.13 and Proposition 1.4.19 that

$$\pi_0^+(\overline{B_0A_0}) = \{X_0(\tau) \mid \tau \in I^+(1)\} \tag{1.4.103}$$

$$\pi_1(\overline{E_1A_1}) = \{X_1(\tau) \mid \tau \in I(1)\}, \tag{1.4.104}$$

where

$$X_0(\tau) = e^{\tilde{\sigma}_0\tau}\begin{pmatrix}\cos\tau & -\sin\tau\\ \sin\tau & \cos\tau\end{pmatrix} x_0(\eta^+(1,\tau),1) \tag{1.4.105}$$

$$X_1(\tau) = e^{-\tilde{\sigma}_1\tau}\begin{pmatrix}\cos\tau & \sin\tau\\ -\sin\tau & \cos\tau\end{pmatrix} x_1(1,\xi(1,\tau))\,. \tag{1.4.106}$$

In the following three lemmas (Lemmas 1.4.22, 1.4.23 and 1.4.24), vectors are projected onto the (x^1, x^2)-plane. For example, $\overrightarrow{OE_1}$ (respectively $\overrightarrow{OE_0}$) should be understood as $\overrightarrow{D_1A_1}$ (respectively $\overrightarrow{C_0E_0}$).

Lemma 1.4.22.

(1) $|A_0|e^{\tilde{\sigma}_0\tau} \geq |X_0(\tau)| \geq |B_0|e^{\tilde{\sigma}_0\tau}$ $\tau \in I^+(1)$, $\beta \in J$;

(2) $|A_1|e^{-\tilde{\sigma}_1\tau} \geq |X_1(\tau)| \geq |E_1|e^{-\tilde{\sigma}_1\tau}$ $\tau \in I(1)$, $\beta \in J$;

(3) $|E_1| \leq |A_1|e^{-\tilde{\sigma}_1\theta_1}$, $\beta \in J$,
where $\theta_1 \geq 0$ *denotes the angle subtended by two vectors,* $\overrightarrow{OE_1}$ *and* $\overrightarrow{OA_1}$ *on the* (x^1, x^2)*-plane;*

(4) $|A_0| \geq |E_0|e^{\tilde{\sigma}_0\theta_0}$, $\beta \in J$,
where $\theta_0 \geq 0$ *is the angle subtended by the two vectors,* $\overrightarrow{OE_0}$ *and* $\overrightarrow{OA_0}$.

Proof.

(1) Proof of (1) is similar to that of (2). We will only prove (2).

(2) It suffices to show that

$$|A_1| \geq |x_1(1, v(1,t))| \geq |E_1|. \tag{1.4.107}$$

Since $x_1(1,v) = \overrightarrow{OE_1} + v\overrightarrow{E_1A_1}$, $v \in [0,1]$, it follows from plane geometry that

$$|x_1(1,v)|^2 = \left\{ v|\overrightarrow{E_1A_1}| + \frac{\langle\overrightarrow{OE_1}, \overrightarrow{E_1A_1}\rangle}{|\overrightarrow{E_1A_1}|}\right\}^2 + |E_1|^2 - \frac{\langle\overrightarrow{OE_1}, \overrightarrow{E_1A_1}\rangle}{|\overrightarrow{E_1A_1}|}\,. \tag{1.4.108}$$

If we can show that

$$\langle\overrightarrow{OE_1}, \overrightarrow{E_1A_1}\rangle > 0 \tag{1.4.109}$$

then (1.4.108) would imply (1.4.107) because $|x_1(1,v)|^2$ is an increasing function of $v \in [0,1]$ and because $|A_1| = |x_1(1,1)|$ and $|E_1| = |x_1(1,0)|$. To prove (1.4.109), we make use of the first two coordinates of E_1 from (1.4.33) and A_1 from (1.4.29) to write

$$\overrightarrow{OE_1} = (\tilde{\gamma}_1(\tilde{\gamma}_1 - \tilde{\sigma}_1 - p_1)/Q_1, \tilde{\gamma}_1\left[1 - p_1(\tilde{\sigma}_1 - \tilde{\gamma}_1)\right]/Q_1) \tag{1.4.110}$$

$$\overrightarrow{E_1A_1} = (\left[\tilde{\sigma}_1(\tilde{\sigma}_1 - \tilde{\gamma}_1) + 1 + \tilde{\gamma}_1 p_1\right]/Q_1, \{p_1\left[\tilde{\sigma}_1(\tilde{\sigma}_1 - \tilde{\gamma}_1) + 1\right] - \tilde{\gamma}_1\}/Q_1). \tag{1.4.111}$$

Computing the inner product between (1.4.110) and (1.4.111), we obtain

$$\langle \overrightarrow{OE_1}, \overrightarrow{E_1A_1} \rangle = -\tilde{\sigma}_1\tilde{\gamma}_1(p_1^2 + 1)/Q_1. \tag{1.4.112}$$

Using (1.4.112) and Lemma 1.4.20 ($\tilde{\sigma}_1 > 0$, $\tilde{\gamma}_1 < 0$), we obtain the desired inequality (1.4.109).

(3) From (1.4.109), it follows that $0 < \theta_1 < \pi/2$ so that

$$\theta_1 < \tan\theta_1. \tag{1.4.113}$$

Recall the monotonicity of $\tilde{\sigma}_1$ with respect to β on J. In [Matsumoto, Chua and Ayaki 1988] verifiable error bounds on the eigenvalues are given. In particular,

$$0.05 \leq \tilde{\sigma}_1 \leq 0.16 \tag{1.4.114}$$

for all $\beta \in J$, and we can show that

$$1 - 2\tilde{\sigma}_1\theta_1 \leq e^{-2\tilde{\sigma}_1\theta_1}. \tag{1.4.115}$$

Since (3) is equivalent to

$$|E_1|^2/|A_1|^2 \leq e^{-2\tilde{\sigma}_1\theta_1}, \tag{1.4.116}$$

it follows from (1.4.115) that it suffices to prove

$$|E_1|^2/|A_1|^2 \leq 1 - 2\tilde{\sigma}_1 \tan\theta_1. \tag{1.4.117}$$

Since A_1 and E_1 are projected onto the plane, we can suppress the x^3-coordinate in (1.4.29) and (1.4.33) and obtain, after simplification,

$$|A_1|^2 = p_1^2 + 1, \ |E_1|^2 = \tilde{\gamma}_1^2(p_1^2 + 1)/Q_1. \tag{1.4.118}$$

Now define the normal vector to E_1 by:

$$E_1^{\perp} := \Big(-\tilde{\gamma}_1(1 - p_1(\tilde{\sigma}_1 - \tilde{\gamma}_1))/Q_1, \tilde{\gamma}_1(\tilde{\gamma}_1 - \tilde{\sigma}_1 - p_1)/Q_1\Big). \tag{1.4.119}$$

Then it follows from (1.4.33) that

$$|E_1| = |E_1^{\perp}| \text{ and } \overrightarrow{OE_1} \perp \overrightarrow{OE_1^{\perp}}. \tag{1.4.120}$$

A straightforward calculation shows

$$\begin{aligned}\tan\theta_1 &= \langle \overrightarrow{OA_1}, \overrightarrow{O_1E_1^{\perp}}\rangle / \langle \overrightarrow{OA_1}, \overrightarrow{O_1E_1}\rangle \\ &= 1/(\tilde{\sigma}_1 - \tilde{\gamma}_1).\end{aligned} \tag{1.4.121}$$

Substituting (1.4.118), (1.4.121), and (1.4.113) into (1.4.117) and solving for $\tilde{\gamma}_1$, we obtain

$$\tilde{\gamma}_1 \le \tilde{\sigma}_1\left(1 + \frac{2}{\tilde{\sigma}_1^2 - 1}\right). \tag{1.4.122}$$

Hence, to prove (3) it is sufficient to prove (1.4.122) holds over the parameter range assumed by $\tilde{\gamma}_1$ and $\tilde{\sigma}_1$ for $\beta \in J$. To verify this, note that the right hand side of (1.4.122) decreases with respect to $\tilde{\sigma}_1$. Using (1.4.114), we see that the right hand side of (1.4.122) is no smaller than $0.16\left[1 + 2/(0.16^2 - 1)\right]$, the latter being no smaller than -0.17. It follows from [Matsumoto, Chua and Ayaki 1988] that $\tilde{\gamma}_1$ satisfies

$$-1.92 \le \tilde{\gamma}_1 \le -1.3$$

so that (1.4.122) holds.

(4) It follows from (1.4.23) and (1.4.27) that (4) is equivalent to

$$(1 + p_0^2) - (1 + \tilde{\sigma}_0^2)e^{2\tilde{\sigma}_0\theta_0} > 0.$$

To prove this inequality, let us define

$$g(t) := 1 + \tan^2(\varphi + t) - (1 + \tilde{\sigma}_0^2)e^{2\tilde{\sigma}_0 t}, \; t \in [0, \theta_0] \tag{1.4.123}$$

where

$$\varphi := \tan^{-1}\tilde{\sigma}_0 \in \left(-\frac{\pi}{2}, 0\right).$$

It is easy to verify that

$$\begin{aligned} g(0) &= 0 \\ g'(t) &= 2\tan(\varphi + t)\left[1 + \tan^2(\varphi + t)\right] \\ &\quad -2\tilde{\sigma}_0(1 + \tilde{\sigma}_0^2)e^{2\tilde{\sigma}_0 t} \\ g'(0) &= 0 \\ g'(t) &\ge 0 \text{ for } 0 < t < \frac{\pi}{2} - \varphi, \end{aligned} \tag{1.4.124}$$

where (1.4.124) follows from $\tilde{\sigma}_0 < 0$ and $-\pi/2 < \varphi < 0$. Since $E_0 = (1, \tilde{\sigma}_0)$, φ is the negative angle between $\overrightarrow{OE_0}$ and the x^1-axis. Hence, $0 < \theta_0 < \pi/2 - \varphi$

falls within the range of t in (1.4.124). Moreover, since $A_0 = (1, p_0)$ and since $\varphi+\theta_0$ is the angle between $\overrightarrow{OA_0}$ and the x^3-axis, it follows that $\tan(\varphi+\theta_0) = p_0$. Hence, letting $t = \theta_0$ in (1.4.123), we obtain

$$g(\theta_0) = (1+p_0^2)-(1+\tilde{\sigma}_0^2)e^{2\tilde{\sigma}_0\theta_0} > 0. \qquad \square$$

Lemma 1.4.23. *In the D_0-unit (Fig. 1.4.1(b)), no trajectory starting on $\overline{A_0E_0}$ hits the $x^1 = -1$ boundary.*

Proof. Suppressing the x^3-coordinate from (1.4.23) and (1.4.24), we write

$$A_0 = (1, p_0), \; p_0 = \tilde{\sigma}_0 + \frac{k_0}{\tilde{\gamma}_0}(\tilde{\sigma}_0^2 + 1).$$

It follows from [Matsumoto, Chua and Ayaki 1988] that

$$-0.78 \le \tilde{\sigma}_0 \le -0.41$$
$$0.38 \le k_0/\tilde{\gamma}_0 \le 0.69.$$

Since $k_0/\tilde{\gamma}_0$ and $\tilde{\sigma}_0$ are monotonic, we see that

$$p_0 \le \max(\tilde{\sigma}_0) + \max\left(\frac{k_0}{\tilde{\gamma}_0}\right)(\max(\tilde{\sigma}_0^2) + 1) < 0.4$$

and hence

$$|A_0|^2 = 1 + p_0^2 < 1.16 \quad \text{and} \quad \varphi_0 := \tan^{-1}(p_0) \in \left(0, \frac{\pi}{4}\right),$$

where φ_0 is the angle between $\overrightarrow{OA_0}$ and the x^1-axis. Now, for $\tau \ge \pi/2 - \varphi_0$

$$\begin{aligned} |X_0(\tau)| &\le |A_0| \exp\left[\tilde{\sigma}_0\left(\frac{\pi}{2} - \varphi_0\right)\right] \\ &\le \sqrt{1.16}\exp\left[\frac{\pi}{4}\max(\tilde{\sigma}_0)\right] \\ &< 1. \end{aligned}$$

Since $0 < \varphi_0 < \pi/4$, it can be shown that the trajectory starting from A_0 remains in the region $x^1 > 0$ for all $0 < \tau < \pi/2 - \varphi_0$. Consequently, $X_0(\tau)$ never strikes the line $x^1 = -1$ for $\tau > 0$, namely,

$$\{X_0(\tau) \mid \tau > 0\} \subset \{(x^1, x^2) \mid x^1 > -1\}.$$

Similarly, it can be shown the trajectory starting from E_0 never reaches the line $x^1 = -1$. Since at any time the flow of a linear system is a linear function of its initial state, the trajectory starting from $\overline{A_0E_0}$ cannot hit $x^1 = -1$. $\square$

Lemma 1.4.24. *Let $C_1 := \Psi_1(C) = (x_C^1, x_C^2)$ and $F_1 := \Psi_1(F) = (x_F^1, x_F^2)$ on the (x^1, x^2)-plane in Fig. 1.4.1(b). Then for every $\beta \in J$,*

$$x_C^1 < x_F^1 < 1 \text{ and } x_C^2 > 0. \tag{1.4.125}$$

Moreover, C_1 is a continuous function of β on J.

Proof. It follows from (1.4.35) that

$$x_F^1 = \tilde{\gamma}_1(\tilde{\gamma}_1 - 2\tilde{\sigma}_1)/Q_1, \; x_F^2 = \tilde{\gamma}_1\left[1 - \tilde{\sigma}_1(\tilde{\sigma}_1 - \tilde{\gamma}_1)\right]/Q_1. \tag{1.4.126}$$

Since $C_1 = \Psi(C_0) = \Phi(0,0)$ when projected onto the (x^1, x^2)-plane, where Φ is the map defined in (1.4.96), we have

$$x_C^1 = 1 - \frac{(\tilde{\sigma}_1^2+1)\left[(\tilde{\sigma}_0 + \tilde{\gamma}_0 k_1)^2 + 1\right]}{(\tilde{\sigma}_0^2+1)Q_1} \tag{1.4.127}$$

$$\begin{aligned} x_C^2 &= \frac{\tilde{\gamma}_1\left[1 - \tilde{\sigma}_1(\tilde{\sigma}_1 - \tilde{\gamma}_1)\right]}{Q_1} - \frac{(\tilde{\sigma}_1^2+1)\tilde{\gamma}_0 k_1}{(\tilde{\sigma}_0^2+1)\tilde{\gamma}_1 Q_1} \\ &\quad \cdot \left\{k_1\tilde{\gamma}_0\left[\tilde{\sigma}_1(\tilde{\sigma}_1 - \tilde{\gamma}_1) + 1\right] + 2\tilde{\sigma}_0\tilde{\gamma}_1(\tilde{\sigma}_1 - \tilde{\gamma}_1)\right\}. \end{aligned} \tag{1.4.128}$$

It follows from Lemma 1.4.20 and the computations in [Matsumoto, Chua and Ayaki 1988] that $\tilde{\gamma}_0 k_1 > 0$ and $\tilde{\sigma}_0 < 0$ for $\beta \in J$. It follows from (1.4.126) and (1.4.128) that

$$x_F^1 - x_C^1 = \frac{\tilde{\sigma}_1^2 + 1}{Q_1(\tilde{\sigma}_0 + 1)}\tilde{\gamma}_0 k_1(\tilde{\gamma}_0 k_1 - 2\tilde{\sigma}_0) > 0\,.$$

Hence, $x_C^1 < x_F^1$. The fact that $x_F^1 < 1$ follows from the geometry of the D_1-unit in Fig. 1.4.1(b), where $\overline{A_1B_1}$ lies on the line $x^1 = 1$. To prove x_C^2 in (1.4.128) is positive, it suffices to show

$$\begin{aligned} &(\tilde{\sigma}_1^2+1)k_1\tilde{\gamma}_0\left\{k_1\tilde{\gamma}_0\left[\tilde{\sigma}_1(\tilde{\sigma}_1 - \tilde{\gamma}_1) + 1\right] + 2\tilde{\sigma}_0\tilde{\gamma}_1(\tilde{\sigma}_1 - \tilde{\gamma}_1)\right\} \\ &\qquad > \left[1 - \tilde{\sigma}_1(\tilde{\sigma}_1 - \tilde{\gamma}_1)\right]\tilde{\gamma}_1^2(\tilde{\sigma}_0^2 + 1) \end{aligned} \tag{1.4.129}$$

because $\tilde{\gamma}_1 < 0$ for $\beta \in J$. We can rewrite (1.4.129) as

$$\begin{aligned} &\frac{(\tilde{\sigma}_1^2+1)\tilde{\sigma}_0^2}{(\tilde{\sigma}_0^2+1)\tilde{\sigma}_1^2}\left\{\tilde{\sigma}_1(\tilde{\sigma}_1 - \tilde{\gamma}_1)\left[\left(\frac{k_1\tilde{\gamma}_0\tilde{\sigma}_1}{\tilde{\gamma}_1\tilde{\sigma}_0} + 1\right)^2 - 1\right] + \left(\frac{k_1\tilde{\gamma}_0\tilde{\sigma}_1}{\tilde{\gamma}_1\tilde{\sigma}_0}\right)^2\right\} \\ &\qquad > 1 - \tilde{\sigma}_1(\tilde{\sigma}_1 - \tilde{\gamma}_1). \end{aligned} \tag{1.4.130}$$

Since for all $\beta \in J$

$$\frac{k_1\tilde{\gamma}_0\tilde{\sigma}_1}{\tilde{\gamma}_1\tilde{\sigma}_0} = -\frac{\sigma_1}{\sigma_0} > 0 \quad \text{and} \quad \tilde{\sigma}_1(\tilde{\sigma}_1 - \tilde{\gamma}_1) > 0\,,$$

we have

$$
\begin{aligned}
&\{\text{left side of (1.4.130)} - \text{right side of (1.4.130)}\} \\
&= \frac{(\tilde{\sigma}_1^2+1)\tilde{\sigma}_0^2}{(\tilde{\sigma}_0^2+1)\tilde{\sigma}_1^2}\left\{\tilde{\sigma}_1(\tilde{\sigma}_1-\tilde{\gamma}_1)\left[\left(1-\frac{\sigma_1}{\sigma_0}\right)^2-1\right]+\left(\frac{\sigma_1}{\sigma_0}\right)^2\right\} \\
&\quad -1+\tilde{\sigma}_1(\tilde{\sigma}_1-\tilde{\gamma}_1) \\
&\geq \frac{(\tilde{\sigma}_1^2+1)\tilde{\sigma}_0^2\sigma_1^2}{(\tilde{\sigma}_0^2+1)\tilde{\sigma}_1^2\sigma_0^2}-1 \\
&= \frac{\sigma_1^2+\omega_1^2}{\sigma_0^2+\omega_0^2}-1 \quad (\text{because } \tilde{\sigma}_i=\sigma_i/\omega_i,\ i=0,1) \\
&= \frac{m_1\gamma_0}{m_0\gamma_1}-1 \quad (\text{because } \gamma_i(\sigma_i^2+\omega_i^2)=-\alpha\beta m_i,\ i=0,1) \\
&\geq \frac{-2\min(\gamma_0)}{\min(\gamma_1)}-1.
\end{aligned}
$$

It follows from [Matsumoto, Chua and Ayaki 1988] that

$$1.81 \leq \gamma_0 \leq 2.14$$

$$-3.60 \leq \gamma_1 \leq -3.27$$

and hence

$$\frac{-2\min(\gamma_0)}{\min(\gamma_1)}-1>0.$$

Since γ_i is a continuous function of β in view of (1.4.101), it follows from (1.4.102) that σ_i, ω_i, and k_i are also continuous functions of β for $i=0,1$. Since $C_1=(x_C^1, x_C^2)$ is given in (1.4.127) and (1.4.128), C_1 is a continuous function of β. □

Lemma 1.4.25. *There is a $\beta \in J$ such that*

$$C_1 \in \pi_1(\overline{E_1A_1})$$

i.e., there is a homoclinic orbit through the origin (Fig. 1.4.6).

Proof. Let us first draw two concentric circles, S_δ and S_ε, with their centers at $D_1=(0,0)$ in the V_1-plane and a radius equal to $|A_1|$ and $|E_1|e^{-2\pi\tilde{\sigma}_1}$, respectively, as shown in Fig. 1.4.7. Let Θ be the horizontal line through D_1 (i.e., the x^1-axis) and Θ' be the vertical line through F_1. Clearly, Θ' is to the left of the $x^1=1$ line in view of Lemma 1.4.24. Let S_δ intersect Θ and Θ' at points δ and δ', respectively. Let S_ε intersect Θ at point ε to the left of D_1. Depending on the value of $|E_1|$ and

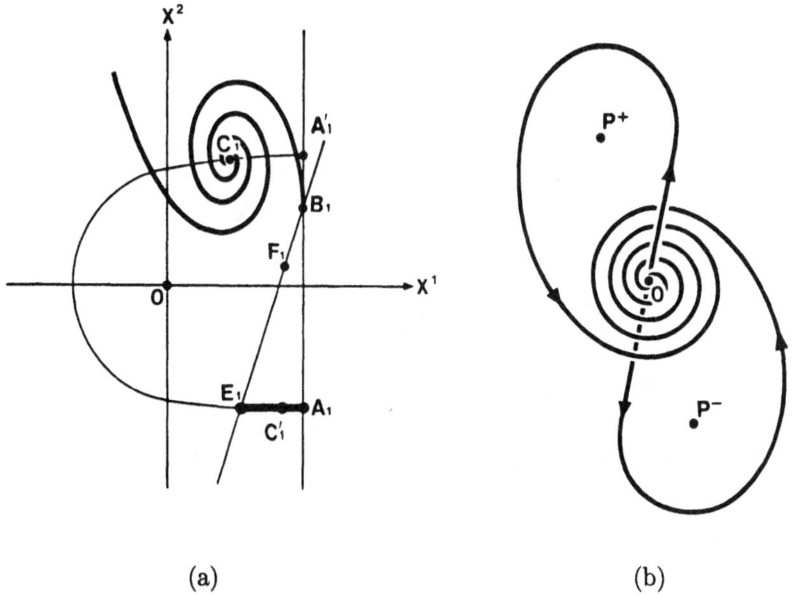

Fig. 1.4.6. The homoclinic orbit. (a) Various sets on the V_1 –plane. (b) Due to symmetry, the homoclinic orbits appear in pair.

$\tilde{\sigma}_1$, S_ε either intersects Θ' at two points, in which case the upper point is labeled ε', otherwise let ε' be the point where S_ε intersects Θ to right of D_1, as shown in Fig. 1.4.7. Let χ be the upper point where S_ε intersects the x^2-axis. Consider next the two logarithmic spirals

$$X_E(\tau) = E_1 \exp\left[-(\tilde{\sigma}_1 + \sqrt{-1})\tau\right], \; \tau \geq 0$$

and

$$X_A(\tau) = A_1 \exp\left[-(\tilde{\sigma}_1 + \sqrt{-1})\tau\right], \; \tau \geq 0.$$

Note that $X_A(\tau)$ and $X_E(\tau)$ correspond to the two *shrinking spirals* $[A_1\kappa''\kappa\kappa']$ (starting from A_1 at $\tau = 0$) and $[E_1\zeta\zeta']$(starting from E_1 at $\tau = 0$), respectively, as shown in Fig. 1.4.7. It follows from Lemma 1.4.22(3) that κ'' lies on the extension of the line $\overline{D_1E_1}$. Since both $|X_E(\tau)|$ and $|X_A(\tau)|$ shrink exponentially with the same rate $\tilde{\sigma}_1$, the time $\tau_{E_1\zeta}$ it takes $X_E(\tau)$ to go from E_1 to ζ (where it first intersects Θ) is equal to the time $\tau_{\kappa''\kappa}$ it takes $X_A(\tau)$ to go from κ'' to κ (where it first intersects Θ). Note that $\tau_{E_1\zeta} = \tau_{\kappa''\kappa} = \angle E_1D_1\kappa$ (in radians), where $\angle E_1D_1\kappa$ is the angle between $\overline{D_1E_1}$ and $\overline{D_1\kappa}$. Since $\angle E_1D_1\kappa < 2\pi$, it follows that κ must lie to the left of ζ, which in turn must lie to the left of ε.

Depending on $\tilde{\sigma}_1$, the continuation of the shrinking spiral from points κ and ζ may either intersect Θ' or Θ. Let this point of intersection be κ' and ζ', respectively.

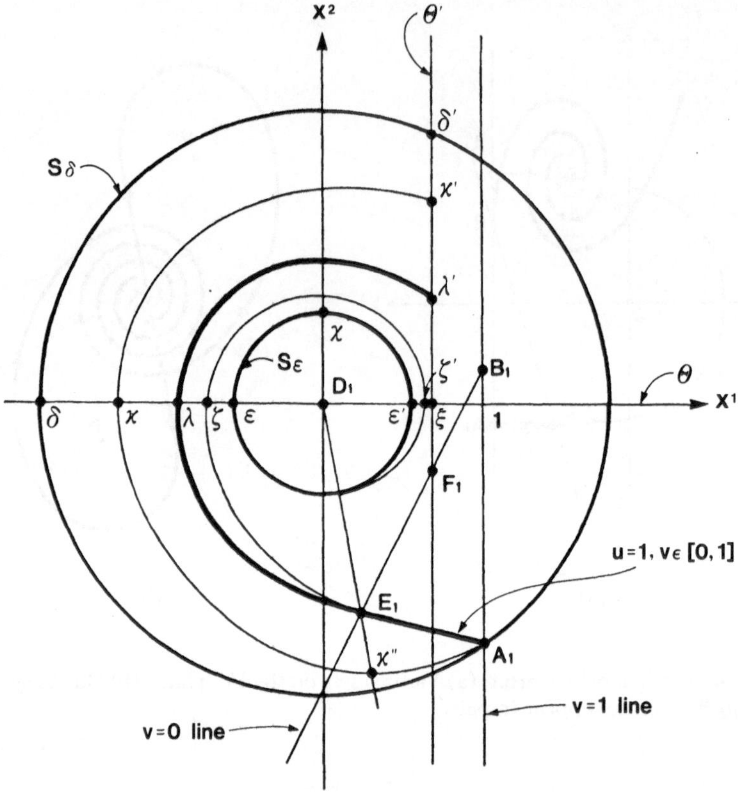

Fig. 1.4.7. The concentric circles S_δ and S_ε. ©1986 IEEE.

Let $\tau_{\kappa''\kappa'}$ denote the time it takes to go from κ'' to κ' and let $\tau_{E_1\zeta'}$ denote the time it takes to go from E_1 to ζ'. Since $\tau_{\kappa''\kappa'} < 2\pi$ and $\tau_{E_1\zeta'} < 2\pi$, both κ' and ζ' must lie outside of S_ε in Fig. 1.4.7, and ζ' must be below κ' in view of Lemma 1.4.22(2). Hence, κ must lie between δ and ζ, whereas κ' must lie between δ' and ζ' in Fig. 1.4.7.

Recall the image under π_1 of the line segment

$$\overline{E_1A_1} = \{(x^1(u,v), x^2(u,v)) \mid u = 1,\ 0 \le v \le 1\} .$$

Its extension beyond $A_1 (v > 1)$ is given by $X_1(\tau)$, defined in (1.4.106). A part of this image is shown by the bold *spiral*$[E_{1\lambda\lambda'}]$ in Fig. 1.4.7. Here $\lambda := X(\tau_1)$ is the point at which $X(\tau)$ first intersects Θ at some time τ_1 and $\lambda' := X(\tau_2)$ is the point at which $X(\tau)$ first intersects either Θ' or Θ to the right of D_1 (if it does not intersect Θ') at some time τ_2. Since both λ and λ' lie to the left of $x^1 = 1$, associated starting point, $X_1(\tau)$, must lie to the left of the $v = 1$ line. Hence

$$X_1(\tau_i) \in \overline{A_1E_1},\ i = 1, 2.$$

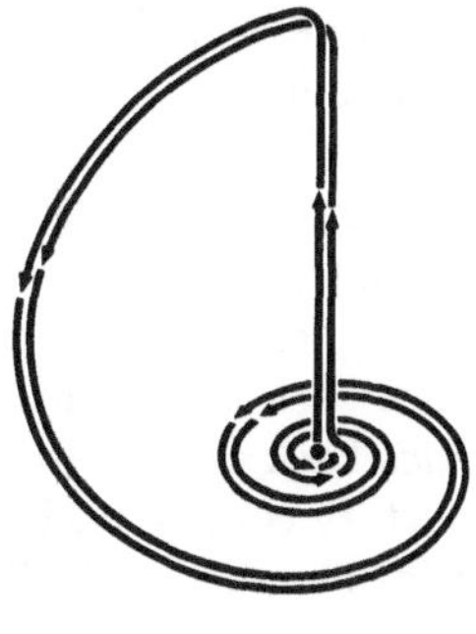

Fig. 1.4.8. Another homoclinicity. ©1987 IEEE.

It follows that λ must lie between ζ and κ, and λ' must lie between ζ' and κ' in Fig. 1.4.7 for all $\beta \in J$.

Our final task is to show that there exists a $\beta \in J$ such that $C_1 := \Psi(C)$ lies on the *bold spiral* $[\lambda\lambda']$. Since C_1 is a function of β (assuming α, m_0, and m_1 are fixed), we will denote this function by $C_1(\beta)$. Lemma 1.4.24 guarantees that $C_1(\beta)$, $\beta \in J$, lies in the simply connected region

$$H = \left\{(x^1, x^2) \mid x^1 \leq x^1_F,\ x^1 \geq 0\right\} .$$

Since $C_1(\beta)$ is continuous in β,

$$\Gamma = \{C_1(\beta) \mid \beta \in J\} \subset H$$

is a plane curve starting from $\beta = 6.5$ and ending at $\beta = 10.5$. It follows from [Matsumoto, Chua and Ayaki 1988] that

$$\begin{aligned} |C_1(\beta = 6.5)| &> |A_1(\beta = 6.5)| \\ |C_1(\beta = 10.5)| &< |E_1(\beta = 10.5)| \exp\left(-4\pi\tilde{\sigma}_1(\beta = 10.5)\right), \end{aligned}$$

which implies that $C_1(\beta = 6.5)$ is outside S_δ while $C_1(\beta = 10.5)$ is inside S_ε, and hence there is a β such that Γ crosses *bold spiral* $[\lambda\lambda']$. □

It follows from [Matsumoto, Chua and Ayaki 1988] that

$$|\sigma_0(\beta)| < \gamma_0(\beta),\ \beta \in J,$$

and hence the conditions of Shil'nikov's theorem are satisfied. □

Remark 1.4.26. Our numerical experiments suggest that at $\beta \approx 8.6$, the homoclinicity of our interest occurs, and the Double Scroll attractor is observed.

Remark 1.4.27. Let β^* be the value of β at the homoclinicity. Note that even though a small change of β would destroy the homoclinicity, the horseshoe is still present, because it is structurally stable.

Remark 1.4.28. It is also worth noting that even though a small change in β may destroy this particular homoclinic trajectory, there are infinitely many values of β near β^* which give rise to other types of homoclinicity. For example, a trajectory starting with O on $E^r(O)$ comes back to a point very close to O but not exactly, makes another round, and comes back exactly to O (see Fig. 1.4.8). Similarly, one can think of a homoclinic trajectory coming back to O after making three rounds, etc. A similar statement holds for heteroclinicity. Therefore, there is an abundance of homoclinicity and heteroclinicity and hence of horseshoes (see Chapter 3).

Remark 1.4.29. Mees and Chapman [Mees and Chapman 1987] also discuss homoclinic/heteroclinic orbits in the Double Scroll circuit.

1.5 Homoclinic Linkage

1.5.1 Introduction

We saw in Section 1.4 that a particular homoclinic orbit plays a crucial role in causing complicated (chaotic) behavior. There are, however, many other homoclinicities/heteroclinicities in (DS 2). This section studies the linkage structure among various homoclinicities/heteroclinicities via a family of periodic orbits. The goal of this section is to explain the following:

(1) The homoclinic/heteroclinic bifurcation sets (the set of parameter values for which homoclinicities/heteroclinicities occur) of interest are all connected with each other via *exactly one family* of periodic orbits;

(2) Moreover, the window (stable periodic orbit) structure of this particular family essentially determines the *global* structure of the periodic windows of (DS2).

1.5.2 Bifurcation Equations

A Normal Form The normal form for proper three-region piecewise-linear vector fields with point symmetry on $\mathbb{R}^3$ (Theorem 2.4.1) of Chapter 2 is directly applicable to (DS2). Namely, such an f is given by

$$f(x) = \begin{cases} Ax & (x \in R_0) \\ B(x-P) & (x \in R_+), \\ B(x+P) & (x \in R_-) \end{cases} \tag{1.5.1}$$

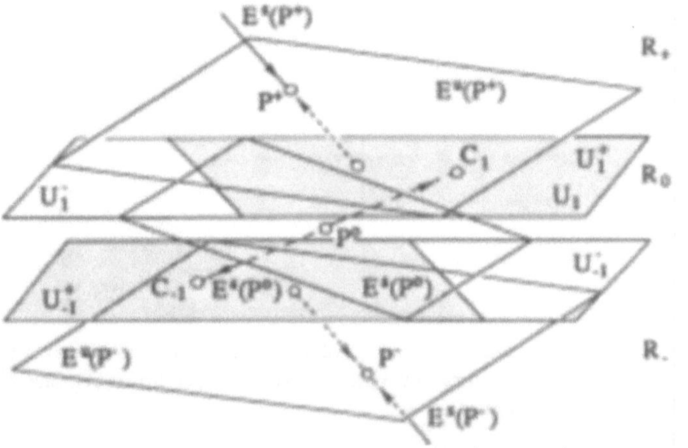

Fig. 1.5.1. Eigenspaces through the equilibrium points $P^{\pm}$ and P^0. ©1991 World Scientific Publishing Company.

where

$$R_{\pm} = \{x \in \mathbb{R}^3 \mid \pm\langle \alpha, x\rangle - 1 > 0\}$$
$$R_0 = \{x \in \mathbb{R}^3 \mid |\langle \alpha, x\rangle| \leq 1\}.$$

The parameters are explicitly given in terms of the eigenvalues of A and B. Define the boundary U_1 and U_{-1} by $\{x \in \mathbb{R}^3 \mid \langle \alpha, x\rangle = \pm 1\}$, and partition U_1 and U_{-1} into two parts respectively:

$$U_1^+ = \{x \in U_1 \mid \langle \alpha, Ax\rangle > 0\},$$
$$U_1^- = \{x \in U_1 \mid \langle \alpha, Ax\rangle < 0\},$$
$$U_{-1}^+ = \{x \in U_{-1} \mid \langle -\alpha, Ax\rangle > 0\},$$
$$U_{-1}^- = \{x \in U_{-1} \mid \langle -\alpha, Ax\rangle < 0\}.$$

For each real eigenvalue μ_i in the middle region R_0, we define the point $C_i \in U_1$ by

$$C_i = \begin{pmatrix} 1 \\ \mu_i \\ \mu_i^2 \end{pmatrix}. \tag{1.5.2}$$

It can be shown that the vector $\overrightarrow{OC_i}$ is an eigenvector of A associated with μ_i. Similarly, for each real eigenvalue ν_i in the outer regions, define the points $D_i \in U_i$ by

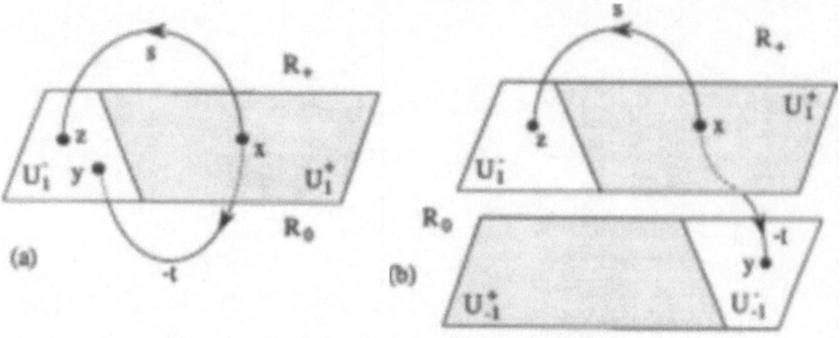

Fig. 1.5.2. Return time coordinates (t, s) of x. (a) Trajectory through x traverses only two linear regions. (b) Trajectory through x traverses three linear regions. ©1991 World Scientific Publishing Company.

$$D_i = \begin{pmatrix} 1 \\ \dfrac{\nu_i a_3}{b_3} \\ \dfrac{\nu_i(\nu_i - c_3)a_3}{b_3} \end{pmatrix}. \tag{1.5.3}$$

It can be shown that the vector $\overrightarrow{P^{\pm} D_i}$ is an eigenvector of B associated with ν_i (see Fig.1.5.1).

B Return Time Coordinates Assume that the matrices A and B are nonsingular.

(1) For $x \in U_1^+$, assume that there are $y, z \in U_1^-$ such that (see Fig.1.5.2(a))

$$\begin{aligned} y &= e^{-At}x \text{ , where} \\ t &= \inf\left\{t' > 0 \mid |\langle \alpha, e^{-At'}x\rangle| = 1\right\}, \\ z &= e^{Bs}(x - P) + P \text{ , where} \\ s &= \inf\left\{s' > 0 \mid |\langle \alpha, e^{Bs'}(x - P) + P\rangle| = 1\right\}. \end{aligned}$$

The pair (t, s) is called the *return time coordinates* of x because there is a one-to-one correspondence between (t, s) and $x \in U_1^+$ (see Chapter 2). Since $Aw = B(w - P)$ for all $w \in U_1$, by the continuity of the vector field one has

$$\begin{aligned} z &= A^{-1}A(e^{Bs}(x - P) + P) = A^{-1}Be^{Bs}(x - P) \\ &= A^{-1}e^{Bs}B(x - P) = A^{-1}e^{Bs}Ax = e^{Cs}x, \end{aligned}$$

where $C = A^{-1}BA$. Since x, y and z all belong to U_1,

$$\alpha^T e^{-At} x = 1, \alpha^T x = 1, \alpha^T e^{Cs} x = 1,$$

or equivalently,

$$[e_1 \alpha^T e^{-At} + e_2 \alpha^T x + e_3 \alpha^T e^{Cs}] x = h_1, \tag{1.5.4}$$

where $e_1 = (1,0,0)^T, e_2 = (0,1,0)^T, e_3 = (0,0,1)^T$ and $h_1 = (1,1,1)^T$. If the matrix inside the brackets on the left-hand side of (1.5.4) is nonsingular, denote its inverse by

$$K(t,s) = [e_1 \alpha^T e^{-At} + e_2 \alpha^T x + e_3 \alpha^T e^{Cs}]^{-1}.$$

Hence,

$$x = K(t,s) h_1.$$

(2) For $x \in U_1^+$, assume that there are $y \in U_{-1}^-$ and $z \in U_1^-$, such that (see Fig.1.5.2(b))

$$\begin{aligned} y &= e^{-At} x \text{ , where} \\ t &= \inf\left\{t' > 0 \mid |\langle \alpha, e^{-At'} x \rangle| = -1\right\}, \\ z &= e^{Bs}(x - P) + P \text{ , where} \\ s &= \inf\left\{s' > 0 \mid |\langle \alpha, e^{Bs'}(x - P) + P \rangle| = 1\right\}. \end{aligned}$$

The pair (t, s) is called the return time coordinates of x for the same reason as before. Since $x \mapsto z$ in Fig.1.5.2(b) is identical to that in Fig.1.5.2(a), one has the same relationship

$$z = e^{Cs} x,$$

where $C = A^{-1} B A$. Since $x, z \in U_1$, and $y \in U_{-1}$,

$$\alpha^T e^{-At} x = -1, \alpha^T x = 1, \alpha^T e^{Cs} x = 1,$$

or equivalently,

$$[e_1 \alpha^T e^{-At} + e_2 \alpha^T x + e_3 \alpha^T e^{Cs}] x = h_2, \tag{1.5.5}$$

where $e_1 = (1,0,0)^T, e_2 = (0,1,0)^T, e_3 = (0,0,1)^T$ and $h_2 = (1,1,1)^T$. Observe that the matrix inside the brackets on the left-hand side of (1.5.5) is the same as that in case (1). Hence, if this matrix is nonsingular, then one can solve for x in (1.5.5) to obtain

$$x = K(t,s) h_2,$$

where the inverse matrix $K(t,s)$ is as defined in case (1).

C Periodic Orbits Characterization of periodic orbits as well as other orbits depends on how the trajectory hits the boundaries U_1 and U_{-1}. Consider first the simplest periodic orbits, namely, those involving intersections with only one boundary, as shown in Fig.1.5.3(a).

Type 1 route: $U_1^+ \to U_1^- \to U_1^+ \to U_1^-$.(Fig.1.5.3(a))

$$\begin{aligned} e^{Cs_1}K(t_1,s_1)h_1 &= e^{-At_2}K(t_2,s_2)h_1 \\ e^{Cs_2}K(t_2,s_2)h_1 &= e^{-At_1}K(t_1,s_1)h_1 \\ \left|\alpha^T e^{-At}K(t_i,s_i)h_1\right| &\neq 1 \text{ for all } t \text{ with } 0<t<t_i \\ \alpha^T e^{Cs}K(t_i,s_i)h_1 &\neq 1 \text{ for all } s \text{ with } 0<s<s_i (i=1,2). \end{aligned}$$

The last two conditions will be called the "open conditions" for the first return times. Note that this particular periodic orbit, which is described in Chapter 2, passes through only two regions. Similar equations can be derived for other periodic orbits, e.g., Fig.1.5.3(b) and (c).

Recall that the matrices A and $C = A^{-1}BA$ are determined by the eigenvalues μ_1, μ_2, μ_3 and ν_1, ν_2, ν_3. Each of the first two equations has three components. However, each side of these equations represents a point lying on the two-dimensional plane U_1. Therefore, there are only two independent equations out of three in each of the above equations, and hence these two vector equations actually contain four independent nonlinear equations in four variables (t_1, s_1, t_2, s_2). Therefore, (t_1, s_1, t_2, s_2) are determined through these equations, and hence the solution set lies in a six-dimensional subset in the ten-dimensional $(\mu_1, \mu_2, \mu_3, \nu_1, \nu_2, \nu_3, t_1, s_1, t_2, s_2)$-space. If $(\mu_1, \mu_2, \mu_3, \nu_1, \nu_2, \nu_3, t_1, s_1, t_2, s_2)$ is a solution, then the point $x = e^{-At_1}K(t_1, s_1)h_1$ is a period-2 point. The quadruple (t_1, s_1, t_2, s_2) is called a *return time sequence* of point x. A similar equation can be derived for a periodic orbit which hits the boundaries many times.

D Bifurcation Conditions for Periodic Orbits Assume x is a periodic point with a return time sequence $(t_1, s_1, t_2, s_2, \ldots, t_n, s_n)$. It follows from (2.5.98) of Chapter 2 that the tangent map of the Poincaré full return map π at x is given by

$$D\pi(x) = \left(I - \frac{Ax\alpha^T}{\alpha^T Ax}\right) e^{Bs_n}e^{At_n}\cdots e^{Bs_2}e^{At_2}e^{Bs_1}e^{At_1}.$$

Furthermore, Theorem 2.5.15 of Chapter 2 says that bifurcation sets are given by

(1) Saddle-node bifurcation: $2 - T + D = 0$;

(2) Period-doubling bifurcation: $T + D = 0$;

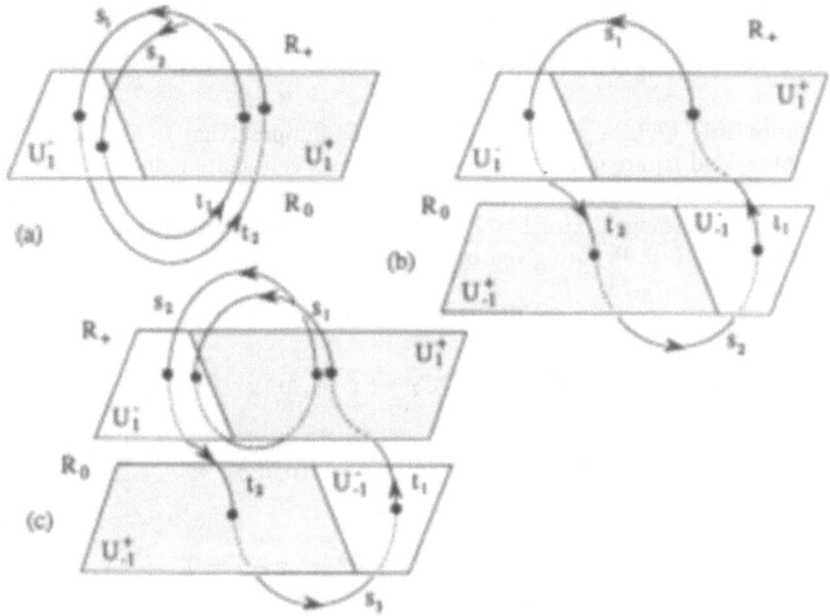

Fig. 1.5.3. Periodic orbits and their return time coordinates. The negative sign for t has been deleted so that all times increase in the direction of the arrows. ©1991 World Scientific Publishing Company.

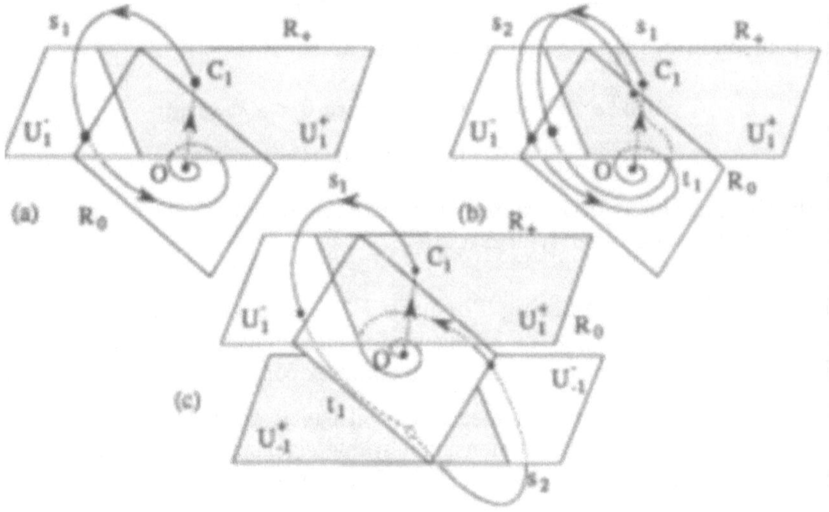

Fig. 1.5.4. Homoclinic orbits passing through O and their return time coordinates. ©1991 World Scientific Publishing Company.

(3) Hopf bifurcation for the Poincaré map: $D - 1 = 0$ and $-1 < T < 3$,

where $T = \mathrm{tr}(M)$, $D = \det(M)$, $M = e^{Bs_n} e^{At_n} e^{Bs_{n-1}} e^{At_{n-1}} \cdots e^{Bs_1} e^{At_1} T$.

E Homoclinic Orbits Passing Through O Suppose that μ_1 is positive real, and that μ_2 and μ_3 are negative reals or a complex conjugate pair with a negative real part. Since the eigenvector associated with μ_i is given by (1.5.2), the two-dimensional stable eigenspace $E^s(O)$ and the one-dimensional unstable eigenspace $E^u(O)$ for $O = (0,0,0)^T$ are given by

$$\begin{aligned} E^u(O) &= \{x \in \mathbb{R}^3 \mid x = r\overrightarrow{OC_1}, |\alpha^T x| \le 1, r \in \mathbb{R}\} \\ E^s(O) &= \{x \in \mathbb{R}^3 \mid x = r\overrightarrow{OC_2} + r'\overrightarrow{OC_3}, |\alpha^T x| \le 1, r, r' \in \mathbb{R}\}. \end{aligned}$$

Set

$$u = (0, (\mu_2 + \mu_3)/(\mu_2\mu_3), -1/(\mu_2\mu_3))^T .$$

Then the intersection $E^u(O) \cap U_1$ is given by

$$E^u(O) \cap U_1 = \{x = (x^1, x^2, x^3) \in \mathbb{R}^3 \mid u^T x - 1 = 0, x = 1\}.$$

Characterization of a homoclinic orbit also depends on how the trajectory hits U_1 and U_{-1}. Consider

Type 1 route : $O \to U_1^+ \to U_1^- \to O$(Fig. 1.5.4(a)).

$$\alpha^T e^{Cs_1} C_1 = 1, u^T e^{Ct} C_1 = 1.$$

The open conditions for the first return times are:

$$\begin{aligned} \alpha^T e^{Cs} C_1 &\ne 1 \quad \text{for all } s \text{ with } 0 < s < s_1 \\ \alpha^T e^{Ct} C_1 &\ne 1 \quad \text{for all } t > 0. \end{aligned}$$

Clearly, there are many other routes which give rise to homoclinic orbits. Fig.1.5.4 (b) and (c) show two other typical routes.

F Homoclinic Orbits Passing Through P^+ Recall matrix B in the dynamics (1.5.1) and let ν_1, ν_2, and ν_3 be its eigenvalues. Assume that ν_1 is negative real, and that ν_2 and ν_3 are positive reals, or a complex conjugate pair with a positive real part. Since the eigenvector associated with ν_i is given by $\overrightarrow{P^\pm D_i}$, where D_i is defined by (1.5.3), the two-dimensional stable eigenspace $E^s(P)$ and the one-dimensional unstable eigenspace $E^u(P)$ at P are given, respectively, by

$$\begin{aligned} E^u(P) &= \{x \in \mathbb{R}^3 \mid x = r\overrightarrow{PD_1} + P, \alpha^T x - 1 \ge 0, r \in \mathbb{R}\} \\ E^s(P) &= \{x \in \mathbb{R}^3 \mid x = r\overrightarrow{PD_2} + r'\overrightarrow{PD_3} + P, \alpha^T x - 1 \ge 0, r, r' \in \mathbb{R}\}. \end{aligned}$$

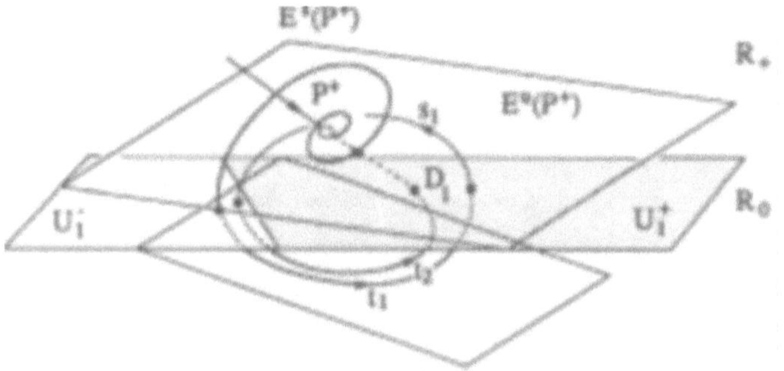

Fig. 1.5.5. Homoclinic orbits passing through P^+ and their return time coordinates. ©1991 World Scientific Publishing Company.

Let

$$v = (0, (\nu_2 + \nu_3 - c_1)b_3/(\nu_2\nu_3a_3), -b_3/(\mu_2\mu_3a_3))^T .$$

Then the intersection $E^u(P) \cap U_1$ is given by

$$E^u(P) \cap U_1 = \{x = (x^1, x^2, x^3) \in \mathbb{R}^3 \; | v^T x - 1 = 0, x^1 = 1\} .$$

Set

$$P^+ = P \text{ and } P^- = -P.$$

Type 1 route: $P^+ \to U_1^+ \to U_1^- \to U_1^+ \to U_1^- \to P^+$(Fig. 1.5.5).

$$v^T e^{-At_1} K(t_1, s_1)h_1 = 1, e^{Cs_1} K(t_1, s_1)h_1 = e^{-At_2} D_1.$$

The open conditions for the first return times are

$$\begin{array}{ll} \alpha^T e^{Cs} e^{-At_1} K(t_1, s_1)h_1 \neq 1 & \text{for all } s > 0 \\ \left|\alpha^T e^{-At} K(t_1, s_1)h_1\right| \neq 1 & \text{for all } t \text{ with } 0 < t < t_1 \\ \alpha^T e^{Cs} K(t_1, s_1)h_1 \neq 1 & \text{for all } s \text{ with } 0 < s < s_1 \\ \left|\alpha^T e^{-At} D_1\right| \neq 1 & \text{for all } t \text{ with } 0 < t < t_2. \end{array}$$

The open conditions for other routes are similar.

G Heteroclinic Orbits Let μ_1, μ_2 and μ_3 be the eigenvalues of A (see (1.5.1)). Assume that μ_1 is positive real, and μ_2 and μ_3 are negative reals, or a complex conjugate pair with a negative real part. Similarly, let ν_1, ν_2 and ν_3 be the eigenvalues of B. Assume also that ν_1 is negative real, and ν_2 and ν_3 are positive reals, or a complex conjugate pair with a positive real part. We will describe heteroclinic orbits with two typical routes.

Type 1 route: $O \to U_1^+ \to P^+$(Fig.1.5.6(a)).

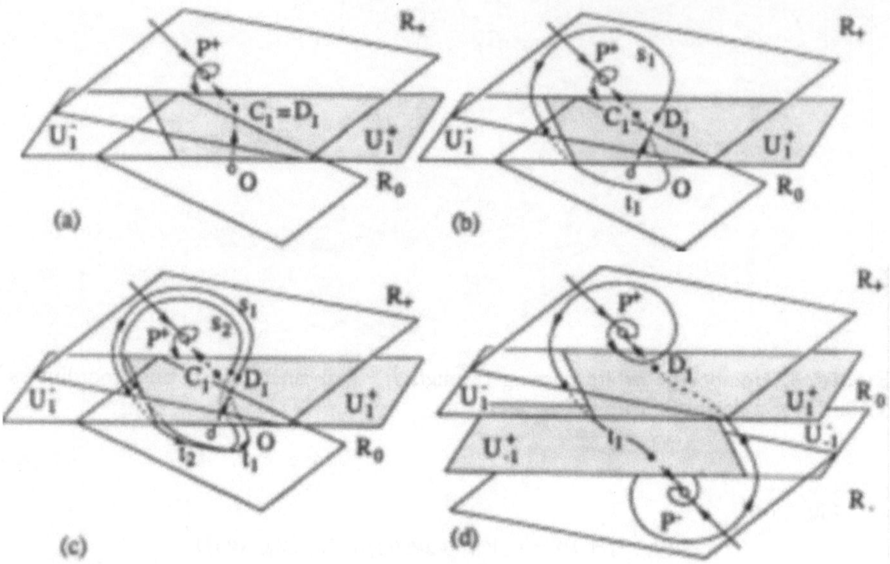

Fig. 1.5.6. Heteroclinic orbits and their return time coordinates. ©1991 World Scientific Publishing Company.

$$C_1 = D_1.$$

Type 2 route: $O \to U_1^+ \to U_1^- \to U_1^+ \to P^+$(Fig.1.5.6(b)).

$$\alpha^T e^{Cs_1} C_1 = 1, e^{Cs_1} C_1 = e^{-At_1} D_1, \alpha^T e^{-At_1} D_1 = 1.$$

The open conditions for the first return times are

$$\begin{aligned} \alpha^T e^{Cs} C_1 &\neq 1 \quad \text{for all } s \text{ with } 0 < s < s_1 \\ \left|\alpha^T e^{-At} D_1\right| &\neq 1 \quad \text{for all } t \text{ with } 0 < t < t_1. \end{aligned}$$

Other types are also possible as shown in Fig.1.5.6(c) and (d).

1.5.3 Global Bifurcations

The bifurcation equations given in the previous subsection will now be used to make a detailed bifurcation analysis. There are two precautions in generating the bifurcation diagrams:

(1) Periods of the periodic orbits near homoclinicity/heteroclinicity become extremely long;

(2) Most of the periodic orbits of interest are saddle-type, i.e., neither attractive nor repelling.

Therefore, the Runge-Kutta, for example, does not work. Those formulas derived in Subsection1.5.2 necessitate no integration scheme for differential equations and are rather powerful for the present purpose.

A Homoclinic/Heteroclinic Bifurcation Sets Let us first look at the bifurcation structures associated with the homoclinic/heteroclinic orbits depicted in Fig.1.5.7 because they are simple and yet they reveal important features common to many other homoclinic/heteroclinic bifurcations. The symbol "Γ" (respectively "Λ") denotes a homoclinic or a heteroclinic orbit passing through two (respectively three) linear regions. The subscript "$_0$" shows that the associated equilibrium point is the origin, while the superscript indicates the "number of rotations with respect to $P^{\pm}$". The subscript "$_p$" implies that the associated equilibrium point is $P^{\pm}$.

Fig. 1.5.8 shows the bifurcation sets associated with the homoclinic/heteroclinic orbits described above in the (α', β)-space, where

$$\alpha' = 80.0\alpha^*/(39.98\beta + 1.6) \quad (1.5.6)$$

$$\alpha^* = \alpha - F(\Gamma_0^1, \beta). \quad (1.5.7)$$

Here $F(\Gamma_0^1, \beta)$ denotes that the values of α for which the homoclinic orbit Γ_0^1 occurs are a function of β. Consequently, $\alpha^* = 0$ for all values of β for which Γ_0^1 exists. The parameter α' is chosen as a bifurcation parameter instead of α because the bifurcation set is easier to draw and is more revealing. In addition to the homoclinic as well as the heteroclinic bifurcation sets, the *Hopf eigenvalue set* (denoted by H) is also plotted. On Hopf eigenvalue set, eigenvalues at equilibrium points are pure imaginary. Whether Hopf bifurcation actually occurs depends on other conditions.

Note that Γ_0^1 corresponds to the vertical axis. Figs. 1.5.8(b) and 1.5.8(c) give enlarged details of the two small rectangles in Fig. 1.5.8(a). Observe that Λ_p, Γ_0^2 and Λ_0^2 are connected to their counterparts $\tilde{\Lambda}_p$, $\tilde{\Gamma}_0^2$ and $\tilde{\Lambda}_0^2$, respectively. Consider next regions R_A and R_B in the diagram. In the parameter region of interest, the eigenvalues at O consist of a real γ_0 and a complex conjugate pair $\sigma_0 \pm \omega_0 i$. These two regions are defined by

$$\text{(1-a)} \qquad R_A = \{(\alpha, \beta) \mid |\sigma_0| < \gamma_0\}, \quad (1.5.8)$$

$$\text{(1-b)} \qquad R_B = \{(\alpha, \beta) \mid |\sigma_0| > \gamma_0\}. \quad (1.5.9)$$

Observe that the hypotheses of Shil'nikov's theorem [Shil'nikov 1965] are satisfied only in region R_A, while one of Shil'nikov's conditions is violated in region R_B. It follows from [Arneodo, Coulett and Tresser 1982] that *each of the five homoclinicities* ($\Gamma_0^1, \Lambda_0^2, \tilde{\Lambda}_0^2, \Gamma_0^2, \tilde{\Gamma}_0^2$) *bifurcates into a pair of stable periodic orbits, where one is symmetric while the other is asymmetric.*

In order to see the structure of this interconnection more clearly, let us look at a typical one-parameter bifurcation diagram located within R_B. To this end, it is more convenient to introduce a new coordinate system. Let

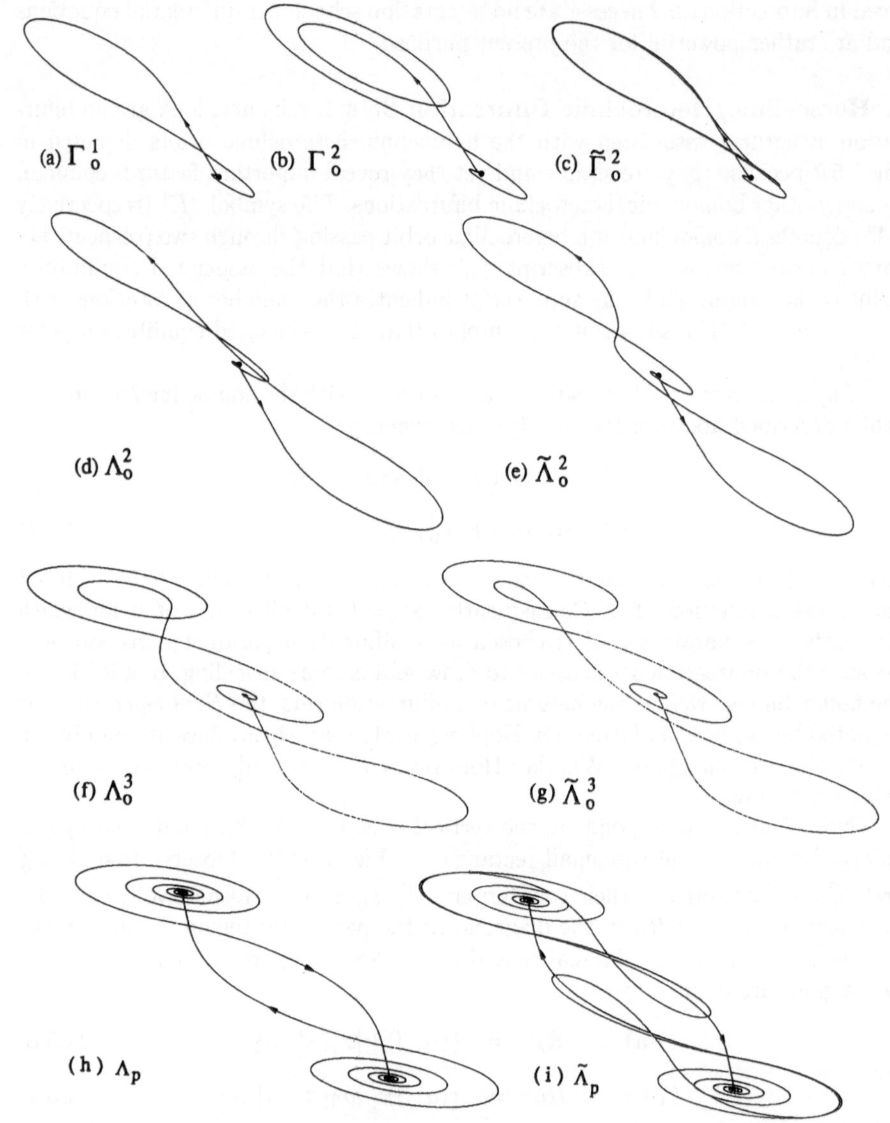

Fig. 1.5.7. Homoclinic orbits (denoted by Γ_o and Λ_o) and heteroclinic orbits (denoted by Λ_p). (a)Γ_o^1. (b)Γ_o^2. (c)$\tilde{\Gamma}_o^2$. (d)Λ_o^2. (e)$\tilde{\Lambda}_o^2$. (f)Λ_o^3. (g)$\tilde{\Lambda}_o^3$. (h)Λ_p. (i)$\tilde{\Lambda}_p$. ©1991 World Scientific Publishing Company.

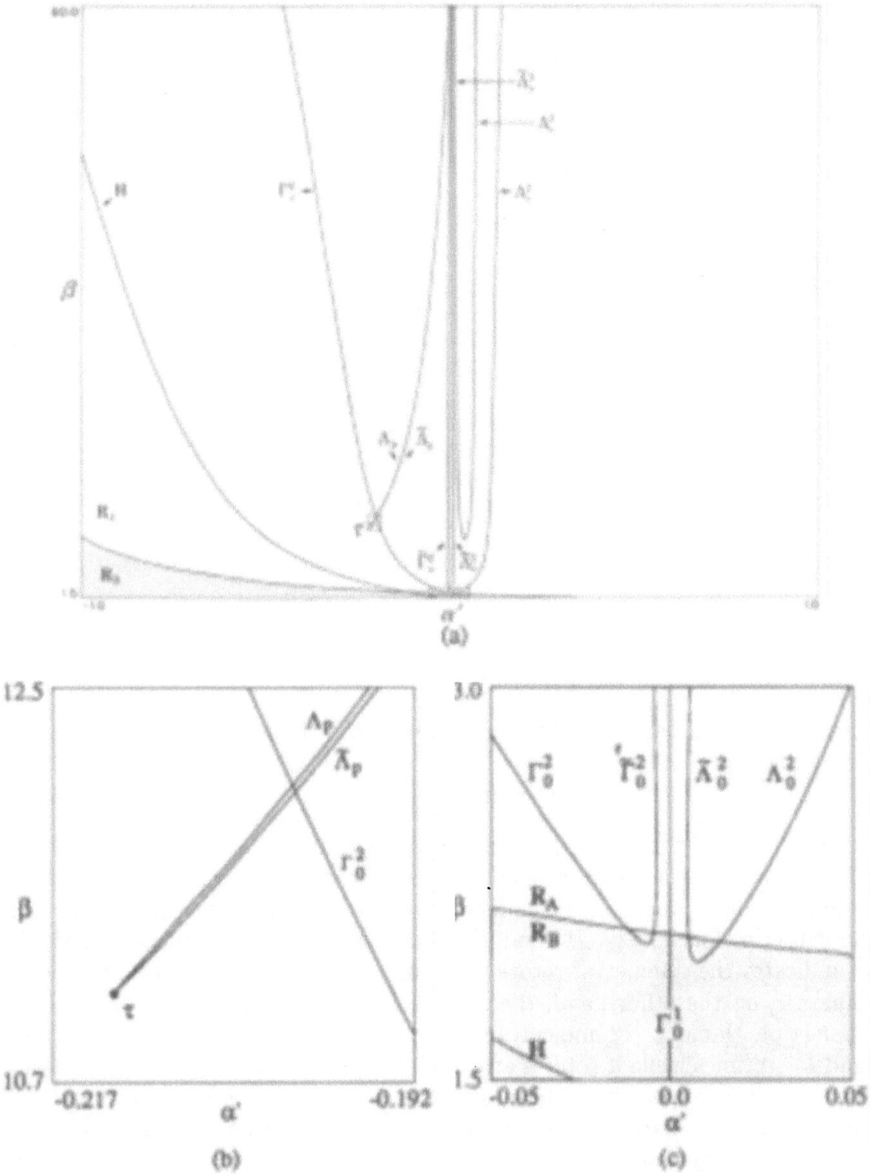

Fig. 1.5.8. Bifurcation diagrams. (a) Global diagram. Note that Γ_o^1 coincides with the vertical axis (α'). The two heteroclinic orbits Λ_p and $\tilde{\Lambda}_p$ are connected at the terminal point τ. These two curves are close to each other that they appear here only as one curve. (b) Enlargement of the rectangle centered at τ. (c) Enlargement of the rectangle near the origin.

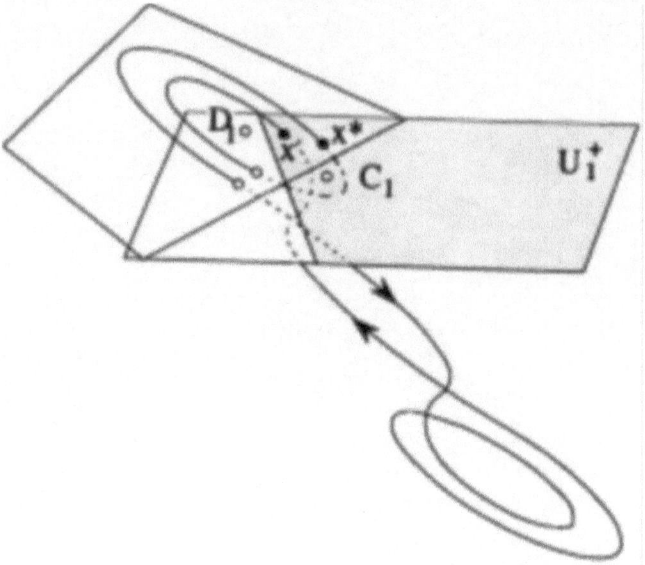

Fig. 1.5.9. Periodic point x^*, which gives rise to an orbit traveling all three regions.

$$x^* = (1, x^{2*}, x^{3*}) \tag{1.5.10}$$

denote a point on a periodic orbit (Fig.1.5.9) on U_1^+ closest to C_1, and define the inner product

$$\xi^* = \langle (0,1,3), x^* \rangle . \tag{1.5.11}$$

Fig. 1.5.10 shows the (α, ξ^*)-bifurcation diagram at $\beta = 2.04$ where each symbol Π indicates the associated family of periodic orbits shown in Fig.1.5.11. If a periodic orbit in question is symmetric with respect to the origin, there will be no subscript for Π, e.g., Π^1 and Π^2. A superscript for a symmetric periodic orbit indicates the number of rotations around P^+ or P^-. If a periodic orbit is asymmetric, on the other hand, the symbol Π will carry a subscript as well as a superscript. Namely, Π_m^n indicates an asymmetric periodic orbit which rotates around P^+ n times while it rotates around P^- m times where $n \geq m$, or an asymmetric periodic orbit which rotates around P^- n times while it rotates around P^+ m times. Since β is fixed here, each homoclinic orbit is defined by a solid point, while a family of symmetric (respectively asymmetric) orbits is defined by a solid (respectively broken) curve. The symbol H indicates Hopf eigenvalue set. Observe that

(1) Families Π_0^1 and Π^1 are terminated by Hopf eigenvalue sets and the homoclinic bifurcation set Γ_0^1;

(2) Family Π_0^2 (respectively $\tilde{\Pi}_0^2$) is terminated by a period-doubling bifurcation set denoted by PD and the homoclinic bifurcation set Γ_0^2 (respectively $\tilde{\Gamma}_0^2$);

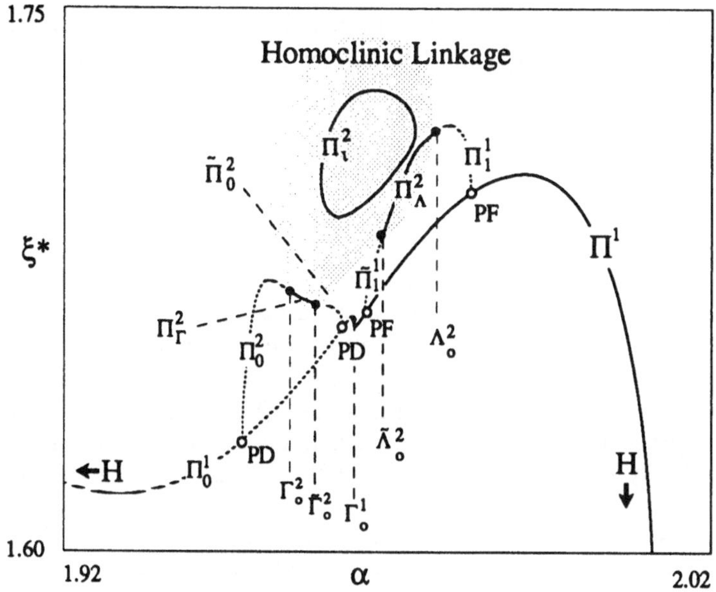

Fig. 1.5.10. (α, ξ^*)-bifurcation diagram for $\beta = 2.04$. ©1991 World Scientific Publishing Company.

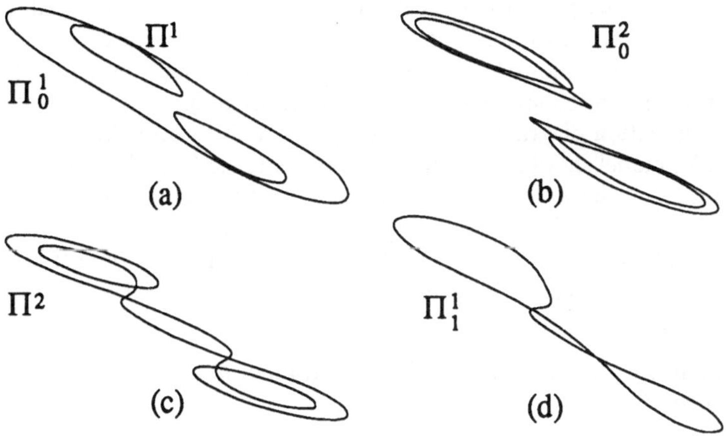

Fig. 1.5.11. Periodic orbit associated with Fig.1.5.10. (a)Π^1 and Π_0^1. (b) Π_0^2. (c) Π^2. (d) Π_1^1. ©1991 World Scientific Publishing Company.

(3) Family Π_1^1 (respectively $\tilde{\Pi}_1^1$) is terminated by a pitchfork bifurcation set denoted by PF and the homoclinic bifurcation set Λ_0^2 (respectively $\tilde{\Lambda}_0^2$);

(4) Family Π_Γ^2 (respectively Π_Λ^2) is terminated by the pair of homoclinic bifurcation sets Γ_0^2 and $\tilde{\Gamma}_0^2$ (respectively Λ_0^2 and $\tilde{\Lambda}_0^2$);

(5) Family Π_l^2 is an isolated loop.

Since the three periodic orbits, Π_l^2, Π_Γ^2 and Π_Λ^2, with two rotations around $P^\pm$ are very similar to each other, only one typical trajectory is shown and labeled as Π^2 without a subscript. A distinct feature discernible in Fig.1.5.10 is that a pair of homoclinicities, e.g., Γ_0^2 and $\tilde{\Gamma}_0^2$ (respectively Λ_0^2 and $\tilde{\Lambda}_0^2$), are connected with each other via the *common* family Π_Γ^2 (respectively Π_Λ^2) of periodic orbits. It seems quite natural, therefore, to call this the $(\Gamma_0^2, \tilde{\Gamma}_0^2)$ (respectively $(\Lambda_0^2, \tilde{\Lambda}_0^2)$)-*homoclinic linkage.* In the rest of this section, we will study the relationships between the family Π^2 and homoclinic bifurcation sets in the half-toned region of Fig.1.5.10.

Now let us look at the typical homoclinic bifurcations in R_A. Fig. 1.5.12 shows the bifurcation structure of the pair of periodic orbits Π^1 and Π_0^1 in the (α^*, ξ)-space where

$$\xi = \langle (0, 2, -1), x^* \rangle , \tag{1.5.12}$$

(compare with (1.5.11)) and $\beta = 30.0$. Observe that this pair of periodic orbits, which are born via (piecewise-linear) Hopf bifurcation, converge to the homoclinic orbit Γ_0^1. This one-parameter bifurcation has been studied in [George 1986]. The spiral structure in the vicinity of Γ_0^1 includes infinitely many folds, namely, saddle-node bifurcation sets, which can be seen more clearly than those in Fig.1.5.10. Consider next what happens when β is varied. The associated bifurcation structures are not easy to see in the (α^*, ξ)-plane. If one embeds it into the (α^*, β, ξ)-space, however, the bifurcation structures are almost transparent. Fig. 1.5.13 shows the (α^*, β, ξ)-bifurcation diagram of Π^1 and Π_0^1, while Fig.1.5.14 shows an enlarged (not-to-scale) model showing only Π_0^1. Note that Fig.1.5.13 and Fig.1.5.14 contain the information of the state variable represented by ξ in addition to the parameters α^* and β. Note that each point (α^*, β, ξ) on the surface represents a periodic orbit belonging to the particular family Π_0^1. Since this is a particular type of bifurcation diagram and since it is extremely informative, we will call Fig.1.5.13 a *bifurcation topography.* One can see that

(1) The surface terminates at a Hopf bifurcation or at a homoclinic/heteroclinic bifurcation set;

(2) Each fold of the surface corresponds to a saddle-node bifurcation;

(3) A new surface is born out of a period-doubling (or a pitchfork) bifurcation.

Observe that as β decreases, the spiral in Fig.1.5.14 gets smaller and that the pair of saddle-node bifurcation sets along the homoclinic orbit Γ_0^1 merge and vanish at a cusp point located in R_B. Consequently, there are infinitely many cusp points

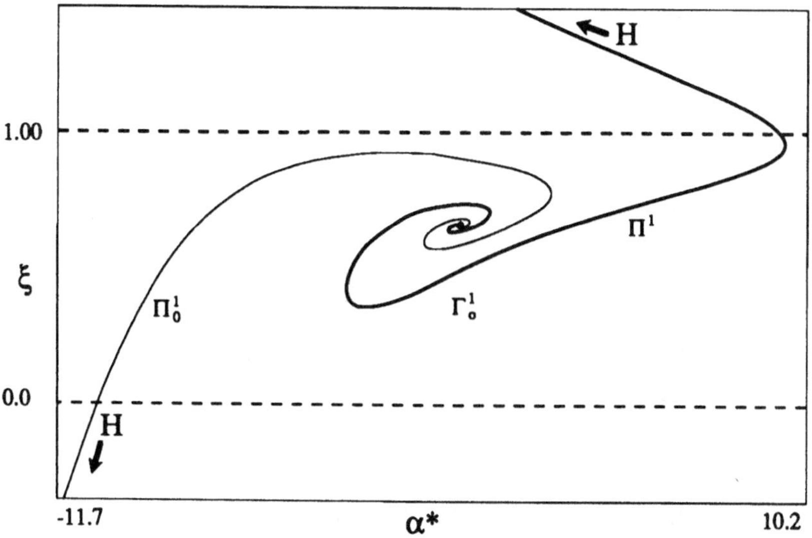

Fig. 1.5.12. (α^*, ξ)-bifurcation diagram for $\beta = 30.0$. ©1991 World Scientific Publishing Company.

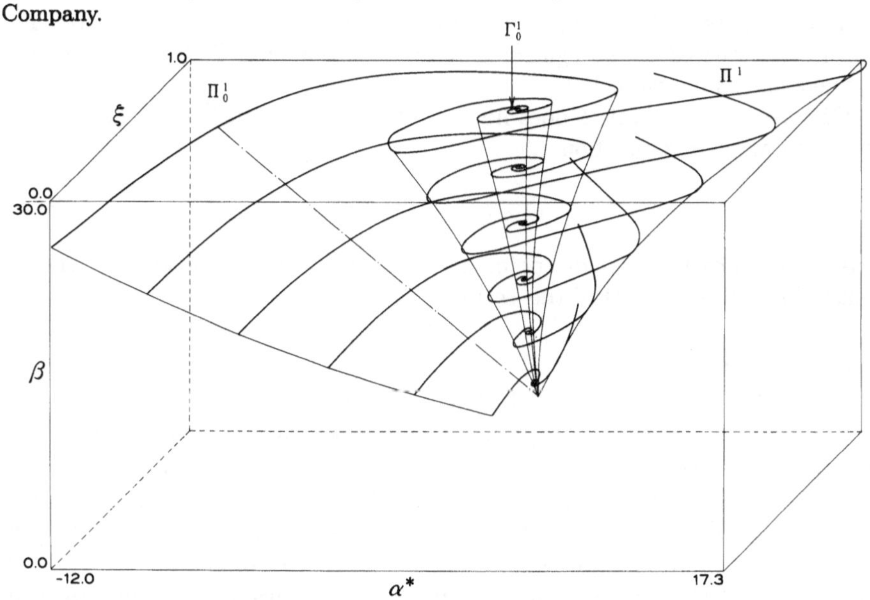

Fig. 1.5.13. (α^*, β, ξ)-bifurcation topography of Π_0^1 and Π^1. ©1991 World Scientific Publishing Company.

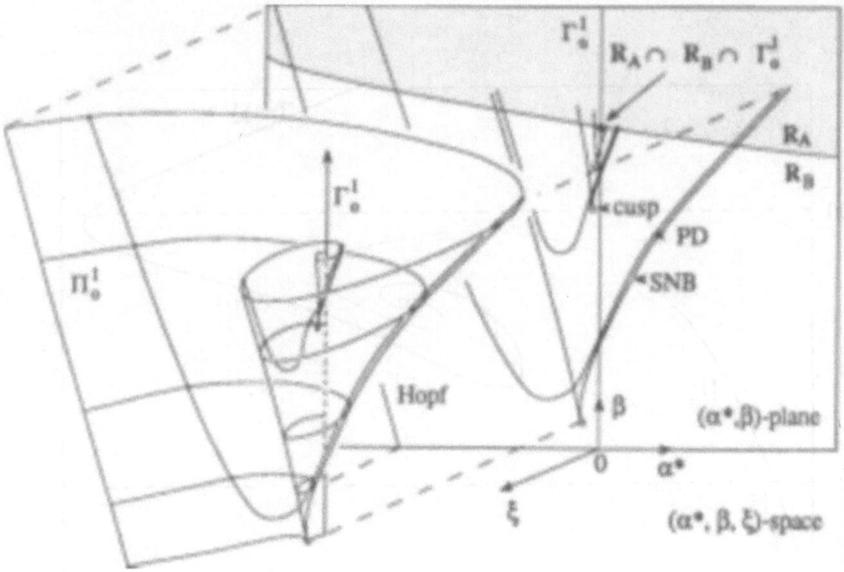

Fig. 1.5.14. Schematic model bifurcation topography of Fig.1.5.13.

which accumulate at the point $\Gamma_0^1 \cap R_A \cap R_B$. Details of the above bifurcation structure are studied in the light of Shil'nikov's theorem in [Fujimoto, Komuro, Tokunaga and Matsumoto 1990].

Of course, the variable ξ does not have to be exactly defined by (1.5.12). It can be any reasonable function

$$\xi = G(x^*), \; G: \mathbb{R}^2 \to \mathbb{R}, \tag{1.5.13}$$

where x^* denotes the periodic point at (α, β). Bifurcation topography will be extensively used in the rest of this section.

B Homoclinic Linkage

B1. Family Π^2 : The $(\Gamma_0^2, \tilde{\Gamma}_0^2)$- and the $(\Lambda_0^2, \tilde{\Gamma}_0^2)$-homoclinic linkages.

Fig. 1.5.15(a) shows the (α^*, β, ξ^*)-bifurcation topography which includes the symmetric periodic orbit Π^2, and the pair of asymmetric periodic orbits Π_0^2 and Π_1^1 (recall Fig.1.5.11), while Fig. 1.5.15 (b) shows the location of each family of periodic orbits, i.e., Π_l^2, Π_Γ^2 and Π_Λ^2. Plate 1.11 shows an equivalent model (not drawn to scale) which emphasizes the important features. Here α^* is defined by (1.5.7) and ξ^* is defined by (1.5.11). Points C and $\tilde{C}$ stand for cusp points while points S and $\tilde{S}$ stand for saddle points. The pair of yellow regions corresponds to the pair of asymmetric periodic orbits, Π_0^2 and Π_1^1. The region with yellow outside and blue inside corresponds to the symmetric periodic orbit Π^2. Now recall the periodic orbits Π_l^2, Π_Γ^2 and Π_Λ^2 in Fig.1.5.10. Examining Fig.1.5.15 and Plate 1.11 one can see how these periodic orbits are deformed as β is varied (see Fig.1.5.15(b)):

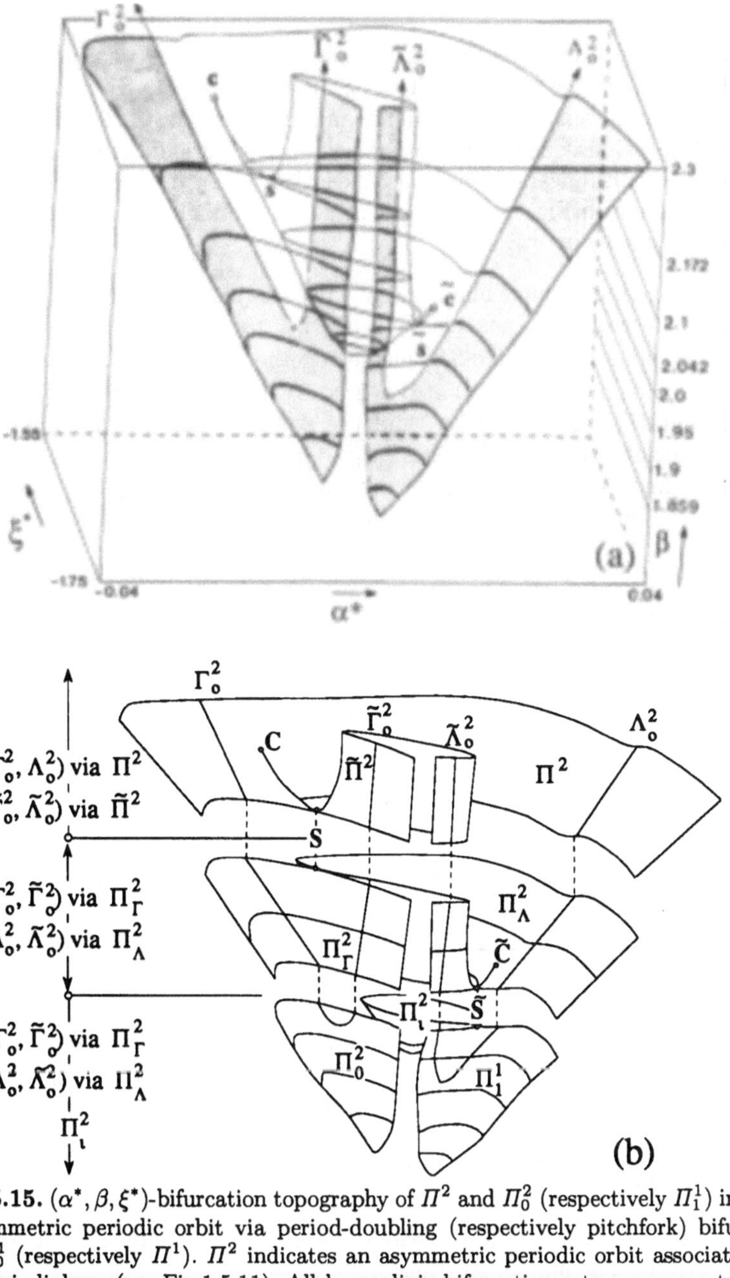

Fig. 1.5.15. (α^*, β, ξ^*)-bifurcation topography of Π^2 and Π_0^2 (respectively Π_1^1) indicates an asymmetric periodic orbit via period-doubling (respectively pitchfork) bifurcation from Π_0^1 (respectively Π^1). Π^2 indicates an asymmetric periodic orbit associated with homoclinic linkage (see Fig.1.5.11). All homoclinic bifurcation sets are connected via a single family of periodic orbits Π^2. (a) Bifurcation topography. (b) Schematic model bifurcation topography. ©1991 World Scientific Publishing Company.

(1) Disappearance of Homoclinic Linkage ($\beta < 2.04$): As β decreases, the distance between the homoclinicities Γ_0^2 and $\tilde{\Gamma}_0^2$ (respectively Λ_0^2 and $\tilde{\Lambda}_0^2$) gets smaller, and eventually, they merge and disappear at $\beta \approx 2.02$ (respectively $\beta \approx 1.96$). Consequently, the pair of symmetric periodic orbits, Π_Γ^2 and Π_Λ^2, eventually disappears. Meanwhile, the isolated loop (the lower-most elliptical cross section in Fig.1.5.10) representing the periodic orbit Π_l^2 gets smaller and disappears at $\beta \approx 1.99$.

(2) Deformation of Homoclinic Linkage ($\beta > 2.04$): As β increases, the isolated loop representing Π_l^2 gets closer to Π_Λ^2, i.e., to the $(\Lambda_0^2, \tilde{\Lambda}_0^2)$-homoclinic linkage. At point $\tilde{S}$, Π_l^2 and Π_Λ^2 merge together and create a new but larger homoclinic linkage $(\Lambda_0^2, \tilde{\Lambda}_0^2)$ at $\beta \approx 2.0428$. Similarly, at point S ($\beta \approx 2.112$), Π_Γ^2 and Π_Λ^2 merge together, and the pair of linkages is deformed into a new pair of homoclinic linkages, $(\Gamma_0^2, \Lambda_0^2)$ and $(\tilde{\Gamma}_0^2, \tilde{\Lambda}_0^2)$ for $\beta > 2.0428$. Let the common symmetric periodic orbit associated with $(\Gamma_0^2, \Lambda_0^2)$ (respectively $(\tilde{\Gamma}_0^2, \tilde{\Lambda}_0^2)$) be denoted by Π^2 (respectively $\tilde{\Pi}^2$).

B2. Family π: The $(\Lambda_p, \tilde{\Lambda}_p)$-heteroclinic and the $(\Lambda_0^3, \tilde{\Lambda}_0^3)$-homoclinic linkages.
For a sufficiently large value of β, another periodic orbit π which is similar to Π^2 is observed. The best way to show this is to make the following change of coordinates:

$$\xi' = \langle T, x^* \rangle, \tag{1.5.14}$$

where

$$T = (0,1,0)\Phi_0$$

$$\Phi_0 = \begin{pmatrix} 1 & \frac{\gamma_0(\gamma_0-\sigma_0-P_0)}{(\sigma_0-\gamma_0)^2+1} & 1 \\ p_0 & \frac{\gamma_0[p_0+(\gamma_0-\sigma_0)]}{(\sigma_0-\gamma_0)^2+1} & \sigma_0 \\ 0 & \frac{1-\gamma_0(\gamma_0-\sigma_0-p_0)}{(\sigma_0-\gamma_0)^2+1} & 0 \end{pmatrix} \times$$

$$\begin{pmatrix} 1 & 1 & 1 \\ \frac{2(\gamma_p-\gamma_0)(\gamma_p+\gamma_0-\beta)+13(\alpha-\beta)}{-2\alpha(\gamma_p-\gamma_0)} & \frac{-33}{7} & \frac{-33}{7} \\ \frac{2(\gamma_p-\gamma_0)(\gamma_p\gamma_0-\beta)+13\gamma_0(\alpha-\beta)}{2\alpha^2(\gamma_p-\gamma_0)} & \frac{77(\gamma_p^2+\gamma_p+1)+66\alpha\gamma_p+91\gamma_p}{14\alpha} & \frac{-7(\gamma_0^2+\gamma_0+\beta)+33\alpha\gamma_0}{7\alpha} \end{pmatrix}^{-1}$$

$$p_0 = \sigma_0 + \frac{k_0}{\gamma_0}(\sigma_0^2+1) \ (k_0 : \text{scaling constant}).$$

Here, γ_p, and $\sigma_p \pm \omega_p i$ denote the real and the complex-conjugate eigenvalues of $P^\pm$, respectively. The matrix Φ_0 transforms the vector field of the linear subspace R_0 (recall Fig.1.5.1) into a normalized vector field (see Chapter 2). Fig. 1.5.16 shows the one-parameter bifurcation diagram in the (α, ξ')-plane at $\beta = 73.7$.

In this bifurcation diagram, there are two pairs of homoclinic linkages: (Λ_p, Λ_0^3) via π and $(\tilde{\Lambda}_p, \tilde{\Lambda}_0^3)$ via $\tilde{\pi}$. As π (respectively $\tilde{\pi}$) approaches the heteroclinic orbit Λ_p (respectively $\tilde{\Lambda}_p$), the number of rotations with respect to $P^\pm$ increases. On the

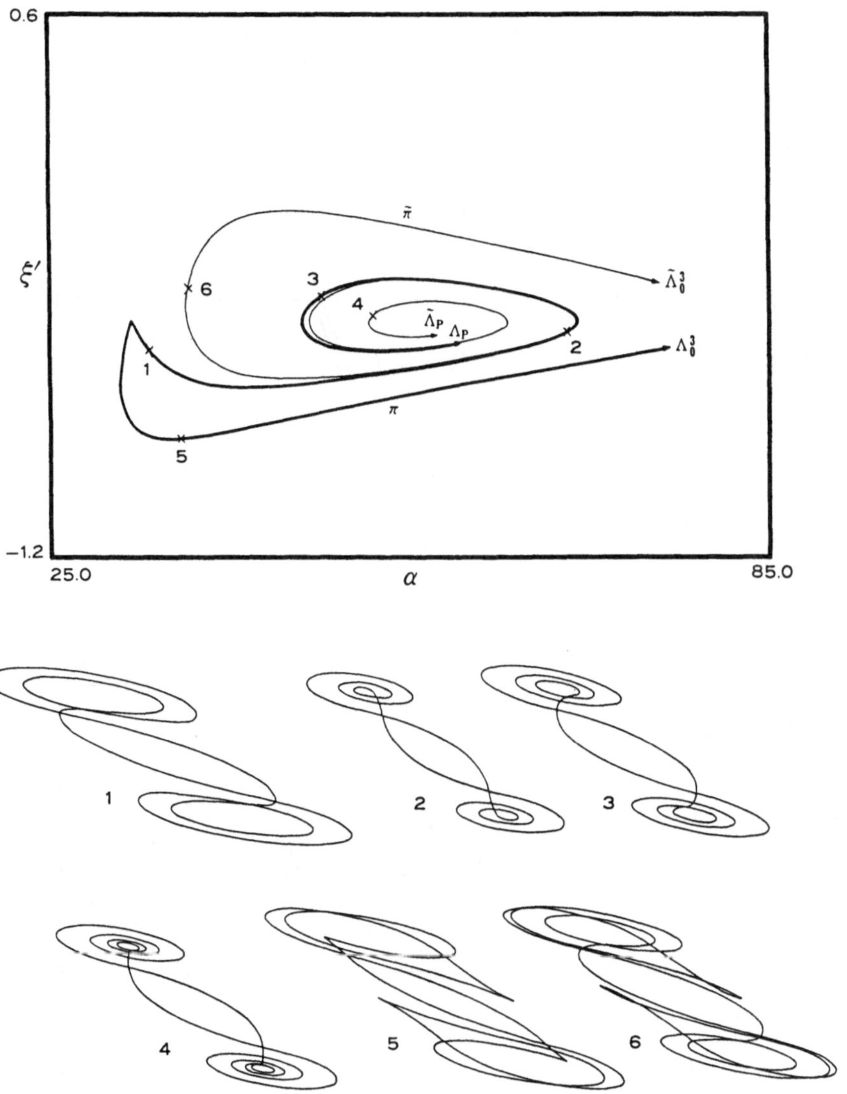

Fig. 1.5.16. (α, ξ')-bifurcation diagram for $\beta = 73.7$.

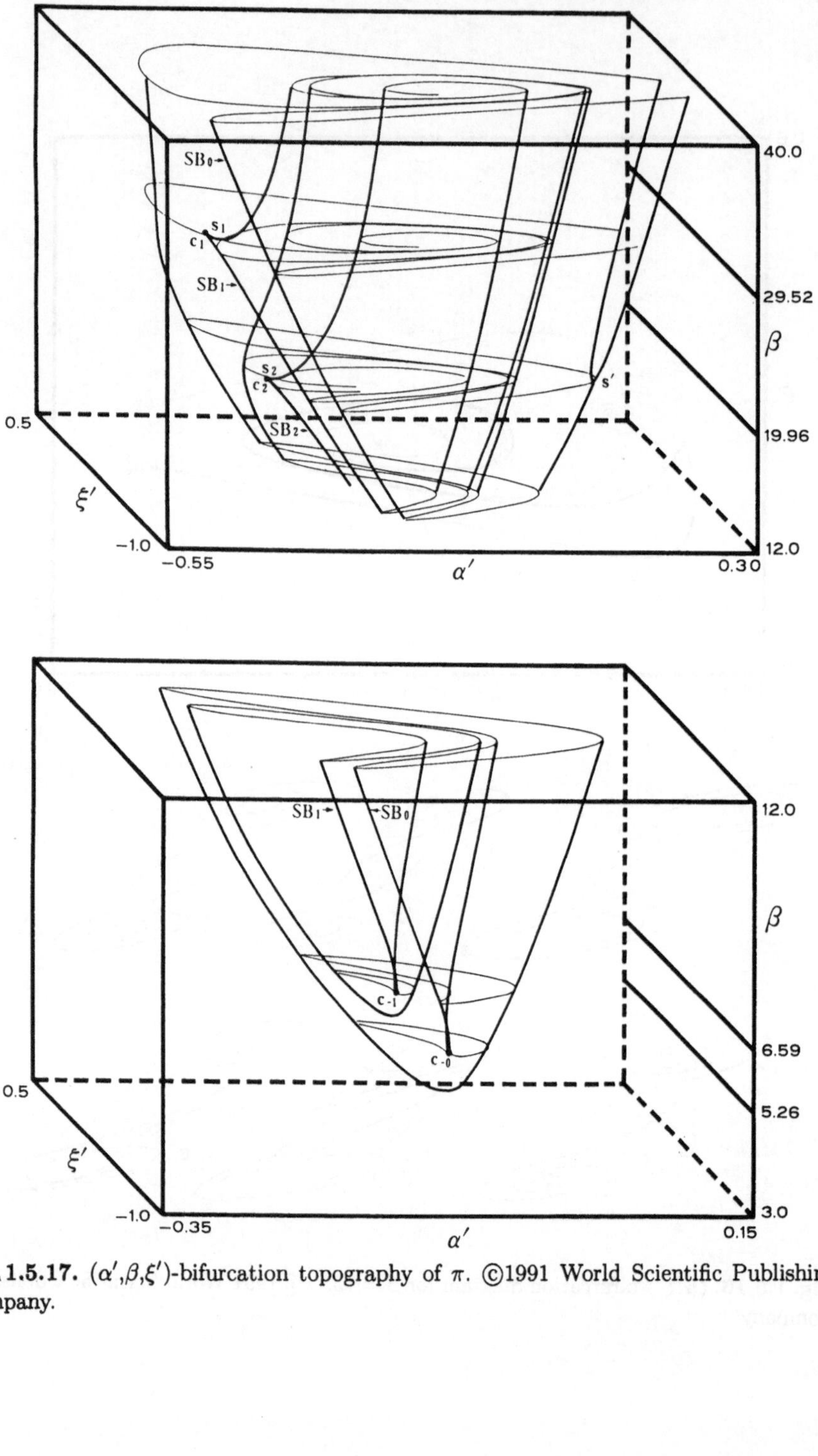

Fig. 1.5.17. (α',β,ξ')-bifurcation topography of π. ©1991 World Scientific Publishing Company.

other hand, as it approaches homoclinic orbit $\tilde{\Lambda}_0^3$ (respectively Λ_0^3), the number of rotations with respect to the unstable (one-dimensional) eigenspace of the origin O increases. Next, let us investigate how the pair of homoclinic linkages are deformed as β is varied.

Fig. 1.5.17 shows the (α', β, ξ')-diagram, while Plate 1.12 shows its schematic model (not drawn to scale). Note that the model is made from a surface whose outside is colored in red and whose inside is colored in gold. (As for the purple region, see C1 below.) One observes:

(1) The (Λ_p, Λ_0^3)- and the $(\tilde{\Lambda}_p, \tilde{\Lambda}_0^3)$-linkage ($29.52 \leq \beta \leq 73.76$): Since there is a cusp point $\tilde{C}_1$ on the surface of π, a pair of folded curves is born. As β decreases, one of these two folds approaches the internal surface of π, and touches it at the saddle point S_1 ($\beta \approx 29.52$). Beneath this level, the pair of homoclinic linkages restructures into the $(\Lambda_p, \tilde{\Lambda}_p)$- and the $(\Lambda_0^3, \tilde{\Lambda}_0^3)$-linkage, respectively. Note that the $(\Lambda, \tilde{\Lambda}_p)$-linkage constitutes a double spiral. Let π_p be the common periodic orbit of this linkage. The common periodic orbit of the $(\Lambda_0^3, \tilde{\Lambda}_0^3)$-homoclinic linkage constitutes a "boomerang"-like curve. Via the saddle point S', this homoclinic linkage is cut off from the boomerang. Since the $(\Lambda_0^3, \tilde{\Lambda}_0^3)$-linkage occupies only a relatively small region, this linkage is not shown in Fig.1.5.17 or in Plate 1.12.

(2) Breakdown of the $(\Lambda_p, \tilde{\Lambda}_p)$-linkage ($11.09 \leq \beta \leq 29.52$): As β decreases, a cusp point C_2 and saddle point S_2 emerge. The boomerang-like loop which consists of four folds is eventually cut off from the double spiral of π_p. At each saddle point S_i, the boomerang is cut off from the double spiral. For $\beta \leq 11.09$, there is no spiral structure. Therefore, this signifies the disappearance of the heteroclinic linkage. This point is denoted by τ at $\beta = 11.09$ in Fig.1.5.8.

(3) Disappearance of the boomerang ($\beta \leq 11.09$): As β decreases further, the distance between the pair of folds of each boomerang decreases. Eventually, these two folds merge and become cusp point C_{-i}. After this process, the boomerang is deformed into a simple loop which includes only a pair of folds. Then, the size of the loop decreases and finally disappears. This disappearance process is repeated from the inner surfaces to the outer surfaces. Finally, the symmetric periodic orbit itself disappears.

B3. Π^2 and π are in the same family Π^*.

We will study the relationships between the two periodic orbits Π^2 (Fig. 1.5.11(c)) and π (Fig.1.5.16). For $\beta \in [29.52, 37.76]$, we find the following four different linkages between various homoclinic orbits (denoted by Γ_0 and Λ_0) and heteroclinic (denoted by Λ_p) orbits:

(a) (Λ_p, Λ_0^3) via π (b) $(\tilde{\Lambda}_p), \tilde{\Lambda}_0^3$ via $\tilde{\pi}$
(c) $(\Gamma_0^2, \Lambda_0^2)$ via Π^2 (d) $(\tilde{\Gamma}_0^2), \tilde{\Lambda}_0^2$ via $\tilde{\Pi}^2$.

Fig. 1.5.18(a) shows the one-parameter bifurcation structure in the (α, ξ')-plane at $\beta = 73.76$, while Fig.1.5.18(b) shows a schematic model of how a pair of linkages "exchanges partners" with each other, thereby forming two new linkages: $(\Gamma_0^2, \Lambda_0^2)$ and (Λ_p, Λ_0^3). Observe that the two linkages intersect at saddle-point S_0 at $\beta = 73.76$. Then, they exchange partners and give birth to the new linkages $(\Gamma_0^2, \Lambda_0^3)$ and (Λ_0^2, Λ_p) for $\beta > 73.76$. Therefore, the two periodic orbits Π^2 and π belong to the *same two-parameter family* of periodic orbits. Let us denote this by $\Pi^* = \{\Pi^2, \pi\}$.

Fig.1.5.19 is obtained by solving bifurcation equations using our numerical methods. The model (Plate 1.13) consists of one surface whose outside is colored in blue and red while the inside is colored in gold. (As for the purple region, see C below.) We summarize our observations as:

The linkage structures of $\Lambda_p, \tilde{\Lambda}_p, \Lambda_0^3, \tilde{\Lambda}_0^3, \Gamma_0^2, \tilde{\Gamma}_0^2, \Lambda_0^2$, and $\tilde{\Lambda}_0^2$ are all determined by only one family. The local structures associated with Π^2 and π are determined by the structures of the homoclinic/heteroclinic orbits.

C Global Bifurcations of Periodic Windows We will add one more important piece of information to our bifurcation diagrams: namely the *stability* of periodic orbits. In Plates 1.12 and 1.13, the purple region corresponds to various periodic windows.[1] Utilizing this new information, more detailed structures of these periodic windows are discernible.

C1. Periodic Windows on the $(\Lambda_p, \tilde{\Lambda}_p)$- and $(\Lambda_0^3, \tilde{\Lambda}_0^3)$-Homoclinic Linkages.

Let us examine first the periodic windows on the $(\Lambda_p, \tilde{\Lambda}_0^3)$-homoclinic linkage. Fig. 1.5.19(b) shows the associated bifurcation diagram where solid curves indicate the homoclinic bifurcation sets and the saddle-node bifurcation sets, while the broken curves indicate the pitchfork bifurcation sets. The small half-toned regions near the center correspond to periodic windows. Note that each of the four band-like periodic windows start from a cusp point C_{+i}. Two follow along the heteroclinic bifurcation sets, Λ_p and $\tilde{\Lambda}_p$, while others connect C_{+i} with C_{-i} and C_{1-i}. In a neighborhood of C_{+i}, a pair of symmetric periodic attractors, Π^{i+2}, coexist. However Π^{i+2} is continuously deformed into Π^{i+3}, if one varies α and β along the saddle-node bifurcation set, SB_i, which connects C_{+i} to C_{-i}. Gaspard et al. call such periodic windows "*fishhooks*" [Gaspard, Kapral and Nicolis 1984] and observe them in Rössler's equation (see Chapter 3 also). They showed, however, only one homoclinic bifurcation set which has one "terminal" point, i.e., a point corresponding to τ in Fig.1.5.8. We conjecture that there is another homoclinic bifurcation set in a neighborhood of the one discussed in [Gaspard et al. 1984], and that the member of homoclinic bifurcation sets are connected with each other as in the present case.

It is clearly seen that the double spiral structure of the $(\Lambda_p, \tilde{\Lambda}_p)$-heteroclinic linkage is responsible for this complicated structure.

[1] In our case, there is no region for periodic repellor on the surface

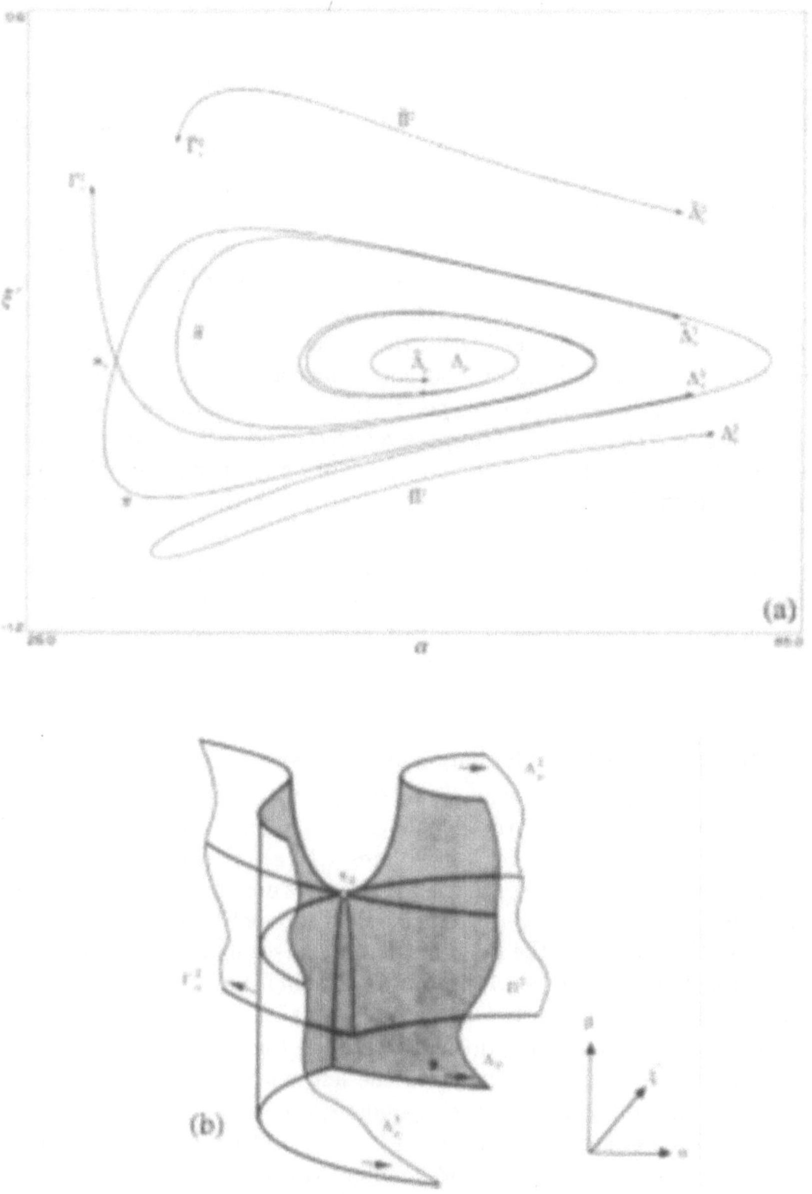

Fig. 1.5.18. Deformation of the homoclinic linkage. (a) (α, ξ')-bifurcation diagram for $\beta = 73.76$. (b) Local bifurcation topography in the (α, β, ξ')-space.

C2. Periodic windows on the $(\Gamma_0^2, \tilde{\Gamma}_0^2)$- and $(\Lambda_0^2, \tilde{\Lambda}_0^2)$-homoclinic linkages

Next, let us examine the periodic windows on the $(\Gamma_0^2, \tilde{\Gamma}_0^2)$- and $(\Lambda_0^2, \tilde{\Lambda}_0^2)$- homoclinic linkages. Fig.1.5.19(c) gives an enlarged picture of the small rectangle encircled by dotted lines near the origin in Fig.1.5.19(a). Note that at $\beta = 73.76$, a cusp point C_0 (see Fig.1.5.18 and Fig.1.5.19(a)) is observed on the $(\Gamma_0^2, \Lambda_0^2)$-homoclinic linkage, and it gives rise to the first fishhook. One of the four band-like periodic windows which start from C_0 descends to C_{-0} along SB_0, while two of these band-like periodic windows rise along Λ_p. The rest of them go down to C^* along the saddle-node bifurcation set SB_Γ. Observe that the sequence of fishhooks is terminated by the cusp point C^* near the center of the dotted rectangle.

Recall Figs. 1.5.13 and 1.5.14, and note that cusp points accumulate at the point $\Gamma_0^2 \cap R_A \cap R_B$. The cusp $\tilde{C}_\Gamma$ (respectively $\tilde{C}_\Lambda$), terminates the pair of saddle-node bifurcations SB_Γ and $\tilde{SB}_\Gamma$ (respectively SB_Λ and $\tilde{SB}_\Lambda$). These bifurcation sets are governed by the homoclinic bifurcation of Γ_0^2 (respectively Λ_0^2) (A similar accumulation point is observed at $\Lambda_0^2 \cap R_A \cap R_B$). On the other hand, cusp C^* terminates the pair of saddle-node bifurcation sets SB^* and $\tilde{SB}^*$. SB^* goes up along SB_0, while $\tilde{SB}^*$ goes up along Λ_p and $\tilde{\Lambda}_p$. It is clear that SB^* and $\tilde{SB}^*$ are governed by the heteroclinicities Λ_p and $\tilde{\Lambda}_p$. Note that the band-like periodic window along SB_Γ approaches at the first fishhook. Consequently, the fishhook periodic windows are terminated by the periodic window which includes C^*.

C3. Π^* captures global information of stability.

We will now examine how the windows of Π^* are related to the set of parameter values for which attractors of the dynamics are observed. In order to see this, let us compare Figs. 1.5.19(a)-(c) with Plates 1.14(a)-(c) which show the set of parameter values in the (α, β)-space where attractors are observed. The parameter ranges of each of Figs.1.5.19 (a)-(c) coincide with those of Plates 1.14 (a)-(c). The color encodes the periods of the periodic attractors (see figure caption). Since the algorithm used to obtain Plate 1.14 is "*carpet bombing*" in the (α, β)-space, all the periodic attractors up to a fairly long period are taken.[1]

In order to observe the relationships between Fig.1.5.19 and Plate 1.14 let us pick a blue point which is located on the left side of Plate 1.14(c). By definition (see the figure caption), it corresponds to a period-1 periodic orbit. More precisely, it will be either Π_0^1 or Π^1 given in Fig.1.5.11(a). If we move "upward" in Plate 1.14(c), we observe a period-doubling (respectively pitchfork) bifurcation and enters the purple region. Namely, Π_0^1 (respectively Π^1) bifurcates into Π_0^2 (respectively Π_1^1) given in Fig.1.5.11(b) (respectively Fig.1.5.11(d)). Similarly, period-doubling cascade and then chaotic attractors are observed. Furthermore, the well-known period-3 window (red region) and its associated period-doubling is observed. Naturally, smaller fishhooks born out of the largest fishhook within the

[1] In order to compute the two-parameter bifurcation diagram which is shown in Plate 1.13 an approximate one-dimensional map (see [Chua, Komuro and Matsumoto 1986]) is exploited.

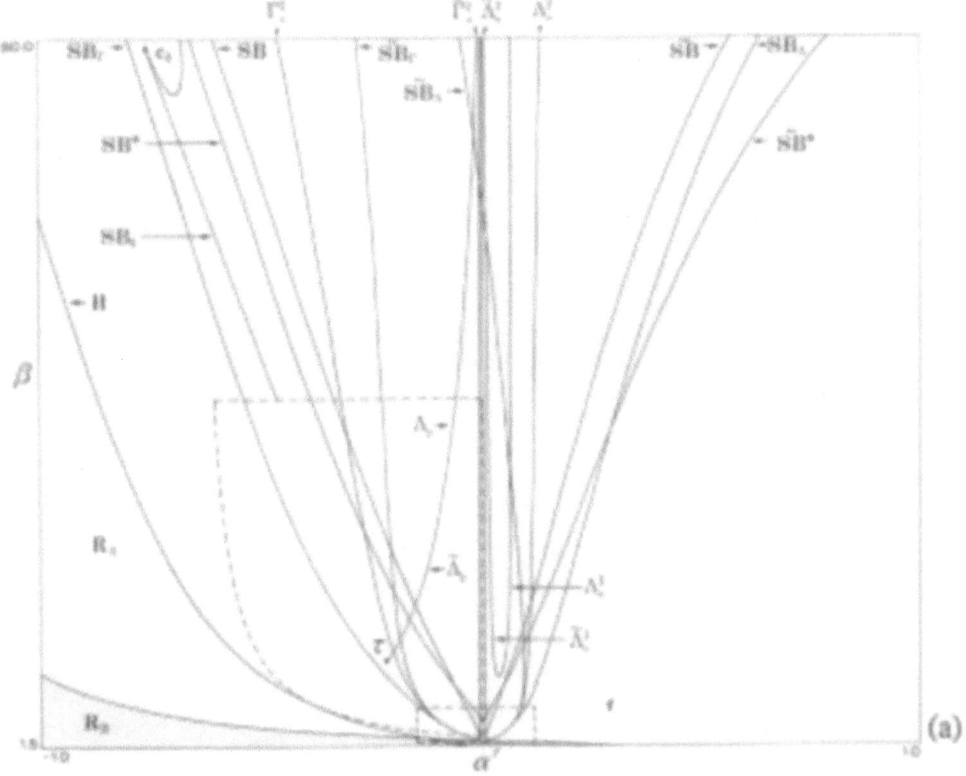

Fig. 1.5.19. (α', β)-bifurcation diagrams which are obtained by solving exact bifurcation equations. (a) Global diagram.©1991 World Scientific Publishing Company.

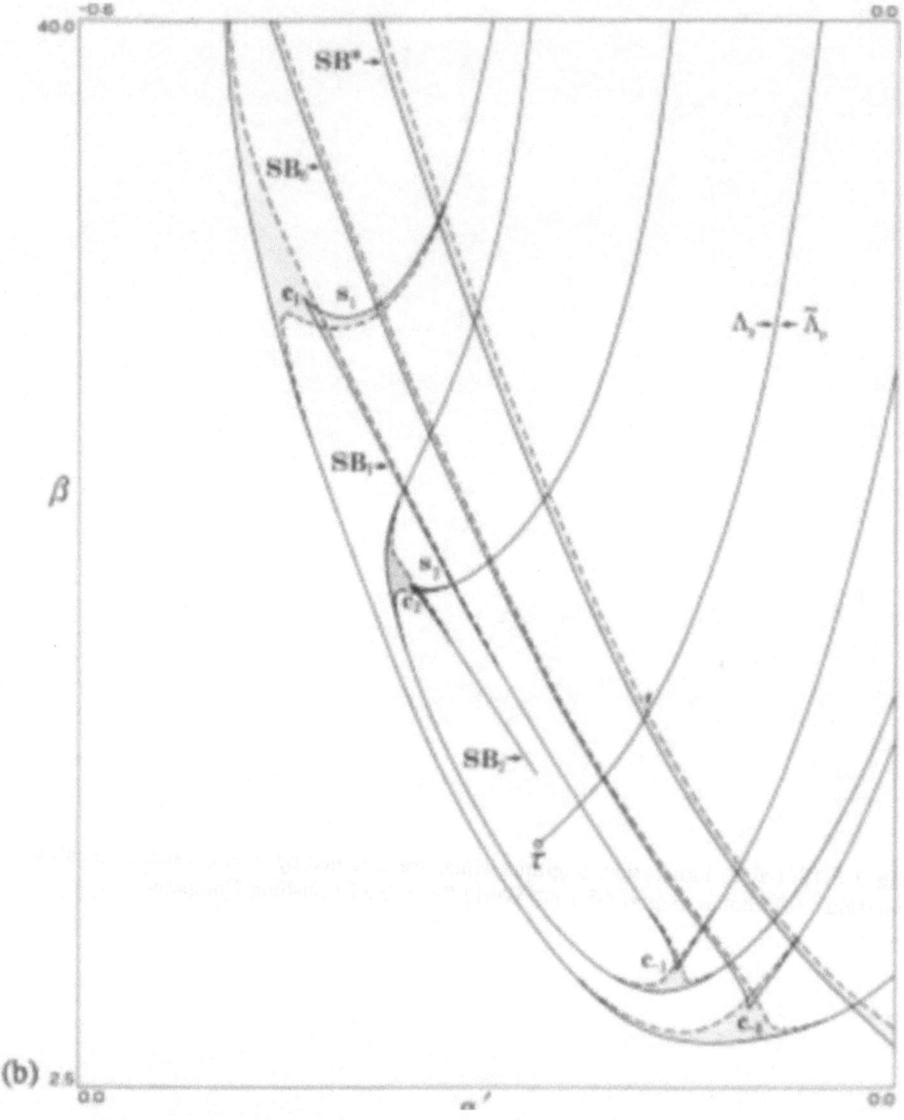

Fig. 1.5.19. continued (b) Enlargement in a neighborhood of τ. ©1991 World Scientific Publishing Company.

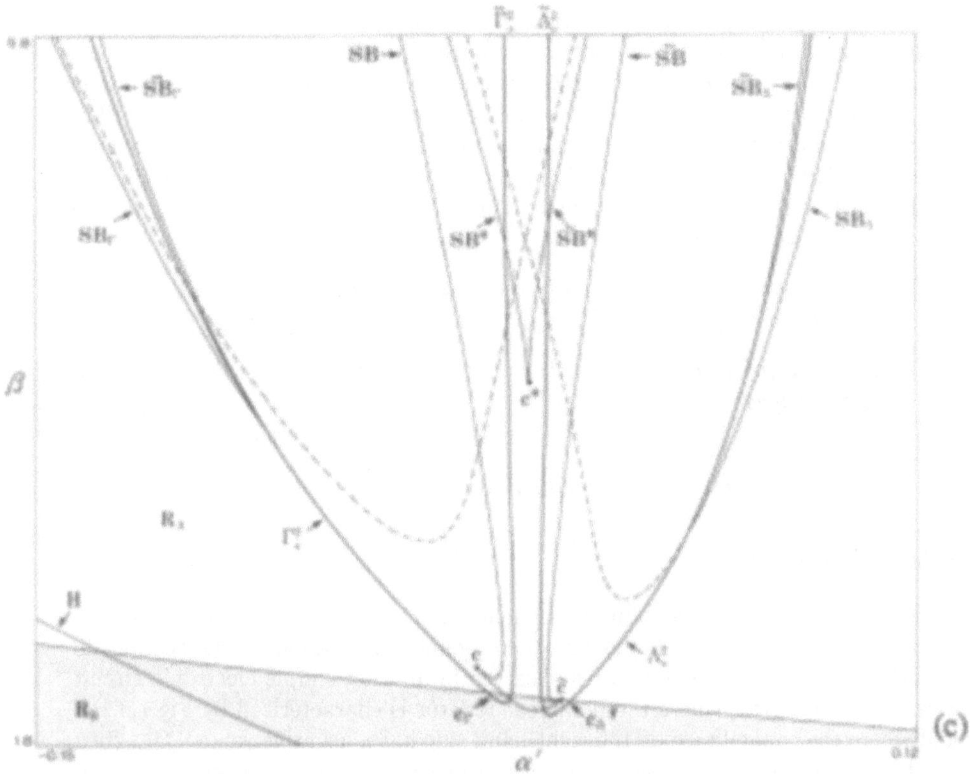

Fig. 1.5.19. continued (c) Enlargement in a neighborhood of C^*. ©1991 World Scientific Publishing Company.

purple region are also discernible. This amounts to the fact that the blue and the purple region play the role of the "trunk" of a tree and essentially determine the shapes of smaller branches. The correspondence of the blue and the purple regions with Fig.1.5.19(c) is clear. Thus, if we were to solve all the bifurcation equations and plot them on Fig.1.5.19(c), we would see the same details as those in Plate 1.14(c). We did not do it simply because it would have consumed unfeasibly large amount of CPU time with any existing computer to date. Similar remarks apply to Fig.1.5.19(a) and (b), and Plate 1.14 (a) and (b). We close this section by summarizing two main facts.

(1) *The homoclinic/heteroclinic bifurcation sets studied here are all linked together via a single family* Π^*;

(2) *The global information of stability is essentially captured by the structure of these linkages.*

1.6 The Torus Breakdown Circuit

1.6.1 Introduction

In Section 1.5, we saw that bifurcation structures of the Double Scroll circuit are essentially determined by homoclinic/heteroclinic bifurcations. This section describes other bifurcation structures observed in another simple electronic circuit, given in Fig.1.6.1(a), where the nonlinear resistor is characterized by Fig.1.6.1(b) and where the capacitance on the right-hand side has a negative value $-C_1$. *Torus breakdown* will be a major phenomenon in this section. There are two major scenarios:

(1) A periodic state and *torus* (quasi-periodic state) appear and disappear alternately many times. Suddenly, however, a periodic state bifurcates into a chaotic attractor, rather than the torus;

(2) After a repeated bifurcation between a periodic and a quasi-periodic state, a periodic state bifurcates into a period-doubling cascade and culminates in a chaotic attractor.

1.6.2 Observations of Torus Breakdown

A The Circuit and its Dynamics This circuit is easily realized by the circuit of Fig.1.6.2. The dynamics of this circuit is governed by the state equation

$$\begin{aligned} C_1 \frac{dv_{C_1}}{dt} &= -g(v_{C_2} - v_{C_1}) \\ C_2 \frac{dv_{C_2}}{dt} &= -g(v_{C_2} - v_{C_1}) - i_L \end{aligned} \qquad (1.6.1)$$

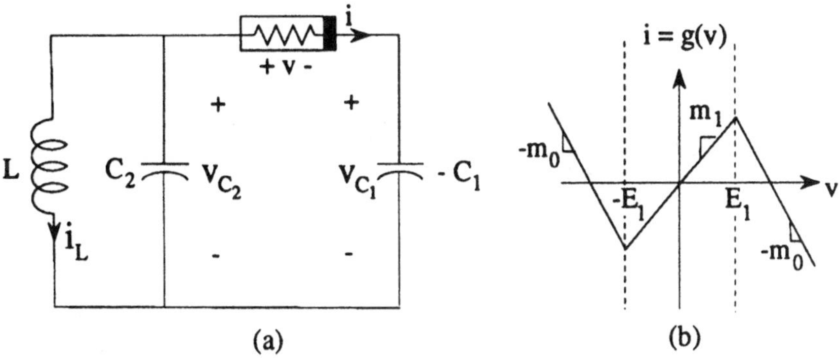

Fig. 1.6.1. A simple third-order autonomous circuit which exhibits a torus breakdown phenomenon. (a) Circuit. (b) Nonlinear resistor *v*-*i* characteristic.

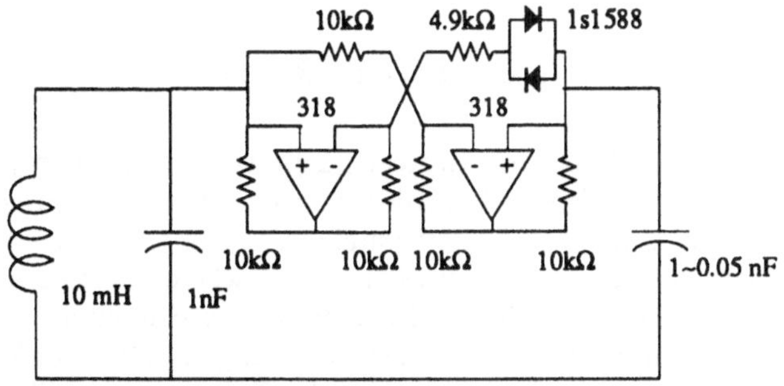

Fig. 1.6.2. Realization of the circuit given in Fig.1.6.1.

$$L\frac{di_L}{dt} = v_{C_2},$$

where v_{C_1},v_{C_2}, and i_L denote, respectively, the voltage across C_1, the voltage across C_2, and the current through L. The function $g(\cdot)$ denotes the v-i characteristic of the nonlinear resistor and is described by

$$g(v) = -m_0 v + 0.5(m_0 + m_1)[\mid v + E_1 \mid - \mid v - E_1 \mid]. \tag{1.6.2}$$

To simplify our analysis, let us transform (1.6.1) into the following dimensionless form:

$$\begin{aligned}\frac{dx^1}{dt} &= -\alpha f(x^2 - x^1)\\ \frac{dx^2}{dt} &= -f(x^2 - x^1) - x^3\\ \frac{dx^3}{dt} &= \beta x^2,\end{aligned} \tag{1.6.3}$$

where

$$\begin{aligned} x^1 &= \frac{v_{C_1}}{E_1}, \quad x^2 = \frac{v_{C_2}}{E_1}, \quad x^3 = \frac{i_L}{C_2 E_1}, \\ \alpha &= \frac{C_2}{C_1}, \quad \beta = \frac{1}{LC_2}, \\ a &= \frac{m_0}{C_2}, \quad b = \frac{m_1}{C_2}, \end{aligned} \tag{1.6.4}$$

$$f(x) = -ax + 0.5(a+b)(|\, x+1 \,| - |\, x-1 \,|). \tag{1.6.5}$$

The dynamics associated with (1.6.3) depends on four parameters: a, b, α, and β. In this subsection, we will choose α as our bifurcation parameter by fixing the other parameters as:

$$a = 0.07, \quad b = 0.1, \quad \beta = 1.0. \tag{1.6.6}$$

B Experiments Plate 1.15 shows thirteen qualitatively distinct trajectories projected onto the (v_{C_2}, v_{C_1})-plane in the order of decreasing C_1 or increasing α. Here, v_{C_1} is the vertical axis and v_{C_2} is the horizontal axis. Plate 1.16 shows thirteen cross sections of the corresponding trajectories at $i_L = 0$, where the cross section in the region $v_{C_2} < 0$ is shown. Figs.1.6.3 and 1.6.4 give digital computer confirmation of Plates 1.15 and 1.16, respectively, obtained by solving (1.6.3). In order to explain these bifurcation phenomena, first note that (1.6.3) has three equilibrium points all located (symmetrically) on the v_{C_1}-axis, namely,

$$O = (0,0,0), P^+ = (1 + \frac{b}{a}, 0, 0), P^- = (-1 - \frac{b}{a}, 0, 0). \tag{1.6.7}$$

Within the range of the parameter values that we are looking at, the eigenvalues at O (respectively $P^\pm$) consist of one real γ_0 (respectively γ_1) and one complex-conjugate pair $\sigma_0 \pm \omega_0 i$ (respectively $\sigma_1 \pm \omega_1 i$). The following observations provide a qualitative explanation of the various bifurcation phenomena shown in Plates 1.15 and 1.16, and in Figs.1.6.3 and 1.6.4.

(1) It can be shown (see Subsection 1.6.3) that $P^\pm$ are always unstable. The equilibrium point O is stable for $\alpha < 0$, but loses its stability at $\alpha = 0$, whereupon a periodic attractor is born (see Plates 1.15 and 1.16, Figs.1.6.3 and 1.6.4(a)). This, however, is *not* a Hopf bifurcation. This bifurcation is more degenerate than Hopf and will not be pursued here since it is outside of the scope of this chapter. The real eigenvalue γ_o becomes positive for $\alpha > 0$, while σ_0 remains negative.

(2) A further increase in the value of α eventually gives rise to a two-torus (see Plates 1.15 and 1.16, Figs.1.6.3 and 1.6.4(b)). To further confirm that this

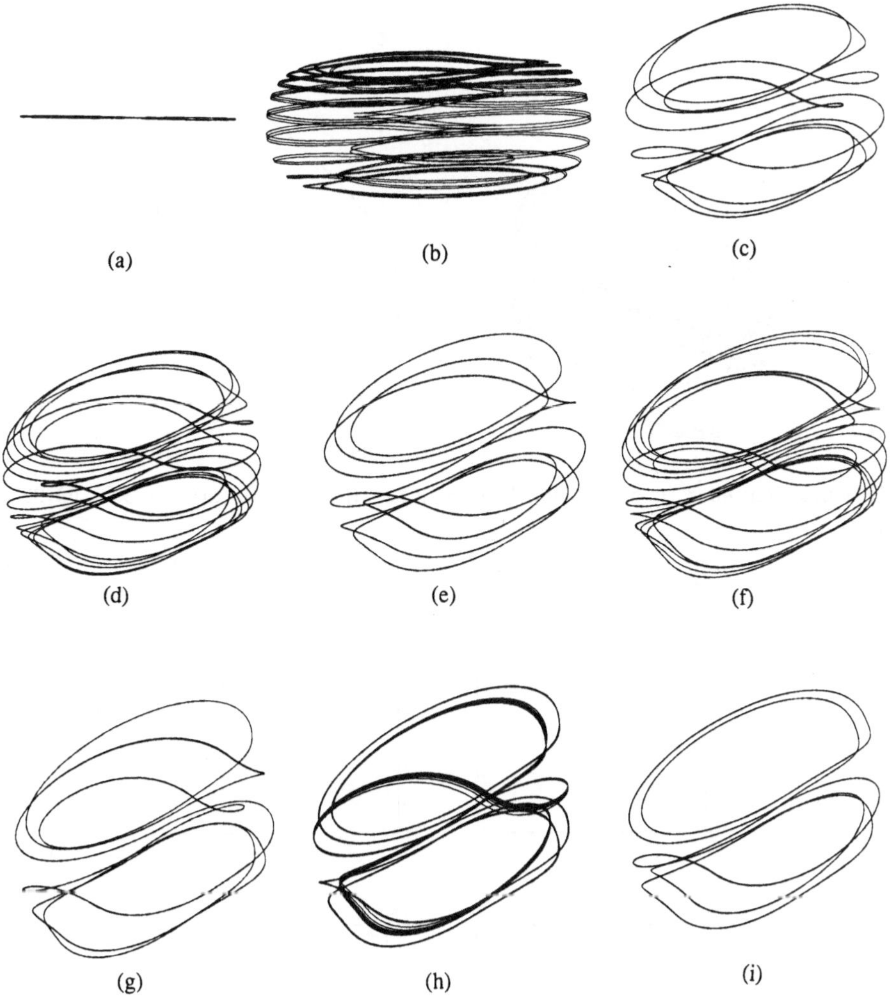

Fig. 1.6.3. Digital computer confirmation of Plate 1.15. (a) $\alpha = 0.5$. (b) $\alpha = 2.0$. (c) $\alpha = 8.0$. (d) $\alpha = 8.8$. (e) $\alpha = 9.6$. (f) $\alpha = 10.8$. (g) $\alpha = 13.0$. (h) $\alpha = 13.4$. (i) $\alpha = 13.4$. ©1987 IEEE.

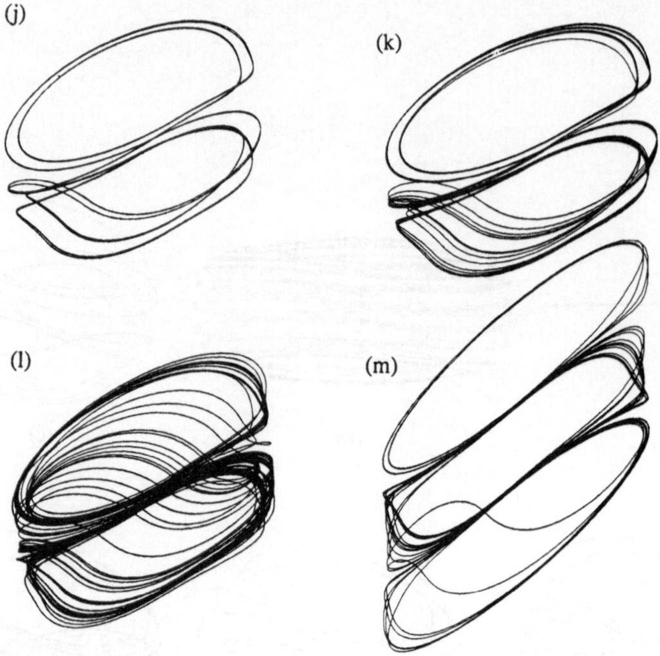

Fig. 1.6.3. continued (j) $\alpha = 13.45$. (k) $\alpha = 13.52$. (l) $\alpha = 15.0$. (m) $\alpha = 33.0$. ©1987 IEEE.

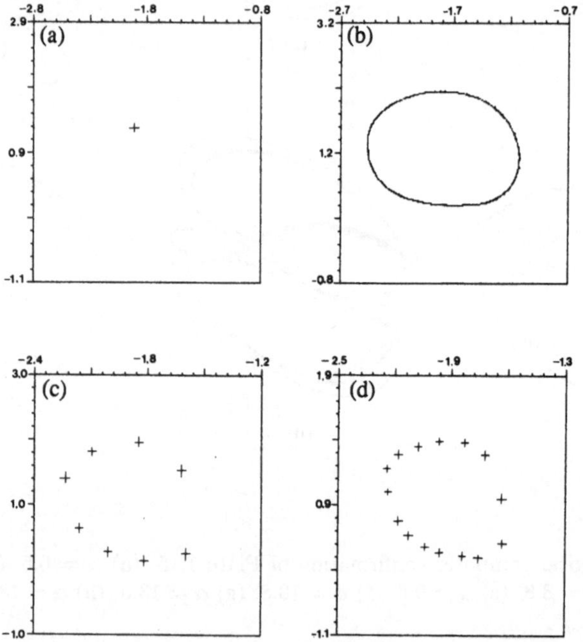

Fig. 1.6.4. Digital computer confirmation of Plate 1.16.©1987 IEEE.

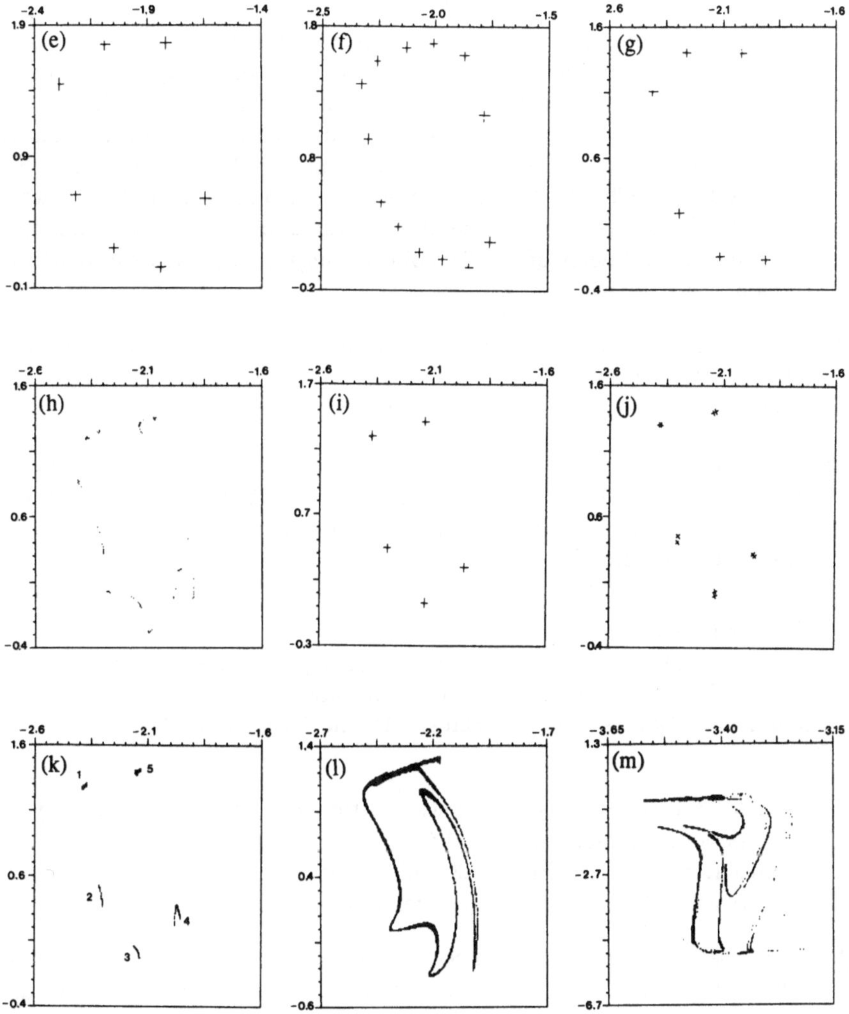

Fig. 1.6.4. continued ©1987 IEEE.

trajectory indeed is a torus, three Lyapunov exponents associated with this trajectory are computed:

$$(\mu_1, \mu_2, \mu_3) \approx (0, 0, -0.00675). \tag{1.6.8}$$

Since no Lyapunov exponent in (1.6.8) is positive, the system is not chaotic. However, only one Lyapunov exponent is negative, and the solution is not a periodic attractor either. The presence of two zero Lyapunov exponents therefore provides a further confirmation that the trajectory in Plate 1.15(b) is indeed the two-torus, namely, quasi-periodic, solution. It appears that the Poincaré map of the periodic attractor born at $\alpha = 0$, has undergone a Hopf bifurcation (see Chapter 2 and Chapter 3), as depicted in Fig.1.6.5, thereby giving birth to a two-torus. The above torus is observed in the half space $v_{C_1} > 0$. The symmetry of (1.6.3), however, suggests the existence of another torus in the half space $v_{C_1} < 0$.

(3) As α is increased further, two-torus and periodic attractor are observed alternatively many times. Plates 1.15 and 1.16, Figs.1.6.3 and 1.6.4 (c)-(f) give several of the periodic attractors in this bifurcation sequence.

(4) For the parameter range $11.0 < \alpha < 13.4$, a period-6 attractor is observed (Plates 1.15 and 1.16, Figs.1.6.3 and 1.6.4(g)). After this, the two-torus does not reappear. After the disappearance (or breakdown) of the torus, two different scenarios are observed .

The First Scenario
At $\alpha \approx 13.4$, *the period-6 attractor suddenly bifurcates into a chaotic attractor.* This is a typical situation where *intermittency* chaos [Pomeau and Manneville 1980] is present. (Plates 1.15 and 1.16, Fig.1.6.3(h))

(5) At $\alpha \approx 13.4$, the situation becomes a little more complicated; namely, another periodic attractor of period-5 is born out of a saddle-node bifurcation (Plates 1.15 and 1.16, Figs.1.6.3 and 1.6.4(i)), and this period-5 attractor *coexists* with the chaotic attractor described above.

(6) Since a saddle-node bifurcation occurs at $\alpha \approx 13.4$, another period-5 periodic orbit of the saddle-type is born simultaneously. Since this orbit is unstable, it is not observable experimentally.

(7) At $\alpha \approx 13.5$, the saddle-type periodic orbit collides with the chaotic attractor and the chaotic attractor disappears. Clearly, this is a boundary crisis [Grebogi, Ott and Yorke 1983].

(8) The period-5 attractor survives after the collision (i.e., for $\alpha > 13.4$), and the two-torus is no longer observed. The following observations provide the

second scenario following the torus breakdown.

The Second Scenario
At $\alpha \approx 13.44$, a period-doubling cascade is initiated (Plates 1.15 and 1.16, Figs.1.6.3 and 1.6.4(j)) and the solution eventually bifurcates into a chaotic attractor consisting of five "islets" (Plates 1.15 and 1.16, Figs.1.6.3 and 1.6.4(k)).

(9) As we increase α beyond $\alpha \approx 13.44$,the islets merge (interior crisis. [Grebogi, Ott and Yorke 1983]) into one chaotic attractor. (Plates 1.15 and 1.16, Figs.1.6.3 and 1.6.4(l)), which appears to be a "folded torus". Its Lyapunov exponents are

$$(\mu_1, \mu_2, \mu_3) \approx (0.027, 0, -0.1134). \tag{1.6.9}$$

Note that the first exponent is positive, as expected. Note also that the symmetry of (1.6.3) implies the existence of another folded torus in the half space $v_{C_1} < 0$.

(10) When α is increased further, two folded tori merge and give rise to an attractor similar to the Double Scroll attractor (Plates 1.15 and 1.16, Figs.1.6.3 and 1.6.4(m)). Recall in Section 1.3 the Double Scroll is born out of a pair of Rössler's screw-type attractors.

(11) When the value of α is increased even further, the Double Scroll-type attractor suddenly disappears. This sudden death signifies a boundary crisis; namely, there is yet another saddle-type periodic orbit which collides with the Double Scroll-type attractor. Beyond this value of α, no further attractor is observed.

C Period-Adding Sequence The periodic state shown in Plates 1.15 and 1.16, and in Figs.1.6.3 and 1.6.4 are related to each other in accordance with a definite law which can be derived (empirically) from the one-dimensional bifurcation diagram given in Fig.1.6.6. This diagram is obtained by numerically solving (1.6.3). Here, the horizontal axis is α, and the vertical axis represents v_{C_1} with $i_L = 0$ and $v_{C_2} < 0$. Fig.1.6.6 shows the following properties.

(1) There is a sequence of periodic states whose period decreases by exactly one: 11, 10, ···, 6.

(2) In between period n and period $n-1$, there is a phase-locked state of period $2n-1$.

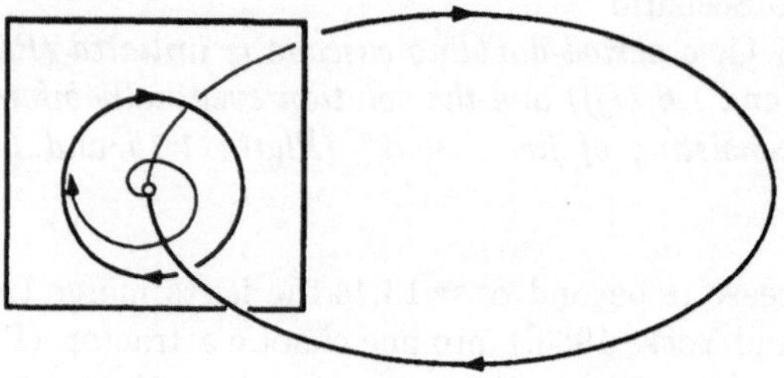

Fig. 1.6.5. Hopf bifurcation for the Poincaré map. ©1987 IEEE.

D Sounds This circuit also behaves within audible frequencies. Since (1.6.1) is again third order, one can view two of the three variables v_{C_1}, v_{C_2} and i_L on an oscilloscope while one can listen to the bifurcations by connecting the remaining variable to a speaker. The sounds of the torus as well as its breakdown are interesting.

1.6.3 Analysis

In this subsection, we will examine some of the phenomena discussed in the previous subsection from a more analytical perspective. Let us partition the state space into three parallel regions R_1, R_0 and R_{-1} separated by boundaries B_1 and B_{-1}, respectively, where

$$\begin{aligned} R_1 &= \{(x^1,x^2,x^3) | x^2 - x^1 < -1\} \\ R_0 &= \{(x^1,x^2,x^3) | \mid x^2 - x^1 \mid < 1\} \\ R_{-1} &= \{(x^1,x^2,x^3) | x^2 - x^1 > 1\} \\ B_1 &= \{(x^1,x^2,x^3) | x^2 - x^1 = -1\} \\ B_{-1} &= \{(x^1,x^2,x^3) | x^2 - x^1 = 1\}. \end{aligned} \tag{1.6.10}$$

The Jacobian matrix associated with (1.6.3) in R_0 (respectively R_1, R_{-1}) is given by

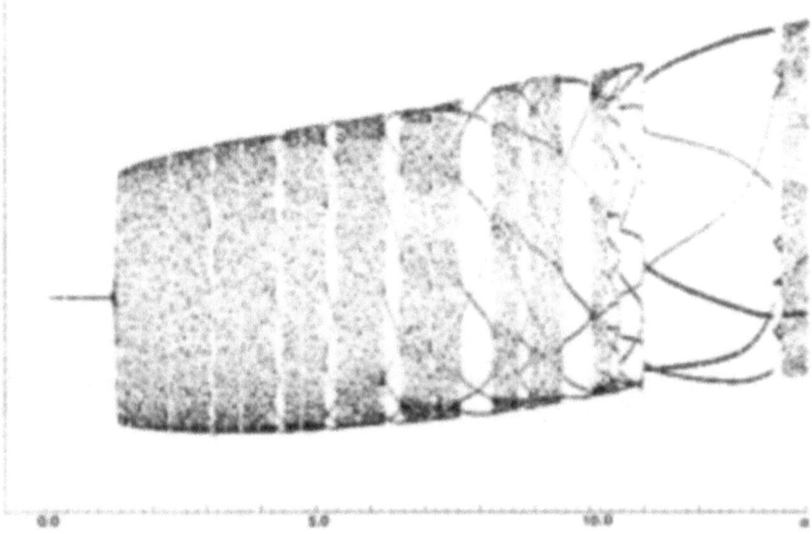

Fig. 1.6.6. Bifurcation diagram where the horizontal axis is α, and the vertical axis is v_{C_1} on the cross section at $i_L = 0$, $v_{C_2} < 0$. ©1987 IEEE.

$$A_0 = \begin{pmatrix} \alpha b & -\alpha b & 0 \\ b & -b & -1 \\ 0 & -\beta & 0 \end{pmatrix},$$

$$\text{respectively } A_1 = \begin{pmatrix} \alpha a & -\alpha a & 0 \\ -a & a & -1 \\ 0 & \beta & 0 \end{pmatrix}. \tag{1.6.11}$$

One can easily show that O (respectively $P^{\pm}$) is stable if and only if

$$\alpha > 0, b(1-\alpha) > 0, b\alpha\beta < 0, (1-\alpha)\beta < 0$$

$$(\text{respectively } \alpha > 0, a(1-\alpha) > 0, a\alpha\beta < 0, (1-\alpha)\beta < 0). \tag{1.6.12}$$

A Divergence Zero Boundary It follows from (1.6.12) that equilibrium point O loses its stability at $\alpha = 0$. This is not a Hopf bifurcation, however, as mentioned earlier, namely, the real eigenvalue γ_0 becomes positive, while the real part σ_0 of the complex-conjugate pair remains negative. There is, however, more to this. Consider the divergence div$f$ of the vector field f associated with (1.6.3),

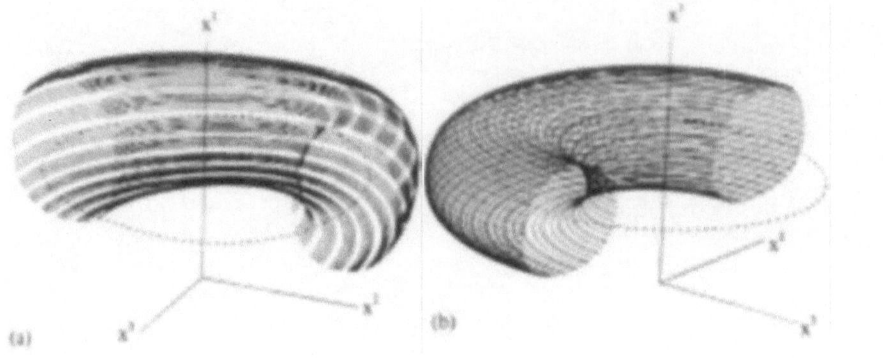

Fig. 1.6.7. (a) Repelling torus in the half space $x^3 < 0$ and period-1 attractor at $\alpha = 0.5$. (b) Attracting torus in the half space $x^3 < 0$ and period-1 repellor at $\alpha = 2.0$. ©1987 IEEE.

$$\mathrm{div} f = \begin{cases} \mathrm{div}_0 f = -b(1-\alpha) & (x_1, x_2, x_3) \in R_0 \\ \mathrm{div}_1 f = a(1-\alpha) & (x_1, x_2, x_3) \in R_\pm . \end{cases} \tag{1.6.13}$$

Note that the $\mathrm{div}_0 f$ and $\mathrm{div}_1 f$ exchange their signs at $\alpha = 1$, namely,

$$\begin{aligned} &for \quad \alpha < 1, \quad \mathrm{div}_0 f < 0, \quad \mathrm{div}_1 f > 0, \\ &for \quad \alpha = 1, \quad \mathrm{div}_0 f = 0, \quad \mathrm{div}_1 f = 0, \\ &for \quad \alpha > 1, \quad \mathrm{div}_0 f > 0, \quad \mathrm{div}_1 f < 0. \end{aligned} \tag{1.6.14}$$

It is observed (by simulation) that, for $\alpha < 1$, a periodic attractor coexists with a *repelling torus* (Fig.1.6.7(a)). At $\alpha = 1$, neither attractor nor repellor is observed as expected since the circuit in this case has no dissipation mechanism. If α is increased, at $\alpha = 1$ an attracting periodic orbit bifurcates into an attracting torus. On the other hand, if α is decreased, a repelling periodic orbit bifurcates into a repelling torus.

B Trajectories on the Torus Recall that the eigenvalues at O (respectively $P^\pm$) consist of one real γ_0 (respectively γ_1) and one complex-conjugate pair $\sigma_0 \pm \omega_0 i$ (respectively $\sigma_1 \pm \omega_1 i$). In particular, at $(\alpha, \beta) = (2, 1)$,

$$\begin{aligned} (\gamma_0, \sigma_0, \omega_0) &= (0.14786, -0.048886, 1.0060), \\ (\gamma_1, \sigma_1, \omega_1) &= (-0.10425, 0.034426, 1.0030). \end{aligned} \tag{1.6.15}$$

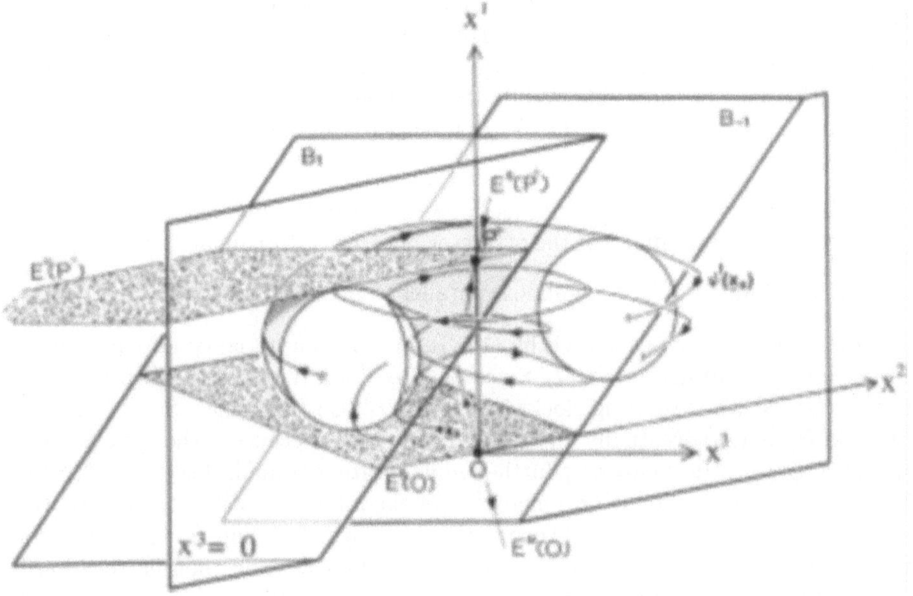

Fig. 1.6.8. The relative positions of the eigenspaces and the two-torus.©1987 IEEE.

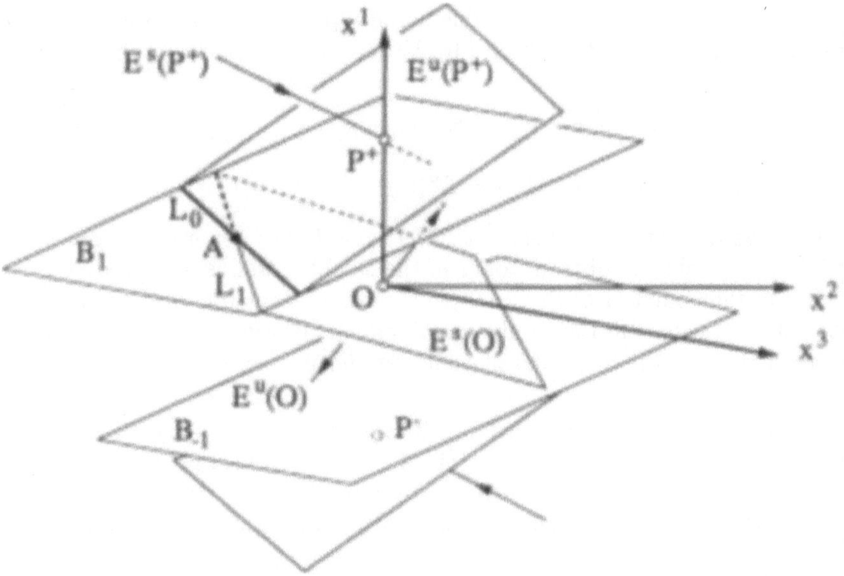

Fig. 1.6.9. The relative positions of the eigenspaces.©1987 IEEE.

Let $E^s(O)$ (respectively $E^u(O)$) denote the eigenspace corresponding to γ_0, (respectively $\sigma_0 \pm \omega_0 i$). Similarly, let $E^u(P^\pm)$ (respectively $E^s(P^\pm)$) denote the eigenspace corresponding to $\sigma_1 \pm \omega_1 i$, (respectively γ_1). Fig.1.6.8 describes the relative positions of these sets. While the relative positions of the eigenvalue in (1.6.15) are identical to those of the Double Scroll circuit, there are two subtle differences:

(1) The magnitude of $|\gamma_1|$ is not as large as in the Double Scroll circuit , and hence "the flattening" of the attractor onto $E^u(P^\pm)$ is relatively weak;

(2) $E^s(P^\pm)$ and $E^u(P^\pm)$ are almost parallel to each other.

Let φ^t be the flow generated by (1.6.3) and pick an initial condition x_0 near O above $E^s(O)$ but not on $E^u(O)$. Since $\gamma_0 > 0$, φ^t starts moving up (with respect to the x^1-axis) while rotating clockwise around $E^u(O)$ (Fig.1.6.8). Since (1.6.3) is linear in R_0, $\varphi^t(x_0)$ eventually hits B_+ and enters R_+. Because of the relative position of $E^s(P^\pm)$, $\varphi^t(x_0)$ moves up further while rotating around, this time, $E^s(P^\pm)$. Since $\sigma_1 > 0$, the solution $\varphi^t(x_0)$ increases its magnitude of oscillation and eventually enters R_0. Then, because of the relative positions of R_0 and R_+, $\varphi^t(x_0)$ starts moving downward (with rotation). It eventually hits B_- and then flattens itself against $E^s(O)$ while rotating around $E^u(O)$. Since $\sigma_0 < 0$, the solution decreases its magnitude of oscillation and moves into the original neighborhood of O. This process then repeats itself *ad infinitum* but never returns to the original point and, hence, the associated loci eventually cover the surface of a two-torus. Next, let us explain the situation where the Double Scroll-type attractor is observed. Typical parameter values are $(\alpha, \beta) = (30.0, 1.0)$. Various sets are shown in Fig.1.6.9 where

$$
\begin{aligned}
&L_{+0} = E^u(P^+) \cap B_+, L_{+1} = E^u(O) \cap B_+, A_+ = L_{+0} \cap L_{+1},\\
&L_{-0} = E^u(P^-) \cap B_-, L_{-1} = E^u(O) \cap B_-, A_- = L_{-0} \cap L_{-1}.
\end{aligned} \tag{1.6.16}
$$

Observe that the angle between $E^s(O)$ and $E^u(P^\pm)$ is much larger than in Fig.1.6.8. The behavior of trajectories in the Double Scroll is described in Section 1.3. Recall that point A of Fig.1.6.9 plays an important role in determining the "fate" of a trajectory in the Double Scroll circuit. In Fig.1.6.8, however, point A is located far from the torus and plays no significant role.

C The Folded Torus and the Double Scroll It should be noted that for the present parameter values, the two points A_+ and A_-, which played an important role in Section 1.3, are located far beyond the region where the torus is observed. However, these points keep getting closer to the attractors as one increases α so that eventually A_+ and A_- touch the folded torus attractors. When this situation occurs, the two folded tori cannot stay away from each other and must therefore merge into one attractor, namely, the Double Scroll-type attractor. This is because if a trajectory (in the attractor) hits A_+ or a point to the left of A_+ as in Fig.1.6.9, then it has to descend and eventually hits B_-.

1.7 The Hyperchaotic Circuit

1.7.1 Introduction

A system is called *hyperchaotic* if there are more than one positive Lyapunov exponents [Rössler 1979a] so that the dynamics expands not only certain line segments but also certain area elements. The circuits discussed so far can have at most one positive Lyapunov exponent because the circuits are third order, autonomous and dissipative. The circuit described below [Matsumoto, Chua and Kobayashi 1986] is probably the first real physical system where hyperchaos is observed. Another experimental observation of hyperchaos is with a p-germanium [Peinke, Muhlbach, Rohricht, Wessely, Mannhart, Parisi and Huebener 1986].

1.7.2 Experiment

A Observation Consider the circuit of Fig. 1.7.1(a), where the nonlinear resistor is characterized by Fig. 1.7.1(b), i.e., a three-segment piecewise-linear v-i characteristic. All other elements are linear and passive except $-R$, which represents a negative resistance. The dynamics is described by

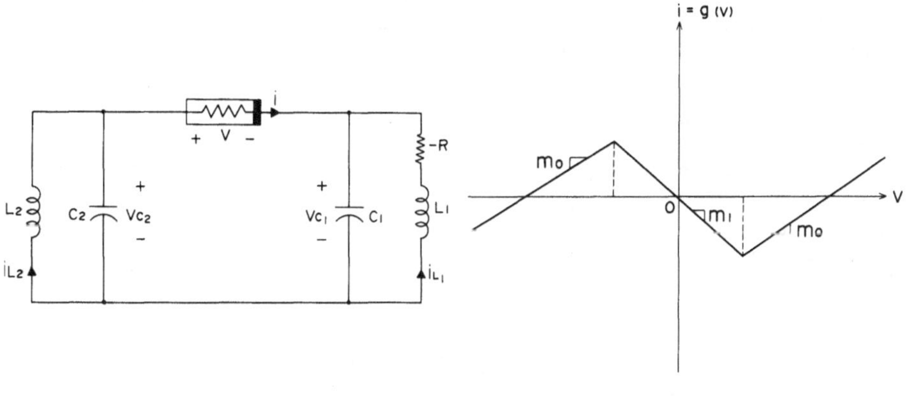

(a) (b)

Fig. 1.7.1. A fourth-order circuit which exhibits hyperchaos. (a) Circuitry. (b) v-i characteristic of N_1, with v and i denoting, respectively, the voltage across the nonlinear resistor and the current through the nonlinear resistor. ©1986 IEEE.

$$\begin{aligned}
C_1\frac{dv_{C_1}}{dt} &= g(v_{C_2}-v_{C_1})-i_{L_1}\\
C_2\frac{dv_{C_2}}{dt} &= -g(v_{C_2}-v_{C_1})-i_{L_2}\\
L_1\frac{di_{L_1}}{dt} &= v_{C_1}+Ri_{L_1}\\
L_2\frac{di_{L_2}}{dt} &= v_{C_2},
\end{aligned} \tag{1.7.1}$$

where $v_{C_1}, v_{C_2}, i_{L_1}$, and i_{L_2} denote, respectively, the voltage across C_1, the voltage across C_2, the current through L_1, and the current through L_2, and $g(\cdot)$ is the piecewise-linear function of Fig. 1.7.1(b), given by

$$g(v)=\begin{cases} m_0 v - m, & v \geq 1\\ m_1, & |v| \leq 1\\ m_0 v + m_1, & v \leq -1\,. \end{cases} \tag{1.7.2}$$

This can easily be realized by the circuit of Fig. 1.7.2(a), where the subcircuit N_1 realizes the nonlinear resistor and N_2 realizes the negative resistance. Fig. 1.7.2(b) shows the measured v-i characteristic of N_1, which is an approximation of Fig. 1.7.1(b) with $m_0 = 3$ and $m_1 = -0.2$.

Plate 1.17 shows the observed attractor from the circuit of Fig. 1.7.2. Plate 1.17(a) is the projection onto the (v_{C_1}, v_{C_2})-plane, while Plate 1.17(b) is the projection onto the (v_{C_1}, i_{L_1})-plane. Plate 1.18 gives the power spectrum of the time waveform $v_{C_1}(t)$. To analyze the attractor further, we measured its cross section on the 3-dimensional hyperplane

$$\Sigma = \{(v_{C_1}, v_{C_2}, i_{L_1}, i_{L_2})|\ v_{C_1} - v_{C_2} = 0\}\,. \tag{1.7.3}$$

Plate 1.13 gives the projection onto the (v_{C_1}, i_{L_1})-plane of the cross section of the attractor on Σ, where $(d/dt)(v_{C_1} - v_{C_2}) < 0$. The picture shows that the cross section still looks fairly "thick" on the three-dimensional hyperplane. This appears to indicate that the attractor exhibits hyperchaos. If there were only one expanding direction, the cross section would look much thinner.

B Sounds The circuit given in Fig. 1.7.2(a) requires no special parts or special parameter values. One can listen to the hyperchaos in addition to being able to see it on an oscilloscope.

1.7.3 Confirmation

In order to confirm our prediction that the above attractor exhibits hyperchaos, we computed four Lyapunov exponents for (1.7.1). The parameter values for (1.7.1)

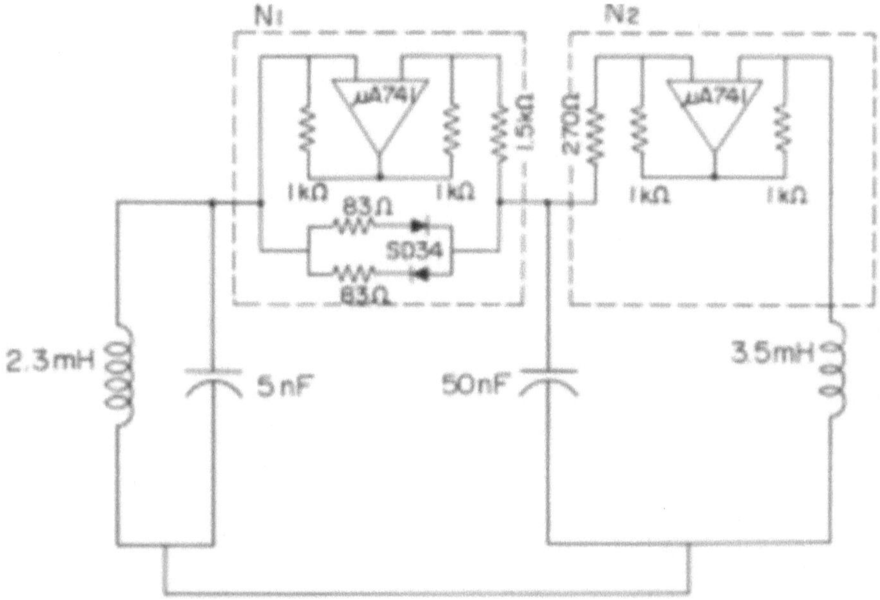

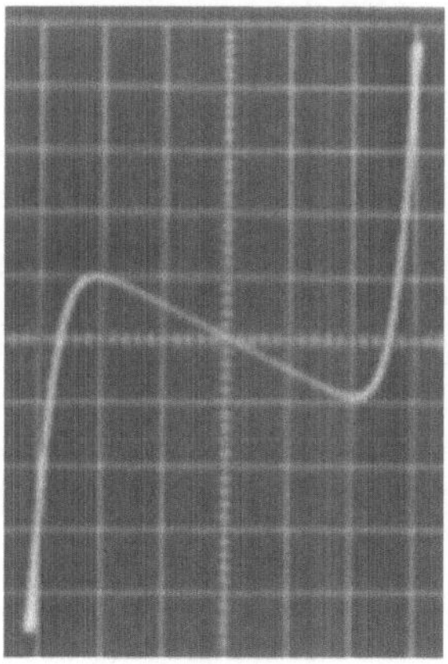

(b)

Fig. 1.7.2. Realization of the circuit of Fig. 1.7.1. (a) Circuitry. (b) Measured characteristic of $N1$. Horizontal scale, $0.2V$/div; vertical scale, $50\mu A$ ©1986 IEEE.

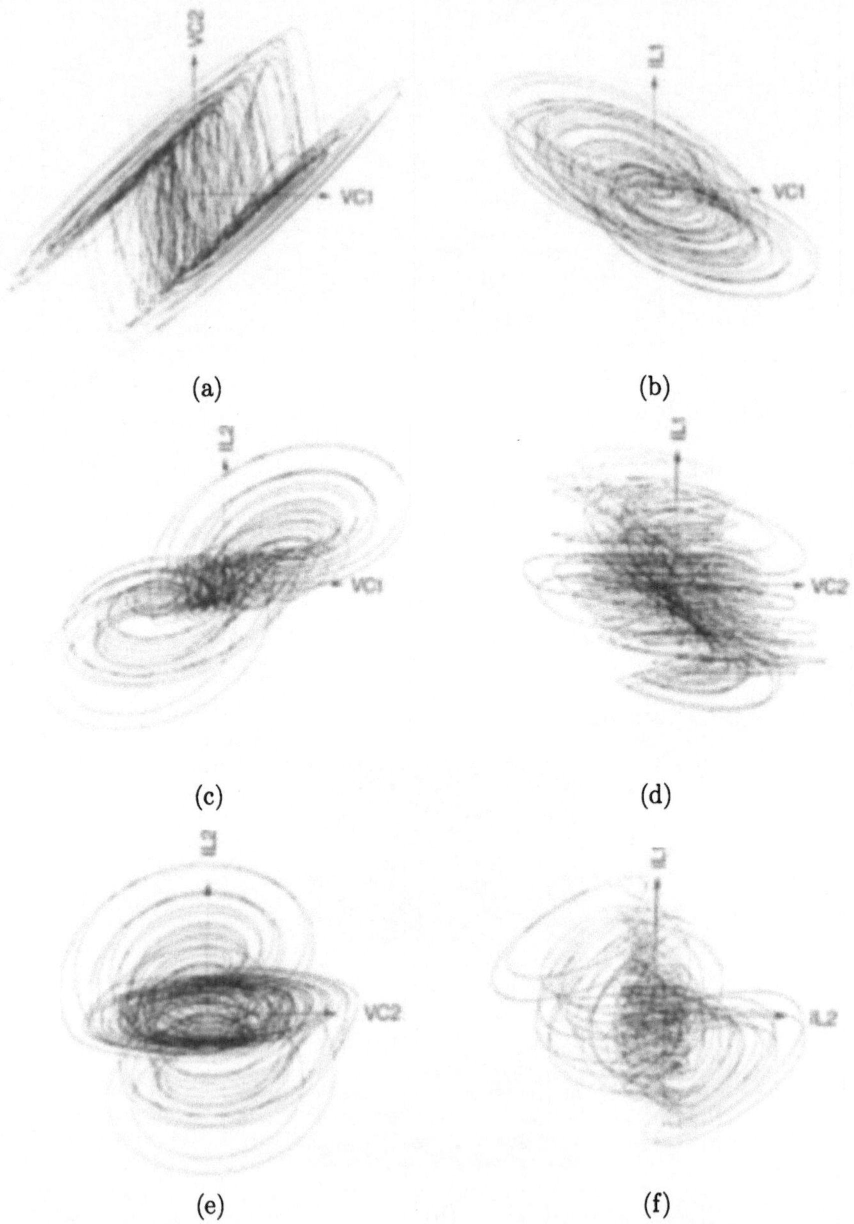

Fig. 1.7.3. Observation by digital computer simulation of (1.7.1). (a) Projection onto the (v_{C_1}, v_{C_2})-plane. (b) Projection onto the (v_{C_1}, i_{L_1})-plane. (c) Projection onto the (v_{C_1}, i_{L_2})-plane. (d) Projection onto the (v_{C_2}, i_{L_1})-plane. (e) Projection onto the (v_{C_2}, i_{L_2})-plane. (f) Projection onto the (i_{L_2}, i_{L_1})-plane. ©1986 IEEE.

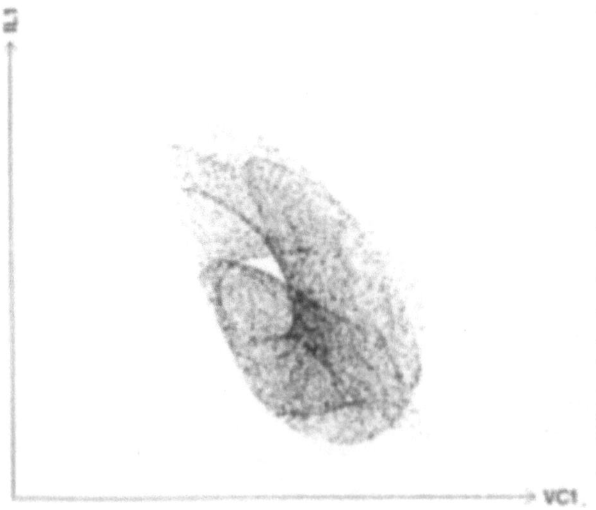

Fig. 1.7.4. Projection onto the (v_{C_1}, i_{L_1})-plane of the cross section of the attractor on $v_{C_1} - v_{C_2} < 0$, observed by digital computer simulation. ©1986 IEEE.

which correspond to the circuit of Fig. 1.7.2 are $1/C_1 = 2.1, 1/C_2 = 20, 1/L_1 = 1, 1/L_2 = 1.5, R = 1, m_0 = 3$, and $m_1 = -0.2$. The Lyapunov exponents are

$$\mu_1 \approx 0.24, \ \mu_2 \approx 0.06, \ \mu_3 \approx 0.00, \text{ and } \mu_4 \approx -53.8 \,.$$

The Lyapunov dimension (see Subsection 1.3.2) given by

$$d_L \approx 3 + 0.3/|-53.8| \approx 3.006,$$

which is a fractal between 3 and 4, agrees with the experimental data given above. Finally, Fig. 1.7.3 gives the projections of the attractor obtained by digital computer simulation of (1.7.1), while Fig. 1.7.4 gives the projection onto the (v_{C_1}, i_{L_1})-plane of the cross section defined by (1.7.3). Correspondence with the experimental data is clear. Note that the dimension of the state space of our circuit is four, which is the minimum in order for a system to exhibit hyperchaos. Recall that the circuit of Fig. 1.7.1 has two active elements: $-R$ and the piecewise-linear resistor. Two would probably be the minimum number of active elements in order to admit hyperchaos, since it has to have at least two expanding directions. For example, the Double Scroll circuit has only one active element and only one expanding direction.

1.8 The Neon Bulb Circuit

1.8.1 Introduction

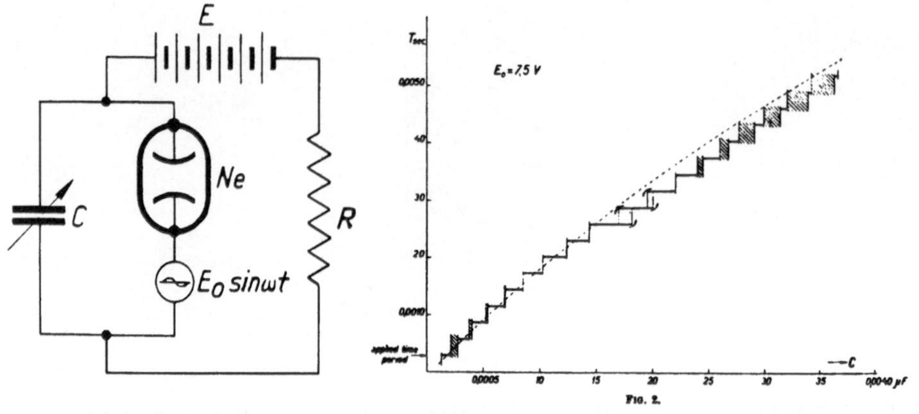

(a) (b)

Fig. 1.8.1. The neon bulb circuit experimented on by Van der Pol and Van der Mark. (a) Circuitry. (b) Bifurcation diagram. ©Macmillan Magazines Ltd.

In 1927, two Dutch engineers, Van der Pol and Van der Mark performed a rather interesting experiment on the circuit given in Fig. 1.8.1(a) [Van der Pol and Van der Mark 1927]. The symbol Ne stands for a neon bulb. They first set $E_0 = 0$ and tuned the parameters in such a way that the autonomous circuit oscillates with frequency f_0 = 1kHz. (E = 200V, R: several MΩ, C: small). Then they applied E_0 = 7.5V, $f = \frac{\omega}{2\pi}$ = 1kHz and varied C up to 3.5nF. The result is shown in Fig. 1.8.1(b), where the vertical axis is the period of the circuit's oscillation. The two electrical engineers were interested in the staircase structure of the graph, in that, the frequency suddenly jumped from f to $1/2f$, to $1/3f$, ..., as they varied the capacitance values. They called this phenomenon "frequency demultiplication" and wrote:

> "while the production of harmonics, as with the frequency multiplication, furnishes us with tone determining the musical major scales, the phenomenon of frequency-division renders the musical minor scale audible. In fact, with a properly chosen fundamental ω, the tuning of the condenser in the region of the third to the sixth subharmonics strongly reminds one of the tunes of a bagpipe."

Note that they used telephone receivers to observe the phenomena. Another very interesting observation they made is that

> "Often an *irregular noise* is heard in the telephone receivers before the frequency jumps to the next lower value" (emphasis by the authors).

The shaded parts in Fig. 1.8.1(b) "Correspond to those settings of condenser where an irregular noise is heard". The irregular noise, in today's term, would probably have been chaos. Van der Pol and Van der Mark, however, did not pursue the irregular noise by saying

> "However, this is a subsidiary phenomenon, the main effect being the regular frequency demultiplication."

Van der Pol appears to have had an interest only in $(m/n)f$, where m and n are integers because of his great amount of knowledge of music. In fact, he tried to explain musical scales in terms of Farey fractions [Van der Pol 1960]. He also had great interest in absolute pitch and performed experiments on measuring the frequency of the note "A" taken from 106 orchestras in Europe in the years around 1940, and found out that the note "A" varied between 430Hz and 447Hz.

1.8.2 Experiment

A Observation Instead of observing capacitance bifurcations, we have performed two parameter bifurcation experiments on the (E_0, f)-space, i.e., the amplitude and the frequency of the input voltage source. The reason for doing this is simply because it is not very interesting to reproduce something which has been obtained already. Plate 1.20 shows the result where the horizontal axis is f/f_0 while the vertical axis is E_0, and

$$1.40 \leq f/f_0 \leq 1.72$$

$$0 \leq E_0 \leq 4\text{V}$$

$$R = 1\text{M}\Omega,\ C = 1\text{nF},\ \text{battery} = 85\text{V},\ \text{and}$$

$$f_0 = 280\text{Hz is the oscillation frequency when } E_0 = 0.$$

At each point (E_0, f) within a region of a particular color corresponds to a particular period of the circuit's oscillation (see the figure caption of Plate 1.22 for color code). Observe the triangular regions touching the $E_0 = 0$ line. The fact that each of these triangular regions has the same color implies that the period of the circuit's oscillation does not change even if one varies (E_0, f), as long as (E_0, f) stays within each triangular region. This is called *phase locking*, and the triangular regions are called *Arnold tongues* after the Russian mathematician Arnold (see the

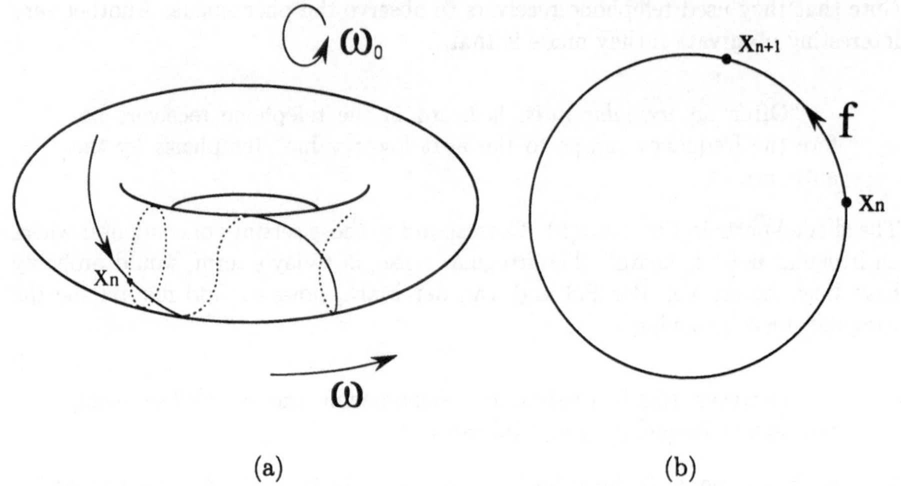

Fig. 1.8.2. Description of the dynamics with two oscillatory components. (a) Dynamics on a torus. (b) Dynamics on a circle.

next subsection). Plate 1.21 is a similar diagram when the driving voltage source is a "saw tooth" wave form. This picture has a nice correspondence, at least partly, with the supercritical sine circle map described in the next subsection (Plate 1.22).

Note that in Plate 1.21, period-doubling and chaotic regions are clearly seen where the input is a saw tooth wave form. In Plate 1.20 with sinusoidal input, however, period-doubling cascade and hence chaos is not observed, while Van der Pol and van der Mark reported period-doubling as well as the "irregular noise". A possible reason would be the differences between the two bulbs used since the two Dutch engineers performed their experiment in 1927, while we did ours in 1987.

B Sounds Since $f_0 = 280$Hz and since $1.40 \leq f/f_0 \leq 1.72$, the circuit behaves well within audible frequencies. This is also true for the circuit of Van der Pol and Van der Mark (Fig. 1.8.1). Recall that the two Dutch engineers listened to the bifurcations through "telephone receiver" instead of viewing the bifurcations.

1.8.3 Arnold Tongues

Recall that the neon bulb circuit (Fig. 1.8.1) can oscillate without the input voltage source with angular frequency, say ω_0. If one applies a nonzero input $E_0 \sin \omega t$, then the behavior can be naturally described on a torus (Fig. 1.8.2(a)). If one "cuts" the torus at a particular ω, one has a circle S^1. If one looks at how the trajectory comes back to S^1, one obtains a *circle map* (Fig. 1.8.2(b))

$$f : S^1 \to S^1. \tag{1.8.1}$$

Although circle maps have a long history [Denjoy 1932, Arnold 1965, Herman 1977], our purpose here is to describe only a few relevant facts. The Russian mathematician Arnold studied the sine circle map

$$f(x) := x + \Omega - \frac{K}{2\pi} \sin 2\pi x \ \text{ mod. } 1. \tag{1.8.2}$$

Note that K corresponds to E_0 while Ω corresponds to $2\pi f/f_0$ in the neon bulb circuit. Plate 1.22 shows the sets of points in the (Ω, K)-space, where each color represents the period of periodic orbit of (1.8.2) where

$$0 \le \Omega \le 1, \ 0 \le K \le 4.$$

For $0 \le K < 1$ (subcritical region), the map (1.8.2) is a diffeomorphism so that there is always a unique stable periodic or a unique quasi-periodic orbit. For $1 > K$ (supercritical region), the map f is non-invertible and there could be more than one attractor. However, it follows from the fact that f has a negative Schwartzian derivative that there are at most two stable periodic orbits [Guckenheimer and Holmes 1983]. The color code for $1 > K$ is obtained by choosing the initial condition $x_0 = \frac{1}{2\pi} \sin^{-1}(2\pi\Omega/K)$ (a local minimum of f) and letting f iterate 1000 times freely, after which the period was checked. For $0 \le K < 1$, the initial condition is $x_0 = 1/2$. Correspondence between the subcritical region of Plate 1.22 and a part of Plate 1.20 is clear. There is also a correspondence at least partly, between the supercritical region of Plate 1.22 and a part of Plate 1.21.

1.8.4 Rotation Numbers

If $f : S^1 \to S^1$ is an orientation preserving diffeomorphism, then it has a unique *lift* $F : \mathbb{R} \to \mathbb{R}$, such that

$$\rho(x_0, F) := \lim_{n\to\infty} \frac{F^n(x_0) - x_0}{n}$$

exists and is independent of x_0. This is the *rotation number* of F and hence of f denoted by $\rho(f)$. If ρ is rational, say P/Q, then there is an x_0 such that

$$F^Q(x_0) = x_0 + P. \tag{1.8.3}$$

If, on the other hand, ρ is irrational, say α, then F is topologically conjugate to the pure rotation

$$R(x) = x + \alpha.$$

It should be noted that in Plate 1.20 as well as in Plate 1.22, Q defined by (1.8.3) is coded by color while P is not. The rotation number P/Q of some of the larger Arnold tongues is indicated in Plate 1.20 and Plate 1.22. For $K > 1$, the map f is not invertible and the rotation number is not unique. In addition, chaotic behavior is possible. Finally $K = 1$ is called the critical line. Almost every point of the critical line gives rise to a rational rotation number. Experimental results on forced Rayleigh-Bénard convection are found in [Glazier and Libchaber 1988] and the references therein. We close this section by remarking that many physicists have been interested in the behavior of f near $K = 1$ with $\rho(f) = (\sqrt{5} - 1)/2$, the reciprocal of the golden mean.

1.9 The *R-L*-Diode Circuit

1.9.1 Experiment 1

Consider the circuit shown in Fig. 1.9.1, where a series connection of R,L and Diode is driven by a sinusoidal voltage source. Plate 1.23 shows the one-dimensional bifurcation diagram of the current i when the amplitude E of the applied sinusoidal voltage source is increased from 0 to 7.7 V, where

$$R = 107\Omega, \ L = 2.5\text{mH}, \ f = 150\text{kHz}, \ E_b = 0 \text{ and diode:3CC13}.$$

Each point in this bifurcation diagram represents a one-dimensional Poincaré section taken at each fundamental period $T = 1/f$ of the sinusoidal source.

There are two striking features here:

(1) A succession of large periodic windows whose periods increase exactly by one as we move from any window to the next window on the right;

(2) A succession of chaotic bands sandwiched between the large periodic windows.

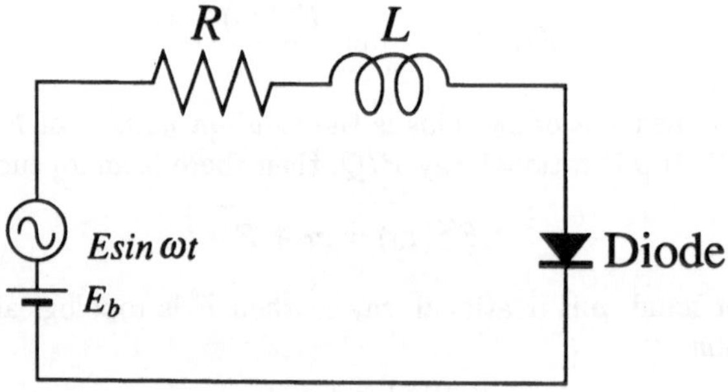

Fig. 1.9.1. An R-L-Diode circuit is driven by a sinusoidal voltage source. ©1987 Elsevier Science Publishers.

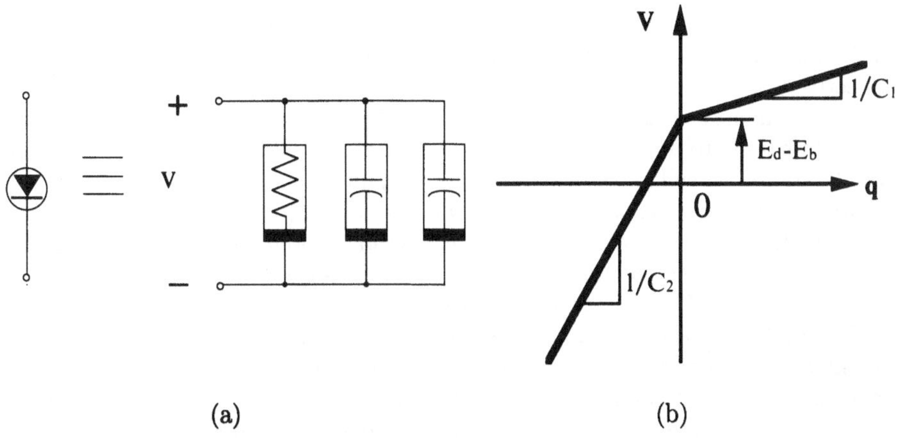

Fig. 1.9.2. Equivalent circuit of a diode. (a) An accurate model is a parallel connection of three nonlinear elements. (b) Simplified capacitor characteristic. ©1987 Elsevier Science Publishers.

Plate 1.24 shows cross sections measured in the voltage-current plane of the diode for four different values of the signal amplitude; namely, $E = 1.5\text{V}, 2.6\text{V}, 5.6\text{V}$ and 6.2V. Note that these values of E are chosen so that each one corresponds to one of the four chaotic bands in Plate 1.23. The Poincaré sections corresponding to the periodic windows between these chaotic bands consist of isolated points: n points for a period-n window.

1.9.2 Analysis 1

A The Dynamics An accurate equivalent circuit of a junction diode is given by a parallel connection of three nonlinear elements [Yang 1987] (see Fig. 1.9.2):

(1) A nonlinear resistor,

$$I_d = I_s\left[\exp(q'v/kT) - 1\right] \; ; \tag{1.9.1}$$

(2) A junction capacitor $C_j(v)$ due to the depletion region,

$$C_j(v) = \frac{C_{j0}}{(1 - v/V_{j0})^{1/2}} \; ; \tag{1.9.2}$$

(3) A diffusion capacitor $C_d(v)$, due to the rearrangement of the minority carrier density,

$$C_d(v) = C_{d0}\exp(q'v/kT), \tag{1.9.3}$$

where $I_s, q', k, T, V_{j0}, C_{j0}$ and C_{d0} are saturation current, electron charge, Boltzmann constant, the absolute temperature, the potential voltage of the pn junction, the junction capacitance at zero bias and the diffusion capacitance at zero bias, respectively.

The dynamics of the circuit with (1.9.1)-(1.9.3) will be extremely difficult, if not impossible to analyze. By several careful measurements and device physics considerations, we will make several drastic simplifications without losing essential features of the experimentally observed bifurcation diagrams.

Note that under reverse bias, the capacitor is dominated by the junction capacitor (1.9.2), whereas under forward bias, the capacitor is dominated by the diffusion capacitor (1.9.3). Through measurements, the capacitance is found to be 90nF at 0.5V (a positive bias) and 235pF at -1.0V (a negative bias). Note that the difference in the capacitance values is more than two orders of magnitude. The diode also exhibits the well-known rectification characteristic (1.9.1): in the reverse bias region the resistance is almost infinite, whereas in the forward bias region the resistance is very small. For example, at 0.5V the resistance is 100Ω. By carefully measuring the impedances of the capacitors and the resistor over a frequency range of more than 25kHz, it was found that the impedances of the capacitors are much smaller than those of the resistor. Therefore, the diode characteristic can be simplified and modeled by the two-segment piecewise-linear capacitor given in Fig. 1.9.2(b) so that the *R-L*-diode circuit can be accurately described by

$$\begin{aligned} \frac{dq}{dt} &= i \\ L\frac{di}{dt} &= -Ri - \left\{ \begin{array}{ll} \frac{1}{C_d} & q \text{ if } q \geq 0 \\ \frac{1}{C_j} & q \text{ if } q < 0 \end{array} \right\} - E_d + E_b + E\sin(2\pi f t), \end{aligned} \quad (1.9.4)$$

where C_d is the diffusion capacitance at 0.5V bias, C_j is the junction capacitance at -1.0V, $E_d = 0.5$V is the break point voltage at which the capacitance value changes between the junction capacitance and the diffusion capacitance, i is the circuit current, and q is the charge of the capacitor.

In order to clarify the effects of circuit parameters on the dynamics, let us perform the following rescaling:

$$Q \leftarrow (Lf^2/E)q, \; I \leftarrow (Lf/E)i, \; \tau \leftarrow ft. \quad (1.9.5)$$

The dynamics is rewritten as

$$\begin{aligned} \frac{dQ}{d\tau} &= I \\ \frac{dI}{d\tau} &= -\frac{R}{fL}I - \left\{ \begin{array}{l} \frac{1}{f^2LC_d}Q \text{ if } Q \geq 0 \\ \\ \frac{1}{f^2LC_j}Q \text{ if } Q < 0 \end{array} \right\} - \frac{E_d - E_b}{E} + \sin(2\pi\tau) \end{aligned} \quad (1.9.6)$$

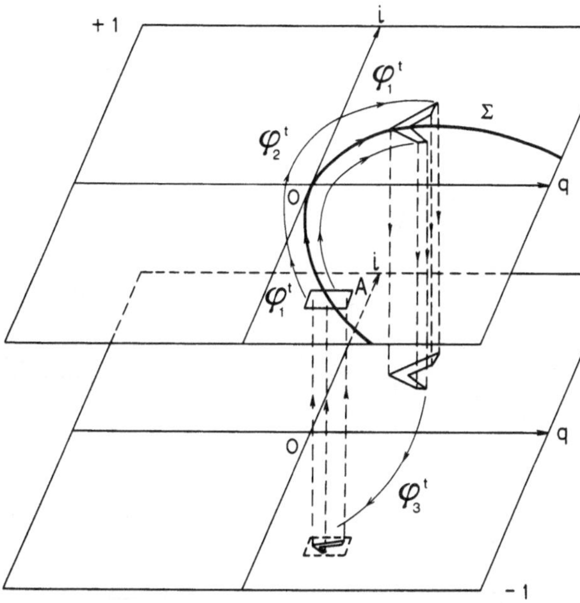

Fig. 1.9.3. Deformation of an initial trapezoid A. ©1987 Elsevier Science Publishers.

We will further simplify (1.9.6) by replacing the sinusoidal voltage source by a rectangular voltage source:

$$\begin{aligned} \frac{dQ}{d\tau} &= I \\ \frac{dI}{dt} &= -kI - \left\{ \begin{array}{c} \alpha Q \text{ if } Q \geq 0 \\ \\ \beta Q \text{ if } Q < 0 \end{array} \right\} \\ & \quad -\frac{E_d - E_b}{E} + \left\{ \begin{array}{ccccc} +1 \text{ if} & n & \leq \tau & < & n + \frac{1}{2} \\ \\ -1 \text{ if} & n + \frac{1}{2} & \leq \tau & < & n+1 \end{array} \right\}, \end{aligned} \tag{1.9.7}$$

where

$$\alpha = 1/LC_d f^2,\ \beta = 1/LC_j f^2,\ k = R/fL.$$

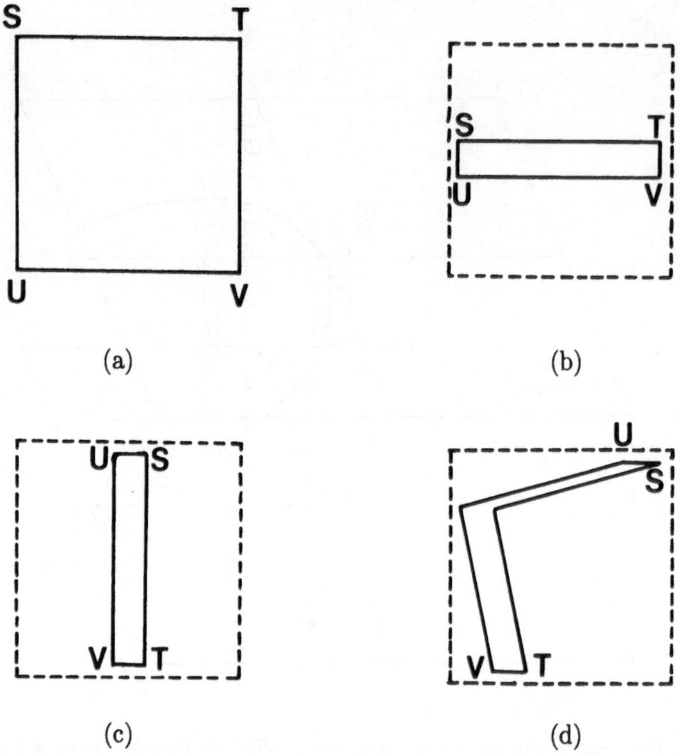

Fig. 1.9.4. Two-dimensional map model. (a) The initial rectangle STUV. (b) The rectangle is compressed in the vertical direction. (c) The compressed rectangle is rotated by 90 degrees. (d) The rectangle is folded. ©1987 Elsevier Science Publishers.

Observe that in Plate 1.23, the current i is shown instead of the charge q of the dynamics (1.9.4), as it is much easier to measure i than to measure q.

Now note that any solution of (1.9.7) is made up of components from the following four *linear autonomous flows on* $\mathbb{R}^2$:

$$\varphi_1^t : Q \geq 0 \text{ and the voltage source is } +1$$

$$\varphi_2^t : Q < 0 \text{ and the voltage source is } +1$$

$$\varphi_3^t : Q \geq 0 \text{ and the voltage source is } -1$$

$$\varphi_4^t : Q < 0 \text{ and the voltage source is } -1.$$

Taking a full advantage of the linearity of each flow, one can capture the basic mechanism of the dynamics. Fig.1.9.3 shows how an initial trapezoid A is deformed, where Σ is the trajectory passing through the origin.

B Two-Dimensional Map Model Based upon Fig.1.9.3, we will propose a simple two-dimensional map (discrete dynamical system) model which mimics the point transformation described in Fig. 1.9.3. In particular, this two-dimensional map exhibits the same bifurcation phenomena as those observed experimentally in the circuit shown in Fig. 1.9.1. Fig. 1.9.4 gives a more precise description of the point transformation mechanism shown in Fig. 1.9.3. Composing the steps described in Fig.1.9.4, one has the discrete dynamical system

$$\begin{aligned} x^1_{n+1} &= x^2_n - 1 + \left\{ \begin{array}{ll} a_1 x^1_n & \text{if } x_n \geq 0 \\ -a_2 x^1_n & \text{if } x^1_n < 0 \end{array} \right\} \\ x^2_{n+1} &= b x^1_n . \end{aligned} \tag{1.9.8}$$

Now let us examine how this simple map captures the essential features of the bifurcation phenomena observed experimentally from our *R-L*-diode circuit. Fig. 1.9.5(a) shows the one-parameter bifurcation diagram of x^1 for (1.9.8), where

$$a_1 = 0.7, \ b = -0.13$$

and a_2 is varied over the range

$$0 \leq a_2 \leq 20.$$

Fig. 1.9.5(b) shows the attractor in the (x^1, x^2)-plane corresponding to

$$a_2 = 18.0 \, .$$

Note that this attractor is qualitatively identical to those obtained experimentally in Section 1.9.1.

C The Bifurcation Scenario Based upon the preceding analysis, we will describe the bifurcation scenario. Fig. 1.9.6 shows the detailed bifurcation mechanisms associated with the period-4 window. Bifurcations associated with the other periodic windows have similar structures.

The sequence drawings in column B of Fig. 1.9.6 shows how the attractor of the two-dimensional map model is deformed as a_2 is increased from its value at the lowest position to a larger value at the top position. Speaking roughly, parameter a_2 corresponds to the amplitude E of the original circuit. (details are omitted). The "snap shots" in Column A show the corresponding experimental observations taken from the original *R-L*-diode circuit as E increases from the bottom. The four insets in Column C are enlarged pictures in a small neighborhood of the periodic point *P4A* (of the two-dimensional map) identified by the solid triangles.

We can give a complete picture of what is happening in the original circuit:

(1) Let us begin with the picture at the bottom of Column B and look at the folded object. The star-like symbol identifies the location of the fixed point of

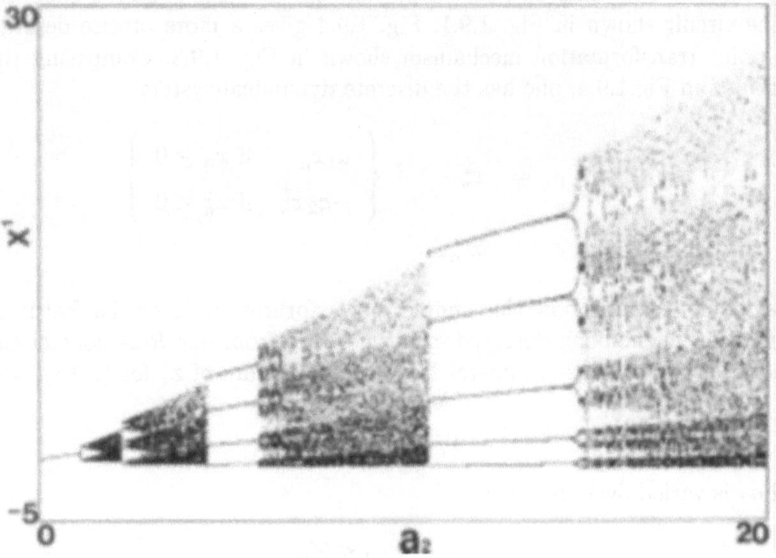

(a)

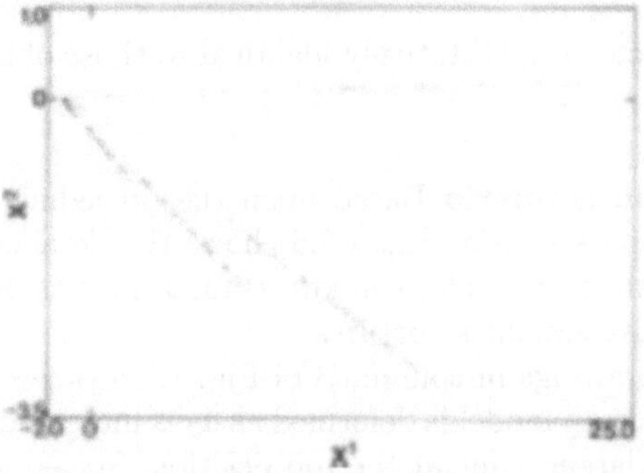

(b)

Fig. 1.9.5. (a) One-parameter bifurcation diagram of x^1 for (1.9.8). (b) Attractor at $a_2 = 18.0$. ©1987 Elsevier Science Publishers.

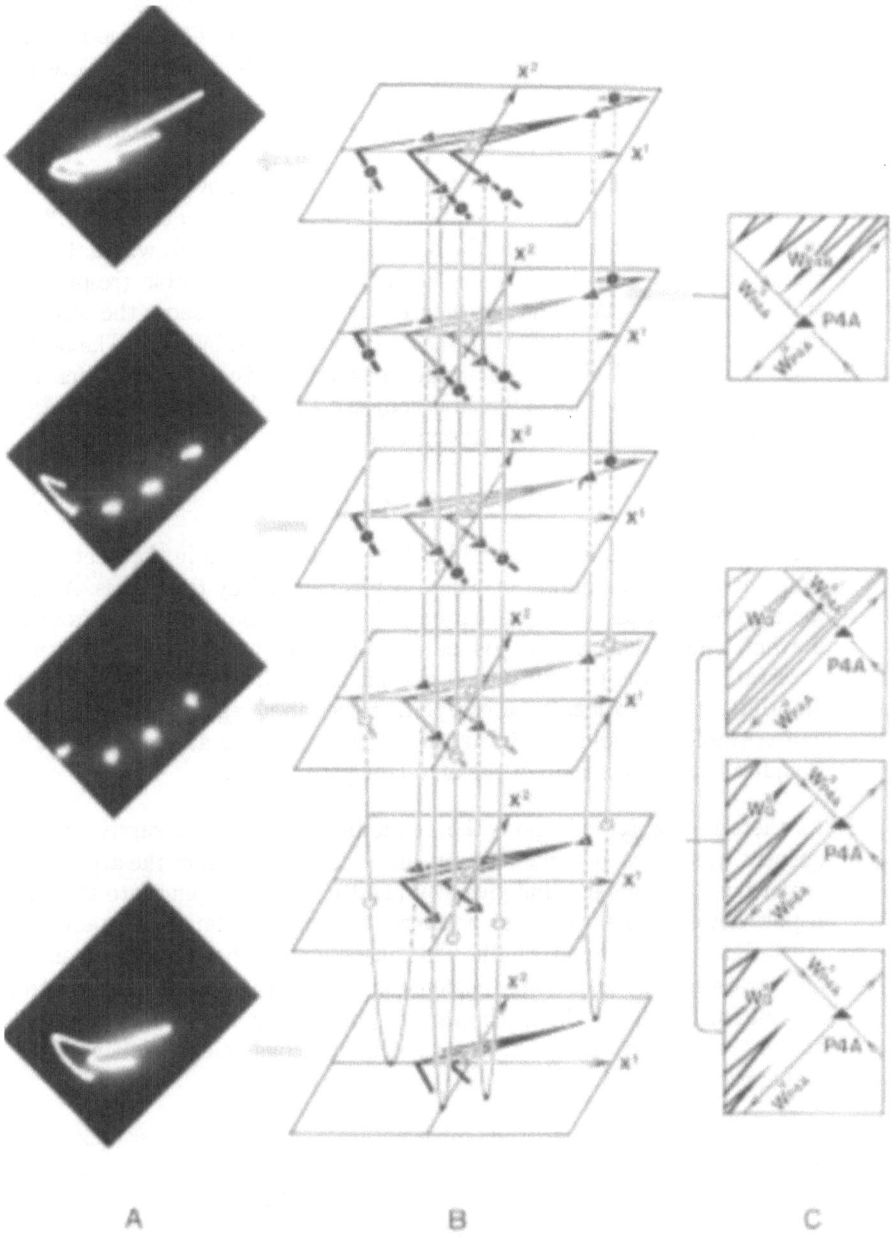

Fig. 1.9.6. Bifurcation mechanisms corresponding to period-4 window. Left column shows experimental observations while the insets in the right column show enlarged pictures around P4A. ©1987 Elsevier Science Publishers.

(1.9.8) which is a saddle for the present parameter range. As we increase the value of a_2 (E in the original circuit), a saddle-node bifurcation of period-4 takes place outside the region where the attractor lies. This period-4 bifurcation has a strong influence on the structure of the attractor. Note that since the bifurcation in this case corresponds to that of a saddle-node, pairs consisting of both a stable and an unstable periodic orbit are born.

(2) As we increase a_2 further, the unstable periodic orbit moves closer and closer to the "crab" attractor, and finally it collides with the attractor. This is depicted in the picture next to the bottom one in Column B, where the solid triangles (respectively open dots) correspond to an unstable (respectively stable) periodic orbit. The three insets in Column C show the situation around the right-most unstable periodic point denoted by $P4A$. The bottom inset in Column C shows the situation before collision, where thick lines indicate W_Q^u (the closure of which is conjectured to be the attractor). As we increase a_2 by an appropriate amount, we see that

$$W_Q^u \text{ collides with } W_{P4A}^s\,, \tag{1.9.9}$$

where W_{P4A}^s is the stable manifold of $P4A$. This is shown in the inset second from the bottom in column C, where W_Q^u is denoted by thick lines. A slight increase of a_2 leads to the situation depicted by the inset third from the bottom, where, this time, W_Q^u is indicated by thick broken lines. The crucial observation in this picture is that the unstable direction of $P4A$ provides an orbit with an exit gate for escape into the outer region. Since the stable and unstable manifolds are invariant, a collision of the attractor with $P4A$ is equivalent to a collision of the attractor with W_{P4A}^s.

(3) As there is now an exit gate, the attractor can no longer survive. Consequently, we observe a sudden disappearance or extinction of the attractor at the critical parameter value given by (1.9.9). After escaping into the outer region, however, the orbit cannot diverge to infinity because the stable periodic orbit is waiting to attract it. This phenomenon, therefore, represents an interior crisis. The situation is depicted in the third picture from the bottom in Column B. This is the mechanism responsible for the extinction (death) of the "two-legged" attractor and the simultaneous emergence (birth) of a stable period-4 orbit.

(4) As we increase a_2 further, the stable period-4 orbit loses its stability via a period-doubling bifurcation. The periodic attractor then deforms itself into a chaotic attractor made up of four islets as depicted in the fourth picture from the bottom in Column B. Note, however, that the period-doubling does not last infinitely many times since (1.9.8) is piecewise-linear. The destabilized periodic points are denoted by four solid dots. Observe that the chaotic attractor in this case is the closure of the unstable manifold of the solid dots rather than that of the star-like symbol (see(1)). Note also that the unstable

period-4 points, represented by the four solid triangles born in the picture below, are still present near the chaotic attractor.

(5) As we increase a_2 even further, the chaotic attractor eventually collides with the stable manifold of the solid triangle, namely,

$$W^u_{P4B} \text{ collides with } W^s_{P4A} \, . \tag{1.9.10}$$

This is depicted in the fifth picture from the bottom in Column B. The corresponding inset in Column C shows the enlarged details around $P4A$. When (1.9.10) occurs, W^u_{P4A} plays bridging role between the chaotic islands, thereby giving birth to the attractor with "three legs" shown in the topmost picture in Column B. Note that the increase in the number of legs (or the number of islands in the chaotic bands) is attributed to the interaction of the attractor with *the other period-4 orbit* which was born earlier via a saddle-node bifurcation.

Remark 1.9.1. Even though (1.9.8) is similar to the Lozi mapping (see Chapter 3), there are two subtle differences: (i) the parameter in (1.9.8) corresponding to a in the Lozi mapping depends on the value of x^1_n, and (ii) $b < 0$. These differences make it impossible to directly apply Misurewicz's idea [Misurewicz 1980] to (1.9.8).

1.9.3 Experiment 2

Fig. 1.9.7 shows another experiment with

$$R = 75\Omega, f = 28\text{kHz}, E_b = -1.0\text{V}, O \le E \le 4.02$$

while other parameters are the same as before. This bifurcation diagram is qualitatively different from Plate 1.23, in that rather than increasing the period of each successive periodic window by one, period-1 windows (e.g. indicated as (a), (b) and (c)) and chaotic bands appear alternatively. Namely, something qualitatively different is happening at lower frequencies.

1.9.4 Analysis 2

The fact that $C_j = 235\text{pf} \ll C_d = 90\text{nF}$ gives rise to two important features of 1.9.6);

(1) The vertical component of the vector field (Q,I) on $Q < 0$ is much faster than that on $Q \ge 0$.

(2) Considering the imaginary part of the eigenvalues on $Q < 0$ (respectively $Q \ge 0$), $\sqrt{4L/C_j - R^2}/2fL$ (respectively $\sqrt{4L/C_d - R^2}/2fL$), the resonant frequency on $Q < 0$ is much higher than that on $Q \ge 0$. This means that there is a strong "twisting" mechanism on $Q < 0$.

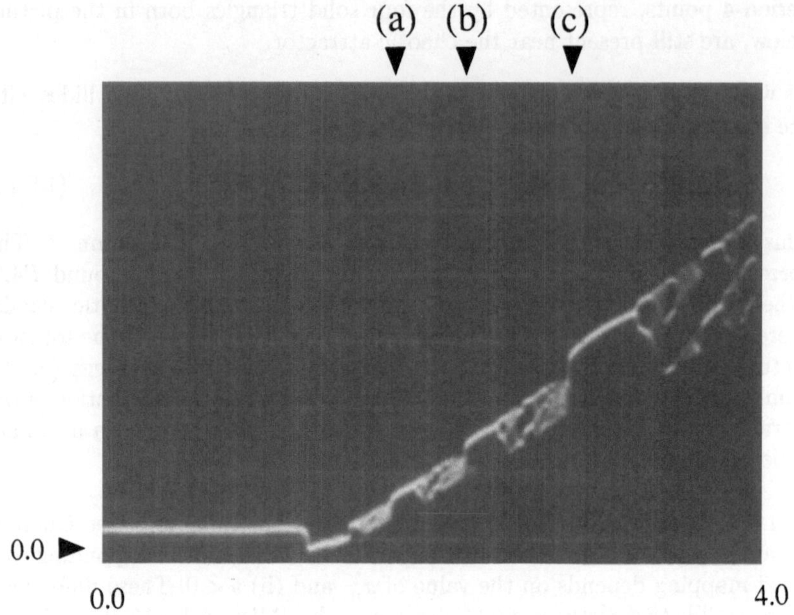

Fig. 1.9.7. Another one-parameter bifurcation diagram with lower frequency and smaller amplitude of the voltage source. Horizontal axis: 0.4V/div. Vertical axis: 2.0mA/div. ©1991 Elsevier Science Publishers.

Plate 1.25(a) shows a Poincaré section of the chaotic attractor with $E = 2.4$V, $f =$ 50kHz at a particular phase of the input voltage source. Naturally, this particular section alone cannot reveal the properties of the Poincaré map. A great amount of information is obtained by looking at other sections in a sequence with various phases between 0 and 2π as shown in Plate 1.25(b)-(f). The horizontal axis is the diode voltage v_d and the vertical axis is the current i. From (1.9.5) and (1.9.6) v_d is given as

$$\begin{aligned} v_d &= E/f^2LC_dQ + E_d, \text{ if } Q \geq 0 \\ &= E/f^2LC_jQ + E_d, \text{ if } Q < 0 \end{aligned}$$

where $C_j = 235\text{pF} \ll C_d = 90\text{nF}$. Therefore the part on $v_d \geq E_d$ is strongly compressed in the horizontal direction. To help our understanding of the transformation process, Fig 1.9.12 gives the geometric structure corresponding to each figure in Plate 1.25, where the small triangle shows the reference orientation, and the vertical line indicates $Q = 0$ ($v_d = E_d$), i.e. the boundary between the diffusion capacitance region (right) and the junction capacitance region (left).

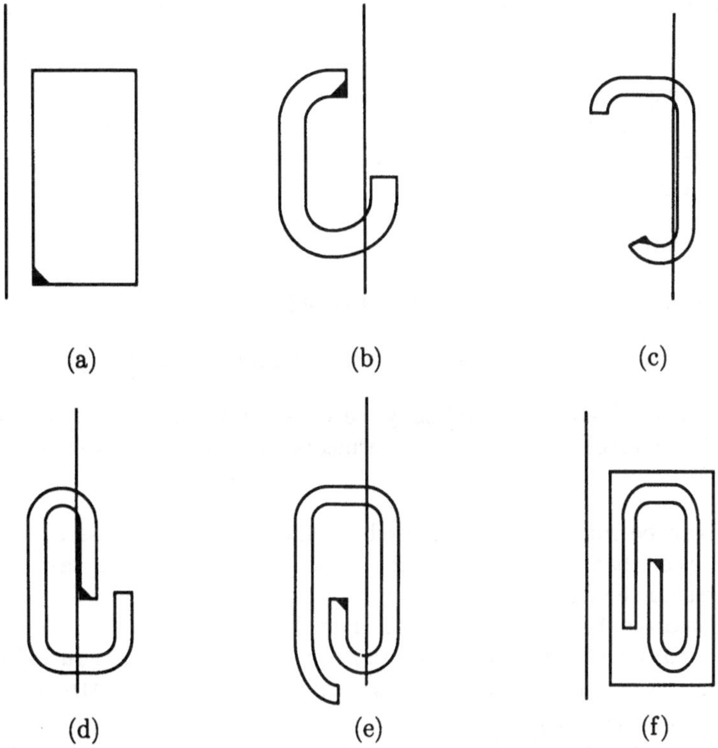

Fig. 1.9.8. Geometric model of the attractor formation. Each figure corresponds to the one in Plate 1.25. ©1991 Elsevier Science Publishers.

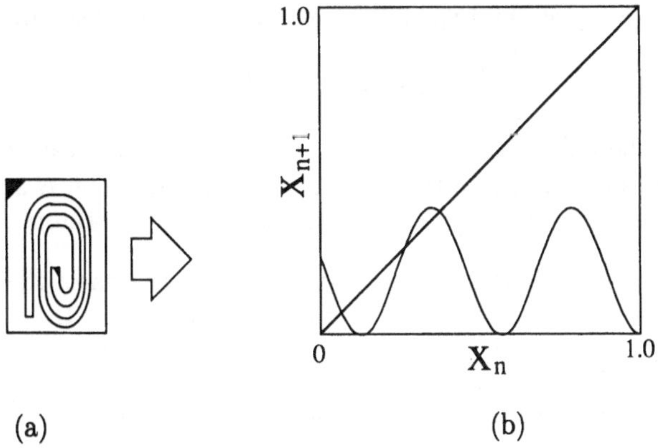

Fig. 1.9.9. Discrete map models. (a) Two-dimensional map. (b) One-dimensional map. ©1991 Elsevier Science Publishers.

(a) The attractor is in the region which is dominated by the diffusion capacitor.

(b) A part of the attractor moves into the region which is dominated by the junction capacitor. It follows from (a) and (b) above that when the attractor moves from the diffusion capacitor region into the junction capacitor region, it is stretched because of the difference between the vector fields. Then it is twisted clockwise.

(c) The attractor is further twisted.

(d) The attractor is stretched again and twisted.

(e) The attractor is further twisted (see Plate 1.26(a) for an enlarged picture).

(f) The attractor is squeezed and finally the attractor returns to the initial region (see Plate 1.26(b) and (c) for two intermediate enlarged pictures between (d) and (f)).

The (a)-(f) can be summarized as: stretching and twisting give rise to a folded object. It is clear that the attractor is folded twice. It is, therefore, easy to infer that the attractor could be twisted many times. This can happen, for instance, when E becomes smaller. This is due to the fact that the bias term in (1.9.6) shifts the dynamics to the left and hence gives rise to a larger time period during which the attractor stays in the $Q < 0$ ($v_d < E_d$) region. Therefore the attractor could be twisted more times ("multi-folding").

The above argument leads us to the two-dimensional map given in Fig. 1.9.9(a). Lut us explain how the "multi-folding" mechanism is responsible for the "alternative appearance of period-one attractors and chaotic attractors". It is rather difficult to obtain an analytically tractable formula for the two-dimensional map described in Fig.1.9.9(a). One can derive, however, a one-dimensional version of Fig. 1.9.9(a), given by Fig.1.9.9(b), which can be modeled by

$$x_{n+1} = a\{1 - \cos[b(1 - x_n)]\}. \tag{1.9.11}$$

In order to show that (1.9.11) captures all the important features of the observed bifurcations, we will discuss the dependencies of a and b on E and f only roughly for our present purpose. Our analysis is based upon laboratory measurements.

(1) *Parameter a.* This parameter controls the extrema of the map (1.9.11). In other words a controls the size of the attractor. From the observations shown in Fig. 1.9.7, the size of the attractor is proportional to E, the amplitude of the voltage source. Moreover, a is inversely proportional to the circuit dissipation which is given by $\exp(R/2fL)$. Therefore,

$$a \propto (E + a_1)\exp(-R/2fL) \tag{1.9.12}$$

would be an appropriate relationship, where a_1 is a parameter.

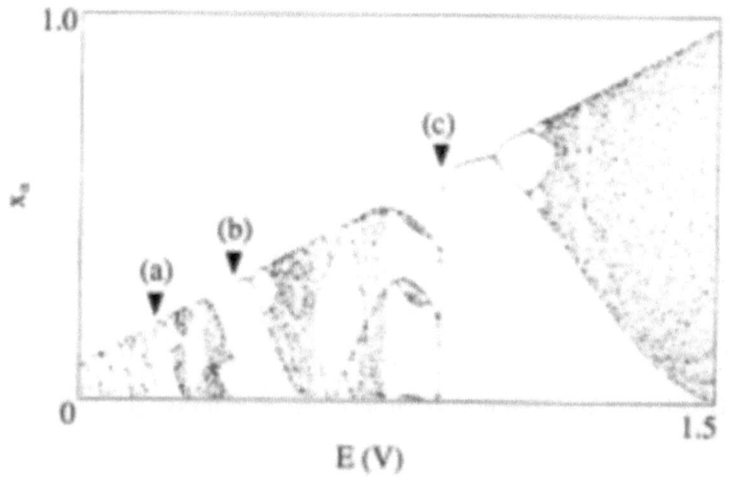

Fig. 1.9.10. One-parameter bifurcation diagram of (1.9.11). ©1991 Elsevier Science Publishers.

(2) *Parameter b*. This parameter controls the number of extrema of (1.9.11). which corresponds to the number of twisting in the junction capacitor region. The latter should be proportional to the imaginary part ω_j of the eigenvalue of the junction capacitor region, and the length t_j of the time interval during which the attractor stays in the junction capacitor region, namely,

$$b \propto \omega_j t_j + \theta, \tag{1.9.13}$$

where θ represents the phase constant. When E is decreased, as mentioned above, the bias term in (1.9.6) shifts the dynamics to the left and hence gives rise to a larger time period t_j. Therefore t_j is inversely proportional to the amplitude E. When E is increased it has been observed that the change of the number of twistings becomes more moderate. This factor is represented by θ and the relationship

$$\theta \propto b_1[1 - b_2/(E + b_2)]. \tag{1.9.14}$$

seems reasonable, where b_1 and b_2 are parameters. Therefore we will write (1.9.13) as

$$b \propto \frac{\sqrt{4L/C_j - R^2}}{2fL} \frac{b_3}{E + b_3} + b_1(1 - \frac{b_2}{E + b_2}) \tag{1.9.15}$$

where b_3 is another parameter.

Fig.1.9.10 shows the bifurcation diagram of (1.9.11) where the horizontal axis is E and the vertical axis is x_n. The parameter values are chosen as: $R = 214\Omega, C_2 =$

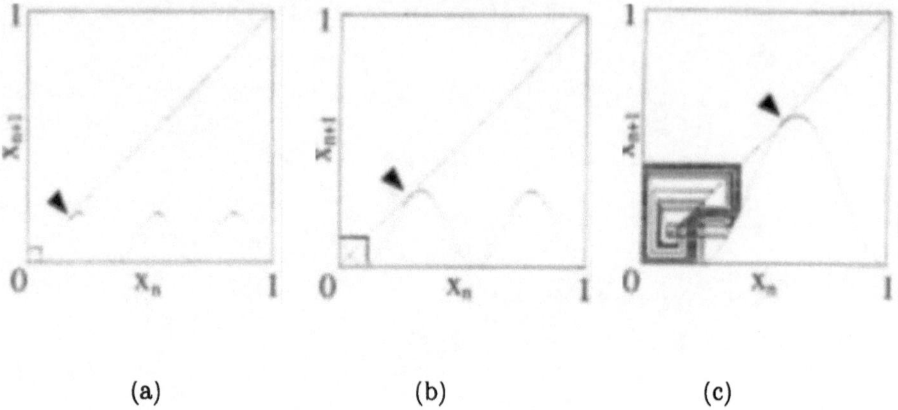

Fig. 1.9.11. Orbits of the one-dimensional map. (a) $E = 0.18$V. (b) $E = 0.36$V. (c) $E = 0.85$V.

235pF, $L = 2.50$mH, $a_1 = 0.15$, $b_1 = 2.4$, $b_2 = 1.0$, $b_3 = 0.2$. The frequency of the voltage source is fixed at $f = 35$kHz, while the amplitude of the voltages varied from 1.5 to 0.0V. Observe that the basic qualitative features of Fig.1.9.7are clearly captured. In particular, period-one windows and chaotic bands appear alternatively. Fig.1.9.11(a)-(c) show orbits of (1.9.11) at the parameter values labeled as (a)-(c), respectively in Fig.1.9.7 and Fig.1.9.9 It is clear that (1.9.11) undergoes a saddle-node bifurcation when it become tangent to the diagonal line. Since the extremum value of (1.9.11) is determined by parameter a and since a is monotonic with respected to E (see(1.9.12)), the only possible reason for (1.9.11) to undergo repeated period-1 saddle-node bifurcations is its multimodality. Namely, the hills and valleys of (1.9.11) become tangent to the diagonal one by one. In terms of the original circuit dynamics, this means that an initial rectangle is mapped into a "multi-folded object".

Remark 1.9.2. Other results on R-L-Diode circuit are found in [Linsay 1981], [Bronson, Dewey and Linsay 1983], [Rollins and Hunt 1982], [Testa, Perez and Jeffries 1982], [Cascais, Dilao and Norondacosta 1983], [Yoon, Song, Shin and Ra 1984], [Tanaka, Higuchi and Matsumoto 1993], [Tanaka, Matsumoto and Chua 1985, 1987], [Matsumoto, Chua and Tanaka 1984], and the references therein.

2. Bifurcations of Continuous Piecewise-Linear Vector Fields

2.1 Introduction

This chapter provides several fundamental theorems for continuous piecewise-linear vector fields. In Section 2.2, we will give a definition of continuous piecewise-linear mappings. And we will give a fundamental representation theorem (Standard Form) for an arbitrary continuous piecewise-linear mapping.

In Section 2.3, we will consider normal forms of two-region continuous piecewise-linear vector fields, which correspond to the Jordan normal forms for linear vector fields. Before considering normal forms of full two-region systems, we will study an affine vector field with a boundary, which is a half of two-region system. We will present two cases. The first one is called the *non-degenerate* case where it is possible to change the affine vector field into a linear vector field by an affine transformation. The second is called the *degenerate* case where it is impossible to change the affine vector field into any linear vector field by any affine transformation. We will consider the non-degenerate case in Subsection 2.3.2, and the degenerate case in Subsection 2.3.3. Using results from these subsections, we will determine the normal forms of two-region systems in Subsection 2.3.4. In Subsection 2.3.5, we will consider the normal forms of proper two-region systems, which are important in the studying bifurcations.

In Section 2.4, we will apply the results from Section 2.3 to multi-region systems, and numerically observe typical attractors. We will derive normal forms of

(1) three-dimensional proper three-region systems with point symmetry (Subsection 2.4.1),

(2) three-dimensional proper four-region systems with axial symmetry (Subsection 2.4.2),

(3) four-dimensional proper three-region systems with point symmetry (Subsection 2.4.3).

From these equations, we can numerically observe various chaotic attractors depending on parameter values. In the three-dimensional three-region system with point symmetry, we will observe (1) spiral-type attractor, (2) Double Scroll-type attractor, (3) double screw-type attractor, (4) toroidal-type attractor, (5) Sparrow-type attractor, and (6) a point symmetric Lorenz-type attractor.

In the three-dimensional four-region system with axial symmetry, we will observe (7) a piecewise-linear Lorenz attractor, and in the four-dimensional three-region system with point symmetry, (8) a piecewise-linear Duffing attractor.

In Section 2.5, we will derive global equations of bifurcation sets (a bifurcation equations) for three-dimensional two-region systems. We will derive bifurcation equations of homoclinic and heteroclinic orbits in Subsection 2.5.4, and bifurcation equations of periodic orbits (that is, saddle-node bifurcation, period-doubling bifurcation, Hopf bifurcation) in Subsection 2.5.5. Subsections 2.5.1-2.5.3 are preparation to derive bifurcation equations. In particular, Subsection 2.5.1 will provide an elementary proof of normal forms for three-dimensional proper two-region systems.

In Section 2.6, we will numerically solve the bifurcation equations and describe global structure of the bifurcation sets. We will describe homoclinic/heteroclinic bifurcation sets in Subsection 2.6.1, and bifurcation sets for periodic orbits in Subsection 2.6.2. Since these result can be obtained without any numerical integration formula, e.g., the Runge-Kutta, the method described is extremely powerful.

2.2 Definition and Standard Forms of Continuous Piecewise-Linear Maps

In this section we will define the piecewise-linear mapping, and give a necessary and sufficient condition for a piecewise-linear mapping to be continuous (Theorem 2.2.4). We will, then, give fundamental representation theorems (Theorems 2.2.6 and 2.2.28) for continuous piecewise-linear mappings.

2.2.1 Definition of Piecewise-Linear Maps

Definition 2.2.1. Define an $(n-1)$-dimensional hyperplane U in n-dimensional euclidian space $\mathbb{R}^n$ by

$$U = U(\alpha, \beta) = \{x \in \mathbb{R}^n \mid \langle \alpha, x \rangle = \beta\},$$

where $\alpha \in \mathbb{R}^n - \{0\}, \beta \in \mathbb{R}$, and $\langle \cdot, \cdot \rangle$ denotes the usual inner product. We suppose that the elements of $\mathbb{R}^n$ are column vectors. For $\alpha_1, \cdots, \alpha_k \in \mathbb{R}^n - \{0\}$ and $\beta_1, \cdots, \beta_k \in \mathbb{R}$, define

$$\tilde{\alpha} = (\alpha_1, \cdots, \alpha_k) \in M(n,k), \quad \tilde{\beta} = (\beta_1, \cdots, \beta_k) \in M(1,k),$$

where $M(m,n)$ denotes the set of all $m \times n$ matrices with real components. Each hyperplane

$$U(\alpha_i, \beta_i) \quad (i = 1, \cdots, k)$$

is called *a boundary* defined by (α_i, β_i), and a union of hyperplanes

$$B = B(\tilde{\alpha}, \tilde{\beta}) = \bigcup_{i=1}^{k} U(\alpha_i, \beta_i)$$

is called *a boundary set* defined by $(\tilde{\alpha}, \tilde{\beta})$. For $(\tilde{\alpha}, \tilde{\beta})$ define a function $\omega : \mathbb{R}^n \to \{0,1\}^k$ by

$$\omega(x) = (\mathrm{sgn}(\langle \alpha_1, x\rangle - \beta_1), \cdots, \mathrm{sgn}(\langle \alpha_k, x\rangle - \beta_k)),$$

where

$$\mathrm{sgn}(t) = \begin{cases} 0 & (t \le 0) \\ 1 & (t > 0), \end{cases}$$

Define a subset of $\{0,1\}^k$ by

$$\Omega = \Omega(\tilde{\alpha}, \tilde{\beta}) = \{\omega \in \{0,1\}^k \mid \omega = \omega(x) \text{ for some } x \in \mathbb{R}^n\}.$$

Then *a polyhedral region* (or simply, *region*) with a sign $\omega \in \Omega$ is

$$R_\omega = \{x \in \mathbb{R}^n \mid \omega(x) = \omega\} \text{ for } \omega \in \Omega.$$

The union $\bigcup\{R_\omega \mid \omega \in \Omega\}$ is a partition of $\mathbb{R}^n$;

$$\mathbb{R}^n = \bigcup_{\omega \in \Omega} R_\omega; \text{ and}$$

$$R_\omega \cap R_{\omega'} = \emptyset \text{ if } \omega \ne \omega'.$$

Definition 2.2.2. A mapping $f : \mathbb{R}^n \to \mathbb{R}^m$ is *piecewise-affine* if there is $(\tilde{\alpha}, \tilde{\beta})$ such that

(1) f is differentiable at all points which do not belong to $B(\tilde{\alpha}, \tilde{\beta})$;

(2) for each $\omega \in \Omega(\tilde{\alpha}, \tilde{\beta})$, the derivative $Df(x)$ is constant in the interior of R_ω, i.e. $x, x' \in \mathrm{int}(R_\omega) \Rightarrow Df(x) = Df(x')$.

If $f : \mathbb{R}^n \to \mathbb{R}^m$ is piecewise-affine, then for each $\omega \in \Omega(\tilde{\alpha}, \tilde{\beta})$, there are $A_\omega \in M(m,n)$ and $q_\omega \in \mathbb{R}^m$ such that

$$\begin{aligned} f(x) &= A_\omega x + q_\omega \text{ for } x \in \mathrm{int}(R_\omega) \\ Df(x) &= A_\omega \text{ for } x \in \mathrm{int}(R_\omega). \end{aligned}$$

When f is piecewise-affine, we will say that f is *piecewise-linear* (*PL*), according to custom. In general, a PL map $f : \mathbb{R}^n \to \mathbb{R}^m$ may be discontinuous at points on B. If f is continuous on B, and hence, on $\mathbb{R}^n$, f is called *a continuous piecewise-linear map* (*CPL map*). A vector field $X : \mathbb{R}^n \to \mathbb{R}^n$ is *a CPL vector field*, or *a CPL system* if there is a CPL map $f : \mathbb{R}^n \to \mathbb{R}^n$ such that

$$X(x) = f(x) \quad (x \in \mathbb{R}^n),$$

where we identify $\mathbb{R}^n$ and a tangent space $T_x\mathbb{R}^n (x \in \mathbb{R}^n)$ under the natural isomorphism. For a vector field $X : \mathbb{R}^n \to \mathbb{R}^n$ and an affine transformation $h : \mathbb{R}^n \to \mathbb{R}^n; h(x) = Hx + p \quad (H \in GL(n, \mathbb{R}), p \in \mathbb{R}^n)$, define a vector field $h_*X : \mathbb{R}^n \to \mathbb{R}^n$ by

$$h_*X(x) = HX(h^{-1}(x)) \quad (x \in \mathbb{R}^n),$$

where $GL(n, \mathbb{R})$ is the set of all $n \times n$ real non-singular matrices. Two CPL systems X and Y are *affine conjugate* (respectively *linearly conjugate*), if there is an affine (respectively linear) transformation $h : \mathbb{R}^n \to \mathbb{R}^n$ such that $h_*(X) = Y$.

2.2.2 Standard Forms of CPL Maps with the Boundary Set in General Position

The simplest PL map is a two-region PL mapping.

Definition 2.2.3. For $\alpha \in \mathbb{R}^n - \{0\}$ and $\beta \in \mathbb{R}$, define

$$\begin{aligned} U &= U(\alpha, \beta) = \{x \in \mathbb{R}^n \mid \langle \alpha, x \rangle = \beta\} \\ R_0 &= \{x \in \mathbb{R}^n \mid \langle \alpha, x \rangle - \beta \leq 0\} \\ R_1 &= \{x \in \mathbb{R}^n \mid \langle \alpha, x \rangle - \beta > 0\} \\ f(x) &= f(x; (A, q_A); (B, q_B); (\alpha, \beta)) \\ &= \begin{cases} Ax + q_A & (x \in R_0) \\ Bx + q_B & (x \in R_1), \end{cases} \end{aligned}$$

where $A, B \in M(m, n), q_A, q_B \in \mathbb{R}^m$. Then $f : \mathbb{R}^n \to \mathbb{R}^m$ is called *two-region PL map* defined by $(A, q_A), (B, q_B)$ and (α, β).

Theorem 2.2.4. (Standard Forms of Two-Region CPL Maps) *Let a two-region PL map*

$$f : \mathbb{R}^n \to \mathbb{R}^m; f(x) = f(x; (A, q_A); (B, q_B); (\alpha, \beta))$$

be given. Then the following conditions are equivalent to each other.

(1) f is continuous.

(2) $Ax + q_A = Bx + q_B$ for all $x \in U(\alpha, \beta)$.

(3) There is an $h \in \mathbb{R}^n$ such that

(i) $B - A = (B - A)h\alpha^T$,

(ii) $q_B - q_A = -\beta(B - A)h$,

(iii) $\langle \alpha, h \rangle = 1$,

where h, α *are column vectors, and* α^T *denotes the transposition of* α.

(4) For any $h \in \mathbb{R}^n$ *with* $\langle \alpha, h \rangle = 1$, *it is satisfied that*

(i) $B - A = (B - A)h\alpha^T$,

(ii) $q_B - q_A = -\beta(B - A)h$,

where h, α *are column vectors, and* α^T *denotes the transposition of* α.

(5) There is a $p \in \mathbb{R}^m$ *such that*

$$f(x) = Ax + q_A + \frac{1}{2}p\{|\langle \alpha, x \rangle - \beta| + (\langle \alpha, x \rangle - \beta)\}.$$

Proof. (1) $\Leftrightarrow$ (2) : clear.

(2) $\Rightarrow$ (4) : Let $h \in \mathbb{R}^n$ with $\langle \alpha, h \rangle = 1$ be given. Since $\langle \alpha, \beta h \rangle = \beta \langle \alpha, h \rangle = \beta$, we have $A(\beta h) + q_A = B(\beta h) + q_B$, hence

$$q_B - q_A = A(\beta h) - B(\beta h) = -\beta(B - A)h. \tag{2.2.1}$$

This proves (ii). Take $y_2, \ldots, y_n \in \mathbb{R}^n$ such that $h, y_2, \cdots, y_n$ are linearly independent, and such that $\langle \alpha, y_i \rangle = 0 \quad (2 \leq i \leq n)$. Define $Y \in GL(n, \mathbb{R})$ by $Y = [h, y_2, \cdots, y_n]$, where h, y_i are column vectors. Since $\langle \alpha, \beta h + y_i \rangle = \beta \langle \alpha, h \rangle + \langle \alpha, y_i \rangle = \beta$ for each $2 \leq i \leq n$, we have

$$A(\beta h + y_i) + q_A = B(\beta h + y_i) + q_B.$$

By (2.2.1), we have

$$(A - B)y_i = q_B - q_A + \beta(B - A)h = 0.$$

Set $d = (A - B)h$, then

$$(A - B)Y = (A - B)[h, y_2, \cdots, y_n] = [d, 0, \cdots, 0] = d(1, 0, \cdots, 0).$$

Since $\alpha^T Y = \alpha^T [h, y_2, \cdots, y_n] = (1, 0, \cdots, 0)$,

$$(A - B)Y = d\alpha^T Y = (A - B)h\alpha^T Y. \tag{2.2.2}$$

Multiplying Y^{-1} from the right hand side of (2.2.2), we have (i).

(4) $\Rightarrow$ (3) : Set $h = \alpha / |\alpha|^2$. Then the statement is clear, where $|\alpha|^2 = \langle \alpha, \alpha \rangle$.

(3) $\Rightarrow$ (5) : Set $p = (B - A)h$, then

$$B = A + p\alpha^T, \quad q_B = q_A - \beta p.$$

Hence

$$f(x) = \begin{cases} Ax + q_A & \text{if } \langle \alpha, x \rangle - \beta \le 0 \\ (A + p\alpha^T)x + q_A - \beta p & \text{if } \langle \alpha, x \rangle - \beta > 0 \end{cases}$$
$$= Ax + q_A + \frac{1}{2}p\{|\langle \alpha, x \rangle - \beta| + (\langle \alpha, x \rangle - \beta)\}.$$

(5) $\Rightarrow$ (1): clear. □

Definition 2.2.5. Given $\tilde{\alpha} = (\alpha_1, \cdots, \alpha_k), \tilde{\beta} = (\beta_1, \cdots, \beta_k), \alpha_i \in \mathbb{R}^n, \beta_i \in \mathbb{R} (1 \le i \le k)$, let

$$U_i = \{x \in \mathbb{R}^n \mid \langle \alpha_i, x \rangle = \beta_i\} (i = 1, \cdots, k).$$

The boundary set $B = B(\tilde{\alpha}, \tilde{\beta})$ is *in general position* if for any subset $\{U_{i_1}, \cdots, U_{i_s}\}$ of $\{U_1, \cdots, U_k\}$,

$$U_{i_1} \cap \cdots \cap U_{i_s} = \emptyset,$$

or

$$U_{i_1} \cap \cdots \cap U_{i_s} \ne \emptyset \quad \text{and}$$

$$\{\alpha_{i_1}, \cdots, \alpha_{i_s}\} \quad \text{are linearly independent.}$$

Theorem 2.2.6. (Standard Forms of CPL Mappings with the Boundary Set in General Position)

Let $B = B(\tilde{\alpha}, \tilde{\beta})$ be a boundary set in general position, where $\tilde{\alpha} = (\alpha_1, \cdots, \alpha_k)$, $\tilde{\beta} = (\beta_1, \cdots, \beta_k)$, $\alpha_i \in \mathbb{R}^n$, $\beta_i \in \mathbb{R} (1 \le i \le k)$. Assume that the sign $0 = (0, \cdots, 0)$ belongs to $\Omega(\tilde{\alpha}, \tilde{\beta})$. Define $f : \mathbb{R}^n \to \mathbb{R}^m$ by

$$f(x) = A_\omega x + q_\omega \text{ for } x \in R_\omega,\ \omega \in \Omega(\tilde{\alpha}, \tilde{\beta}),$$

where $A_\omega \in M(m, n)$, $q_\omega \in \mathbb{R}^m$. Then f is continuous if and only if there are $p_1, \cdots, p_k \in \mathbb{R}^m$ such that

$$f(x) = A_0 x + q_0 + \frac{1}{2}\sum_{i=1}^{k} p_i\{|\langle \alpha_i, x \rangle - \beta_i| + (\langle \alpha_i, x \rangle - \beta_i)\},$$

where $A_0 = A_{(0,\cdots,0)}$, $q_0 = q_{(0,\cdots,0)}$.

Lemma 2.2.7. *Let $\alpha_1, \alpha_2 \in \mathbb{R}^n - \{0\}$ and $\beta_1, \beta_2 \in \mathbb{R}$ be given such that α_1 is not parallel to α_2. Set*

$$f(x) = A_\omega x + q_\omega \text{ for } x \in R_\omega,\ \omega \in \{0, 1\}^2,$$

where $R_\omega = \{x \in \mathbb{R}^n \mid \mathrm{sgn}(\langle \alpha_i, x \rangle - \beta_i) = \omega_i\ (i = 1, 2)\}$, $\omega = (\omega_1, \omega_2)$. If f is continuous, then

$$A_{10} - A_{00} = A_{11} - A_{01}.$$

Proof. Without loss of generality, we may assume $|\alpha_1| = |\alpha_2| = 1$. Since α_1 is not parallel to α_2, there are $a \in \mathbb{R}$ and $b \in \mathbb{R} - \{0\}$ such that

$$\alpha_1 = a\alpha_2 + b\alpha_2',$$

where $\alpha_2' \in \mathbb{R}^n$ is a vector with $\langle \alpha_2, \alpha_2' \rangle = 0$ and $|\alpha_2'| = 1$. Set $h_1 = (1/b)\alpha_2'$ and $h_2 = \alpha_2$, then

$$\langle \alpha_1, h_1 \rangle = 1, \quad \langle \alpha_2, h_1 \rangle = 0, \quad \langle \alpha_2, h_2 \rangle = 1.$$

Set

$$\begin{aligned}
f_1(x) &= f(x; (A_{00}, q_{00}); (A_{10}, q_{10}); (\alpha_1, \beta_1)) \\
f_2(x) &= f(x; (A_{01}, q_{01}); (A_{11}, q_{11}); (\alpha_1, \beta_1)) \\
f_3(x) &= f(x; (A_{00}, q_{00}); (A_{01}, q_{01}); (\alpha_2, \beta_2)) \\
f_4(x) &= f(x; (A_{10}, q_{10}); (A_{11}, q_{11}); (\alpha_2, \beta_2))
\end{aligned}$$

The two-region map f_1 coincides f on $R_{00} \cup R_{10}$ and f is continuous. Hence f_1 is continuous. Similarly f_2, f_3 and f_4 are continuous. By Theorem 2.2.4(4), it follows that

$$A_{10} - A_{00} = (A_{10} - A_{00})h_1\alpha_1^T, \quad A_{11} - A_{01} = (A_{11} - A_{01})h_1\alpha_1^T,$$

$$A_{01} - A_{00} = (A_{01} - A_{00})h_2\alpha_2^T, \quad A_{11} - A_{10} = (A_{11} - A_{10})h_2\alpha_2^T.$$

Hence, using $\alpha_2^T h_1 = \langle \alpha_2, h_1 \rangle = 0$, we obtain

$$(A_{10} - A_{00}) - (A_{11} - A_{01}) = (A_{10} - A_{00} - A_{11} + A_{01})h_1\alpha_1^T$$

$$= ((A_{10} - A_{11}) - (A_{00} - A_{01}))h_2\alpha_2^T h_1\alpha_1^T = 0.$$

□

Lemma 2.2.8. *Let $\alpha_i \in \mathbb{R}^n - \{0\}$ and $\beta_i \in \mathbb{R} (i = 1, \cdots, k)$ be given. Assume $\alpha_1, \cdots, \alpha_k$ are linearly independent. Set*

$$f(x) = A_\omega x + q_\omega \text{ for } x \in R_\omega, \ \omega \in \{0,1\}^k,$$

where $R_\omega = \{x \in \mathbb{R}^n \mid \mathrm{sgn}(\langle \alpha_i, x \rangle - \beta_i) = \omega_i \ (1 \le i \le k)\}$, $\omega = (\omega_1, \cdots, \omega_k)$.
Let $\omega, \omega', \sigma, \sigma' \in \{0,1\}^k$ be given such that

$$\omega_i = \omega_i' (i \ne k), \omega_k = 0 \quad \text{and} \quad \omega_k' = 1,$$

and

$$\sigma_i = \sigma_i' (i \ne k), \sigma_k = 0 \quad \text{and} \quad \sigma_k' = 1.$$

If f is continuous, then

$$A_\omega - A_{\omega'} = A_\sigma - A_{\sigma'}.$$

Proof. In $(\omega_1, \cdots, \omega_{k-1})$ and $(\sigma_1, \cdots, \sigma_{k-1})$, if there is only one different component, i.e.

$$\omega_s \neq \sigma_s, \quad \text{and} \quad \omega_i = \sigma_i \quad (i \neq s),$$

then, by applying Lemma 2.2.7 to U_k and U_s, we have the conclusion.

In $(\omega_1, \cdots, \omega_{k-1})$ and $(\sigma_1, \cdots, \sigma_{k-1})$, if there are l different components $(1 \leq l \leq k-1)$, then, by applying the above fact repeatedly l times, we have the conclusion.

Proof of Theorem 2.2.6. The "if "-statement is clear. To prove the "only if "-statement, we use an induction with respect to k. Without loss of generality, we may assume $|\alpha_i| = 1$ for $i = 1, \cdots, k$. Take

$$\omega, \omega' \in \Omega \text{ such that } \omega_i = \omega_i' \quad (i \neq k), \omega_k' = 1 \text{ and } \omega_k = 0.$$

Set $p_k = (A_{\omega'} - A_\omega)\alpha_k$. Since f is continuous, we have

$$A_{\omega'} - A_\omega = p_k \alpha_k^T.$$

Define

$$\begin{aligned} g(x) &= \begin{cases} (p_k \alpha_k^T)x - \beta_k p_k, & \text{if } \langle \alpha_k, x \rangle - \beta_k > 0 \\ 0 & \text{if } \langle \alpha_k, x \rangle - \beta_k \leq 0 \end{cases} \\ &= \frac{1}{2} p_k \{ |\langle \alpha_k, x \rangle - \beta_k| + (\langle \alpha_k, x \rangle - \beta_k) \}, \text{ and} \\ h(x) &= f(x) - g(x). \end{aligned}$$

Let $\sigma, \sigma' \in \Omega$ be given such that $\sigma_i = \sigma_i'(i \neq k), \sigma_k = 0$ and $\sigma_k' = 1$. If we repeatedly apply Lemma 2.2.8, we have $A_{\sigma'} - A_\sigma = A_{\omega'} - A_\omega$, hence $p_k \alpha_k^T = A_{\sigma'} - A_\sigma$. Since $Dh(x) = A_{\sigma'} - p_k \alpha_k' = A_{\sigma'} - (A_{\sigma'} - A_\sigma) = A_\sigma$ at $x \in R_{\sigma'}$, we have $Dh(x) = A_\sigma$ (constant) on $R_\sigma \cup R_{\sigma'}$. Hence, by assumption of induction, there are $p_1, \cdots, p_{k-1} \in \mathbb{R}^n$ such that

$$h(x) = A_0 x + q_0 + \frac{1}{2} \sum_{i=1}^{k-1} p_i \{ |\langle \alpha_i, x \rangle - \beta_i| + (\langle \alpha_i, x \rangle - \beta_i) \}.$$

Since $f(x) = h(x) + g(x)$, we obtain the conclusion. □

2.2.3 Standard Forms of CPL Functions

In this subsection, we will consider representing a continuous piecewise-linear mapping $f : \mathbb{R}^n \to \mathbb{R}^m$ by using the absolute value function $|\cdot| : \mathbb{R} \to \mathbb{R}$;

$$|x| = \begin{cases} x & (x \geq 0) \\ -x & (x < 0) \end{cases}$$

Definition 2.2.9. A continuous piecewise-linear map from $\mathbb{R}^n$ to $\mathbb{R}$ is called *a continuous piecewise-linear function*(CPL function) of $\mathbb{R}^n$. The set of all CPL functions of $\mathbb{R}^n$ is denoted by $\text{CPL}(\mathbb{R}^n)$.

If we denote a continuous piecewise-linear map $f : \mathbb{R}^n \to \mathbb{R}^m$ by

$$f(x) = (f_1(x), \cdots, f_m(x)), \quad x \in \mathbb{R}^n,$$

then each f_i is a CPL function of $\mathbb{R}^n$. Hence, to represent CPL maps by the absolute function, it is enough to represent CPL functions by the absolute value function.

Definition 2.2.10. Define a set of formal expressions with variable $x \in \mathbb{R}^n$, $L_k(\mathbb{R}^n)$ $(k \geq 0)$, inductively as follows:

$$\begin{aligned} L_0(\mathbb{R}^n) &= \{\langle a, x\rangle + b \mid a \in \mathbb{R}^n, b \in \mathbb{R}\} \\ L_k(\mathbb{R}^n) &= \{f_0(x) + \sum_{i=1}^{N} \varepsilon_i |f_i(x)| \mid f_i(x) \in L_{k-1}(\mathbb{R}^n) \quad (0 \leq i \leq N), \\ &\qquad \varepsilon_i \in \{-1, 1\} \quad (1 \leq i \leq N), \quad N \geq 0\}, \end{aligned}$$

where $N = 0$ means that the summation is not taken. Then the following holds:

$$L_0(\mathbb{R}^n) \subset L_1(\mathbb{R}^n) \subset \cdots \subset L_k(\mathbb{R}^n) \subset \cdots.$$

Hence $L_k(\mathbb{R}^n)$ is a set of expressions with at most a k-ply absolute value function. Define

$$L_\infty(\mathbb{R}^n) = \bigcup_{k=0}^{\infty} L_k(\mathbb{R}^n).$$

An element of $L_\infty(\mathbb{R}^n)$ is called an *expression* of the CPL function of $\mathbb{R}^n$.

Definition 2.2.11. Define a mapping $\mathcal{S}$ from $L_\infty(\mathbb{R}^n)$ to $CPL(\mathbb{R}^n)$ by

$$\mathcal{S}(f)(x) = F(x) \quad \text{for} \quad f(x) \in L_\infty(\mathbb{R}^n)$$

where $F(x) \in \mathbb{R}$ is a value that a formal expression $f(x)$ takes when $x \in \mathbb{R}^n$ is substituted to $f(x)$. If $f_1(x)$ and $f_2(x)$ are different elements of $L_\infty(\mathbb{R}^n)$, and if $f_1(x) = f_2(x)$ for all $x \in \mathbb{R}^n$, then we say that they are *different expressions of same CPL function.*

Example 2.2.12. Two expressions $f_1(x) = 1 - |x| + |1 - |x||$ and $f_2(x) = |x + 1| + |2x| + |x - 1|$ for $x \in \mathbb{R}$ are considered to be two different elements of $L_2(\mathbb{R})$. However, if we substitute any $x \in \mathbb{R}$ to them, we have $f_1(x) = f_2(x)$, hence they are the same function as an element of CPL($\mathbb{R}$), i.e., $\mathcal{S}(f_1)(x) = \mathcal{S}(f_2)(x)$. Hence they are different expressions of the same CPL function.

From now on, if there is no confusion, we use simply $f(x)$ in stead of $\mathcal{S}(f)(x)$ for $f(x) \in L_\infty(\mathbb{R}^n)$, and comment that we consider $f(x)$ as a function of $x \in \mathbb{R}^n$.

Definition 2.2.13. For $f(x) = \langle a, x\rangle + b \in L_0(\mathbb{R}^n)$, the $b \in \mathbb{R}$ is called *a constant term* of $f(x)$. Inductively, for $f(x) \in L_k(\mathbb{R}^n)$, if

$$f(x) = f_0(x) + \sum_{i=1}^{N} \varepsilon_i |f_i(x)|, \quad f_i(x) \in L_{k-1}(\mathbb{R}^n) \quad (0 \le i \le N),$$

each constant term of $f_i(x)$ is called *a constant term* of $f(x)$.

Definition 2.2.14. For $f(x) \in L_\infty(\mathbb{R}^n)$ and $g(y) \in L_\infty(\mathbb{R})$, define an expression $F(x,y) \in L_\infty(\mathbb{R}^{n+1})$ for $(x,y) \in \mathbb{R}^n \times \mathbb{R} = \mathbb{R}^{n+1}$ by multiplying $g(y)$ by all constant terms of $f(x)$. Define a operator

$$\mathcal{F} : L_\infty(\mathbb{R}^n) \times L_\infty(\mathbb{R}) \to L_\infty(\mathbb{R}^{n+1})$$

by

$$\mathcal{F}(f(x), g(y)) = F(x,y) \quad \text{for} \quad (f(x), g(y)) \in L_\infty(\mathbb{R}^n) \times L_\infty(\mathbb{R}).$$

Definition 2.2.15. For $f(x) \in L_k(\mathbb{R}^n)$, define an expression $\bar{f}(x,y)$ by multiplying $-y \in \mathbb{R}$ by all constant terms of $f(x)$:

$$\bar{f}(x,y) = \mathcal{F}(f(x), -y).$$

Clearly $\bar{f}(x,y)$ has at most k-ply absolute value function, hence

$$\bar{f}(x,y) \in L_k(\mathbb{R}^{n+1}), \quad (x,y) \in \mathbb{R}^n \times \mathbb{R} = \mathbb{R}^{n+1}.$$

Define a function $F_{k,n}$ from $L_k(\mathbb{R}^n)$ to $L_k(\mathbb{R}^{n+1})$ by

$$F_{k,n}(f) = \bar{f}.$$

Definition 2.2.16. Define a new notation $[y]$ for $y \in \mathbb{R}$ by

$$[y] = \frac{1}{2}\{y + |y|\}.$$

For $f(x) \in L_k(\mathbb{R}^n)$, define an expression $\tilde{f}(x,y)$ by multiplying $[y]$ by all constant terms of $f(x)$:

$$\tilde{f}(x,y) = \mathcal{F}(f(x), [y]).$$

Clearly $\tilde{f}(x,y)$ has at most $(k+1)$-ply absolute value function, hence

$$\tilde{f}(x,y) \in L_{k+1}(\mathbb{R}^{n+1}), \quad (x,y) \in \mathbb{R}^n \times \mathbb{R} = \mathbb{R}^{n+1}.$$

Define a function $G_{k,n}$ from $L_k(\mathbb{R}^n)$ to $L_{k+1}(\mathbb{R}^{n+1})$ by

$$G_{k,n}(f) = \tilde{f}.$$

Example 2.2.17. Set $f(x) = 1 - |x| + |1 - |x||$. Then

$$\begin{aligned}\bar{f}(x,y) &= -y - |x| + |-y - |x||,\\ \tilde{f}(x,y) &= [y] - |x| + |[y] - |x||\\ &= \frac{1}{2}\{y + |y|\} - |x| + |\frac{1}{2}\{y + |y|\} - |x||.\end{aligned}$$

Definition 2.2.18. Using two functions $F_{k,n}$ and $G_{k,n}$, we define a function $T_{k,n}$ as follows:

$$T_{k,n} : L_k(\mathbb{R}^n) \times L_k(\mathbb{R}^n) \to L_{k+1}(\mathbb{R}^{n+1});$$

$$T_{k,n}(f,g) = F_{k,n}(f) + G_{k,n}(g).$$

Definition 2.2.19. Define subsets $L_n^a(\mathbb{R}^n)$, $L_n^b(\mathbb{R}^n)$ and $L_n^c(\mathbb{R}^n)$ of $L_n(\mathbb{R}^n)$ inductively as follows:

$$L_1^a(\mathbb{R}) := \{ax + \frac{b}{2}\{x + |x|\} \mid a, b, x \in \mathbb{R}\}$$

$$L_1^c(\mathbb{R}) := \{c + \sum_{i=1}^{N} f_i(x - x_i) \mid f_i(x) \in L_1^a(\mathbb{R}), c \in \mathbb{R}, x_i \in \mathbb{R}, N \geq 1\}$$

$$L_1^b(\mathbb{R}) := \{f(x) \in L_1^c(\mathbb{R}) \mid \mathcal{S}(\tilde{f})(x,y) = 0 \quad \text{for all } x \in \mathbb{R} \quad \text{and} \quad y = 0\},$$

where $\tilde{f}(x,y) = \mathcal{F}(f(x), [y])$.

$$L_2^a(\mathbb{R}^2) := T_{1,1}(L_1^c(\mathbb{R}), L_1^b(\mathbb{R}))$$

$$L_2^c(\mathbb{R}^2) := \{c + \sum_{i=1}^{N} f_i(x - x_i) \mid f_i(x) \in L_2^a(\mathbb{R}^2), c \in \mathbb{R}, x_i \in \mathbb{R}^2, N \geq 1\}$$

$$L_2^b(\mathbb{R}^2) := \{f(x) \in L_2^c(\mathbb{R}^2) \mid \mathcal{S}(\tilde{f})(x,y) = 0 \quad \text{for all } x \in \mathbb{R}^2 \quad \text{and} \quad y = 0\},$$

where $\tilde{f}(x,y) = \mathcal{F}(f(x), [y])$.

$$L_n^a(\mathbb{R}^n) := T_{n-1,n-1}(L_{n-1}^c(\mathbb{R}^{n-1}), L_{n-1}^b(\mathbb{R}^{n-1}))$$

$$L_n^c(\mathbb{R}^n) := \{c + \sum_{i=1}^{N} f_i(x - x_i) \mid f_i(x) \in L_n^a(\mathbb{R}^n), c \in \mathbb{R}, x_i \in \mathbb{R}^n, N \geq 1\}$$

$$L_n^b(\mathbb{R}^n) := \{f(x) \in L_n^c(\mathbb{R}^n) \mid \mathcal{S}(\tilde{f})(x,y) = 0 \quad \text{for all } x \in \mathbb{R}^n \quad \text{and} \quad y = 0\},$$

where $\tilde{f}(x,y) = \mathcal{F}(f(x), [y])$.

Example 2.2.20. Define a new notation $[x]^\varepsilon$ for $x \in \mathbb{R}$ and $\varepsilon \in \{0, 1\}$ by

$$[x]^\varepsilon = \begin{cases} \frac{1}{2}\{x + |x|\} & (\varepsilon = 1) \\ x & (\varepsilon = 0) \end{cases}$$

Assume that $a_i, b_i, c_i, d_i, e_i, f_i, g_i$, and h_i belong to $\mathbb{R}$, and $\varepsilon, \varepsilon_i, \varepsilon_{ij}$ belong to $\{0, 1\}$.

(1) $L_1^a(\mathbb{R})$ consists of all expressions with following form:

$$a_0 x + b_0 [x]^\varepsilon \quad \text{for} \quad x \in \mathbb{R}.$$

$L_1^c(\mathbb{R})$ consists of all expressions with following form:

$$a_0 + \sum_{i=1}^{N} b_i [x + c_i]^{\varepsilon_i} \quad \text{for} \quad x \in \mathbb{R}.$$

Clearly

$$L_1(\mathbb{R}) = L_1^c(\mathbb{R}).$$

(2) $L_2^a(\mathbb{R}^2)$ consists of all expressions with following form:

$$\sum_{i=1}^{N} \{a_i [y]^{\varepsilon_{i2}} + b_i [x + c_i [y]^{\varepsilon_{i2}}]^{\varepsilon_{i1}}\} \quad \text{for} \quad (x, y) \in \mathbb{R}^2.$$

$L_2^c(\mathbb{R}^2)$ consists of all expressions with following form:

$$\sum_{i=1}^{N} \{a_i [y + e_i]^{\varepsilon_{i2}} + b_i [x + d_i + c_i [y + e_i]^{\varepsilon_{i2}}]^{\varepsilon_{i1}}\} \quad \text{for} \quad (x, y) \in \mathbb{R}^2.$$

(3) $L_3^a(\mathbb{R}^3)$ consists of all expressions with following form:

$$\sum_{i=1}^{N} \{a_i [y + e_i [z]^{\varepsilon_{i3}}]^{\varepsilon_{i2}} + b_i [x + d_i [z]^{\varepsilon_{i3}} + c_i [y + e_i [z]^{\varepsilon_{i3}}]^{\varepsilon_{i2}}]^{\varepsilon_{i1}}\}$$

$$\text{for} \quad (x, y, z) \in \mathbb{R}^3.$$

$L_3^c(\mathbb{R}^n)$ consists of all expressions with following form:

$$\sum_{i=1}^{N} \{a_i [y + g_i + e_i [z + h_i]^{\varepsilon_{i3}}]^{\varepsilon_{i2}}$$

$$+ b_i [x + f_i + d_i [z + h_i]^{\varepsilon_{i3}} + c_i [y + g_i + e_i [z + h_i]^{\varepsilon_{i3}}]^{\varepsilon_{i2}}]^{\varepsilon_{i1}}\}$$

$$\text{for} \quad (x, y, z) \in \mathbb{R}^3.$$

Our goal in this subsection is to prove that any CPL function $f(x)$ of $\mathbb{R}^n$ is represented by an element of $L_n^c(\mathbb{R}^n)$ (Theorem 2.2.28). To prove Theorem 2.2.28, we will prepare some lemmas.

Lemma 2.2.21. *If $f_1(x), f_2(x) \in L_n^c(\mathbb{R}^n)$ and $c \in \mathbb{R}^n$, then*

(1) $f_1(x+c) \in L_n^c(\mathbb{R}^n)$,

(2) $f_1(x) + f_2(x) \in L_n^c(\mathbb{R}^n)$.

Proof. Proof immediately follows from definition of $L_n^c(\mathbb{R}^n)$. □

Proposition 2.2.22. *Let $f(x) \in L_k(\mathbb{R}^n)$ be given, and define $\hat{f}(x,y)$ by multiplying $y \in \mathbb{R}$ by all constant terms of $f(x)$, that is*

$$\hat{f}(x,y) = \mathcal{F}(f(x), y).$$

Define $f^-(x) \in L_k(\mathbb{R}^n)$ by

$$f^-(x) = \hat{f}(x,-1) \quad \textit{for} \quad x \in \mathbb{R}^n.$$

Then we have, for any $x \in \mathbb{R}^n$ and any $y \neq 0$,

$$\hat{f}(x,y) \quad = \quad \begin{cases} yf(\frac{x}{y}) & (y>0) \\ -yf^-(-\frac{x}{y}) & (y<0) \end{cases}$$

Moreover, if we consider $\hat{f}(x,y)$ as a function of $(x,y) \in \mathbb{R}^{n+1}$, $\hat{f}(x,y)$ is a CPL function all boundaries of which pass through the origin.

Proof. To prove the first statement, we use an induction with respect to $k \geq 0$.

(i) For $f(x) = \langle a, x\rangle + b \in L_0(\mathbb{R}^n)$, if $y > 0$, we have

$$\hat{f}(x,y) = \langle a, x\rangle + by = y(\langle a, \frac{x}{y}\rangle + b) = yf(\frac{x}{y}),$$

and if $y < 0$,

$$\hat{f}(x,y) = \langle a, x\rangle + by = -y(\langle a, -\frac{x}{y}\rangle - b) = -yf^-(-\frac{x}{y}).$$

(ii) Assume for a $k \geq 0$, the statement holds. Then, for any

$$f(x) = f_0(x) + \sum_{i=1}^{N} \varepsilon_i |f_i(x)| \in L_{k+1}(\mathbb{R}^n),$$

$$f_i(x) \in L_k(\mathbb{R}^n) \quad (0 \leq i \leq N), \quad \varepsilon_i \in \{1,-1\},$$

if $y > 0$, we have

$$\begin{aligned}\hat{f}(x,y) &= \hat{f}_0(x,y)+\sum_{i=1}^{N}\varepsilon_i|\hat{f}_i(x,y)| \\ &= yf_0(\frac{x}{y})+\sum_{i=1}^{N}\varepsilon_i|yf_i(\frac{x}{y})| \\ &= y\{f_0(\frac{x}{y})+\sum_{i=1}^{N}\varepsilon_i|f_i(\frac{x}{y})|\} \\ &= yf(\frac{x}{y}),\end{aligned}$$

and if $y<0$,

$$\begin{aligned}\hat{f}(x,y) &= \hat{f}_0(x,y)+\sum_{i=1}^{N}\varepsilon_i|\hat{f}_i(x,y)| \\ &= -yf_0^-(-\frac{x}{y})+\sum_{i=1}^{N}\varepsilon_i|-yf_i^-(-\frac{x}{y})| \\ &= -y\{f_0^-(-\frac{x}{y})+\sum_{i=1}^{N}\varepsilon_i|f_i^-(-\frac{x}{y})|\} \\ &= -yf^-(-\frac{x}{y}).\end{aligned}$$

Therefore, for all $k\geq 0$, the statement holds.

To prove the second statement, let

$$U_i=\{x\in \mathbb{R}^n \mid \langle\alpha_i,x\rangle-\beta_i=0\},\quad (i=1,\cdots,m)$$

be boundaries of $f(x)$, and let $R_j\subset\mathbb{R}^n$ $(j=1,\cdots,l)$ be regions of $f(x)$. Also let

$$U_i^-=\{x\in \mathbb{R}^n \mid \langle\alpha_i^-,x\rangle-\beta_i^-=0\},\quad (i=1,\cdots,m)$$

be boundaries of $f^-(x)$, and let $R_j^-\subset\mathbb{R}^n$ $(j=1,\cdots,l)$ be regions of $f^-(x)$.

Then all boundaries of $\hat{f}(x,y)$ are given by

$$V_i=\{(x,y)\in\mathbb{R}^{n+1}\mid\langle\alpha_i,x\rangle-\beta_i y=0\},$$
$$V_i^-=\{(x,y)\in\mathbb{R}^{n+1}\mid\langle\alpha_i^-,x\rangle-\beta_i^-(-y)=0\},\quad (i=1,\cdots,m)$$

and

$$V_0=\{(x,y)\in\mathbb{R}^{n+1}\mid y=0\}.$$

To prove this, assume $(x_0, y_0) \in \mathbb{R}^{n+1}$ does not belong to $V_0 \cup \bigcup_{i=1}^{m}(V_i \cap V_i^-)$. It is enough to show that there is a neighborhood on which the derivative $D\hat{f}(x,y)$ is constant.

If $y > 0$, there exists R_j such that $\frac{x_0}{y_0} \in R_j$. Since $f(x)$ is piecewise-linear, there are a $1 \times n$ matrix A_j and $q_j \in \mathbb{R}$ such that $f(x) = A_j x + q_j$ for all $x \in R_j$. Define

$$S_j = \{(x,y) \in \mathbb{R}^{n+1} \mid \frac{x}{y} \in R_j, \quad y > 0\},$$

which is a neighborhood of (x_0, y_0). Since $\hat{f}(x,y) = yf(\frac{x}{y})$ for $y > 0$,

$$\begin{aligned} D_x\hat{f}(x,y) &= Df(\frac{x}{y}) = A_j, \\ D_y\hat{f}(x,y) &= f(\frac{x}{y}) - Df(\frac{x}{y})\frac{x}{y} = q_j. \end{aligned}$$

Hence

$$D\hat{f}(x,y) = [A_j, q_j] \quad \text{for all } (x,y) \in S_j.$$

If $y < 0$, there exists R_j^- such that $-\frac{x_0}{y_0} \in R_j^-$. Since $f^-(x)$ is also piecewise-linear, there are a $1 \times n$ matrix A_j^- and $q_j^- \in \mathbb{R}$ such that $f^-(x) = A_j^- x + q_j^-$ for all $x \in R_j^-$. Define

$$S_j^- = \{(x,y) \in \mathbb{R}^{n+1} \mid -\frac{x}{y} \in R_j^-, \quad y < 0\},$$

which is a neighborhood of (x_0, y_0). Since $\hat{f}(x,y) = -yf^-(-\frac{x}{y})$ for $y < 0$,

$$\begin{aligned} D_x\hat{f}(x,y) &= Df^-(-\frac{x}{y}) = A_j^-, \\ D_y\hat{f}(x,y) &= -f^-(-\frac{x}{y}) + Df^-(-\frac{x}{y})(-\frac{x}{y}) = -q_j^-. \end{aligned}$$

Hence

$$D\hat{f}(x,y) = [A_j, -q_j] \quad \text{for} \quad \text{all } (x,y) \in S_j^-.$$

Therefore, V_i, V_i^- $(i = 1, \cdots, m)$ and V_0 are all boundaries of $\hat{f}(x,y)$, which clearly pass through the origin. □

Proposition 2.2.23. *Let $f(x) \in L_k(\mathbb{R}^n)$ be given, and define $\bar{f}(x,y)$ by multiplying $-y \in \mathbb{R}$ by all constant terms of $f(x)$, that is*

$$\bar{f}(x,y) = \mathcal{F}(f(x), -y).$$

Then if $y < 0$, we have for any $x \in \mathbb{R}^n$,

$$\bar{f}(x,y) = -yf(-\frac{x}{y}).$$

Moreover, if we consider $\bar{f}(x,y)$ as a function of $(x,y) \in \mathbb{R}^{n+1}$, $\bar{f}(x,y)$ is a CPL function all boundaries of which pass through the origin.

Proof. Since $-y > 0$, the proof immediately follows from Proposition 2.2.22. □

Proposition 2.2.24. *Assume that for an $n \geq 1$, any CPL function of $\mathbb{R}^n$ has an expression in $L_n^c(\mathbb{R}^n)$. Let $f(x,y)$ be a CPL function of $\mathbb{R}^{n+1}$:*

$$f(x,y) \in \mathrm{CPL}(\mathbb{R}^{n+1}), \quad (x,y) \in \mathbb{R}^{n+1}$$

such that all boundaries of $f(x,y)$ pass through the origin O and $f(0,0) = 0$.

Since $f(x,-1)$ is a CPL function of $\mathbb{R}^n$, by the assumption there is its expression $f_-(x) \in L_n^c(\mathbb{R}^n)$. Define $\bar{f}_-(x,y)$ by multiplying $-y \in \mathbb{R}$ by all constant terms of $f_-(x)$:

$$\bar{f}_-(x,y) = \mathcal{F}(f_-(x), -y).$$

Since $g(x) = f(x,1) - \bar{f}_-(x,1)$ is a CPL function of $\mathbb{R}^n$, by the assumption there is its expression $g_1(x) \in L_n^c(\mathbb{R}^n)$. Define $\tilde{g}_1(x,y)$ by multiplying $[y] = \frac{1}{2}\{y + |y|\}$ by all constant terms of $g_1(x)$:

$$\tilde{g}_1(x,y) = \mathcal{F}(g_1(x), [y]).$$

Then we have

(1) $f(x,y) = \bar{f}_-(x,y)$ *for all* $x \in \mathbb{R}^n$ *and all* $y \leq 0$,

(2) $f(x,y) = \bar{f}_-(x,y) + \tilde{g}_1(x,y)$ *for all* $(x,y) \in \mathbb{R}^{n+1}$,

(3) $g_1(x) \in L_n^b(\mathbb{R}^n)$,

(4) $f(x,y)$ *has an expression* $\bar{f}_-(x,y) + \tilde{g}_1(x,y)$ *in* $L_n^a(\mathbb{R}^{n+1})$.

Proof. (1) Since $f(x,y)$ is a CPL function all boundaries of which pass through the origin, and since $f(0,0) = 0$, for any $c > 0$,

$$cf(x,y) = f(cx,cy)$$

holds. Hence, for $y < 0$, we have

$$\begin{aligned} f(x,y) &= -yf(-\frac{x}{y}, -1) \\ &= -yf_-(-\frac{x}{y}) = \bar{f}_-(x,y) \quad \text{(by Proposition 2.2.23)}. \end{aligned}$$

For $y = 0$, we have

$$\bar{f}_-(x,0) = \lim_{y \to 0-\varepsilon} \bar{f}_-(x,y) = \lim_{y \to 0-\varepsilon} f(x,y) = f(x,0).$$

(2) By Proposition 2.2.23, $\bar{f}_-(x,y)$ is a CPL function all boundaries of which pass through the origin, and $\bar{f}_-(0,0) = 0$. Since $[y] = y$ for $y > 0$, by Proposition 2.2.22, $\tilde{g}_1(x,y)$ is also a CPL function all boundaries of which pass through the origin, and $\tilde{g}_1(0,0) = 0$. Hence, if $y > 0$,

$$f(x,y) - \bar{f}_-(x,y) = yf(\frac{x}{y},1) - y\bar{f}_-(\frac{x}{y},1)$$

$$= y\{f(\frac{x}{y},1) - \bar{f}_-(\frac{x}{y},1)\}$$

$$= yg(\frac{x}{y}) = yg_1(\frac{x}{y}) = \tilde{g}_1(x,y).$$

If $y \leq 0$, since $[y] = 0$, by (1),

$$\tilde{g}_1(x,y) = \tilde{g}_1(x,0) = 0 = f(x,y) - \bar{f}_-(x,y).$$

Hence (2) holds.

(3) From (1) and (2), we have

$$\tilde{g}_1(x,0) = f(x,0) - \bar{f}_-(0,0) = 0,$$

that is, $g_1(x) \in L_n^b(\mathbb{R}^n)$.

(4) Since $f_-(x) \in L_n^c(\mathbb{R}^n)$ by assumption, and since $g_1(x) \in L_n^b(\mathbb{R}^n)$ by (3), $f(x,y)$ has an expression $\bar{f}_-(x,y) + \tilde{g}_1(x,y)$ in $L_n^a(\mathbb{R}^{n+1})$. □

Definition 2.2.25. Let $f(x) \in \mathrm{CPL}(\mathbb{R}^n)$, and U be a boundary of $f(x)$, and $p \in U$. By $\{U_{i_1}, \cdots, U_{i_r}\}$ denote the set of all boundaries of $f(x)$ which pass through the point p. Then a function $g(x) \in \mathrm{CPL}(\mathbb{R}^n)$ is a *germ* of $f(x)$ at the point p, if $g(x)$ has only boundaries $\{U_{i_1}, \cdots, U_{i_r}\}$, and $g(x)$ coincides $f(x)$ on a neighbourhood of p.

We denote by $\{V_1, \cdots, V_N\}$, the set of all $(n-2)$-dimensional planes on U which are intersection sets of U and other boundaries:

$$V_i = U \cap U_j \neq \emptyset \quad \text{for some} \quad j$$

A point $x_0 \in V_i$ is a *redundant point* if $f(x)$ takes value 0 at all point of a neighbourhood of the point x_0, that is, the germ $g(x)$ of $f(x)$ at x_0 is 0 function. Denote all connected components of the set $\{x \in V_i | x \quad \text{is not redundant point}\}$ by

$$\{V_{i1}, \cdots, V_{ik_i}\}.$$

Then the set $\{V_{ij} \mid 1 \leq j \leq k_i, \quad 1 \leq i \leq N\}$ is called $(n-2)$-*dimensional intersection set* on U.

We denote by $\{W_1, \cdots, W_M\}$, the set of all $(n-3)$-dimensional planes on U which are intersection of two planes in $\{V_1, \cdots, V_N\}$:

$$W_i = V_j \cap V_k \quad \text{for some} \quad j,k \quad (j \neq k).$$

Denote all connected components of the set

$$\{x \in W_i \mid x \quad \text{is not redundant point}\}$$

by

$$\{W_{i1}, \cdots, W_{ik_i}\}.$$

Then the set $\{W_{ij} \mid 1 \leq j \leq k_i, \quad 1 \leq i \leq N\}$ is called $(n-3)$-*dimensional intersection set* on U.

In this way, we can inductively define k-*dimensional intersection set* on U for $k = n-2, n-3, \cdots, 1, 0$.

Proposition 2.2.26. *Assume that any element of* $\mathrm{CPL}(\mathbb{R}^n)$ *has an expression in* $L_n^c(\mathbb{R}^n)$ *for some* $n \geq 1$. *Let* $f(x,y) \in \mathrm{CPL}(\mathbb{R}^{n+1})$, U *be a boundary of* $f(x,y)$, $p \in U$, *and* $g(x,y)$ *be the germ of* $f(x,y)$ *at* p, *where* $(x,y) \in \mathbb{R}^n \times \mathbb{R}$. *Then* $g(x,y)$ *has an expression in* $L_{n+1}^c(\mathbb{R}^{n+1})$.

Proof. Let $H : \mathbb{R}^{n+1} \to \mathbb{R}^{n+1}$ be a parallel translation such that

$$H(O) = p,$$

where O is the origin. Then $h(x,y) = g(H(x,y))$ is a CPL function all boundaries of which pass through the origin. By Proposition 2.2.24, $h(x,y) - h(0,0) \in \mathrm{CPL}(\mathbb{R}^{n+1})$ has an expression in $L_{n+1}^a(\mathbb{R}^{n+1}) \subset L_{n+1}^c(\mathbb{R}^{n+1})$. Since $L_{n+1}^c(\mathbb{R}^{n+1})$ is invariant under parallel translations by Lemma 2.2.21, $g(x,y)$ has also an expression in $L_{n+1}^c(\mathbb{R}^{n+1})$. □

Proposition 2.2.27. *Any CPL function of* $\mathbb{R}$, $f(x) \in \mathrm{CPL}(\mathbb{R})$, *has an expression in* $L_1^c(\mathbb{R})$.

Proof. Denote boundaries of $f(x)$ in order by

$$x_1 < x_2 < \cdots < x_l.$$

Define $m_i \in \mathbb{R} \quad (0 \leq i \leq l)$ by

$$Df(x) = \begin{cases} m_0 & (x < x_1) \\ m_i & (x_i < x < x_{i+1}, \quad 1 \leq i < l) \\ m_l & (x_l < x). \end{cases}$$

Then

$$f_1(x) = f(x_1) + m_0(x - x_1) + \frac{m_1 - m_0}{2}\{x - x_1 + |x - x_1|\} +$$

$$\frac{m_2 - m_1}{2}\{x - x_2 + |x - x_2|\} +$$

$$\cdots + \frac{m_l - m_{l-1}}{2}\{x - x_l + |x - x_l|\}$$

$$= f(x_1) + m_0(x - x_1) + \sum_{i=1}^{l} \frac{m_i - m_{i-1}}{2}\{x - x_i + |x - x_i|\}$$

is an expression in $L_1^c(\mathbb{R})$ for $f(x)$. □

Theorem 2.2.28. (Standard Forms of CPL Functions) *An arbitrary* $f(x) \in \mathrm{CPL}(\mathbb{R}^n)$ *has an expression in* $L_n^c(\mathbb{R}^n)$.

Proof. Induction with respect to $n \geq 1$.

When $n = 1$, the statement is valid by Proposition 2.2.27. Assume that the statement is valid for $n - 1 \geq 1$. Then we use an induction with respect to $l \geq 1$, where l is the number of boundaries of $f(x)$.

(1) When $l = 1$, since $f(x)$ is a two-region CPL function, $f(x)$ has an expression in $L_1(\mathbb{R}^n) \subset L_n^c(\mathbb{R}^n)$ as shown in Theorem 2.2.4.

(2) Assume that the statement is valid when $l - 1 \geq 1$, and assume $f(x)$ has l boundaries:

$$U_1, \cdots, U_l.$$

Denote the zero-dimensional intersection set on U_l by $\{p_1, \cdots, p_N\}$. By Proposition 2.2.26, the germ of $f(x)$ at p_N has an expression $g_N(x) \in L_n^c(\mathbb{R}^n)$. If we define $f_N(x)$ by $f_N(x) = f(x) - g_N(x)$, since $f_N(x)$ takes value 0 at a neighbourhood of the point p_N, p_N does not belong to the zero-dimensional intersection set on U_l of $f_N(x)$. Hence the zero-dimensional intersection set on U_l of $f_N(x)$ consists of less than or equal to $N - 1$ points. Rewrite this by $\{p_1, \cdots, p_{N'}\}$ $(N' \leq N - 1)$. Take the germ $g_{N'}(x) \in L_n^c(\mathbb{R}^n)$ of $f_N(x)$ at the point $p_{N'}$, and define $f_{N'}(x) = f_N(x) - g_{N'}(x)$. Then the zero-dimensional intersection set on U_l of $F_{N'}(x)$ consists of less than or equal to $N' - 1$ points. Repeating this operation, the zero-dimensional intersection set on U_l becomes empty. Then the one-dimensional intersection set on U_l has no intersection between each other. If we subtract a germ at a point of one-dimensional line in the one-dimensional intersection set, and if we repeat the operation as above, we can remove the one-dimensional intersection set on U_l. Then the two-dimensional intersection set has no intersection between each other. If we subtract a germ at a point of two-dimensional plane in the two-dimensional intersection set, and if we repeat the operation as above, we can remove the two-dimensional intersection set on U_l.

Repeating this operation, we can remove the $(n-1)$-dimensional intersection set. Finally, by subtracting a germ at a point of U_l, we can remove the boundary U_l. Since each germ subtracted to remove intersection sets belongs to $L_n^c(\mathbb{R}^n)$ by Proposition 2.2.26, there is an $h(x) \in L_n^c(\mathbb{R}^n)$ such that $f(x) - h(x)$ has $l - 1$ boundaries. By our assumption, since $f(x) - h(x)$ has an expression in $L_n^c(\mathbb{R}^n)$, $f(x)$ has also an expression in $L_n^c(\mathbb{R}^n)$. □

2.2.4 Examples of CPL functions

Recall the notation $[\cdot]$ defined by

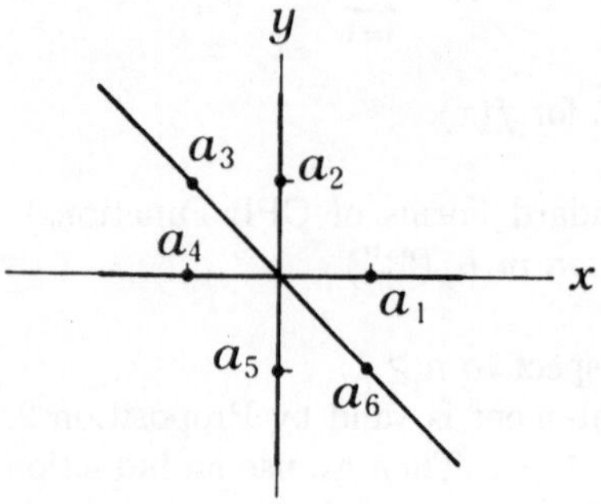

Fig. 2.2.1. $F(x, y; a), \quad a = (a_1, \cdots, a_6)$.

$$[x] = \frac{1}{2}\{x + |x|\}.$$

Example 2.2.29.

We will derive a standard form of two dimensional CPL function $f(x, y)$ having three boundaries

$$V_1 : x = 0, \quad V_2 : y = 0, \quad V_3 : x + y = 0$$

as in Fig.2.2.1. Set

$$f(0,0) = 0, \quad f(1,0) = a_1, \quad f(0,1) = a_2, \quad f(-1,1) = a_3,$$
$$f(-1,0) = a_4, \quad f(0,-1) = a_5, \quad f(1,-1) = a_6.$$

Since the section $f(x, -1)$ of $f(x, y)$ at $y = -1$ is a one dimensional CPL function, it has an expression $f_-(x)$ in $L_1^c(\mathbb{R})$:

$$f_-(x) = -a_4 x + a_5 + (a_6 - a_5 + a_4)[x] + (a_1 - a_6 + a_5)[x - 1].$$

Define $\bar{f}_-(x, y)$ by multiplying $-y$ by all constant terms of $f_-(x)$:

$$\begin{aligned} \bar{f}_-(x, y) &= \mathcal{F}(f_-(x), -y) \\ &= -a_4 x - a_5 y + (a_6 - a_5 + a_4)[x] + (a_1 - a_6 + a_5)[x + y]. \end{aligned}$$

The section $f(x, 1)$ of $f(x, y)$ at $y = 1$ has an expression

$$f_+(x) = -a_4(x + 1) + a_3 + (a_2 - a_3 + a_4)[x + 1] + (a_1 - a_2 + a_3)[x],$$

which belongs to $L_1^c(\mathbb{R})$. Hence $g(x) = f(x, 1) - \bar{f}_-(x, 1)$ has an expression in $L_1^c(\mathbb{R})$:

$$
\begin{aligned}
g_1(x) &= f_+(x) - \bar{f}_-(x,1) \\
&= a_3 - a_4 + a_5 + (a_1 - a_2 + a_3 - a_4 + a_5 - a_6)([x] - [x+1])
\end{aligned}
$$

Since

$$
\begin{aligned}
\tilde{g}_1(x,y) &= \mathcal{F}(g_1(x),[y]) \\
&= (a_3 - a_4 + a_5)[y] + (a_1 - a_2 + a_3 - a_4 + a_5 - a_6)([x] - [x + [y]]),
\end{aligned}
$$

and

$$
\tilde{g}_1(x,0) = (a_1 - a_2 + a_3 - a_4 + a_5 - a_6)([x] - [x]) = 0 \quad \forall x \in \mathbb{R},
$$

$g_1(x)$ belongs to $L_1^b(\mathbb{R})$. The function $f(x,y)$ has an expression in $L_2^a(\mathbb{R}^2)$:

$$
\begin{aligned}
f(x,y) &= \bar{f}_-(x,y) + \tilde{g}_1(x,y) \\
&= -a_4 x - a_5 y + (a_1 - a_2 + a_3)[x] + (a_3 - a_4 + a_5)[y] + (a_1 + a_5 - a_6)[x+y] \\
&\quad +(-a_1 + a_2 - a_3 + a_4 - a_5 + a_6)[x + [y]].
\end{aligned}
$$

Denote this expression by

$$
F(x,y;a), \quad a = (a_1, \cdots, a_6).
$$

Example 2.2.30. We will derive a standard form of two-dimensional CPL function $f(x,y)$ having five boundaries

$$
\begin{aligned}
&V_1 : x = -1, \quad V_2 : y = 1, \quad V_3 : x + y = 0 \\
&V_4 : x = 1, \quad V_5 : y = -1,
\end{aligned}
$$

as in Fig.2.2.2. Set

$$
\begin{aligned}
&f(-2,2) = c_1, \quad f(-1,2) = c_2, \quad f(-2,1) = c_3, \quad f(-1,1) = c_4, \\
&f(0,1) = c_5, \quad f(-1,0) = c_6, \quad f(0,0) = c_7, \quad f(2,-1) = c_8, \\
&f(1,-2) = c_9, \quad f(2,-2) = c_{10}.
\end{aligned}
$$

Using the expression $F(x,y;a)$ in Example 2.2.29, a germ of $f(x,y)$ at a point $p_1 = (-1,1)$ has an expression

$$
g_1(x,y) = F(x+1, y-1; a) + f(-1,1), \quad a = (a_1, \cdots, a_6)
$$

where

$$
\begin{aligned}
&a_1 = c_5 - c_4, \quad a_2 = c_2 - c_4, \quad a_3 = c_1 - c_4, \\
&a_4 = c_3 - c_4, \quad a_5 = c_6 - c_4, \quad a_6 = c_7 - c_4.
\end{aligned}
$$

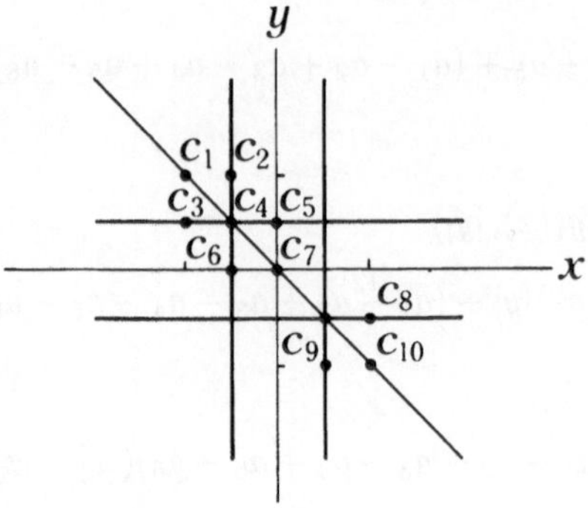

Fig. 2.2.2. $F_2(x, y; c), \quad c = (c_1, \cdots, c_{10})$.

The germ of $h(x, y) = f(x, y) - g_1(x, y)$ at a point $p_2 = (1, -1)$ has an expression $h_1(x, y) \in L_2^c(\mathbb{R}^2)$:

$$h_1(x, y) = F(x - 1, y + 1; b), \quad b = (b_1, \cdots, b_6),$$

where

$$b_1 = h(2, -1) = c_8 - 2c_7 - c_5 + 2c_4, \quad b_2 = h(1, 0) = 0, \quad b_3 = h(0, 0) = 0,$$

$$b_4 = h(0, 0) = 0, \quad b_5 = h(1, -2) = c_9 - 2c_7 - c_6 + 2c_4,$$

$$b_6 = h(2, -2) = c_{10} - 3c_7 + 2c_4.$$

By Proposition 2.2.24, $f(x, y)$ has an expression in $L_2^c(\mathbb{R}^2)$ as follows:

$$\begin{aligned}
f(x, y) &= g_1(x, y) + h_1(x, y) \\
&= F(x + 1, y - 1; a) + F(x - 1, y + 1; b) + f(-1, 1) \\
&= (c_4 - c_3)x + (-c_9 + 2c_7 - c_4)y + (c_5 - c_4 - c_2 + c_1)[x + 1] \\
&\quad +(c_8 - 2c_7 - c_5 + 2c_4)[x - 1] + (c_9 - 2c_7 - c_6 + 2c_4)[y + 1] \\
&\quad +(c_6 - c_4 - c_3 + c_1)[y - 1] + (-c_{10} + c_9 + c_8 - 2c_7 + c_4)[x + y] \\
&\quad +(c_7 - c_6 - c_5 + c_3 + c_2 - c_1)[x + 1 + [y - 1]] \\
&\quad +(c_{10} - c_9 - c_8 + c_7 + c_6 + c_5 - 2c_4)[x - 1 + [y + 1]] \\
&\quad +(-c_9 + 2c_7 + 2c_6 - c_4 - c_3).
\end{aligned}$$

Denote this expression by

$$F_2(x, y; c), \quad c = (c_1, \cdots, c_{10}).$$

Example 2.2.31. We will derive a condition under which the expression $F_2(x, y; c)$ in Example 2.2.30 becomes an element of $L_2^b(\mathbb{R}^2)$. For $f(x, y) = F_2(x, y; c)$, we have

$$\begin{aligned}
&\tilde{f}(x, y, 0) = \mathcal{F}(f(x, y), 0) \\
&= (c_4 - c_3)x + (-c_9 + 2c_7 - c_4)y + (c_8 - 2c_7 + c_4 - c_2 + c_1)[x] \\
&\quad +(c_9 - 2c_7 + c_4 - c_3 + c_1)[y] + (-c_{10} + c_9 + c_8 - 2c_7 + c_4)[x + y] \\
&\quad +(c_{10} - c_9 - c_8 + 2c_7 - 2c_4 + c_3 + c_2 - c_1)[x + [y]].
\end{aligned}$$

Since $\tilde{f}(x, y, 0) = 0$ for all $(x, y) \in \mathbb{R}^2$, we have

$$c_1 = c_2 = c_3 = c_4, \quad c_8 = c_9 = c_{10} = 2c_7 - c_1.$$

Hence if $f(x, y)$ belongs to $L_2^b(\mathbb{R}^2)$, it should be written by

$$\begin{aligned}
f(x, y) &= (c_6 - c_1)([y - 1] - [y + 1] + 2) + (c_5 - c_1)([x + 1] - [x - 1]) \\
&\quad +(-c_7 + c_6 + c_5 - c_1)([x - 1 + [y + 1]] - [x + 1 + [y - 1]]),
\end{aligned}$$

where

$$\begin{aligned}
&c_1 = f(-2, 2) = f(-1, 2) = f(-2, 1) = f(-1, 1), \\
&f(0, 1) = c_5, \quad f(-1, 0) = c_6, \quad f(0, 0) = c_7.
\end{aligned}$$

Example 2.2.32. We will derive a standard form of three-dimensional CPL function $f(x, y, z)$ having six boundaries:

$$\begin{aligned}
&V_1 : z = 0, \quad V_2 : y + z = 0, \quad V_3 : y - z = 0 \\
&V_4 : x + z = 0, \quad V_5 : x - z = 0, \quad V_6 : x + y = 0
\end{aligned}$$

as in Fig.2.2.3. Set

$$\begin{aligned}
&c_1 = f(-2, 2, -1), \quad c_2 = f(-1, 2, -1), \quad c_3 = f(-2, 1, -1), \\
&c_4 = f(-1, 1, -1), \quad c_5 = f(0, 1, -1), \quad c_6 = f(-1, 0, -1), \\
&c_7 = f(0, 0, -1), \quad c_8 = f(2, -1, -1), \quad c_9 = f(1, -2, -1), \\
&c_{10} = f(2, -2, -1), \quad c_{11} = f(-2, 2, 1), \quad c_{12} = f(0, 1, 1), \\
&c_{13} = f(-1, 0, 1), \quad c_{14} = f(0, 0, 1), \quad f(0, 0, 0) = 0.
\end{aligned}$$

The section $f(x, y, -1)$ of $f(x, y, z)$ at $z = -1$ has an expression $f_-(x, y)$ in $L_2^c(\mathbb{R}^2)$:

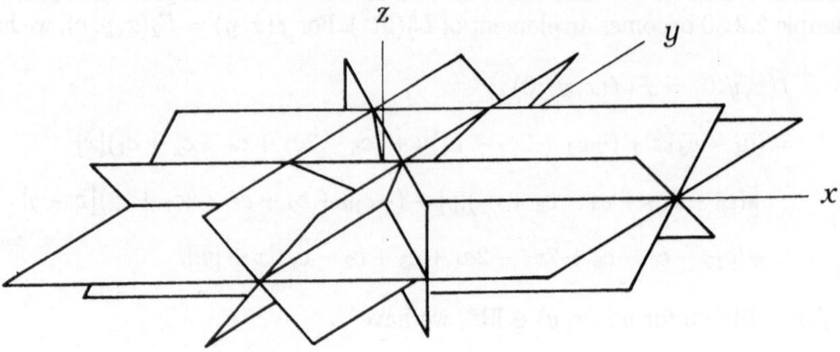

Fig. 2.2.3. Boundaries of $f(x, y, z)$

$$f_-(x,y) = F_2(x,y;c), \quad c = (c_1, \cdots, c_{10}),$$

where $F_2(x, y)$ is an expression derived in Example 2.2.30. Define $\bar{f}_-(x, y, z)$ by multiplying $-z$ by all constant terms of $f_-(x, y)$:

$$\begin{aligned}
&\bar{f}_-(x,y,z) = \mathcal{F}(f_-(x,y), -z) \\
&= (c_4 - c_3)x + (-c_9 + 2c_7 - c_4)y + (c_5 - c_4 - c_2 + c_1)[x - z] \\
&\quad +(c_8 - 2c_7 - c_5 + 2c_4)[x + z] + (c_9 - 2c_7 - c_6 + 2c_4)[y - z] \\
&\quad +(c_6 - c_4 - c_3 + c_1)[y + z] + (-c_{10} + c_9 + c_8 - 2c_7 + c_4)[x + y] \\
&\quad +(c_7 - c_6 - c_5 + c_3 + c_2 - c_1)[x - z + [y + z]] \\
&\quad +(c_{10} - c_9 - c_8 + c_7 + c_6 + c_5 - 2c_4)[x + z + [y - z]] \\
&\quad -(-c_9 + 2c_7 + 2c_6 - c_4 - c_3)z.
\end{aligned}$$

By Proposition 2.2.24, $g(x, y) = f(x, y, 1) - \bar{f}_-(x, y, 1)$ has an expression $g_1(x, y) \in L_2^b(\mathbb{R}^2)$, which is given as follows by Example 2.2.31. Set

$$\begin{aligned}
d_1 &= g(-2,2) = f(-2,2,1) - \bar{f}_-(-2,2,1) \\
&= c_{11} - 3(-c_4 + c_1) \\
d_5 &= g(0,1) = f(0,1,1) - \bar{f}_-(0,1,1) \\
&= c_{12} - (c_8 - 2c_7 - c_6 + c_4 + c_2 + c_1) \\
d_6 &= g(-1,0) = f(-1,0,1) - \bar{f}_-(-1,0,1) \\
&= c_{13} - (c_9 + c_8 - 4c_7 - 2c_6 + 2c_4 + c_3 + c_1)
\end{aligned}$$

$$\begin{aligned} d_7 &= g(0,0) = f(0,0,1) - \bar{f}_-(0,0,1) \\ &= c_{14} - (c_{10} - 3c_7 - c_6 + c_5 + c_4 + c_1). \end{aligned}$$

Then

$$\begin{aligned} &g_1(x,y) = (d_6 - d_1)([y-1] - [y+1] + 2) + (d_5 - d_1)([x+1] - [x-1]) \\ &\quad +(d_5 + d_6 - d_7 - d_1)([x - 1 + [y+1]] - [x + 1 + [y-1]]) \\ &= (c_{13} - c_{11} - c_9 - c_8 + 4c_7 + 2c_6 - 5c_4 - c_3 + 2c_1)([y-1] - [y+1] + 2) \\ &\quad +(c_{12} - c_{11} - c_8 + 2c_7 + c_6 - 4c_4 - c_2 + 2c_1)([x+1] - [x-1]) \\ &\quad +(c_{12} + c_{13} - c_{14} - c_{11} + c_{10} - c_9 - 2c_8 + 3c_7 + 2c_6 + c_5 - 2c_4 \\ &\quad -c_3 - c_2 + 2c_1) \times ([x - 1 + [y+1]] - [x + 1 + [y-1]]). \end{aligned}$$

Hence $f(x,y,z)$ has an expression in $L_3^c(\mathbb{R}^3)$ as follows:

$$\begin{aligned} &f(x,y,z) = \bar{f}_-(x,y,z) + \mathcal{F}(g_1(x,y), [z]) \\ &= (c_4 - c_3)x + (-c_9 + 2c_7 - c_4)y + (c_5 - c_4 - c_2 + c_1)[x - z] \\ &\quad +(c_8 - 2c_7 - c_5 + 2c_4)[x + z] + (c_9 - 2c_7 - c_6 + 2c_4)[y - z] \\ &\quad +(c_6 - c_4 - c_3 + c_1)[y + z] + (-c_{10} + c_9 + c_8 - 2c_7 + c_4)[x + y] \\ &\quad +(c_7 - c_6 - c_5 + c_3 + c_2 - c_1)[x - z + [y + z]] \\ &\quad +(c_{10} - c_9 - c_8 + c_7 + c_6 + c_5 - 2c_4)[x + z + [y - z]] \\ &\quad -(-c_9 + 2c_7 + 2c_6 - c_4 - c_3)z \\ &\quad +(c_{13} - c_{11} - c_9 - c_8 + 4c_7 + 2c_6 - 5c_4 - c_3 + 2c_1) \\ &\qquad \times([y - [z]] - [y + [z]] + 2[z]) \\ &\quad +(c_{12} - c_{11} - c_8 + 2c_7 + c_6 - 4c_4 - c_2 + 2c_1) \\ &\qquad \times([x + [z]] - [x - [z]]) \\ &\quad +(c_{12} + c_{13} - c_{14} - c_{11} + c_{10} - c_9 - 2c_8 + 3c_7 + 2c_6 + c_5 - 2c_4 \\ &\quad -c_3 - c_2 + 2c_1) \times ([x - [z] + [y + [z]]] - [x + [z] + [y - [z]]]). \end{aligned}$$

2.3 Normal Forms of Piecewise-Linear Vector Fields

In this section we will derive normal forms for affine conjugate classes of n-dimensional two-region systems, which are the most fundamental piecewise-linear vector fields. More general piecewise-linear systems are combinations of two-region systems, which are dealt with in the next section.

Define an n-dimensional two-region piecewise-linear vector field $X : \mathbb{R}^n \to \mathbb{R}^n$ by

$$X(x) = \begin{cases} Ax + q_A & \text{if} \quad \langle \alpha, x \rangle - \beta < 0 \\ Bx + q_B & \text{if} \quad \langle \alpha, x \rangle - \beta \geq 0. \end{cases}$$

where A and B are $n \times n$ real matrices, and $q_A, q_B \in \mathbb{R}^n$, $\alpha \in \mathbb{R}^n - \{0\}$ and $\beta \in \mathbb{R}$. Since X is determined by $(A, B, q_A, q_B, \alpha, \beta)$, it has $2n^2 + 3n + 1$ parameters. As proved in Theorem 2.2.4, the continuity condition of X restricts them to n^2+3n+1 parameters. Then, using n-dimensional parameter $p \in \mathbb{R}^n$, a continuous X is written by

$$X(x) = Ax + q_A + \frac{1}{2}p\{|\langle \alpha, x \rangle - \beta| + \langle \alpha, x \rangle - \beta\}.$$

By suitable linear transformation, we can change A into its real Jordan matrix J:

$$X(x) = Jx + q + \frac{1}{2}p\{|\langle \alpha, x \rangle - \beta| + \langle \alpha, x \rangle - \beta\}. \tag{2.3.1}$$

Our goal is to change q, p, α and β into the simplest form under suitable affine transformation which preserves J.

Before considering the normal forms of full two-region systems, we will study an affine vector field with a boundary, which is a half of a two-region system. We will present two cases. In the first case which is called the *non-degenerate* case, it is possible to change q into a zero vector by a suitable affine transformation. In the second case which is called the *degenerate* case, it is impossible to change the q into zero vector by any affine transformation. In Subsection 2.3.2, we will derive the normal forms of non-degenerate linear vector fields with a boundary. Namely, (α, β) will be changed into the simplest form (i.e., it has as many zero elements as possible, and non-zero elements take as many values 1 as possible) under an affine transformation which preserves J (Theorem 2.3.21). In Subsection 2.3.3, we will derive normal forms of degenerate affine vector fields with a boundary. Since we can use the simplest form of (α, β) derived in Subsection 2.3.2, our task is to change q into the simplest form under an affine transformation which preserves J and (α, β) (Theorem 2.3.32). The simplest α will be called a *canonical normal vector* for J (Definition 2.3.22), and the simplest q will be called a *canonical constant vector* for (J, α) (Definition 2.3.33). Also, we will show that β can take the value 0 or 1.

In Subsection 2.3.4, we will consider normal forms of full two-region systems. Combining Equation (2.3.1) and the results of Subsections 2.3.2 and 2.3.3, the remaining task will be to change p into the simplest form under an affine transformation which preserves (J, α) in a non-degenerate case (Theorem 2.3.41), or (J, α, q) in a degenerate case (Theorem 2.3.43). The simplest p will be called a *canonical parameter* for (J, α) or (J, α, q) corresponding to the non-degenerate case (Definition 2.3.40) or the degenerate case (Definition 2.3.42), respectively. Consequently, the normal form of continuous two-region piecewise-linear vector fields are determined by:

(1) a real Jordan matrix J,

(2) a canonical normal vector α for J,

(3) a scalar $\beta \in \{0, 1\}$,

(4) a canonical constant vector q for (J, α), and

(5) a canonical parameter p for (J, α) or (J, α, q).

In Subsection 2.3.5, we will derive normal form of proper two-region piecewise-linear vector fields (Theorem 2.3.50), which is special class of non-degenerate two-region systems. Since the proper condition (i.e., any invariant subspaces are not parallel to the boundary) is generic, the proper systems are an important class for studying bifurcations of piecewise-linear vector fields. Theorem 2.3.50 will show us that normal form of proper systems is determined by the values of elementary symmetric functions of eigenvalues in each region.

2.3.1 Notations

In this subsection, we prepare notations and some lemmas.

Definition 2.3.1. Define the following notations.

(1)

$$\mathbb{R} : \quad \text{the set of all real numbers.}$$

$$\mathbb{C} : \quad \text{the set of all complex numbers.}$$

$$K_d := \begin{cases} \mathbb{R} & \text{if} \quad d = 1 \\ \mathbb{C} & \text{if} \quad d = 2 \end{cases}$$

$$\mathbb{R}^n : \quad \text{the direct product of } \mathbb{R} \text{ by } n \text{ times.}$$

For $\omega \in \mathbb{C}$, $\bar{\omega}$ denotes the complex-conjugate of ω. Elements of $\mathbb{R}^n$ are *column vectors* unless otherwise stated.

(2)

$$\mathbb{N} := \{1, 2, 3, \cdots\} : \text{the set of all natural numbers.}$$

$$\mathbb{Z}_+ := \mathbb{N} \cup \{0\} : \text{the set of all non-negative integers.}$$

$$\mathbb{N}^r = \mathbb{N}(r) := \{(n_1, \cdots, n_r) \mid n_i \in \mathbb{N}, 1 \leq i \leq r\}$$

$$: \text{the direct product of } \mathbb{N} \text{ by } r \text{ times.}$$

$$\mathbb{N}^r_< = \mathbb{N}_<(r) := \{(n_1, \cdots, n_r) \in \mathbb{N}^r \mid n_1 < n_2 < \cdots < n_r\}$$

$$\mathbb{Z}^r_+ = \mathbb{Z}_+(r) := \{(n_1, \cdots, n_r) \mid n_i \in \mathbb{Z}_+, 1 \leq i \leq r\}$$

$$: \text{the direct product of } \mathbb{Z}_+ \text{ by } r \text{ times.}$$

Elements of $\mathbb{N}^r$ or $\mathbb{Z}^r_+$ are *row vectors* unless otherwise stated.

(3)

$$\begin{aligned}
M(k,l) &\quad : \text{the set of all } k \times l \text{ matrices},\\
M(k) &:= M(k,k), \quad GL(k,\mathbb{R}) := \{A \in M(k) \mid \det(A) \neq 0\},\\
M_1(k,l) &:= M(k,l), \quad M_2(k,l) := M(2k,2l),\\
M_1(k) &:= M(k), \quad M_2(k) := M(2k).
\end{aligned}$$

We will often identify $M(n,1) = M_1(n,1) = \mathbb{R}^n$.

(4) For $A_{ij} \in M(k_i, l_i), 1 \le i \le l$ and $1 \le j \le m$, denote

$$(A_{ij}) = \begin{pmatrix} A_{11} & \cdots & A_{1m} \\ \vdots & & \vdots \\ A_{l1} & \cdots & A_{lm} \end{pmatrix} \in M((\sum_{i=1}^{l} k_i), (\sum_{j=1}^{m} l_j)).$$

(5) For $A_i \in M(k_i), 1 \le i \le r$, denote

$$\begin{aligned}
\dot{\sum}_{i=1}^{r} A_i &= A_1 \dot{+} \cdots \dot{+} A_r\\
&= \begin{pmatrix} A_1 & & & 0 \\ & A_2 & & \\ & & \ddots & \\ 0 & & & A_r \end{pmatrix} \in M(\sum_{i=1}^{r} k_i).
\end{aligned}$$

(6) For $p_i \in M(k_i, 1), 1 \le i \le r$, denote

$$\dot{\prod}_{i=1}^{r} p_i = p_1 \dot{\times} \cdots \dot{\times} p_r = \begin{pmatrix} p_1 \\ \vdots \\ p_r \end{pmatrix} \in M((\sum_{i=1}^{r} k_i), 1).$$

(7) The zero matrix of $k \times l$ size is denoted by $O(k,l)$. The zero matrix of $k \times k$ size is denoted by $O(k)$. They are simply denoted by O if there is no confusion.

(8) The identity matrix of $k \times k$ size is denoted by $I(k)$. It is simply denoted by I if there is no confusion.

(9) For $A = (a_{ij}) \in M(k)$ and $l \in \mathbb{Z}_+$, define

$$[OA] := (b_{ij}) \in M(k, (k+l)),$$

where

$$b_{ij} = \begin{cases} 0 & \text{if} \quad j \leq l \\ a_{i,j-l} & \text{if} \quad j > l. \end{cases}$$

Also define

$$A\#O := (c_{ij}) \in M((k+l), k),$$

where

$$c_{ij} = \begin{cases} a_{ij} & \text{if} \quad i \leq k \\ 0 & \text{if} \quad i > k. \end{cases}$$

(10) For $\tilde{k} = (k_1, \cdots, k_r), \tilde{m} = (m_1, \cdots, m_r) \in \mathbb{N}^r$ and $d = 1$ or 2, define

$$\tilde{k} \cdot \tilde{m} := \sum_{i=1}^{r} k_i m_i, \quad d\tilde{k} \cdot \tilde{m} := d \sum_{i=1}^{r} k_i m_i.$$

Definition 2.3.2. Let $k, m \in \mathbb{N}, \tilde{k} = (k_1, \cdots, k_r), \tilde{m} = (m_1, \cdots, m_r) \in \mathbb{N}^r, \lambda \in \mathbb{R}$ and $\omega = a + \sqrt{-1}b \in \mathbb{C}$ be given. Then define the following notations.

(1)

$$J_1(\lambda; k) := \begin{pmatrix} \lambda & 1 & & \\ & \ddots & \ddots & \\ & & \ddots & 1 \\ \mathbf{0} & & & \lambda \end{pmatrix} \in M(k),$$

i.e., the $k \times k$ Jordan block with eigenvalue λ.

(2)

$$J_1(\lambda; k, m) := J_1(\lambda; k) \dot{+} \cdots \dot{+} J_1(\lambda; k),$$

the direct sum of $J_1(\lambda; k)$ by m times.

(3)

$$J_1(\lambda; \tilde{k}, \tilde{m}) := \dot{\sum}_{i=1}^{r} J_1(\lambda; k_i, m_i) \in M(\tilde{k} \cdot \tilde{m}).$$

(4)

$$J_2(\omega; k) := \begin{pmatrix} A & I & & \\ & \ddots & \ddots & \\ & & \ddots & I \\ \mathbf{0} & & & A \end{pmatrix} \in M(2k),$$

i.e., the $2k \times 2k$ Jordan block with the eigenvalue $a \pm \sqrt{-1}b$, where

$$A = \begin{pmatrix} a & -b \\ b & a \end{pmatrix}, \quad I = \begin{pmatrix} 1 & 0 \\ 0 & 1 \end{pmatrix}$$

(5)

$$J_2(\omega; k, m) := J_2(\omega; k) \dot{+} \cdots \dot{+} J_2(\omega; k),$$

the direct sum of $J_2(\omega; k)$ by m times.

(6)

$$J_2(\omega; \tilde{k}, \tilde{m}) := \dot{\sum}_{i=1}^{r} J_2(\omega; k_i, m_i) \in M(2\tilde{k} \cdot \tilde{m}).$$

(7)

$$R(1) := \mathbb{R}, R(2) := \{(a_{ij}) \in M(2) \mid a_{11} = a_{22}, -a_{12} = a_{21}\}.$$

Example 2.3.3.

$$J_1(\lambda; 2, 3) = \begin{pmatrix} \lambda & 1 & & & & \\ 0 & \lambda & & & & \\ & & \lambda & 1 & & \\ & & 0 & \lambda & & \\ & & & & \lambda & 1 \\ & & & & 0 & \lambda \end{pmatrix}$$

$$J_1(\lambda; (1, 2), (2, 2)) = \begin{pmatrix} \lambda & 0 & & & & \\ 0 & \lambda & & & & \\ & & \lambda & 1 & & \\ & & 0 & \lambda & & \\ & & & & \lambda & 1 \\ & & & & 0 & \lambda \end{pmatrix}$$

Definition 2.3.4. Let $k, m \in \mathbb{N}$, and

$$\tilde{k} = (k_1, \cdots, k_r),\ \tilde{m} = (m_1, \cdots, m_r) \in \mathbb{N}^r$$

be given. Then define the following notations.

(1) $C_1(k)$ is a subset of $M_1(k)$ defined by

$$(a_{ij}) \in C_1(k)\ (1 \le i, j \le k)$$

$$\Longleftrightarrow$$

$$a_{ij} \in \mathbb{R}\ (1 \le i, j \le k),\ a_{ij} = 0\ (i > j) \quad \text{and}$$
$$a_{ij} = a_{i+1, j+1}\ (1 \le i, j < k).$$

(2) $C_2(k)$ is a subset of $M_2(k)$ defined by

$$(A_{ij}) \in C_2(k)\ (1 \le i, j \le k)$$

$$\Longleftrightarrow$$

$$A_{ij} \in R(2)\ (1 \le i, j \le k),\ A_{ij} = O\ (i > j) \quad \text{and}$$

$$A_{ij} = A_{i+1,j+1}\ (1 \le i, j < k).$$

(3) $C_1(\tilde{k})$ (respectively $C_2(\tilde{k})$) is a subset of $M_1(\sum_{i=1}^{r} k_i)$ (respectively $M_2(\sum_{i=1}^{r} k_i)$) defined by

$$(A_{ij}) \in C_1(\tilde{k}) \quad (\text{respectively } (A_{ij}) \in C_2(\tilde{k}))\ (1 \le i, j \le r)$$

$$\Longleftrightarrow$$

$$A_{ij} \in M_1(k_i, k_j) \quad (\text{respectively } M_2(k_i, k_j)) \quad \text{and}$$

$$A_{ij} = [OB_{ij}] \quad \text{or} \quad B_{ij}\#O,$$

where $B_{ij} \in C_1(\min\{k_i, k_j\})$ (respectively $C_2(\min\{k_i, k_j\})$).

(4) Set $\tilde{h} = (k, \cdots, k) \in \mathbb{N}^m$, then define

$$C_1(k, m) := C_1(\tilde{h}), \quad C_2(k, m) := C_2(\tilde{h}).$$

(5) Set $\tilde{h} = (\tilde{h}_1, \cdots, \tilde{h}_r)$, where $\tilde{h}_i = (k_i, \cdots, k_i) \in \mathbb{N}(m_i)(1 \le i \le r)$, then define

$$C_1(\tilde{k}, \tilde{m}) := C_1(\tilde{h}), \quad C_2(\tilde{k}, \tilde{m}) := C_2(\tilde{h}).$$

(6) Define nilpotent matrices as follows:

$$N_1(k) := J_1(0; k), N_2(k) := J_2(0; k),$$

$$N_1(k; j) := \begin{cases} I(k) & (\text{if j=0}), \\ N_1(k)^j & (\text{if } 1 \le j < k), \\ O(k) & (\text{if } j > k). \end{cases}$$

$$N_2(k; j) := \begin{cases} I(2k) & (\text{if j=0}), \\ N_2(k)^j & (\text{if } 1 \le j < k), \\ O(2k) & (\text{if } j > k). \end{cases}$$

(7) For $a_i, c_{ij} \in \mathbb{R}(1 \le i, j \le n)$, define

$$\Delta_1(a_1, \cdots, a_n) = (c_{ij}) \in C_1(n)$$

$$\Longleftrightarrow$$

$$c_{1j} = a_j (1 \le j \le n),$$

i.e.,

$$\Delta_1(a_1, \cdots, a_n) = \begin{pmatrix} a_1 & a_2 & \cdots & \cdots & a_n \\ & a_1 & a_2 & & \vdots \\ & & \ddots & \ddots & \\ & & & \ddots & a_2 \\ 0 & & & & a_1 \end{pmatrix}.$$

(8) For $\tilde{a}_i, \tilde{c}_{ij} \in R(2)(1 \leq i, j \leq n)$, define

$$\Delta_2(\tilde{a}_1, \cdots, \tilde{a}_n) = (\tilde{c}_{ij}) \in C_2(n)$$

$$\Longleftrightarrow$$

$$\tilde{c}_{1j} = \tilde{a}_j (1 \leq j \leq n),$$

i.e.,

$$\Delta_2(\tilde{a}_1, \cdots, \tilde{a}_n) = \begin{pmatrix} \tilde{a}_1 & \tilde{a}_2 & \cdots & \cdots & \tilde{a}_n \\ & \tilde{a}_1 & \tilde{a}_2 & & \vdots \\ & & \ddots & \ddots & \\ & & & \ddots & \tilde{a}_2 \\ & & & & \tilde{a}_1 \end{pmatrix}.$$

Example 2.3.5.

(1)

$$\begin{pmatrix} a & b & c & d \\ 0 & a & b & c \\ 0 & 0 & a & b \\ 0 & 0 & 0 & a \end{pmatrix} \quad \text{an element of } C_1(4)$$

(2)

$$\left(\begin{array}{c|cc|ccc} a & 0 & b & 0 & 0 & c \\ \hline d & e & f & 0 & g & h \\ 0 & 0 & e & 0 & 0 & g \\ \hline k & l & m & n & p & q \\ 0 & 0 & l & 0 & n & p \\ 0 & 0 & 0 & 0 & 0 & n \end{array}\right) \quad \text{an element of } C_1((1,2,3))$$

(3)

$$\left(\begin{array}{ccc|ccc} a & b & c & d & e & f \\ 0 & a & b & 0 & d & e \\ 0 & 0 & a & 0 & 0 & d \\ \hline g & h & k & m & n & p \\ 0 & g & h & 0 & m & n \\ 0 & 0 & g & 0 & 0 & m \end{array}\right) \quad \text{an element of } C_1(3,2)$$

(4)

$$\left(\begin{array}{c|c|cc|cc} a & b & 0 & c & 0 & d \\ \hline e & f & 0 & g & 0 & h \\ \hline k & l & m & n & p & q \\ 0 & 0 & 0 & m & 0 & p \\ \hline r & s & t & u & v & w \\ 0 & 0 & 0 & t & 0 & v \end{array}\right) \quad \text{an element of } C_1((1,2),(2,2)).$$

Definition 2.3.6. Let $n, k \in \mathbb{N}, u \in \mathbb{Z}_+$, $\tilde{k} = (k_1, \cdots, k_r) \in \mathbb{N}^r_<$, $\tilde{m} = (m_1, \cdots, m_r) \in \mathbb{N}^r$ be given.

(1)

$$e_1(n; u) := (a_1, \cdots, a_n)^T \in M(n, 1),$$

where

$$a_i = \begin{cases} 1 & (\text{if} \quad i = u) \\ 0 & (\text{if} \quad i \neq u). \end{cases}$$

In particular, $e_1(n; 0) = O(n, 1)$.

$$e_2(n; u) := (\tilde{a}_1, \cdots, \tilde{a}_n)^T \in M(2n, 1),$$

where

$$\tilde{a}_i = \begin{cases} \tilde{1} = (1, 0) & (\text{if} \quad i = u) \\ \tilde{0} = (0, 0) & (\text{if} \quad i \neq u). \end{cases}$$

In particular, $e_2(n; 0) = O(2n, 1)$.

(2) For $1 \leq s_* \leq m$ and $0 \leq u_* \leq k$,

$$e_1(k, m; (s_*, u_*)) := \dot{\prod}_{j=1}^{m} u_j, \quad u_j \in M(k, 1),$$

where

$$u_j = \begin{cases} e_1(k; u_*) & (\text{if} \quad j = s_*) \\ O(k, 1) & (\text{if} \quad j \neq s_*). \end{cases}$$

In particular, $e_1(k, m; (s_*, 0)) = O(km, 1)$.

$$e_2(k, m; (s_*, u_*)) := \dot{\prod}_{j=1}^{m} u_j, \quad u_j \in M(2k, 1),$$

where

$$u_j = \begin{cases} e_2(k; u_*) & (\text{if} \quad j = s_*) \\ O(2k, 1) & (\text{if} \quad j \neq s_*). \end{cases}$$

In particular, $e_2(k, m; (s_*, 0)) = O(2km, 1)$.

(3) For $1 \le i_* \le r, 1 \le s_* \le m_{i_*}$ and $0 \le u_* \le k_{i_*}$,

$$e_1(\tilde{k}, \tilde{m}; (i_*, s_*, u_*)) := \dot{\prod}_{i=1}^{r} \dot{\prod}_{j=1}^{m_i} u_{ij}, \quad u_{ij} \in M(k_i, 1),$$

where

$$u_{ij} = \begin{cases} e_1(k_*; u_*) & (\text{if} \quad i = i_*, j = s_*) \\ O(k_i, 1) & (\text{otherwise}). \end{cases}$$

In particular, $e_1(\tilde{k}, \tilde{m}; (i_*, s_*, 0)) = O(\tilde{k} \cdot \tilde{m}, 1)$.

$$e_2(\tilde{k}, \tilde{m}; (i_*, s_*, u_*)) := \dot{\prod}_{i=1}^{r} \dot{\prod}_{j=1}^{m_i} u_{ij}, \quad u_{ij} \in M(2k_i, 1),$$

where

$$u_{ij} = \begin{cases} e_2(k_*; u_*) & (\text{if} \quad i = i_*, j = s_*) \\ O(2k_i, 1) & (\text{otherwise}). \end{cases}$$

In particular, $e_2(\tilde{k}, \tilde{m}; (i_*, s_*, 0)) = O(2\tilde{k} \cdot \tilde{m}, 1)$.

(4) Let $p \in M(\tilde{k} \cdot \tilde{m}, 1)$ be given. If

$$p = (p(1), \cdots, p(r))^T, \quad p(i)^T \in M(k_i m_i, 1) \quad (1 \le i \le r),$$

then $p(i)^T$ is called *the i-block of p*. If

$$p(i) = (p(i,1), \cdots, p(i, m_i)), \quad p(i,s)^T \in M(k_i, 1)$$
$$(1 \le i \le r, 1 \le s \le m_i),$$

then $p(i,s)^T$ is called *the (i,s)-block of p*. If

$$p(i,s) = (p(i,s,1), \cdots, p(i,s,k_i)), \quad p(i,s,u) \in \mathbb{R}$$
$$(1 \le i \le r, 1 \le s \le m_i, 1 \le u \le k_i),$$

then $p(i,s,u)$ is called *the (i,s,u)-component of p*.

(5) Let $p \in M(2\tilde{k} \cdot \tilde{m}, 1)$ be given. If

$$p = (p(1), \cdots, p(r))^T, \quad p(i)^T \in M(2k_i m_i, 1) \quad (1 \le i \le r),$$

then $p(i)^T$ is called *the i-block of p*. If

$$p(i) = (p(i,1), \cdots, p(i, m_i)), \quad p(i,s)^T \in M(2k_i, 1)$$
$$(1 \le i \le r, 1 \le s \le m_i),$$

then $p(i,s)^T$ is called *the (i,s)-block of p*. If

$$p(i,s) = (p(i,s,1), \cdots, p(i,s,k_i)), \quad p(i,s,u)^T \in M(2,1)$$

$$(1 \le i \le r, 1 \le s \le m_i, 1 \le u \le k_i),$$

then $p(i,s,u)^T$ is called *the (i,s,u)-component of p*. If

$$p(i,s,u) = (a,b), \quad a,b \in \mathbb{R},$$

then a is called *the first (i,s,u)-component*, and b *the second (i,s,u)-component* of p.

(6) Let $A \in M_1(\tilde{k} \cdot \tilde{m})$ be given. If

$$A = (A_{ij}), \quad A_{ij} \in M_1(k_i m_i, k_j m_j) \quad (1 \le i,j \le r),$$

then A_{ij} is called *the (i,j)-block* of A. If

$$A_{ij} = (A_{ij,st}), \quad A_{ij,st} \in M_1(k_i, k_j)$$

$$(1 \le s \le m_i, 1 \le t \le m_j),$$

then $A_{ij,st}$ is called *the ((i,s),(j,t))-block* of A. If

$$A_{ij,st} = (a_{ij,st,uv}), \quad a_{ij,st,uv} \in \mathbb{R}$$

$$(1 \le u \le k_i, 1 \le v \le k_j),$$

then $a_{ij,st,uv}$ is called *the ((i,s,u),(j,t,v))-component* of A.

(7) Let $A \in M_2(\tilde{k} \cdot \tilde{m})$ be given. If

$$A = (A_{ij}), \quad A_{ij} \in M_2(k_i m_i, k_j m_j) \quad (1 \le i,j \le r),$$

then A_{ij} is called *the (i,j)-block* of A. If

$$A_{ij} = (A_{ij,st}), \quad A_{ij,st} \in M_2(k_i, k_j)$$

$$(1 \le s \le m_i, 1 \le t \le m_j),$$

then $A_{ij,st}$ is called *the ((i,s),(j,t))-block* of A. If

$$A_{ij,st} = (\tilde{a}_{ij,st,uv}), \quad \tilde{a}_{ij,st,uv} \in M(2)$$

$$(1 \le u \le k_i, 1 \le v \le k_j),$$

then $\tilde{a}_{ij,st,uv}$ is called *the ((i,s,u),(j,t,v))-component* of A.

(8) For $1 \le i_* \le r$, $1 \le s_* \le m_{i_*}$, and $1 \le u_* \le k_{i_*}$, set $\alpha = e_1(\tilde{k}, \tilde{m}; (i_*, s_*, u_*))$. Then, $C_1(\tilde{k}, \tilde{m}; \alpha)$ is a subset of $C_1(\tilde{k}, \tilde{m})$ defined by

$$A = (a_{ij}) \in C_1(\tilde{k}, \tilde{m}; \alpha),$$

$$a_{ij} \in \mathbb{R} \quad and \quad 1 \le i, j \le \tilde{k} \cdot \tilde{m}$$

$$\Longleftrightarrow$$

there is an $a \in \mathbb{R}$ such that

the $((i_*, s_*, u_*), (j, t, v))$-component

$$= \begin{cases} a & \text{if} \quad (j, t, v) = (i_*, s_*, u_*) \\ 0 & \text{otherwise,} \end{cases}$$

where $1 \le j \le r$, $1 \le t \le m_j$ and $1 \le v \le k_j$. Also, $C_2(\tilde{k}, \tilde{m}; \alpha)$ is a subset of $C_2(\tilde{k}, \tilde{m})$ defined by

$$A = (\tilde{a}_{ij}) \in C_2(\tilde{k}, \tilde{m}; \alpha),$$

$$\tilde{a}_{ij} \in R(2) \quad and \quad 1 \le i, j \le \tilde{k} \cdot \tilde{m}$$

$$\Longleftrightarrow$$

there is an $\tilde{a} \in R(2)$ such that

the $((i_*, s_*, u_*), (j, t, v))$-component

$$= \begin{cases} \tilde{a} & \text{if} \quad (j, t, v) = (i_*, s_*, u_*) \\ O(2) & \text{otherwise,} \end{cases}$$

where $1 \le j \le r$, $1 \le t \le m_j$ and $1 \le v \le k_j$.

Example 2.3.7. An element of $C_1(\tilde{k}, \tilde{m}; \alpha)$ with $\tilde{k} = (2, 4)$, $\tilde{m} = (2, 2)$ and $\alpha = e_1(\tilde{k}, \tilde{m}; (2, 1, 2))$:

$$\left(\begin{array}{cc|cc|cccc|cccc}
a & b & c & d & 0 & 0 & e & f & 0 & 0 & g & h \\
0 & a & 0 & c & 0 & 0 & 0 & e & 0 & 0 & 0 & g \\
\hline
i & j & k & l & 0 & 0 & m & n & 0 & 0 & p & q \\
0 & i & 0 & k & 0 & 0 & 0 & m & 0 & 0 & 0 & p \\
\hline
0 & r & 0 & s & t & 0 & 0 & u & 0 & 0 & 0 & v \\
0 & 0 & 0 & 0 & 0 & t & 0 & 0 & 0 & 0 & 0 & 0 \\
0 & 0 & 0 & 0 & 0 & 0 & t & 0 & 0 & 0 & 0 & 0 \\
0 & 0 & 0 & 0 & 0 & 0 & 0 & t & 0 & 0 & 0 & 0 \\
\hline
w & x & y & z & a' & b' & c' & d' & e' & f' & g' & h' \\
0 & w & 0 & y & 0 & a' & b' & c' & 0 & e' & f' & g' \\
0 & 0 & 0 & 0 & 0 & 0 & a' & b' & 0 & 0 & e' & f' \\
0 & 0 & 0 & 0 & 0 & 0 & 0 & a' & 0 & 0 & 0 & e'
\end{array}\right)$$

Remark 2.3.8. The subscript 1 of $J_1(\lambda;k)$, $C_1^{\cdot}(k)$, $N_1(k)$, $\Delta_1(a_1,\cdots,a_n)$ or $e_1(n;u)$ in Definitions 2.3.2 - 2.3.6 will be removed if there is no confusion. Similarly, we will denote

$$J(\lambda;k,m) := J_1(\lambda;k,m), \quad J(\lambda;\tilde{k},\tilde{m}) := J_1(\lambda;\tilde{k},\tilde{m})$$
$$C(\tilde{h}) := C_1(\tilde{h}), \quad C(k,m) := C_1(k,m), \quad C(\tilde{k},\tilde{m}) := C_1(\tilde{k},\tilde{m}),$$
$$N(k;j) := N_1(k;j), \quad e(k,m;(s_*,u_*)) := e_1(k,m;(s_*,u_*)),$$
$$e(\tilde{k},\tilde{m};(i_*,s_*,u_*)) := e_1(\tilde{k},\tilde{m};(i_*,s_*,u_*)).$$

However, the subscript 2 of $J_2(\lambda;k)$, $C_2(k)$, $N_2(k)$, $\Delta_2(\tilde{a}_1,\cdots,\tilde{a}_n)$, $e_2(n;u)$, $M_2(n)$ etc., cannot be always removed.

Lemma 2.3.9. *Let $\tilde{k},\tilde{m} \in \mathbb{N}^r (r \geq 1), \tilde{h},\tilde{n} \in \mathbb{N}^s (s \geq 1)$ and $\theta,\varphi \in K_d$ be given, where d=1 or 2.*

(1) If $J = J_d(\theta;\tilde{k},\tilde{m})$ and $G \in M_d(\tilde{k}\cdot\tilde{m})$, then

$$JG = GJ \iff G \in C_d(\tilde{k},\tilde{m}).$$

(2) If $J = J_d(\theta;\tilde{k},\tilde{m}), K = J_d(\varphi;\tilde{h},\tilde{n})$ and $\varphi \neq \theta,\bar{\theta}$, then

$$JB = BK \iff B = O,$$

where $B \in M_d(\tilde{k}\cdot\tilde{m},\tilde{h}\cdot\tilde{n})$.

(3) If $J = J_1(\lambda;\tilde{k},\tilde{m}), K = J_2(\omega;\tilde{h},\tilde{n}), \lambda \in \mathbb{R}, \omega \in \mathbb{C}$ and $\lambda \neq \omega,\bar{\omega}$, then

$$JB = BK \iff B = O$$
$$CJ = KC \iff C = O,$$

where $B \in M(\tilde{k}\cdot\tilde{m},2\tilde{h}\cdot\tilde{n})$ and $C \in M(2\tilde{k}\cdot\tilde{m},\tilde{h}\cdot\tilde{n})$.

Proof. Theorems 1 and 2 of Chapter VIII in [Gantmacher 1977] (pp.215 - 222). □

Lemma 2.3.10. *Assume*

$$\tilde{k}_i,\tilde{m}_i \in \mathbb{N}(r(i)) \quad (1 \leq i \leq r), \quad \tilde{h}_j,\tilde{n}_j \in \mathbb{N}(s(j)) \quad (1 \leq j \leq s),$$
$$\lambda_i \in \mathbb{R} \quad (1 \leq i \leq r), \quad \lambda_i \neq \lambda_j \quad (i \neq j), \quad \text{and}$$
$$\omega_i \in \mathbb{C} \quad (1 \leq i \leq s), \quad \omega_i \neq \omega_j,\bar{\omega}_j \quad (i \neq j).$$

If

$$J = \dot{\sum}_{i=1}^{r} J_1(\lambda_i; \tilde{k}_i, \tilde{m}_i) \dot{+} \dot{\sum}_{j=1}^{s} J_2(\omega_i; \tilde{h}_i, \tilde{n}_i) \quad \textit{and} \quad G \in M(n),$$

where $n = \sum_{i=1}^{r} \tilde{k}_i \cdot \tilde{m}_i + 2\sum_{j=1}^{s} \tilde{h}_j \cdot \tilde{n}_j$, *then*

$$JG = GJ \iff G = \dot{\sum}_{i=1}^{r} A_i \dot{+} \dot{\sum}_{j=1}^{s} B_j,$$

where $A_i \in C_1(\tilde{k}_i, \tilde{m}_i) \quad (1 \le i \le r)$ *and* $B_j \in C_2(\tilde{h}_j, \tilde{n}_j) \quad (1 \le j \le s)$.

Proof. The proof follows from Lemma 2.3.9. □

Lemma 2.3.11. *Assume d=1 or 2. Then the following holds:*

(1) $N_d(n; i) e_d(n; j) = e_d(n; j - i) \quad (0 \le j - i \le j, 1 \le j \le n)$.

(2) $N_d(n; i)^T e_d(n; j) = e_d(n; j + i) \quad (j \le j + i \le n, 1 \le j \le n)$.

(3) If $N = N_d(n)$ *and* $e = e_d(n; n)$, *then*

$$(N^{j+1})^T N^j e = 0 \quad (0 \le j \le n).$$

Proof. Since $N_d(n; i) = J_d(0; n)^i$, the above statements are immediately verified. □

Lemma 2.3.12.

(1) If $G = \Delta_1(a_1, \cdots, a_n)$ *and* $e = e_1(n; i) \quad (1 \le i \le n)$, *then*

$$G^T e = (0, \cdots, 0, a_1, \cdots, a_{n-i+1}).$$

(2) If $G = \Delta_2(\tilde{a}_1, \cdots, \tilde{a}_n)$, $\tilde{a}_j = \begin{pmatrix} a_j & b_j \\ -b_j & a_j \end{pmatrix} \quad (1 \le j \le n)$ *and* $e = e_2(n; i) \quad (1 \le i \le n)$, *then*

$$G^T e = (0, \cdots, 0, a_1, b_1, a_2, b_2, \cdots, a_{n-i+1}, b_{n-i+1}).$$

Proof. (1) If $G = (c_{ij})$, by Definition 2.3.4(1) and (7),

$$c_{ij} = 0 (i > j), \quad c_{ij} = c_{i+1,j+1} \quad (1 \le i, j < n) \quad \text{and}$$

$$a_j = c_{1j} \quad (1 \le j \le n).$$

Hence

$$\begin{aligned} G^T e &= (c_{i1}, c_{i2}, \cdots, c_{in})^T = (0, \cdots, 0, c_{ii}, \cdots, c_{in})^T \\ &= (0, \cdots, 0, c_{11}, c_{12}, \cdots, c_{1,n-i+1})^T \\ &= (0, \cdots, 0, a_1, \cdots, a_{n-i+1})^T. \end{aligned}$$

(2) Notice that $\tilde{a}_j^T \begin{pmatrix} 1 \\ 0 \end{pmatrix} = \begin{pmatrix} a_j \\ b_j \end{pmatrix}$. Then the statement is similarly obtained as in (1). □

Lemma 2.3.13. *Set* $d = 1$ *or* 2.

(1) If $G \in C_d(k) \cap GL(dk, \mathbb{R}) \quad (k \in \mathbb{N})$ *and*

$$e_d(k,i)^T G = e_d(k;j)^T \quad (1 \le i, j \le k),$$

then $i = j$.

(2) If $G \in C_d(k,m) \cap GL(dkm, \mathbb{R}) \quad (k, m \in \mathbb{N})$ *and*

$$e_d(k,m;(1,i))^T G = e_d(k,m;(1,j))^T \quad (1 \le i, j \le k),$$

then $i = j$.

(3) If $\tilde{k} = (k_1, \cdots, k_r) \in \mathbb{N}^r_<, \quad \tilde{m} = (m_1, \cdots, m_r) \in \mathbb{N}^r, G \in C_d(\tilde{k}, \tilde{m}) \cap GL(d\tilde{k} \cdot \tilde{m}, \mathbb{R})$ *and*

$$e_d(\tilde{k}, \tilde{m}; (i,1,j))^T G = e_d(\tilde{k}, \tilde{m}; (i',1,j'))^T$$
$$(1 \le j \le k_i, 1 \le j' \le k_{i'}, 1 \le i, i' \le r)$$

then $i = i'$ *and* $j = j'$.

Proof. Case of $d = 1$.

(1) If $G \in C_1(k) \cap GL(j, \mathbb{R})$, there exist $a_1, \cdots, a_k \in \mathbb{R}$ such that

$$G = \Delta_1(a_1, \cdots, a_k) \quad \text{and} \quad a_1 \neq 0.$$

Since $e_1(k;i)^T G = (0, \cdots, 0, a_1, \cdots, a_{k-i+1})$, if $e_1(k;i)^T G = e_1(k;j)^T$, we have $a_1 = 1$ and $i = j$.

(2)(respectively (3)) Apply the same argument to the $(1,i)$- (respectively the $(i,1,j)$-) block of G.

Case of $d = 2$. Proof follows *mutatis mutandis.* □

Lemma 2.3.14. *Assume* $\tilde{k} = (k_1, \cdots, k_r) \in \mathbb{N}^r_<, \quad \tilde{m} = (m_1, \cdots, m_r) \in \mathbb{N}^r$.

(1) $\alpha = e_1(\tilde{k}, \tilde{m}; (i,s,u))$ *and* $H \in C_1(\tilde{k}, \tilde{m})$, *then*

$$\alpha^T H = a\alpha^T \quad (a \in \mathbb{R})$$

$$\iff$$

$$H \in C_1(\tilde{k}, \tilde{m}; \alpha) \quad \textit{and "the } ((i,s,u),(i,s,u))\textit{-component of } H\textit{"} = a.$$

(2) $\alpha = e_2(\tilde{k}, \tilde{m}; (i, s, u))$ *and* $H \in C_2(\tilde{k}, \tilde{m})$, *then*

$$\alpha^T H = a\alpha^T (a \in \mathbb{R})$$

$$\Longleftrightarrow$$

$$H \in C_2(\tilde{k}, \tilde{m}; \alpha) \text{ and "the } ((i,s,u),(i,s,u))\text{-component of } H\text{"} = aI(2).$$

Proof. (1) Notice that all components of α are zero except the (i, s, u)-component, which is 1 (Definition 2.3.6(3)), and that the $((i, s, u), (j, t, v))$-component of H is zero except the case of $(i, s, u) = (j, t, v)$ (Definition 2.3.6(7)). From this, the statement is immediately obtained.

(2) Proof is similar to that for (1). □

Lemma 2.3.15. *Assume* $\tilde{k} = (k_1, \cdots, k_r) \in \mathbb{N}^r_<$, $\tilde{m} = (m_1, \cdots, m_r) \in \mathbb{N}^r$ *and* $H \in C_d(\tilde{k}, \tilde{m})$ *(*$d = 1$ *or* 2*). Denote by* $a_{ij,st,uv} \in R(d)$ *the* $((i, s, u), (j, t, v))$*-component of* H*. For* $1 \le i_* \le r$ *and* $1 \le s_* \le m_{i_*}$*, assume*

$$a_{ij,st,uv} \neq 0 \quad \text{if } (i,s,u) = (j,t,v); \text{ and}$$

$$a_{ij,st,uv} = 0 \quad \text{if } (i,s) \neq (i_*, s_*) \text{ and } (i,s,u) \neq (j,t,v).$$

Then H *belongs to* $GL(d\tilde{k} \cdot \tilde{m}, \mathbb{R})$.

Proof. Case of $d = 1$. All diagonal components of H are nonzero, and offdiagonal components $a_{ij,st,uv}$ are zero if $(i, s) \neq (i_*, s_*)$. From this we have the rank$(H) = \tilde{k} \cdot \tilde{m}$. Hence $H \in GL(\tilde{k} \cdot \tilde{m}, \mathbb{R})$.

Case of $d = 2$. Proof follows similarly to the case of $d = 1$. □

Lemma 2.3.16. *Assume* $\tilde{k} = (k_1, \cdots, k_r) \in \mathbb{N}^r_<$, $\tilde{m} = (m_1, \cdots, m_r) \in \mathbb{N}^r$ *and* $p = e_d(\tilde{k}, \tilde{m}; (i_0, s_0, u_0))$ $(1 \le i_0 \le r, 1 \le s_0 \le m_{i_0}, 1 \le u_0 \le k_{i_0})$*, where* $d = 1$ *or* 2.

(1) Let $i_0 \le j_0 \le r$ *and* $1 \le t_0 \le m_{j_0}$ *be given. If there are* $H \in C_d(\tilde{k}, \tilde{m})$ *and* $1 \le v_0 \le k_{j_0}$ *such that*

$$\text{"the } (j_0, t_0, v_0)\text{-component of } Hp\text{"} \neq 0,$$

then $v_0 \le u_0$.

(2) Let $i_0 \le j_0 \le r$, $1 \le t_0 \le m_{j_0}$, $1 \le v_0 \le u_0$ *and* $c \in M(d, 1)$ *be given. Then there are* $H \in C_d(\tilde{k}, \tilde{m}) \cap GL(d\tilde{k} \cdot \tilde{m}, \mathbb{R})$ *such that*

$$\text{"the } (j_0, t_0, v_0)\text{-component of } Hp\text{"} = c.$$

(3) *Let* $1 \le j_0 \le i_0$ *and* $1 \le t_0 \le m_{j_0}$ *be given. If there are* $H \in C_d(\tilde{k}, \tilde{m})$ *and* $1 \le v_0 \le k_{j_0}$ *such that*

$$\text{“the } (j_0, t_0, v_0)\text{-component of } Hp\text{”} \ne 0,$$

then $k_{j_0} - v_0 \le k_{i_0} - u_0$.

(4) *Let* $1 \le j_0 \le i_0, \quad 1 \le t_0 \le m_{j_0}, \quad 1 \le v_0 \le u_0 - k_{i_0} + k_{j_0}$ *and* $c \in M(d, 1)$ *be given. Then there is an* $H \in C_d(\tilde{k}, \tilde{m})$ *such that*

$$\text{“the } (j_0, t_0, v_0)\text{-component of } Hp\text{”} = c.$$

Proof. (1) The $((j_0, t_0), (i_0, s_0))$-block of $H \in C_d(\tilde{k}, \tilde{m})$ has the form

$$H_{j_0 i_0, t_0 s_0} = A\#O \in M(dk_{j_0}, dk_{i_0}),$$

$$A = \Delta_d(a_1, \cdots, a_{k_{i_0}}), a_1, \cdots, a_{k_{i_0}} \in R(d).$$

This implies that, if $u > u_0$, then the (j_0, t_0, u)-component of Hp is equal to 0.

(2) Define $H = (H_{ij,st}) \in C_d(\tilde{k}, \tilde{m}) \cap GL(d\tilde{k} \cdot \tilde{m}, \mathbb{R})$ as follows:

$$H_{ij,st} = \begin{cases} I(dk_i) & (\text{if } (i, s) = (j, t)) \\ N\#O & (\text{if } (i, s) = (j_0, t_0) \quad \text{and} \quad (j, t) = (i_0, s_0)) \\ O & (\text{otherwise}), \end{cases}$$

where

$$N = \Delta_d(a_1, \cdots, a_{k_{i_0}}),$$

$$a_i = \begin{cases} A & (\text{if } i = u_0 - v_0 + 1) \\ O & (\text{otherwise}), \end{cases}$$

$$A = c \quad (\in \mathbb{R}) \quad (\text{if } d = 1)$$

$$A = \begin{pmatrix} c_1 & -c_2 \\ c_2 & c_1 \end{pmatrix} \quad (\text{where } c = (c_1, c_2)^T) \ (\text{if } d = 2).$$

(3) The $((j_0, t_0), (i_0, s_0))$-block of $H \in C_d(\tilde{k}, \tilde{m})$ has the form

$$H_{j_0 i_0, t_0 s_0} = [OA] \in M(dk_{j_0}, dk_{i_0}),$$

$$A = \Delta_d(a_1, \cdots, a_{k_{i_0}}), a_1, \cdots, a_{k_{i_0}} \in R(d).$$

This implies that, if $k_{j_0} - u > k_{i_0} - u_0$, then the (j_0, t_0, u)-component of Hp is equal to 0.

(4) Define $H = (H_{ij,st}) \in C_d(\tilde{k}, \tilde{m}) \cap GL(d\tilde{k} \cdot \tilde{m}, \mathbb{R})$ as follows:

$$H_{ij,st} = \begin{cases} I(dk_i) & (\text{if } (i, s) = (j, t)) \\ [ON] & (\text{if } (i, s) = (j_0, t_0) \quad \text{and} \quad (j, t) = (i_0, s_0)) \\ O & (\text{otherwise}), \end{cases}$$

where

$$N = \Delta_d(a_1, \cdots, a_{k_{i_0}}),$$

$$a_i = \begin{cases} A & (\text{if } i = k_{j_0} - v_0 - k_{i_0} + u_0 + 1) \\ O & (\text{otherwise}), \end{cases}$$

$$A = c (\in \mathbb{R}) \quad (\text{if } d = 1)$$

$$A = \begin{pmatrix} c_1 & -c_2 \\ c_2 & c_1 \end{pmatrix} \quad (\text{where } c = (c_1, c_2)^T) \ (\text{if } d = 2).$$

□

2.3.2 Normal Forms of Linear Vector Fields with a Boundary

Define an affine vector field $X_{(A,q)} : \mathbb{R}^n \to \mathbb{R}^n$ by

$$X_{(A,q)}(x) = Ax + q \quad (x, q \in \mathbb{R}^n, A \in M(n))$$

and $(n-1)$-dimensional hyperplane U (called a boundary) by

$$U = U(\alpha, \beta) := \{x \in \mathbb{R}^n \mid \langle \alpha, x \rangle = \beta\} \quad (\alpha \in \mathbb{R}^n - \{0\}, \beta \in \mathbb{R}).$$

The pair $(X_{(A,q)}, U)$ is called *an affine vector field with a boundary.* (For simplicity, $\{x \in \mathbb{R}^n \mid \langle \alpha, x \rangle = \beta\}$ will be denoted by $\{\langle \alpha, x \rangle = \beta\}$.) Two affine vector fields with a boundary; (X, U) and (X', U') are *affine conjugate* if there is an affine transformation $g : \mathbb{R}^n \to \mathbb{R}^n; g(x) = Gx + p$ $(G \in GL(n, \mathbb{R}), p \in \mathbb{R}^n)$ such that

$$GX(x) = X'(g(x)) \quad (x \in \mathbb{R}^n) \quad \text{and } g(U) = U'.$$

An affine vector field $X_{(A,q)}$ is *degenerate* if

$$X_{(A,q)}(x) = Ax + q \neq 0 \quad \text{for all } x \in \mathbb{R}^n.$$

Clearly, this is equivalent to the condition:

$$q \notin \{Ax | x \in \mathbb{R}^n\}.$$

An affine vector field $X_{(A,q)}$ is *non-degenerate* if it is not degenerate. $(X_{(A,q)}, U)$ is called *degenerate* (respectively *non-degenerate*) if $X_{(A,q)}$ is degenerate (respectively non-degenerate). A non-degenerate $(X_{(A,q)}, U)$ is *regular* if there is an $x_0 \notin U$ such that

$$X_{(A,q)}(x_0) = Ax_0 + q = 0.$$

A non-degenerate $(X_{(A,q)}, U)$ is *singular* if it is not regular. If $(X_{(A,q)}, U)$ is singular, there is an $x_0 \in U$ such that

$$X_{(A,q)}(x_0) = Ax_0 + q = 0.$$

Hence, if $X_{(A,q)}$ is non-degenerate, by a parallel translation $h(x) = x - x_0$, $X_{(A,q)}$ is transformed to a linear vector field

$$X_A : \mathbb{R}^n \to \mathbb{R}^n; X_A(x) = Ax.$$

Then $U = U(\alpha, \beta)$ is transformed to

$$U(\alpha, \beta') = \{\langle \alpha, x \rangle = \beta'\} \quad (\beta' \neq 0)$$

if $(X_{(A,q)}, U(\alpha, \beta))$ is regular, and to

$$U(\alpha, 0) = \{\langle \alpha, x \rangle = 0\}$$

if $(X_{(A,q)}, U(\alpha, \beta))$ is singular. Therefore, we can assume that affine vector fields with a boundary have the following forms, without loss of generality:

$$\text{degenerate} \iff (X_{(A,q)}, U(\alpha, \beta)), \quad q \notin A(\mathbb{R}^n)$$

non-degenerate :

$$\text{regular} \iff (X_A, U(\alpha, \beta)), \quad \beta \neq 0$$

$$\text{singular} \iff (X_A, U(\alpha, 0)).$$

A non-degenerate affine vector field with a boundary is considered a linear vector field with a boundary. In this subsection we consider the normal forms of linear vector fields with a boundary under linear conjugacy. The normal forms of degenerate affine vector fields with a boundary will be considered in the next subsection.

It is well known that any linear vector field $X_A : \mathbb{R}^n \to \mathbb{R}^n; X_A(x) = Ax$ is transformed by a suitable linear transformation $G : \mathbb{R}^n \to \mathbb{R}^n (G \in GL(n, \mathbb{R}))$ into a vector field X_J, defined by a real Jordan matrix J:

$$GX_AG^{-1} = X_{GAG^{-1}} = X_J.$$

When we consider the normal form of a linear vector field with a boundary $(X_A, U(\alpha, \beta))$, we can assume that A is a real Jordan matrix, without loss of generality. Hence our goal in this subsection is to simplify $\alpha \in \mathbb{R}^n - \{0\}$ and $\beta \in \mathbb{R}$ as much as possible under linear transformation by which the real Jordan matrix is invariant. We will identify a linear transformation and an element of $GL(n, \mathbb{R})$ with respect to the natural basis of $\mathbb{R}^n$, if there is no confusion.

Lemma 2.3.17. *Assume $\theta \in K_d$ and $d = 1$ or 2. Let*

$$J = J_d(\theta; n) \quad \text{and} \quad \alpha = (\alpha_1, \cdots, \alpha_n)^T \neq 0$$

be given, where $\alpha_1, \cdots, \alpha_n \in M(1, d)$. Set

$$i_* = \min\{1 \leq i \leq n \mid \alpha_i \neq 0\}, \qquad \alpha' = e_d(n; i_*).$$

Then there is a $G \in GL(dn, \mathbb{R})$ *such that*

$$(i)\quad JG = GJ, \qquad (ii)\quad (\alpha')^T G = \alpha^T.$$

Proof. Case of $d = 1$. Set

$$G = \Delta_1(\alpha_{i_*}, \cdots, \alpha_n, 0, \cdots, 0) \in C_1(n).$$

Since $\alpha_{i_*} \neq 0$, we have $G \in GL(n, \mathbb{R})$. By Lemma 2.3.9(1) (setting $\tilde{k} = n, \tilde{m} = 1$), $JG = GJ$. Also, by Lemma 2.3.12(1),

$$G^T\alpha' = G^T e(n; i_*) = (0, \cdots, 0, \alpha_{i_*}, \cdots, \alpha_n)^T = \alpha,$$

hence $(\alpha')^T G = \alpha^T$.

Case of $d = 2$. For $\alpha_i = (\alpha_{i1}, \alpha_{i2}) \in M(1,2)(1 \leq i \leq n)$, set

$$\begin{aligned} G &= \Delta_2(A_{i_*}, \cdots, A_n, 0, \cdots, 0) \in C_2(n) \\ A_i &= \begin{pmatrix} \alpha_{i1} & \alpha_{i2} \\ -\alpha_{i2} & \alpha_{i1} \end{pmatrix} \quad (1 \leq i \leq n). \end{aligned}$$

Since $\alpha_{i_*} \neq (0,0)$, we have $G \in GL(2n, \mathbb{R})$. (i) follows from Lemma 2.3.9(2), and (ii) follows from Lemma 2.3.12(2). □

Lemma 2.3.18. *Assume* $k, m \in \mathbb{N}, \theta \in K_d$ *and* $d = 1$ *or* 2. *Let*

$$J = J_d(\theta; k, m), \quad \alpha = \dot{\prod}_{i=1}^{m} u_i \neq 0$$

be given, where $u_i = e_d(k; j(i))(0 \leq j(i) \leq k, 1 \leq i \leq m)$. *Set*

$$j_* = \min\{j(i) \neq 0 \mid 1 \leq i \leq m\}, \alpha' = e_d(k, m; (1, j_*)).$$

Then, there is a $G \in GL(dkm, \mathbb{R})$ *such that*

$$(i)\quad JG = GJ, \qquad (ii)\quad (\alpha')^T G = \alpha^T.$$

Proof. Without loss of generality, we can assume that $j_* = j(1)$. Define $G \in GL(dkm, \mathbb{R})$ by

$$G^{-1} = \begin{pmatrix} I & N_2 & N_3 & \cdots & \cdots & N_m \\ & I & 0 & \cdots & \cdots & 0 \\ & & I & 0 & \cdots & 0 \\ & & & & \ddots & \\ \mathbf{0} & & & & & I \end{pmatrix},$$

where $I = I(dk), N_i = -N_d(k; j(i) - j_*)(2 \le i \le m)$. Since $G^{-1} \in C_d(k,m)$, (i) follows from Lemma 2.3.9(1), setting $\tilde{k} = (k)$ and $\tilde{m} = (m)$. By Lemma 2.3.11(2), since

$$\begin{aligned} u_1^T N_i &= -e_d(k; j_*)^T N_d(k; j(i) - j_*) \\ &= -e_d(k; j(i))^T = -u_i^T \quad (2 \le i \le m), \end{aligned}$$

we have

$$\begin{aligned} \alpha^T G^{-1} &= (u_1^T, \cdots, u_m^T) G^{-1} \\ &= (u_1^T, u_1^T N_2 + u_2^T, \cdots, u_1^T N_m + u_m^T) \\ &= (u_1^T, 0, \cdots, 0) = e_d(k, m; (1, j_*)) = (\alpha')^T, \end{aligned}$$

hence $(\alpha')^T G = \alpha^T$. □

Lemma 2.3.19. *Assume* $\tilde{k} = (k_1, \cdots, k_r) \in \mathbb{N}_<^r, \tilde{m} = (1, \cdots, 1) \in \mathbb{N}^r, n = \sum_{i=1}^r k_i, \theta \in K_d$ *and* $d = 1$ *or* 2. *Let*

$$J = J_d(\theta; \tilde{k}, \tilde{m}), \quad \alpha = \dot{\prod}_{i=1}^m u_i \neq 0$$

be given, where $u_i = e_d(k_i; j(i))(0 \le j(i) \le k_i, 1 \le i \le r)$. *Set*

$$j_* = \min\{j(i) \neq 0 \mid 1 \le i \le r\}, \quad i_* = \max\{1 \le i \le r \mid j_* = j(i)\}$$

$$\alpha' = e_d(\tilde{k}, \tilde{m}; (i_*, 1, j_*)).$$

Then, there is $G \in GL(dn, \mathbb{R})$ *such that*

$$\text{(i)} \quad JG = GJ, \quad \text{(ii)} \quad (\alpha')^T G = \alpha^T.$$

Proof. Define $G \in GL(dn, \mathbb{R})$ by

$$G^{-1} = (A_{ij}), \quad A_{ij} \in M_d(k_i, k_j) \quad (1 \le i, j \le r),$$

where all $A_{ij} = O$ except

$$A_{ii} = I(dk_i)(1 \le i \le r),$$

$$A_{i_* j} = \begin{cases} [ON_j] & (\text{if} \quad j > i_*) \\ N_j \# O & (\text{if} \quad j < i_*) \end{cases}$$

$$N_i = -N_d(k_{i_*}; j(i) - j_*) \quad (1 \le i \le r).$$

It is easy to see rank$G^{-1} = n$, hence G and G^{-1} are well-defined. Since it is easy to verify $G^{-1} \in C_d(\tilde{k}, \tilde{m})$, (i) follows from Lemma 2.3.9(1). By Lemma 2.3.11(2), if $j \neq i_*$, then, for $L = [ON_j]$ or $N_j \# O$,

$$\begin{aligned} u_{i_*}^T A_{i_*j} &= -e_d(k_{i_*}; j_*)^T L \\ &= -e_d(k_j; j(i))^T = -u_j^T. \end{aligned}$$

Set $\alpha^T G^{-1} = ((\alpha_1')^T, \cdots, (\alpha_r')^T), \alpha_i' \in M(k_i, 1) \quad (1 \le i \le r)$, then

$$(\alpha_j')^T = u_j^T + u_{i_*}^T A_{i_*j} = 0 \quad \text{if} \quad j \neq i_*;$$

$$(\alpha_j')^T = u_{i_*}^T A_{i_*j} = u_{i_*}^T \quad \text{if} \quad j = i_*.$$

That is, $\alpha^T G^{-1} = e_d(\tilde{k}, \tilde{m}; (i_*, 1, j_*))^T = (\alpha')^T$. □

Lemma 2.3.20. *Assume* $\tilde{k} = (k_1, \cdots, k_r) \in \mathbb{N}_<^r$, $\tilde{m} = (m_1, \cdots, m_r) \in \mathbb{N}^r$, $n = \tilde{k} \cdot \tilde{m}, \theta \in K_d$ *and* $d = 1$ *or* 2. *Let*

$$J = J_d(\theta; \tilde{k}, \tilde{m}), \quad \alpha = \dot{\prod}_{i=1}^r \dot{\prod}_{s=1}^{m_i} e_d(k_i; j(i,s)) \neq 0$$

be given, where $0 \le j(i,s) \le k_i, 1 \le s \le m_i$ *and* $1 \le i \le r$. *Set*

$$j_* = \min\{j(i,s) \neq 0 \mid 1 \le s \le m_i, 1 \le i \le r\},$$

$$i_* = \max\{1 \le i \le r \mid j_* = j(i,s) \quad \textit{for some } 1 \le s \le m_i\},$$

$$\alpha' = e_d(\tilde{k}, \tilde{m}; (i_*, 1, j_*)).$$

Then, there is a $G \in GL(dn, \mathbb{R})$ *such that*

$$(i) \quad JG = GJ, \quad (ii) \quad (\alpha')^T G = \alpha^T.$$

Proof. Applying Lemma 2.3.18 to each $J_d(\theta; k_i, m_i) \quad (1 \le i \le r)$, we have a $G' \in GL(dn, \mathbb{R})$ such that $JG' = G'J$ and $(\alpha'')^T G = \alpha^T$, where

$$\alpha'' = \dot{\prod}_{i=1}^r \dot{\prod}_{s=1}^{m_i} v_{is},$$

$$v_{is} = \begin{cases} e_d(k_i; j(i)_*) & (\text{ if } s = 1) \\ O(dk_i, 1) & (\text{ if } s \neq 1) \end{cases}$$

$$j(i)_* = \min\{j(i,s) \mid 1 \le s \le m_i\} \quad (1 \le i \le r).$$

That is, $\alpha'' = \dot{\prod}_{i=1}^r e_d(k_i, m_i; (1, j(i)_*))$. Moreover, apply Lemma 2.3.19 to the direct sum of $J_d(\theta; k_i)$ corresponding to $v_{i1} \quad (1 \le i \le r)$, i.e.,

$$J_d(\theta; k_1) \dot{+} \cdots \dot{+} J_d(\theta; k_r),$$

then the conclusion follows. □

Theorem 2.3.21. (Normal Forms of Linear Vector Fields with a Boundary) *Let a real Jordan matrix J and a boundary U be given as follows:*

$$J = \dot{\sum}_{i=1}^{r} J_1(\lambda_i; \tilde{k}_i, \tilde{m}_i) \dot{+} \dot{\sum}_{i=1}^{s} J_2(\omega_i; \tilde{h}_i, \tilde{n}_i)$$

where

$$\tilde{k}_i = (k_{i1}, \cdots, k_{ir(i)}) \in \mathbb{N}_<(r(i)) \quad (1 \le i \le r)$$

$$\tilde{h}_i = (h_{i1}, \cdots, h_{is(i)}) \in \mathbb{N}_<(s(i)) \quad (1 \le i \le s)$$

$$\tilde{m}_i = (m_{i1}, \cdots, m_{ir(i)}) \in \mathbb{N}(r(i)) \quad (1 \le i \le r)$$

$$\tilde{n}_i = (n_{i1}, \cdots, n_{is(i)}) \in \mathbb{N}(s(i)) \quad (1 \le i \le s)$$

$$U = \{x \in \mathbb{R}^n \mid \langle \alpha, x \rangle = \beta\}, \quad \alpha \in \mathbb{R}^n - \{0\},\ \beta \in \mathbb{R},$$

and $n = \sum_{i=1}^{r} \tilde{k}_i \cdot \tilde{m}_i + 2\sum_{i=1}^{s} \tilde{h}_i \cdot \tilde{n}_i$. Let (X_J, U) be the linear vector field X_J with a boundary U. Then the following hold:

There is a $G \in GL(n, \mathbb{R})$ and $\alpha' \in \mathbb{R}^n - \{0\}$ such that

(1) $JG = GJ$,

(2) $G(U) = \{\langle \alpha', x \rangle = \varepsilon\}$, *where*

$$\varepsilon = \varepsilon(\beta) = \begin{cases} 1 & (\beta \ne 0, \quad \text{i.e. } (X_J, U) \text{ is regular}) \\ 0 & (\beta = 0, \quad \text{i.e. } (X_J, U) \text{ is singular}) \end{cases},$$

(3) $\alpha' = \dot{\prod}_{i=1}^{r} u_i' \dot{\times} \dot{\prod}_{j=1}^{s} v_j'$, *where* $u_i' \in M(\tilde{k}_i \cdot \tilde{m}_i, 1)$ *and* $v_j' \in M(2\tilde{h}_j \cdot \tilde{n}_j, 1)$,

(4) for each $1 \le i \le r$, *there are* $0 \le k(i) \le r(i)$ *and* $1 \le u(i) \le k_{ik(i)}$ *such that*

$$u_i' = e_1(\tilde{k}_i, \tilde{m}_i; (k(i), 1, u(i))),$$

where $u_i' = 0$ *if* $k(i) = 0$,

(5) for each $1 \le j \le s$, *there are* $0 \le h(j) \le s(j)$ *and* $1 \le v(j) \le h_{jh(j)}$ *such that*

$$v_j' = e_2(\tilde{h}_i, \tilde{n}_i; (h(j), 1, v(j))),$$

where $v_j' = 0$ *if* $h(j) = 0$.

Also, let $G', G'' \in GL(n, \mathbb{R})$ and $\alpha', \alpha' \in \mathbb{R}^n - \{0\}$ be given. If (G', α') and (G'', α'') satisfy the above (1) - (5), then $\alpha' = \alpha''$. In this sense, α' is uniquely determined for (X_J, U).

Proof. Let α divide into blocks:

$$\alpha = \dot{\prod}_{i=1}^{r} u_i \dot{\times} \dot{\prod}_{j=1}^{s} v_j, \quad u_i \in M(\tilde{k}_i \cdot \tilde{m}_i, 1), \quad v_j \in M(2\tilde{h}_j \cdot \tilde{n}_j, 1)$$

$$u_i = \dot{\prod}_{1 \le j \le r(i)} \dot{\prod}_{1 \le t \le m_{ij}} u_i(j,t), \quad u_i(j,t) \in M(k_j, 1) \quad (1 \le i \le r)$$

$$v_i = \dot{\prod}_{1 \le j \le s(i)} \dot{\prod}_{1 \le t \le n_{ij}} v_i(j,t), \quad v_i(j,t) \in M(2h_j, 1) \quad (1 \le i \le s)$$

Apply Lemma 2.3.17 to all non-zero $u_i(j,t)$, and apply Lemma 2.3.20 to all non-zero u_i. Then we have a $G_i \in C_1(\tilde{k}_i, \tilde{m}_i) \cap GL(\tilde{k}_i \cdot \tilde{m}_i, \mathbb{R})$ such that $(u_i')^T G_i = u_i^T$, if $u_i \neq 0$, where u_i' is a vector in (4). If $u_i = 0$, set $G_i = I(\tilde{k}_i \cdot \tilde{m}_i)$. Also, apply Lemma 2.3.17 to all non-zero $v_i(j,t)$, and apply Lemma 2.3.20 to all non-zero v_i. Then we have an $H_i \in C_2(\tilde{h}_i, \tilde{n}_i) \cap GL(2\tilde{h}_i \cdot \tilde{n}_i, \mathbb{R})$ such that $(v_i')^T H_i = v_i^T$, if $v_i \neq 0$, where v_i' is a vector in (5). If $v_i = 0$, set $H_i = I(\tilde{k}_i \cdot \tilde{m}_i)$. Then the matrix

$$F = \dot{\sum}_{i=1}^{r} G_i \dot{+} \dot{\sum}_{j=1}^{s} H_j \in GL(n, \mathbb{R})$$

satisfies (i) $JF = FJ$, and (ii) $(\alpha')^T F = \alpha^T$. For

$$\delta = \begin{cases} \beta & (\text{if } \beta \neq 0) \\ 1 & (\text{if } \beta = 0) \end{cases},$$

set $G = (1/\delta)F$. Then clearly (1) holds. (2) also holds, because

$$\begin{aligned} y \in G(U) \quad &\Longleftrightarrow \quad \langle \alpha, G^{-1}y \rangle = \beta \quad \Longleftrightarrow \quad \langle (G^{-1})^T \alpha, y \rangle = \beta \\ &\Longleftrightarrow \quad \langle \delta (F^{-1})^T \alpha, y \rangle = \beta \quad \Longleftrightarrow \quad \langle \alpha', y \rangle = \varepsilon, \end{aligned}$$

where $\varepsilon = 0$ if $\beta = 0$; $\varepsilon = 1$ if $\beta \neq 0$.

To prove the second statement, suppose (G', α') and (G'', α'') satisfy conditions (1) - (5). Set $G = G'(G'')^{-1}$. Since $JG' = G'J$ and $JG'' = G''J$, we have $JG = GJ$. Hence, by Lemma 2.3.11, G is written as follows:

$$G = G_1 \dot{+} G_2$$

$$G_1 = (A_{ij}), \quad A_{ij} \in M(\tilde{k}_i \cdot \tilde{m}_i, \tilde{k}_j \cdot \tilde{m}_j)$$

$$A_{ij} = O \quad (i \neq j), \quad A_{ii} \in C_1(\tilde{k}_i, \tilde{m}_i)$$

$$G_2 = (B_{ij}), \quad B_{ij} \in M_2(\tilde{h}_i \cdot \tilde{n}_i, \tilde{h}_j \cdot \tilde{n}_j)$$

$$B_{ij} = O \quad (i \neq j), \quad B_{ii} \in C_2(\tilde{h}_i, \tilde{n}_i)$$

As was done in (3), divide α' and α'' into blocks:

$$\alpha' = \dot{\prod}_{i=1}^{r} u_i' \dot{\times} \dot{\prod}_{j=1}^{s} v_j'$$

$$\alpha'' = \dot{\prod}_{i=1}^{r} u_i'' \dot{\times} \dot{\prod}_{j=1}^{s} v_j''$$

Since
$$(\alpha')^T G' = \delta' \alpha^T \quad \text{and} \quad (\alpha'')^T G'' = \delta'' \alpha^T$$
for some $\delta', \delta'' \in \mathbb{R} - \{0\}$, we have
$$(\alpha')^T G = (\alpha')^T G' (G'')^{-1} = \delta' \alpha^T (G'')^{-1} = (\delta'/\delta'')(\alpha'')^T.$$
Hence we have, for $\delta = \delta''/\delta'$,
$$(u_i')^T(\delta A_{ii}) = (u_i'')^T \quad (1 \leq i \leq r), \quad \text{and}$$
$$(v_j')^T(\delta B_{jj}) = (v_j'')^T \quad (1 \leq j \leq s).$$
By Lemma 2.3.13(3), we obtain $u_i' = u_i''$ $(1 \leq i \leq r)$ and $v_j' = v_j''$ $(1 \leq j \leq s)$, that is $\alpha' = \alpha''$. □

Definition 2.3.22. The normal vector $\alpha' \in \mathbb{R}^n - \{0\}$ determined from Theorem 2.3.21 is called *a canonical normal vector* corresponding to the real Jordan matrix J (or, the linear vector field X_J). A boundary $U = U(\alpha', \varepsilon)(\varepsilon = 0$ or $1)$ is called *a canonical boundary* for J (or X_J).

Example 2.3.23. (Two-Dimensional Canonical Normal Vectors)
The two-dimensional real Jordan matrices have the following three types:
$$(1) \begin{pmatrix} \lambda & 1 \\ 0 & \lambda \end{pmatrix}, \quad (2) \begin{pmatrix} \lambda & 0 \\ 0 & \mu \end{pmatrix}, \quad (3) \begin{pmatrix} a & b \\ -b & a \end{pmatrix}.$$
The canonical normal vector α corresponding to each Jordan matrix J is given as follow:

Case(1). $J = J_1(\lambda; 2)$;
$$\alpha = (1,0)^T, \ (0,1)^T.$$

Case(2). If $\lambda \neq \mu$, $J = J_1(\lambda; 1) \dot{+} J_1(\mu; 1)$:
$$\alpha = (1,0)^T, \ (1,1)^T.$$

If $\lambda = \mu$, $J = J_1(\lambda; 1, 2)$:
$$\alpha = (1,0)^T.$$

Case(3). $J = J_2(a + \sqrt{-1}b; 1)$;
$$\alpha = (1,0)^T.$$

Example 2.3.24. (Three-Dimensional Canonical Normal Vectors)
The three-dimensional real Jordan matrices have the following four types:

$$(1) \begin{pmatrix} \lambda & 1 & 0 \\ 0 & \lambda & 1 \\ 0 & 0 & \lambda \end{pmatrix}, \quad (2) \begin{pmatrix} \lambda & 0 & 0 \\ 0 & \mu & 1 \\ 0 & 0 & \mu \end{pmatrix},$$

$$(3) \begin{pmatrix} \lambda & 0 & 0 \\ 0 & \mu & 0 \\ 0 & 0 & \nu \end{pmatrix}, \quad (4) \begin{pmatrix} \lambda & 0 & 0 \\ 0 & a & b \\ 0 & -b & a \end{pmatrix}.$$

The canonical normal vector α corresponding to each Jordan matrix J is given as follow:

Case(1). $J = J_1(\lambda; 3)$;

$$\alpha = (1,0,0)^T,\ (0,1,0)^T,\ (0,0,1)^T.$$

Case(2). If $\lambda \neq \mu$, $J = J_1(\lambda; 1) \dot{+} J_1(\mu; 2)$;

$$\alpha = (1,1,0)^T,\ (1,0,1)^T,\ (1,0,0)^T,\ (0,1,0)^T,\ (0,0,1)^T.$$

If $\lambda = \mu$, $J = J_1(\lambda; (1,2), (1,1))$;

$$\alpha = (1,0,0)^T,\ (0,1,0)^T,\ (0,0,1)^T.$$

Case(3). If $\lambda \neq \mu \neq \nu \neq \lambda$, $J = J_1(\lambda; 1) \dot{+} J_1(\mu; 1) \dot{+} J_1(\nu; 1)$;

$$\alpha = (1,1,1)^T,\ (0,1,1)^T,\ (0,0,1)^T.$$

If $\lambda \neq \mu = \nu$, $J = J_1(\lambda; 1) \dot{+} J_1(\mu; 1, 2)$;

$$\alpha = (1,1,0)^T,\ (1,0,0)^T,\ (0,1,0)^T.$$

If $\lambda = \mu = \nu$, $J = J_1(\lambda; 1, 3)$;

$$\alpha = (1,1,0)^T.$$

Case(4). $J = J_1(\lambda; 1) \dot{+} J_2(a + \sqrt{-1}b; 1)$;

$$\alpha = (1,1,0)^T,\ (1,0,0)^T,\ (0,1,0)^T.$$

Example 2.3.25. (Four-Dimensional Canonical Normal Vectors)
The four-dimensional real Jordan matrices have the following nine types:

$$(1) \begin{pmatrix} \lambda & 1 & 0 & 0 \\ 0 & \lambda & 1 & 0 \\ 0 & 0 & \lambda & 1 \\ 0 & 0 & 0 & \lambda \end{pmatrix}, \quad (2) \begin{pmatrix} \lambda & 0 & 0 & 0 \\ 0 & \mu & 1 & 0 \\ 0 & 0 & \mu & 1 \\ 0 & 0 & 0 & \mu \end{pmatrix}, \quad (3) \begin{pmatrix} \lambda & 1 & 0 & 0 \\ 0 & \lambda & 0 & 0 \\ 0 & 0 & \mu & 1 \\ 0 & 0 & 0 & \mu \end{pmatrix},$$

$$(4)\begin{pmatrix} \lambda & 0 & 0 & 0 \\ 0 & \mu & 0 & 0 \\ 0 & 0 & \nu & 1 \\ 0 & 0 & 0 & \nu \end{pmatrix}, \quad (5)\begin{pmatrix} \lambda & 0 & 0 & 0 \\ 0 & \mu & 0 & 0 \\ 0 & 0 & \nu & 0 \\ 0 & 0 & 0 & \eta \end{pmatrix}, \quad (6)\begin{pmatrix} \lambda & 1 & 0 & 0 \\ 0 & \lambda & 0 & 0 \\ 0 & 0 & a & b \\ 0 & 0 & -b & a \end{pmatrix},$$

$$(7)\begin{pmatrix} \lambda & 0 & 0 & 0 \\ 0 & \mu & 0 & 0 \\ 0 & 0 & a & b \\ 0 & 0 & -b & a \end{pmatrix}, \quad (8)\begin{pmatrix} a & b & 0 & 0 \\ -b & a & 0 & 0 \\ 0 & 0 & c & d \\ 0 & 0 & -d & c \end{pmatrix}, \quad (9)\begin{pmatrix} a & b & 1 & 0 \\ -b & a & 0 & 1 \\ 0 & 0 & a & b \\ 0 & 0 & -b & a \end{pmatrix}.$$

The canonical normal vector α corresponding to each Jordan matrix J is given as follow:

Case(1). $J = J_1(\lambda; 4)$;

$$\alpha = (1,0,0,0)^T,\ (0,1,0,0)^T,\ (0,0,1,0)^T,\ (0,0,0,1)^T.$$

Case(2). If $\lambda \neq \mu$, $J = J_1(\lambda; 1) \dot{+} J_1(\mu; 3)$;

$$\begin{aligned} \alpha &= (1,1,0,0)^T,\ (1,0,1,0)^T,\ (1,0,0,1)^T,\ (1,0,0,0)^T, \\ &\quad (0,1,0,0)^T,\ (0,0,1,0)^T,\ (0,0,0,1)^T. \end{aligned}$$

If $\lambda = \mu$, $J = J_1(\lambda; (1,3),(1,1))$;

$$\alpha = (1,0,0,0)^T,\ (0,1,0,0)^T,\ (0,0,1,0)^T,\ (0,0,0,1)^T.$$

Case(3). If $\lambda \neq \mu$, $J = J_1(\lambda; 2) \dot{+} J_1(\mu; 2)$;

$$\alpha = (1,0,1,0)^T,\ (1,0,0,1)^T,\ (1,0,0,0)^T,\ (0,1,0,1)^T,\ (0,1,0,0)^T.$$

If $\lambda = \mu$, $J = J_1(\lambda; 2, 2)$:

$$\alpha = (1,0,0,0)^T,\ (0,1,0,0)^T.$$

Case(4). If $\lambda \neq \mu \neq \nu \neq \lambda$, $J = J_1(\lambda; 1) \dot{+} J_1(\mu; 1) \dot{+} J_1(\nu; 2)$;

$$\begin{aligned} \alpha &= (1,1,1,0)^T,\ (1,1,0,1)^T,\ (1,1,0,0)^T,\ (1,0,1,0)^T,\ (1,0,0,1)^T, \\ &\quad (1,0,0,0)^T,\ (0,0,1,0)^T,\ (0,0,0,1)^T. \end{aligned}$$

If $\lambda \neq \mu = \nu$, $J = J_1(\lambda; 1) \dot{+} J_1(\mu; (1,2),(1,1))$:

$$\begin{aligned} \alpha &= (1,1,0,0)^T,\ (1,0,1,0)^T,\ (1,0,0,1)^T,\ (1,0,0,0)^T,\ (0,1,0,0)^T, \\ &\quad (0,0,1,0)^T,\ (0,0,0,1)^T. \end{aligned}$$

If $\lambda = \mu \neq \nu$, $J = J_1(\lambda; 1, 2) \dot{+} J_1(\nu; 2)$;

$$\alpha = (1,0,1,0)^T,\ (1,0,0,1)^T,\ (1,0,0,0)^T,\ (0,0,1,0)^T,\ (0,0,0,1)^T.$$

If $\lambda = \mu = \nu$, $J = J_1(\lambda; (1,2), (2,1))$;

$$\alpha = (1,0,0,0)^T,\ (0,0,1,0)^T,\ (0,0,0,1)^T.$$

Case(5). If all eigenvalues are distinct, $J = J_1(\lambda; 1) \dot{+} J_1(\mu; 1) \dot{+} J_1(\nu; 1) \dot{+} J_1(\eta; 1)$;

$$\alpha = (1,1,1,1)^T,\ (0,1,1,1)^T,\ (0,0,1,1)^T,\ (0,0,0,1)^T.$$

If $\lambda = \mu \neq \nu \neq \eta \neq \mu$, $J = J_1(\lambda; 1, 2) \dot{+} J_1(\nu; 1) \dot{+} J_1(\eta; 1)$;

$$\alpha = (1,0,1,1)^T,\ (1,0,1,0)^T,\ (1,0,0,0)^T,\ (0,0,1,1)^T,\ (0,0,1,0)^T.$$

If $\lambda = \mu \neq \nu = \eta$, $J = J_1(\lambda; 1, 2) \dot{+} J_1(\nu; 1, 2)$;

$$\alpha = (1,0,1,0)^T,\ (1,0,0,0)^T.$$

If $\lambda = \mu = \nu \neq \eta$, $J = J_1(\lambda; 1, 3) \dot{+} J_1(\eta; 1)$;

$$\alpha = (1,0,0,1)^T,\ (1,0,0,0)^T,\ (0,0,0,1)^T.$$

If $\lambda = \mu = \nu = \eta$, $J = J_1(\lambda; 1, 4)$;

$$\alpha = (1,0,0,0)^T.$$

Case(6). $J = J_1(\lambda; 2) \dot{+} J_2(a + \sqrt{-1}b; 1)$;

$$\alpha = (1,0,1,0)^T,\ (1,0,0,0)^T,\ (0,1,1,0)^T,\ (0,1,0,0)^T,\ (0,0,1,0)^T.$$

Case(7). If $\lambda \neq \mu$, $J = J_1(\lambda; 1) \dot{+} J_1(\mu; 1) \dot{+} J_2(a + \sqrt{-1}b; 1)$;

$$\alpha = (1,1,1,0)^T,\ (1,1,0,0)^T,\ (1,0,1,0)^T,\ (1,0,0,0)^T,\ (0,0,1,0)^T.$$

If $\lambda = \mu$, $J = J_1(\lambda; 1, 2) \dot{+} J_2(a + \sqrt{-1}b; 1)$;

$$\alpha = (1,0,1,0)^T,\ (1,0,0,0)^T,\ (0,0,1,0)^T.$$

Case(8). If $a + \sqrt{-1}b \neq c \pm \sqrt{-1}d$, $J = J_2(a + \sqrt{-1}b; 1) \dot{+} J_2(c + \sqrt{-1}d; 1)$;

$$\alpha = (1,0,1,0)^T,\ (1,0,0,0)^T.$$

If $a + \sqrt{-1}b = c \pm \sqrt{-1}d$, $J = J_2(a + \sqrt{-1}b; 2)$;

$$\alpha = (1,0,0,0)^T.$$

Case(9). $J = J_2(a + \sqrt{-1}b; 2)$;

$$\alpha = (1,0,0,0)^T,\ (0,0,1,0)^T.$$

2.3.3 Normal Forms of Degenerate Affine Vector Fields with a Boundary

In this subsection, we consider normal forms of degenerate affine vector fields with a boundary, under affine conjugacy. As stated in Subsection 2.3.2, we can assume that a degenerate affine vector field and a boundary have the following forms;

$$\begin{aligned} &X_{(A,q)}(x) = Ax + q \quad (A \in M(n), q \in \mathbb{R}^n) \\ &q \notin A(\mathbb{R}^n) \\ &U = U(\alpha, \beta) = \{\langle \alpha, x \rangle = \beta\} \quad (\alpha \in \mathbb{R}^n - \{0\}, \beta \in \mathbb{R}). \end{aligned} \tag{2.3.2}$$

It follows from Equation (2.3.2) that the rank of A is less than n, hence A has zero eigenvalues. Under suitable linear transformation, A is transformed to a real Jordan matrix J, and by Theorem 2.3.21, the normal vector α is transformed to a canonical normal vector for J:

$$\begin{aligned} &X_{(J,q)}(x) = Jx + q \\ &J = J_* \dot{+} J_1(0; \tilde{k}, \tilde{m}), \quad \tilde{k} \in \mathbb{N}^r_<, \tilde{m} \in \mathbb{N}, \\ &J_* \quad \text{is a real Jordan matrix with } \det J_* \neq 0 \\ &q \notin J(\mathbb{R}^n) \\ &U = U(\alpha, \beta) = \{\langle \alpha, x \rangle = \beta\}, \\ &\alpha \quad \text{is a canonical normal vector for } J\,. \end{aligned}$$

Our goal in this subsection is to simplify $q \in \mathbb{R}^n$ and $\beta \in \mathbb{R}$ as much as possible, under affine transformation by which J and α are invariant.

Standing Assumptions.

(1) As stated in Remark 2.3.8, we will remove the subscript 1 of $J_1(\lambda; k), e_1(n; k)$ etc., in this subsection, if there is no confusion.

(2) We will assume

$$\tilde{k} = (k_1, \cdots, k_r) \in \mathbb{N}^r_<, \quad \tilde{m} = (m_1, \cdots, m_r) \in \mathbb{N}^r,$$

unless otherwise stated.

Definition 2.3.26.

(1) For $1 \leq i \leq r$, $e(\tilde{k}, \tilde{m}; (i, m_i, k_i))$ is simplified as $e(\tilde{k}, \tilde{m}; (i, \mathrm{end}))$.

(2) For $1 \le i, l \le r$, and $\tau_1, \tau_2 = 0$ or 1, define

$$\begin{aligned} & e(\tilde{k}, \tilde{m}; (i, 1, k_i), \tau_2; (l, \mathrm{end}), \tau_1) \\ &\qquad = \tau_2 e(\tilde{k}, \tilde{m}; (i, 1, k_i)) + \tau_1 e(\tilde{k}, \tilde{m}; (l, \mathrm{end})). \end{aligned}$$

I.e., all components are 0, except the $(i, 1, k_i)$-component $= \tau_2$ and the (l, m_l, k_l)-component $= \tau_1$.

(3) For $1 \le u \le k_i, 1 \le i, l \le r$, and $\tau_1, \tau_2, \tau_3 = 0$ or 1, define

$$\begin{aligned} & e(\tilde{k}, \tilde{m}; (i, 1, u-1), \tau_3; (i, 1, k_i), \tau_2; (l, \mathrm{end}), \tau_1) \\ &\qquad = \tau_3 e(\tilde{k}, \tilde{m}; (i, 1, u-1)) + e(\tilde{k}, \tilde{m}; (i, 1, k_i), \tau_2; (l, \mathrm{end}), \tau_1). \end{aligned}$$

Lemma 2.3.27. *Assume*

$$J = J_* \dot{+} J(0; \tilde{k}, \tilde{m}), \quad J_* \in GL(n_1, \mathbb{R}), \quad n_2 = \tilde{k} \cdot \tilde{m}.$$

Set

$$X(x) = Jx + q, \quad q \notin J(\mathbb{R}^n), \quad U = \{\langle \alpha, x \rangle = \beta\},$$

where $n = n_1 + n_2$, and α is a canonical normal vector for J. Then there is an affine transformation

$$h : \mathbb{R}^n \to \mathbb{R}^n; h(x) = Hx + p \quad (H \in GL(n, \mathbb{R}), \quad p \in \mathbb{R}^n)$$

such that

$$h_* X(x) = Jx + q', \quad h(U) = \{\langle \alpha, x \rangle = \beta'\} \ (\beta' \in \mathbb{R}),$$

where $q' = (q'(i, s, u)) \in \mathbb{R}^n \quad (1 \le i \le r, 1 \le s \le m_i, 1 \le u \le k_i)$, and

$$q'(i, s, u) = \begin{cases} 0 \text{ or } 1 & \text{if } u = k_i \\ 0 & \text{if } u \ne k_i. \end{cases}$$

Proof. Set $J_2 = J(0; \tilde{k}, \tilde{m})$. Split $\mathbb{R}^n$ into the direct sum:

$$\mathbb{R}^n = \mathbb{R}^{n_1} \dot{+} J_2(\mathbb{R}^{n_2}) \dot{+} \mathrm{Ker}(J_2),$$

where $\mathrm{Ker}(J_2) = \{x \in \mathbb{R}^{n_2} \mid J_2 x = 0\}$ and split q into

$$q = q_1 \dot{\times} q_2 \dot{\times} q_3, \quad q_1 \in \mathbb{R}^{n_1}, q_2 \in J_2(\mathbb{R}^{n_2}), q_3 \in \mathrm{Ker}(J_2).$$

Also, split $\alpha \in \mathbb{R}^n - \{0\}$ into

$$\alpha = \alpha_1 \dot{\times} \alpha_2, \quad \alpha_1 \in \mathbb{R}^{n_1}, \alpha_2 \in \mathbb{R}^{n_2}.$$

Since α is canonical, we have, for some $0 \le i_* \le r$ and $1 \le u_* \le k_{i_*}$,

$$\alpha_2 = e(\tilde{k}, \tilde{m}; (i_*, 1, u_*)),$$

where $\alpha_2 = 0$ if $i_* = 0$ (i.e. $e(\tilde{k}, \tilde{m}; (0, 1, u_*)) = 0$). Since $q_1 \in J_*(\mathbb{R}^{n_1})$ and $q_2 \in J_2(\mathbb{R}^{n_2})$, there is a $p' \in \mathbb{R}^n$ such that

$$Jp' = q_1 \dot{\times} q_2 \dot{\times} O.$$

Define an affine transformation $h_1 : \mathbb{R}^n \to \mathbb{R}^n$ by $h_1(x) = x + p'$. Then

$$h_{1*}X(x) = Jx - Jp' + q = Jx + (O \dot{\times} q_3)$$

$$h_1\{\langle \alpha, x \rangle = \beta\} = \{\langle \alpha, x \rangle = \langle \alpha, p' \rangle + \beta\} = \{\langle \alpha, x \rangle = \beta''\},$$

where $\beta'' = \langle \alpha, p' \rangle + \beta$. Since $q_3 \in \mathrm{Ker}(J_2)$, there are $q_{ij} \in \mathbb{R}$ $(1 \le i \le r, 1 \le j \le m_i)$ such that

$$q_3 = \sum_{i=1}^{r} \sum_{j=1}^{m_i} q_{ij} e(\tilde{k}, \tilde{m}; (i, j, k_i)).$$

Set, for $1 \le i \le r, 1 \le j \le m_i$,

$$\delta(i,j) = \begin{cases} 1/q_{ij} & (q_{ij} \ne 0) \\ 1 & (q_{ij} = 0), \end{cases}$$

Define $G \in GL(n, \mathbb{R})$ as follows:
(i) If $i_* = 0$ (i.e. $\alpha_2 = 0$), then

$$G = I(n_1) \dot{+} \dot{\sum}_{i=1}^{r} \dot{\sum}_{j=1}^{m_i} \delta(i,j) I(k_i).$$

(ii) If $1 \le i_* \le r$ (i.e. $\alpha_2 \ne 0$), then

$$G = \delta(i_*, 1) I(n_1) \dot{+} \dot{\sum}_{i=1}^{r} \dot{\sum}_{j=1}^{m_i} \delta(i,j) I(k_i).$$

Then we have

$$G_* h_{1*} X(x) = Jx + q',$$

where

$$q' = O(n_1, 1) \dot{+} q_3', \quad q_3' = \sum_{i=1}^{r} \sum_{j=1}^{m_i} q_{ij}' e(\tilde{k}, \tilde{m}; (i, j, k_i)), \quad q_{ij}' = 0 \text{ or } 1.$$

If $i_* = 0$, since $\alpha_2 = 0$, we have $\alpha^T G = \alpha^T$. If $1 \le i_* \le r$, by the definition of G, we have

$$\alpha^T G = \delta(i_*, 1) \alpha^T.$$

Hence,

$$Gh_1(U) = \{\langle (G^{-1})^T\alpha, x\rangle = \beta''\} = \{\langle \alpha, x\rangle = \delta\beta''\} = \{\langle \alpha, x\rangle = \beta'\},$$

where $\beta' = \delta\beta''$ and

$$\delta = \begin{cases} \delta(i_*, 1) & (\text{if } i_* \neq 0) \\ 1 & (\text{if } i_* = 0). \end{cases}$$

Therefore $h = Gh_1$ is a required transformation. □

Lemma 2.3.28. *Assume*

$$\alpha = e(\tilde{k}, \tilde{m}; (i_*, 1, u_*)) \quad (1 \le i_* \le r, 1 \le u_* \le k_{i_*}).$$

Let

$$p = (p(i, s, u)) \in \mathbb{R}^n, \quad n = \tilde{k}\cdot\tilde{m}$$

be given, where

$$p(i, s, u) = \begin{cases} 0 \quad \text{or } 1 & (\text{ if } u = k_i) \\ 0 & (\text{ otherwise}). \end{cases}$$

Set

$$j_* = \max\{1 \le i \le r \mid p(i, s, k_i) \neq 0 \quad \text{for some } 1 \le s \le m_i\}$$

$$s_* = \max\{1 \le s \le m_{j_*} \mid p(j_*, s, k_{j_*}) \neq 0\}.$$

(1) There is a $G \in C_1(\tilde{k}, \tilde{m}) \cap GL(n, \mathbb{R})$ such that

$$Gp = e,$$

where $e = e(\tilde{k}, \tilde{m}; (j_*, m_{j_*}, k_{j_*}))$.

(2) If $(j_*, s_*) \neq (i_*, 1)$, then there is a $G \in C_1(\tilde{k}, \tilde{m}; \alpha) \cap GL(n, \mathbb{R})$ such that $\alpha^T G = \alpha^T$ and

$$Gp = e + \tilde{p},$$

where

$$e = e(\tilde{k}, \tilde{m}; (j_*, m_{j_*}, k_{j_*})), \qquad \tilde{p} = \varepsilon e(\tilde{k}, \tilde{m}; (i_*, 1, k_{i_*}),$$

$$\varepsilon = \begin{cases} 0 & (\text{if } p(i_*, 1, k_{i_*}) = 0) \\ 1 & (\text{if } p(i_*, 1, k_{i_*}) = 1). \end{cases}$$

(3) If $j_* = i_*$ and $p(i_*, 1, k_{i_*}) = 1$, then there is a $G \in C_1(\tilde{k}, \tilde{m}; \alpha) \cap GL(n, \mathbb{R})$ such that $\alpha^T G = \alpha^T$ and

$$Gp = e,$$

where $e = e(\tilde{k}, \tilde{m}; (i_*, 1, k_{i_*}))$.

Proof. (1) Clearly there is a linear transformation in $C_1(\tilde{k},\tilde{m})\cap GL(n,\mathbb{R})$ which exchanges the (j_*,s_*)-block of p and the (j_*,m_{j_*})-block of p. Hence we can assume $s_*=m_{j_*}$, without loss of generality. Define $G=(G_{ij,st}), G_{ij,st}\in M(k_i,k_j)(1\le i,j\le r, 1\le s\le m_i, 1\le t\le m_j)$ as follows: all $G_{ij,st}=O$ except

$$G_{ii,ss}=I(k_i)\quad(1\le i\le r, 1\le s\le m_i),$$

and

$$G_{ii,st}=-p(j,t,k_j)[OI(k_j)]\quad\text{if } (i,s)=(j_*,m_{j_*})\quad\text{and } 1\le t\le m_j,$$

for $1\le j\le j_*$. It is easy to verify that $G\in C_1(\tilde{k},\tilde{m})\cap GL(n,\mathbb{R})$ and $Gp=e(\tilde{k},\tilde{m};(j_*,m_{j_*},k_{j_*}))$.

(2) Suppose $(j_*,s_*)\ne(i_*,1)$. Then there is a linear transformation in $C_1(\tilde{k},\tilde{m};\alpha)\cap GL(n,\mathbb{R})$ which exchanges the (j_*,s_*)-block of p and the (j_*,m_{j_*})-block of p. Hence we can assume $s_*=m_{j_*}$, without loss of generality.

Define $G=(G_{ij,st}), G_{ij,st}\in M(k_i,k_j)(1\le i,j\le r, 1\le s\le m_i, 1\le t\le m_j)$ as follows; all $G_{ij,st}=O$ except

$$G_{ii,ss}=I(k_i)\quad(1\le i\le r, 1\le s\le m_i),$$

and

$$G_{ij,st}=-p(j,t,k_j)[OI(k_j)]\quad\text{if } (i,s)=(j_*,m_{j_*})\quad\text{and } (j,t)\ne(i_*,1),$$

for $1\le j\le j_*$. It is easy to verify that $G\in C_1(\tilde{k},\tilde{m})\cap GL(n,\mathbb{R})$ and G satisfies the conclusion.

(3) Suppose $j_*=i_*$ and $p(i_*,1,k_{i_*})=1$. Define $G=(G_{ij,st}), G_{ij,st}\in M(k_i,k_j)(1\le i,j\le r, 1\le s\le m_i, 1\le t\le m_j)$ as follows: all $G_{ij,st}=O$ except

$$G_{ii,ss}=I(k_i)\quad(1\le i\le r, 1\le s\le m_i),$$

and

$$G_{ij,st}=-p(j,t,k_j)[OI(k_j)]\quad\text{if } (i,s)=(i_*,m_{i_*})\quad\text{and } (j,t)\ne(i_*,1),$$

for $1\le j\le i_*$. It is easy to verify that $G\in C_1(\tilde{k},\tilde{m})\cap GL(n,\mathbb{R})$ and G satisfies the conclusion. □

Lemma 2.3.29. *Assume*

$$J=J_*\dot{+}J(0;\tilde{k},\tilde{m}),\quad J_*\in GL(n_1,\mathbb{R}),\quad n_2=\tilde{k}\cdot\tilde{m}.$$

Set

$$X(x)=Jx+q,\quad q\notin J(\mathbb{R}^n),\quad n=n_1+n_2,$$

$$U=\{\langle\alpha,x\rangle=\beta\},\quad \beta\in\mathbb{R},$$

where

$$q = O(n_1, 1)\overset{\bullet}{\times}(\sum_{i=1}^{r}\sum_{j=1}^{m_i} q_{ij} e(\tilde{k}, \tilde{m}; (i, j, k_i))),$$

$$q_{ij} = 0 \quad or\ 1 \quad (1 \leq i \leq r, 1 \leq j \leq m_i),$$

$\alpha = \alpha_1 \times \alpha_2$ $(\alpha_1 \in \mathbb{R}^{n_1}, \alpha_2 \in \mathbb{R}^{n_2})$ *is a canonical normal vector for* J,

$$\alpha_2 = e(\tilde{k}, \tilde{m}; (i_*, 1, u_*)), \quad 0 \leq i_* \leq r,\ 1 \leq u_* \leq k_{i_*}.$$

Then there is an $H \in GL(n, \mathbb{R})$ *such that*

$$H_* X(x) = Jx + q',$$

$$q' = O(n_1, 1)\overset{\bullet}{\times}$$

$$e(\tilde{k}, \tilde{m}; (i_*, 1, k_{i_*}), \varepsilon; (l_*, end), \tau)$$

$$H(U) = U,$$

where

$$l_* = \max\{1 \leq i \leq r \mid q_{ij} \neq 0 \quad for\ some 1 \leq j \leq m_i\}$$

$$\varepsilon = \begin{cases} 0 & (if\ i_* = 0 \quad or\ q_{i_*1} = 0) \\ 1 & (otherwise) \end{cases}$$

$$\tau = \begin{cases} 0 & (if\ \varepsilon = 1 \quad and\ l_* = i_*) \\ 1 & (otherwise). \end{cases}$$

Proof. (1) Case of $i_* = 0$ (i.e. $\alpha_2 = 0$). Define $H \in GL(n, \mathbb{R})$ by

$$H = I(n_1)\overset{\bullet}{+}G,$$

where $G \in C_1(\tilde{k}, \tilde{m}) \cap GL(n_2, \mathbb{R})$ is the matrix determined by Lemma 2.3.28(1). Then H satisfies $JH = HJ$ and $Hq = O(n_1, 1)\overset{\bullet}{\times}e(\tilde{k}, \tilde{m}; (l_*, \text{end}))$ (i.e., $\varepsilon = 0$ and $\tau = 1$). Since $\alpha = \alpha_1 \overset{\bullet}{\times} O$ and $\alpha^T H = (\alpha_1^T I, O^T G) = \alpha^T$,

$$H(U) = H\{\langle \alpha, x\rangle = \beta\} = \{\langle (H^{-1})^T \alpha, x\rangle = \beta\} = U.$$

(2) Case of $1 \leq i_* \leq r$ and $q_{i_*1} = 0$. Define $H \in GL(n, \mathbb{R})$ by

$$H = I(n_1)\overset{\bullet}{+}G,$$

where $G \in C_1(\tilde{k}, \tilde{m}; \alpha_2) \cap GL(n_2, \mathbb{R})$ is the matrix determined by Lemma 2.3.28(2). Then H satisfies $JH = HJ$ and $Hq = O(n_1, 1)\overset{\bullet}{\times}e(\tilde{k}, \tilde{m}; (l_*, \text{end}))$ (i.e., $\varepsilon = 0$ and $\tau = 1$). Since $\alpha_2^T G = \alpha_2^T$ and $\alpha^T H = (\alpha_1^T I, \alpha_2^T G) = \alpha^T$,

$$H(U) = H\{\langle \alpha, x\rangle = \beta\} = \{\langle (H^{-1})^T\alpha, x\rangle = \beta\} = U.$$

(3) Case of $1 \le i_* \le r$ and $q_{i_*1} = 1$ and $l_* = i_*$. Define $H \in GL(n, \mathbb{R})$ by

$$H = I(n_1)\dot{+}G,$$

where $G \in C_1(\tilde{k}, \tilde{m}; \alpha_2) \cap GL(n_2, \mathbb{R})$ is the matrix determined by Lemma 2.3.28(3). Then H satisfies $JH = HJ$ and $Hq = O(n_1, 1)\dot{\times}e(\tilde{k}, \tilde{m}; (i_*, 1, k_{i_*}))$ (i.e., $\varepsilon = 1$ and $\tau = 0$). Similar to (ii), we have $H(U) = U$.

(4) Case of otherwise, i.e., $1 \le i_* \le r$ and $q_{i_*1} = 1$ and $l_* > i_*$. Define $H \in GL(n, \mathbb{R})$ by

$$H = I(n_1)\dot{+}G,$$

where $G \in C_1(\tilde{k}, \tilde{m}; \alpha_2) \cap GL(n_2, \mathbb{R})$ is the matrix determined by Lemma 2.3.28(2). Then H satisfies $JH = HJ$ and

$$Hq = O(n_1, 1)\dot{\times}(e(\tilde{k}, \tilde{m}; (i_*, 1, k_{i_*})) + e(\tilde{k}, \tilde{m}; (l_*, \mathrm{end})))$$

(i.e., $\varepsilon = \tau = 1$). Similar to Case (2), we have $H(U) = U$. □

Lemma 2.3.30. *Assume*

$$J = J_*\dot{+}J(0; \tilde{k}, \tilde{m}), \quad J_* \in GL(n_1, \mathbb{R}), \quad n_2 = \tilde{k} \cdot \tilde{m}.$$

Set

$$X(x) = Jx + q, \quad q \notin J(\mathbb{R}^n), \quad n = n_1 + n_2,$$
$$U = \{\langle \alpha, x\rangle = \beta\}, \quad \beta \in \mathbb{R},$$

where

$$q = O(n_1, 1)\dot{\times}e(\tilde{k}, \tilde{m}; (i_*, 1, k_{i_*}), \tau_2; (l_*, \mathit{end}), \tau_1))$$

$$\tau_2 = \begin{cases} 0 & (\text{if } i_* = 0) \\ 1 & (\text{otherwise}) \end{cases}$$

$$\tau_1 = \begin{cases} 0 & (\text{if } \tau_2 = 1 \text{ and } l_* \le i_*) \\ 1 & (\text{otherwise}) \end{cases}$$

$\alpha = \alpha_1\dot{\times}\alpha_2$ $(\alpha_1 \in \mathbb{R}^{n_1}, \alpha_2 \in \mathbb{R}^{n_2})$ *is a canonical normal vector for* J,

$$\alpha_2 = e(\tilde{k}, \tilde{m}; (i_*, 1, u_*)), \quad 0 \le i_* \le r, \quad 1 \le u_* \le k_{i_*}.$$

(1) If $i_* = 0$ *(i.e.,* $\alpha_2 = O(n_2, 1)$*), then there is an* $H \in GL(n, \mathbb{R})$ *such that*

$$H_*X(x) = Jx + q,$$

$$H(U) = \{\langle \alpha, x\rangle = \varepsilon\},$$

$$\varepsilon = \varepsilon(\beta) = \begin{cases} 0 & (\text{if } \beta = 0) \\ 1 & (\text{if } \beta \ne 0). \end{cases}$$

(2) If $i_* \neq 0$ (i.e., $\alpha_2 = e(\tilde{k}, \tilde{m}; (i_*, 1, u_*)), 1 \leq i_* \leq r, 1 \leq u_* \leq k_{i_*}$), then there is an affine transformation

$$h : \mathbb{R}^n \to \mathbb{R}^n, \quad h(x) = Hx + p \quad (H \in GL(n, \mathbb{R}), \quad p \in \mathbb{R}^n)$$

such that

$$h_* X(x) = Jx + q',$$
$$h(U) = \{\langle \alpha, x \rangle = 0\},$$

where

$$q' = O(n_1, 1) \overset{\bullet}{\times} \{e(\tilde{k}, \tilde{m}; (i_*, 1, k_{i_*}), \tau_2; (l_*, end), \tau_1) + \tau_3 e(\tilde{k}, \tilde{m}; (i_*, 1, u_* - 1))\}$$

$$\tau_3 = \begin{cases} 1 & (\textit{if } \tau_2 = 0 \textit{ and } \tau_1 = 1 \textit{ and } i_* > l_* \textit{ and } k_{l_*} < u_* - 1) \\ 0 & (\textit{otherwise}) \end{cases}$$

Proof. (1) Case of $i_* = 0$. Set

$$\delta = \delta(\beta) = \begin{cases} 1/\beta & (\text{if } \beta \neq 0) \\ 1 & (\text{if } \beta = 0) \end{cases}$$

$$H = (\delta I(n_1)) \overset{\bullet}{+} I(n_2) \in GL(n, \mathbb{R}).$$

Since

$$HJH^{-1} = J, \quad Hq = q, \quad \alpha^T H = \delta \alpha^T,$$

we have

$$H_* X(x) = HJH^{-1}x + Hq = Jx + q,$$
$$H\{\langle \alpha, x \rangle = \beta\} = \{\langle (H^{-1})^T \alpha, x \rangle = \beta\}$$
$$= \{\delta^{-1} \langle \alpha, x \rangle = \beta\} = \{\langle \alpha, x \rangle = \varepsilon\},$$

where

$$\varepsilon = \varepsilon(\beta) = \begin{cases} 0 & (\text{if } \beta = 0) \\ 1 & (\text{if } \beta \neq 0). \end{cases}$$

(2) Case of $i_* \neq 0$. Set

$$p = O(n_1, 1) \overset{\bullet}{\times} (-\beta) e(\tilde{k}, \tilde{m}; (i_*, 1, u_*)), \quad \text{and}$$
$$h_1 : \mathbb{R}^n \to \mathbb{R}^n, \quad h_1(x) = x + p.$$

Since $\langle \alpha, p \rangle = -\beta$, we have

$$h_1\{\langle\alpha,x\rangle=\beta\}=\{\langle\alpha,x\rangle=\langle\alpha,p\rangle+\beta\}=\{\langle\alpha,x\rangle=0\}.$$

Since $Jp=O(n_1,1)\stackrel{\bullet}{\times}(-\beta)J(0;\tilde{k},\tilde{m})e(\tilde{k},\tilde{m};(i_*,1,u_*))$, we have

$$\begin{aligned} h_{1*}X(x) &= Jx-Jp+q \\ &= Jx+O(n_1,1)\stackrel{\bullet}{\times}\{e(\tilde{k},\tilde{m};(i_*,1,k_{i_*}),\tau_2;(l_*,\text{end}),\tau_1) \\ &\quad +\beta e(\tilde{k},\tilde{m};(i_*,1,u_*-1))\}. \end{aligned}$$

Define a linear transformation $G=(G_{ij,st})\in GL(n_2,\mathbb{R}),G_{ij,st}\in M(k_i,k_j)$ as follows:

(i) Case of $\tau_2=1$. All $G_{ij,st}=O$, except

$$\begin{aligned} G_{ii,ss}&=I(k_i)\quad((i,s)\neq(i_*,1)) \\ G_{i_*i_*,11}&=(-\beta)N(k_{i_*};k_{i_*}-u_*+1)+I(k_{i_*}). \end{aligned}$$

(ii) Case of $\tau_2=0$ and $i_*=l_*$. All $G_{ij,st}=O$, except

$$\begin{aligned} G_{ii,ss}&=I(k_i)\quad(1\le i\le r,1\le s\le m_i) \\ G_{i_*i_*,1m_*}&=(-\beta)N(k_{i_*};k_{i_*}-u_*+1), \end{aligned}$$

where $m_*=m_{i_*}$.

(iii) Case of $\tau_2=0$ and $i_*<l_*$. All $G_{ij,st}=O$, except

$$\begin{aligned} G_{ii,ss}&=I(k_i)\quad(1\le i\le r,1\le s\le m_i) \\ G_{i_*l_*,1m_*}&=[O\ (-\beta)N(k_{i_*};k_{i_*}-u_*+1)], \end{aligned}$$

where $m_*=m_{i_*}$.

(iv) Case of $\tau_2=0$ and $l_*<i_*$ and $k_{l_*}\ge u_*-1$. All $G_{ij,st}=O$, except

$$\begin{aligned} G_{ii,ss}&=I(k_i)\quad(1\le i\le r,1\le s\le m_i) \\ G_{i_*l_*,1m_*}&=B\#O,\quad B=(-\beta)N(k_{i_*};k_{i_*}-u_*+1), \end{aligned}$$

where $m_*=m_{i_*}$.

(v) Case of $\tau_2=0$ and $l_*<i_*$ and $k_{l_*}<u_*-1$. All $G_{ij,st}=O$, except

$$\begin{aligned} G_{ii,ss}&=I(k_i)\quad((i,s)\neq(i_*,1)) \\ G_{i_*i_*,11}&=\delta I(k_i), \end{aligned}$$

where

$$\delta=\begin{cases}1/\beta & (\text{if }\beta\neq0)\\ 1 & (\text{if }\beta=0).\end{cases}$$

In Cases (i)–(iv), we can verify that

$$G\{e(\tilde{k},\tilde{m};(i_*,1,k_{i_*}),\tau_2;(l_*,\mathrm{end}),\tau_1)+\beta e(\tilde{k},\tilde{m};(i_*,1,u_*-1))\}$$
$$=e(\tilde{k},\tilde{m};(i_*,1,k_{i_*}),\tau_2;(l*,\mathrm{end}),\tau_1),$$

and, in Case (v),

$$G\{e(\tilde{k},\tilde{m};(i_*,1,k_{i_*},\tau_2;(l_*,\mathrm{end}),\tau_1)+\beta e(\tilde{k},\tilde{m};(i_*,1,u_*-1))\}$$
$$=e(\tilde{k},\tilde{m};(i_*,1,k_{i_*}),\tau_2;(l*,\mathrm{end}),\tau_1)+\varepsilon e(\tilde{k},\tilde{m};(i_*,1,u_*-1)),$$

where

$$\varepsilon=\begin{cases}0 & (\text{if } \beta=0)\\ 1 & (\text{if } \beta\neq 0).\end{cases}$$

In any case, setting $F=I(n_1)\dot{+}G$, we have

$$FJF^{-1}=J \quad \text{and} \quad F\{\langle\alpha,x\rangle=0\}=\{\langle\alpha,x\rangle=0\}.$$

Therefore, the affine transformation

$$h:\mathbb{R}^n\to\mathbb{R}^n,\quad h(x)=F(x+p)$$

is a required one. □

Lemma 2.3.31. *Assume*

$$H\in C_1(\tilde{k},\tilde{m}),\ \ J=J(0;\tilde{k},\tilde{m})\ \ \textit{and}\ \ p\in\mathbb{R}^n\ \ (n=\tilde{k}\cdot\tilde{m}).$$

(1) If $q=e(\tilde{k},\tilde{m};(l,t,k_1))$ $(1\le t\le m_l)$, and $i>l$, then for all $1\le s\le m_i$,

"the (i,s,k_i)-component of Hq" $=0$.

(2) Let $q=e(\tilde{k},\tilde{m};(l,m_l,k_l))$ and $q'=e(\tilde{k},\tilde{m};(l',s,k_{l'}))(1\le s\le m_{l'})$. If $H\in GL(n,\mathbb{R})$ and $q'=Jp+Hq$, then $l=l'$.

(3) Let

$$q=e(\tilde{k},\tilde{m};(i,1,k_i))+e(\tilde{k},\tilde{m};(l,m_l,k_l))(i<l)$$
$$q'=e(\tilde{k},\tilde{m};(i,1,k_i))+e(\tilde{k},\tilde{m};(l',m_{l'},k_{l'}))(i<l').$$

If $H\in GL(n,\mathbb{R})$ and $q'=Jp+Hq$, then $l=l'$.

Proof. Denote the $((i,s),(j,t))$-block of $H\in M(n)$ by $A_{ij,st}\in M(k_i,k_j)$.

(1) Since $i>l$ (i.e., $k_i>k_l$), the $((i,s),(l,t))$-block of H has the following form:

$$A_{il,st}=A\#O,\quad A\in C_1(k_l).$$

From this, it follows that the (i, s, k_i)-component of Hq is zero.

(2) Since $H \in C_1(\tilde{k}, \tilde{m}) \cap GL(n, \mathbb{R})$,

$$q' = Jp + Hq \quad \text{and} \quad q = -JH^{-1}p + H^{-1}q'.$$

Notice that the (l, m_l, k_l)-, and the $(l', s, k_{l'})$-components of Jp are zero, because $J = J(0; \tilde{k}, \tilde{m})$. Similarly, the (l, m_l, k_l)- and the $(l', s, k_{l'})$-components of $JH^{-1}p$ are zero. If $l < l'$, by (1), we have $q' \neq Jp + Hq$. If $l > l'$, by (1), we have $q \neq -JH^{-1}p + H^{-1}q'$. Hence, we have $l = l'$ if $q' = Jp + Hq$.

(3) If $l < l'$, by (1),

$$\text{“the } (l', m_{l'}, k_{l'})\text{-component of } Jp + Hq\text{”} = 0,$$

hence $q' \neq Jp + Hq$. If $l > l'$, by (2),

$$\text{“the } (l, m_l, k_l)\text{-component of } -JH^{-1}p + H^{-1}q'\text{”} = 0,$$

hence $q \neq -JH^{-1}p + H^{-1}q'$. Therefore, if $q' = Jp + Hq$, we have $l = l'$. □

Theorem 2.3.32. *Assume* $\tilde{k} = (k_1, \cdots, k_r) \in \mathbb{N}^r_<, \quad \tilde{m} = (m_1, \cdots, m_r) \in \mathbb{N}^r$.

$$J = J_* \dot{+} J(0; \tilde{k}, \tilde{m}), \quad J_* \in GL(n_1, \mathbb{R}), \quad n_2 = \tilde{k} \cdot \tilde{m}.$$

$$X(x) = Jx + q, \quad q \notin J(\mathbb{R}^n), \quad n = n_1 + n_2,$$

$$U = \{\langle \alpha, x \rangle = \beta\}, \quad \beta \in \mathbb{R},$$

and $\alpha = \alpha_1 \dot{\times} \alpha_2 \quad (\alpha_1 \in \mathbb{R}^{n_1}, \alpha_2 \in \mathbb{R}^{n_2})$ *is a canonical normal vector for* J.

(1) If $\alpha_2 = O(n_2, 1)$*, then there is an affine transformation*

$$h : \mathbb{R}^n \to \mathbb{R}^n, \quad h(x) = Hx + p \quad (H \in GL(n, \mathbb{R}), \quad p \in \mathbb{R}^n)$$

such that

$$h_* X(x) = Jx + q',$$

$$q' = O(n_1, 1) \dot{\times} e(\tilde{k}, \tilde{m}; (l_*, end)) \quad (1 \leq l_* \leq r),$$

$$h(U) = \{\langle \alpha, x \rangle = \varepsilon\} \quad (\varepsilon = 0, 1).$$

(2) If $\alpha_2 = e(\tilde{k}, \tilde{m}; (i_*, 1, u_*))(1 \leq i_* \leq r, 1 \leq u_* \leq k_{i_*})$*, then there is an affine transformation*

$$h : \mathbb{R}^n \to \mathbb{R}^n, \quad h(x) = Hx + p \quad (H \in GL(n, \mathbb{R}), \quad p \in \mathbb{R}^n)$$

such that

$$h_*X(x) = Jx + q',$$

$$q' = O(n_1, 1)\dot{\times} e(\tilde{k}, \tilde{m}; (i_*, 1, u_* - 1), \tau_3; (i_*, 1, k_{i_*}), \tau_2; (l_*, end), \tau_1),$$

$$h(U) = \{\langle \alpha, x\rangle = 0\},$$

where τ_1, τ_2 and τ_3 satisfy the following Rule (τ):

$$\begin{cases} \tau_2 = 0 \Rightarrow \tau_1 = 1 \text{ and } \begin{cases} \tau_3 = 0 \text{ or } 1 & \text{if } i_* > l_* \text{ and } k_{l_*} < u_* - 1 \quad (A^\circ) \\ \tau_3 = 0 & \text{otherwise} \quad (A') \end{cases} \\ \tau_2 = 1 \Rightarrow \tau_3 = 0 \text{ and } \begin{cases} \tau_1 = 1 & \text{if } i_* < l_* \quad (B^\circ) \\ \tau_1 = 0 & \text{otherwise (i.e. } i_* \geq l_*) \quad (B'). \end{cases} \end{cases}$$

(3) Let two affine transformations be given by

$$h_i(x) = H_i x + p_i \quad (H_i \in GL(n, \mathbb{R}), \quad p_i \in \mathbb{R}^n, i = 1, 2).$$

If both h_1 and h_2 satisfy the conditions for h in (1)-(2), then

$$h_{1*}X(x) = h_{2*}X(x) \quad (x \in \mathbb{R}^n),$$

and $h_1(U) = h_2(U)$.

In this sense, if J and α are fixed, q' and $h(U)$ in (1)-(2) are uniquely determined by the affine conjugate class.

Definition 2.3.33. In (2), define, for $i_* = 0$,

$$\begin{aligned} q' &= O(n_1, 1)\dot{\times} e(\tilde{k}, \tilde{m}; (i_*, 1, u_* - 1), \tau_3; (i_*, 1, k_{i_*}), \tau_2; (l_*, \text{end}), \tau_1) \\ &:= O(n_1, 1)\dot{\times} e(\tilde{k}, \tilde{m}; (l_*, \text{end})), \end{aligned}$$

which is equal to the q' in (1). Thus, we will call

$$q' = O(n_1, 1)\dot{\times} e(\tilde{k}, \tilde{m}; (i_*, 1, u_* - 1), \tau_3; (i_*, 1, k_{i_*}), \tau_2; (l_*, \text{end}), \tau_1)$$

a canonical constant vector for (J, α), including the case of $i_* = 0$. Also, we will call the $h(U)$ determined from (1) and (2) *a canonical boundary* for (J, α).

Proof of Theorem 2.3.32. (1) and (2) are derived immediately from Lemmas 2.3.27, 2.3.28, 2.3.29 and 2.3.30.

To prove (3), denote

$$H = H_* \dot{+} H_0, \quad H_* \in M(n_1), \quad H_0 \in M(n_2),$$

$$p = p_* \dot{\times} p_0, \quad p_* \in \mathbb{R}^{n_1}, \quad p_0 \in \mathbb{R}^{n_2},$$

$$q = q_* \dot{\times} q_0, \quad q_* \in \mathbb{R}^{n_1}, \quad q_0 \in \mathbb{R}^{n_2}.$$

Notice that $JH = HJ$ implies

$$J_* H_* = H_* J_*, \quad J_0 H_0 = H_0 J_0,$$

where $J = J_* \dot{+} J_0$, $J_0 = J(0; \tilde{k}, \tilde{m})$.

Case 1; $\alpha_2 = O(n_2, 1)$. Set

$$X(x) = Jx + q,$$

$$q = O(n_1, 1) \dot{\times} e(\tilde{k}, \tilde{m}; (l_*, m_{l_*}, k_{l_*})) \quad (1 \le l_* \le r)$$

$$U = \{\langle \alpha, x \rangle = \varepsilon\} \quad (\varepsilon = 0, 1).$$

Assume there exists an affine transformation

$$h : \mathbb{R}^n \to \mathbb{R}^n, \quad h(x) = Hx + p \quad (H \in GL(n, \mathbb{R}), \quad p \in \mathbb{R}^n)$$

such that

$$h_* X(x) = Jx + q'$$

$$q' = O(n_1, 1) \dot{\times} e(\tilde{k}, \tilde{m}; (l'_*, m_{l'_*}, k_{l'_*})) \quad (1 \le l'_* \le r)$$

$$h(U) = \{\langle \alpha, x \rangle = \varepsilon'\} \quad (\varepsilon' = 0, 1).$$

Then it is enough to show that $l_* = l'_*$ and $\varepsilon = \varepsilon'$. Since $q' = -Jp + Hq$, we have $q'_0 = -J_0 p_0 + H_0 q_0$. By Lemma 2.3.31(2), we obtain $l_* = l'_*$. Since $h\{\langle \alpha, x \rangle = \varepsilon\} = \{\langle \alpha, x \rangle = \varepsilon'\}$, there is a $k \neq 0$ such that

$$(H^{-1})^T \alpha = k\alpha \quad \text{and} \quad \varepsilon' = \langle \alpha, p \rangle + \varepsilon / k.$$

Since $q'_* = q_* = O(n_1, 1)$,

$$O(n_1, 1) = -J_* p_* + H_* O(n_1, 1).$$

Since $J_* \in GL(n_1, \mathbb{R})$, we have $p_* = O(n_1, 1)$. Since $\alpha_2 = O(n_2, 1)$,

$$\langle \alpha, p \rangle = \langle \alpha_1, p_* \rangle + \langle \alpha_2, p_0 \rangle = 0,$$

hence $\varepsilon' = \varepsilon / k$. Since $\varepsilon, \varepsilon' = 0$ or 1, we obtain $\varepsilon = \varepsilon'$.

Case 2; $\alpha_2 = e(\tilde{k}, \tilde{m}; (i_*, 1, u_*))$ $(1 \le i_* \le r, 1 \le u_* \le k_{i_*})$. Set

$$X(x) = Jx + q,$$

$$q = O(n_1, 1) \dot{\times} e(\tilde{k}, \tilde{m}; (i_*, 1, u_* - 1), \tau_3; (i_*, 1, k_{i_*}), \tau_2; (l_*, \text{end}), \tau_1)$$

$$U = \{\langle \alpha, x \rangle = 0\}.$$

Assume there exists an affine transformation

$$h : \mathbb{R}^n \to \mathbb{R}^n, \quad h(x) = Hx + p \quad (H \in GL(n, \mathbb{R}), \quad p \in \mathbb{R}^n)$$

such that

$$h_* X(x) = Jx + q'$$

$$q' = O(n_1, 1) \overset{\bullet}{\times} e(\tilde{k}, \tilde{m}; (i_*, 1, u_* - 1), \tau_3'; (i_*, 1, k_{i_*}), \tau_2'; (l_*, \mathrm{end}), \tau_1')$$

$$h(U) = \{\langle \alpha, x \rangle = 0\},$$

where τ_1', τ_2 and τ_3' satisfy Rule (τ). Then it is enough to prove that $q = q'$. We will consider the following five cases:

(i) $\tau_2 = 1$ and $l_* \leq i_*$

(ii) $\tau_2 = 1$ and $i_* < l_*$

(iii) $\tau_2 = 0$ and $i_* \leq l_*$

(iv) $\tau_2 = 0$ and $i_* > l_*$ and $k_{l_*} \geq u_* - 1$

(v) $\tau_2 = 0$ and $i_* > l_*$ and $k_{l_*} < u_* - 1$

(i) Case of $\tau_2 = 1$ and $l_* \leq i_*$. From Rule (τ), we have

$$\tau_3 = 0 \quad \text{and} \quad q_0 = e(\tilde{k}, \tilde{m}; (i_*, 1, k_{i_*})).$$

Notice that

$$\begin{aligned} q_0' &= -J_0 p_0 + H_0 q_0 \\ &= e(\tilde{k}, \tilde{m}; (i_*, 1, u_* - 1), \tau_3'; (i_*, 1, k_{i_*}), \tau_2'; (l_*, \mathrm{end}), \tau_1'). \end{aligned}$$

If $l_*' > i_*$, the $(l_*', m_{l_*'}, k_{l_*'})$-component of $-J_0 p_0 + H_0 q_0$ is zero by Lemma 2.3.31(1). Hence $\tau_1' = 0$, which contradicts Rule (τ). Thus the following two cases are possible:

(i-a) Case of $\tau_2' = 1$ and $l_*' \leq i_*$ (then $\tau_1' = \tau_3' = 0$)

(i-b) Case of $\tau_2' = 0$ and $l_*' \leq i_*$ (then $\tau_1' = 1, \tau_3' = 0$)

In Case (i-a), since $q_0' = e(\tilde{k}, \tilde{m}; (i_*, 1, k_{i_*}))$, we have $q_0 = q_0'$, that is, $q = q'$. In Case (i-b), since $q_0' = e(\tilde{k}, \tilde{m}; (l_*', m_{l_*'}, k_{l_*'}))$, by Lemma 2.3.31(2), we have $l_*' = i_*$. That is, $q_0' = e(\tilde{k}, \tilde{m}; (i_*, m_{i_*}, k_{i_*}))$. Since $\alpha_2 = e(\tilde{k}, \tilde{m}; (i_*, 1, u_*))$,

$$\alpha_2 = J_0^w q_0 \quad \text{and} \quad (J_0^w)^T \alpha_2 = q_0,$$

where $w = k_{i_*} - u_*$. From $(H^{-1})^T \alpha = k\alpha$ and $(H_0^{-1})^T \alpha_2 = k\alpha_2$, we have

$$\begin{aligned} q_0^T H_0 q_0 &= q_0^T H_0 (J_0^w)^T \alpha_2 = \alpha_2^T J_0^w H_0 (J_0^w)^T \alpha_2 \\ &= \alpha_2^T H_0 J_0^w (J_0^w)^T \alpha_2 \quad \text{(by Lemma 2.3.11(1),(2))} \\ &= \alpha_2^T H_0 \alpha_2 = (1/k) \alpha_2^T \alpha_2. \end{aligned}$$

Hence, the $(i_*, 1, k_{i_*})$-component of $-J_0p_0 + H_0q_0$ is equal to $1/k$. Note that q_0' has non-zero component only at (i_*, m_{i_*}, k_{i_*}), which is equal to 1. Therefore, we have $m_{i_*} = 1$ and $k = 1$, i.e., $q' = q$.

(ii) Case of $\tau_2 = 1$ and $i_* < l_*$. From Rule (τ), we have

$$\tau_3 = 0 \quad \text{and} \quad q_0 = e(\tilde{k}, \tilde{m}; (i_*, 1, k_{i_*})) + e(\tilde{k}, \tilde{m}; (l_*, m_{l_*}, k_{l_*})).$$

Let us consider the following cases:

(ii-a) $\tau_2' = 1$ and $l_*' \leq i_*$ (then $\tau_1' = \tau_3' = 0$)
(ii-b) $\tau_2' = 1$ and $i_* < l_*'$ (then $\tau_1' = 1$ and $\tau_3' = 0$)
(ii-c) $\tau_2' = 0$ and $l_*' > i_*$ (then $\tau_1' = 1$)
(ii-d) $\tau_2' = 0$ and $i_* \geq l_*'$ (then $\tau_1' = 1$)

Case (ii-a). Since $q_0' = e(\tilde{k}, \tilde{m}; (i_*, 1, k_{i_*}))$, by Lemma 2.3.31(1),

$$\text{"the } (l_*, m_{l_*}, k_{l_*})\text{-component of } J_0H_0^{-1}p_0 + H_0^{-1}q_0'\text{"} = 0.$$

Hence, $q_0 \neq J_0H_0^{-1}p_0 + H_0^{-1}q_0'$, which contradicts with $q' = Jp + Hq$. Namely, Case (ii-a) does not happen.

Case (ii-b). Since

$$q_0' = e(\tilde{k}, \tilde{m}; (i_*, 1, k_{i_*})) + e(\tilde{k}, \tilde{m}; (l_*', m_{l_*'}, k_{l_*'})),$$

by Lemma 2.3.31(3), we have $q_0 = q_0'$, i.e., $q = q'$.

Case (ii-c). Since $q_0' = e(\tilde{k}, \tilde{m}; (l_*', m_{l_*'}, k_{l_*'}))$ and $q_0 = J_0H_0^{-1}p_0 + H_0^{-1}q_0'$, by Lemma 2.3.31(1), we have $l_* \leq l_*'$, i.e., $l_* = l_*'$. Similar to Case (i-b), from $(H_0^{-1})^T\alpha_2 = k\alpha_2$, it follows that the $(i_*, 1, k_{i_*})$-component of $-J_0p_0 + H_0q_0$ is equal to $1/k$. Since q_0' has a non-zero component at only $(l_*', m_{l_*'}, k_{l_*'})$, we have $i_* = i_*'$, which contradicts $i_* < l_*'$. Hence, Case (ii-c) does not happen.

Case (ii-d). Since

$$q_0' = e(\tilde{k}, \tilde{m}; (l_*', m_{l_*'}, k_{l_*'})) + \tau_3' J_0\alpha_2$$

and $q_0 = J_0H_0^{-1}p_0 + H_0^{-1}q_0'$, we have

$$q_0 = J_0(H_0^{-1}p_0 + \tau_1' H_0^{-1}\alpha_2) + H_0^{-1}e(\tilde{k}, \tilde{m}; (l_*', m_{l_*'}, k_{l_*'})).$$

Since $l_*' \leq i_* < l_*$, this contradicts Lemma 2.3.31(1), i.e., Case (ii-d) does not happen. Therefore, only Case (ii-b) is possible, and then we have $q = q'$.

(iii) Case of $\tau_2 = 0$ and $i_* \leq l_*$. From Rule (τ), we have

$$\tau_1 = 1, \quad \tau_3 = 0 \quad \text{and} \quad q_0 = e(\tilde{k}, \tilde{m}; (l_*, m_{l_*}, k_{l_*})).$$

Let us consider the following cases:

(iii-a) $\tau_2' = 1$ and $l_*' \leq i_*$ (then $\tau_1' = \tau_3' = 0$)
(iii-b) $\tau_2' = 1$ and $i_* < l_*'$ (then $\tau_1' = 1$ and $\tau_3' = 0$)
(iii-c) $\tau_2' = 0$ and $l_*' \geq i_*$ (then $\tau_1' = 1$ and $\tau_3' = 0$)
(iii-d) $\tau_2' = 0$ and $i_* > l_*'$ (then $\tau_1' = 1$)

From $q_0' = -J_0p_0 + H_0q_0$, it follows that

$$q_0' = -J_0p_0 + H_0e(\tilde{k}, \tilde{m}; (l_*, m_{l_*}, k_{l_*})), \quad \text{and} \tag{2.3.3}$$

$$e(\tilde{k}, \tilde{m}; (l_*, m_{l_*}, k_{l_*})) = J_0H_0^{-1}p_0 + H_0^{-1}q_0'. \tag{2.3.4}$$

In Case (iii-a) and Case (iii-d), from Lemma 2.3.31(5) it follows that Equation (2.3.4) does not hold. Hence Cases (iii-a) and (iii-d) do not happen. In Case (iii-b), similar to Case (ii-c), $l_* = l_*' = i_*$ follows, which contradicts $i_* < l_*'$. Hence Case (iii-b) does not happen. In Case (iii-c), since $q_0' = e(\tilde{k}, \tilde{m}; (l_*', m_{l_*'}, k_{l_*'}))$, substituting it into (2.3.3) and applying Lemma 2.3.31(2), we have $l_* = l_*'$, i.e., $q = q'$.

(iv) Case of $\tau_2 = 0$ and $i_* > l_*$ and $k_{l_*} \geq u_* - 1$. From Rule (τ),

$$\tau_1 = 1, \quad \tau_3 = 0 \quad \text{and} \quad q_0 = e(\tilde{k}, \tilde{m}; (l_*, m_{l_*}, k_{l_*})).$$

Then, similar to Case (iii), we can obtain $q = q'$.

(v) Case of $\tau_2 = 0$ and $i_* > l_*$ and $k_{l_*} < u_* - 1$. From Rule (τ),

$$\tau_1 = 1, \quad q_0 = e(\tilde{k}, \tilde{m}; (l_*, m_{l_*}, k_{l_*})) + \tau_3J_0\alpha_2, \quad \text{and} \quad \tau_3 = 0 \;\text{ or }\; 1.$$

Let us consider the following cases:

(v-a) $\tau_2' = 1$ and $l_*' \leq i_*$ (then $\tau_1' = \tau_3' = 0$)
(v-b) $\tau_2' = 1$ and $i_* < l_*'$ (then $\tau_1' = 1$ and $\tau_3' = 0$)
(v-c) $\tau_2' = 0$ and $l_*' \geq i_*$ (then $\tau_1' = 1$ and $\tau_3' = 0$)
(v-d) $\tau_2' = 0$ and $i_* > l_*'$ and $k_{l_*'} \geq u_* - 1$(then $\tau_1' = 1$ and $\tau_3' = 0$)
(v-e) $\tau_2' = 0$ and $i_* > l_*'$ and $k_{l_*'} < u_* - 1$(then $\tau_1' = \tau_3' = 1$)

From $q_0' = -J_0p_0 + H_0q_0$, it follows that

$$q_0' = J_0(-p_0 + \tau_3H_0\alpha_0) + H_0e(\tilde{k}, \tilde{m}; (l_*, m_{l_*}, k_{l_*})), \quad \text{and} \tag{2.3.5}$$

$$e(\tilde{k}, \tilde{m}; (l_*, m_{l_*}, k_{l_*})) = J_0(H_0^{-1}p_0 - \tau_3\alpha_2) + H_0^{-1}q_0'. \tag{2.3.6}$$

In Cases (v-a),(v-b) and (v-c), however, from Lemma 2.3.31(1) it follows that Equation (2.3.6) does not hold. Hence Cases (v-a),(v-b) and (v-c) do not happen. In Case (v-d), since $q_0' = e(\tilde{k}, \tilde{m}; (l_*', m_{l_*'}, k_{l_*'}))$, substituting it into (2.3.5) and applying Lemma 2.3.31(2), we have $l_* = l_*'$, which contradicts $k_{l_*} < u_* - 1 \leq k_{l_*'}$. Hence Case (v-d) does not happen. In Case (v-e), since

$$q_0' = e(\tilde{k}, \tilde{m}; (l_*', m_{l_*'}, k_{l_*'})) + \tau_3'J_0\alpha_0,$$

substituting it into (2.3.5), we have

$$e(\tilde{k}, \tilde{m}; (l_*', m_{l_*'}, k_{l_*'})) = J_0(-p_0 + \tau_3H_0\alpha_2 - \tau_3'\alpha_2) + H_0e(\tilde{k}, \tilde{m}; (l_*, m_{l_*}, k_{l_*})).$$

By Lemma 2.3.31(2), we can obtain $l_* = l_*'$. Since $l_* < i_*$ and $\alpha_2^TH_0 = (1/k)\alpha_2^T$, by Lemma 2.3.31(1) and the definition of $C_1(\tilde{k}, \tilde{m}; \alpha_2)$, we have, for any u with $u_* \leq u \leq k_{i_*}$,

'the $(i_*, 1, u)$-component of $H_0e(\tilde{k}, \tilde{m}; (l_*, m_{l_*}, k_{l_*}))$" $= 0$.

Hence, we have

$$\text{“the } (i_*, 1, u+1)\text{-component of } -p_0 + \tau_3 H_0 \alpha_2 - \tau_3' \alpha_2\text{”} = 0$$

for all $u_* \leq u \leq k_{i_*} - 1$. By Lemma 2.3.31(1), since

$$H_0\alpha_2 = H_0 e(\tilde{k}, \tilde{m}; (i_*, 1, u_*)) = (1/k) e(\tilde{k}, \tilde{m}; (i_*, 1, u_*)) = (1/k)\alpha_2,$$

and since $\langle p_0, \alpha_2 \rangle = 0$, we have

$$\text{“the } (i_*, 1, u)\text{-component of } p_0\text{”} = 0.$$

Hence, the $(i_*, 1, u)$-component of the vector

$$-p_0 + \tau_3 H_0 \alpha_2 - \tau_3' \alpha_2 = -p_0 + (\tau_3/k - \tau_3')\alpha_2$$

is equal to $\tau_3/k - \tau_3'$, which is equal to 0. Since $\tau_3, \tau_3' = 0$ or 1, we have $\tau_3 = \tau_3'$, namely $q = q'$. □

Example 2.3.34. (Canonical Constant Vectors)
Assume

$$J = J_* \dot{+} J(0; \tilde{k}, \tilde{m}), \quad \tilde{k} = (2, 5),$$

$$\tilde{m} = (2, 2), \quad n_2 = \tilde{k} \cdot \tilde{m} = 14,$$

$$\alpha = \alpha_1 \dot{\times} \alpha_2; \quad \text{a canonical normal vector,}$$

$$q' = O(n_1, 1) \dot{\times} q; \quad \text{a canonical constant vector.}$$

When $\alpha_2 = O(n_2, 1)$, q is given by

$$e(\tilde{k}, \tilde{m}; (1,2,2)) = (0,0, \quad 0,1, \quad 0,0,0,0,0, \quad 0,0,0,0,0)^T,$$

$$e(\tilde{k}, \tilde{m}; (2,2,5)) = (0,0, \quad 0,0, \quad 0,0,0,0,0, \quad 0,0,0,0,1)^T.$$

When $\alpha_2 = e(\tilde{k}, \tilde{m}; (i_*, 1, u_*))$, we have the cases as shown in Table 2.3.1; By Rule (τ),

$$q = e(\tilde{k}, \tilde{m}; (i_*, 1, u_* - 1), \tau_3; (i_*, 1, k_{i_*}), \tau_2; (l_*, m_{l_*}, k_{l_*}), \tau_1)$$

is given as follows:

$$(1) \quad e(\tilde{k}, \tilde{m}; (1,1,0), 0; \quad (1,1,2), 0; (1,2,2), 1)$$
$$= (0,0, \quad 0,1, \quad 0,0,0,0,0, \quad 0,0,0,0,0)^T$$
$$e(\tilde{k}, \tilde{m}; (1,1,0), 0; \quad (1,1,2), 1; (1,2,2), 0)$$
$$= (0,1, \quad 0,0, \quad 0,0,0,0,0, \quad 0,0,0,0,0)^T$$

Table 2.3.1.

k_{l_*}		2	5
i_*	u_* \ l_*	1	2
1	1	(1) A', B'	(2) A', B°
	2	(3) A', B'	(4) A', B°
2	1	(5) A', B'	(6) A', B'
	2	(7) A', B'	(8) A', B'
	3	(9) A', B'	(10) A', B'
	4	(11) A°, B'	(12) A', B'
	5	(13) A°, B'	(14) A', B'

(2) $e(\tilde{k}, \tilde{m}; (1,1,0), 0; \quad (1,1,2), 0; (2,2,5), 1)$

$$= (0,0, \quad 0,0, \quad 0,0,0,0,0, \quad 0,0,0,0,1)^T$$

$e(\tilde{k}, \tilde{m}; (1,1,0), 0; \quad (1,1,2), 1; (2,2,5), 1)$

$$= (0,1, \quad 0,0, \quad 0,0,0,0,0, \quad 0,0,0,0,1)^T$$

(3) $e(\tilde{k}, \tilde{m}; (1,1,1), 0; \quad (1,1,2), 0; (1,2,2), 1)$

$$= (0,0, \quad 0,1, \quad 0,0,0,0,0, \quad 0,0,0,0,0)^T$$

$e(\tilde{k}, \tilde{m}; (1,1,1), 0; \quad (1,1,2), 1; (1,2,2), 0)$

$$= (0,1, \quad 0,0, \quad 0,0,0,0,0, \quad 0,0,0,0,0)^T$$

(4) $e(\tilde{k}, \tilde{m}; (1,1,1), 0; \quad (1,1,2), 0; (2,2,5), 1)$

$$= (0,0, \quad 0,0, \quad 0,0,0,0,0, \quad 0,0,0,0,1)^T$$

$e(\tilde{k}, \tilde{m}; (1,1,1), 0; \quad (1,1,2), 1; (2,2,5), 1)$

$$= (0,1, \quad 0,0, \quad 0,0,0,0,0, \quad 0,0,0,0,1)^T$$

(5) $e(\tilde{k}, \tilde{m}; (2,1,0), 0; \quad (2,1,5), 0; (1,2,2), 1)$

$$= (0,0, \quad 0,1, \quad 0,0,0,0,0, \quad 0,0,0,0,0)^T$$

$e(\tilde{k}, \tilde{m}; (2,1,0), 0; \quad (2,1,5), 1; (1,2,2), 0)$

$$= (0,0, \quad 0,0, \quad 0,0,0,0,1, \quad 0,0,0,0,0)^T$$

(6) $e(\tilde{k}, \tilde{m}; (2,1,0), 0; \quad (2,1,5), 0; (2,2,5), 1)$

$$= (0,0, \quad 0,0, \quad 0,0,0,0,0, \quad 0,0,0,0,1)^T$$

$$e(\tilde{k},\tilde{m};(2,1,0),0;\quad (2,1,5),1;(2,2,5),0)$$
$$= (0,0,\quad 0,0,\quad 0,0,0,1,0,\quad 0,0,0,0,0)^T$$

(7) $$e(\tilde{k},\tilde{m};(2,1,1),0;\quad (2,1,5),0;(1,2,2),1)$$
$$= (0,0,\quad 0,1,\quad 0,0,0,0,0,\quad 0,0,0,0,0)^T$$

$$e(\tilde{k},\tilde{m};(2,1,1),0;\quad (2,1,5),1;(1,2,2),0)$$
$$= (0,0,\quad 0,0,\quad 0,0,0,0,1,\quad 0,0,0,0,0)^T$$

(8) $$e(\tilde{k},\tilde{m};(2,1,1),0;\quad (2,1,5),0;(2,2,5),1)$$
$$= (0,0,\quad 0,0,\quad 0,0,0,0,0,\quad 0,0,0,0,1)^T$$

$$e(\tilde{k},\tilde{m};(2,1,1),0;\quad (2,1,5),1;(2,2,5),0)$$
$$= (0,0,\quad 0,0,\quad 0,0,0,0,1,\quad 0,0,0,0,0)^T$$

(9) $$e(\tilde{k},\tilde{m};(2,1,2),0;\quad (2,1,5),0;(1,2,2),1)$$
$$= (0,0,\quad 0,1,\quad 0,0,0,0,0,\quad 0,0,0,0,0)^T$$

$$e(\tilde{k},\tilde{m};(2,1,2),0;\quad (2,1,5),1;(1,2,2),0)$$
$$= (0,0,\quad 0,0,\quad 0,0,0,0,1,\quad 0,0,0,0,0)^T$$

(10) $$e(\tilde{k},\tilde{m};(2,1,2),0;\quad (2,1,5),0;(2,2,5),1)$$
$$= (0,0,\quad 0,0,\quad 0,0,0,0,0,\quad 0,0,0,0,1)^T$$

$$e(\tilde{k},\tilde{m};(2,1,2),0;\quad (2,1,5),1;(2,2,5),0)$$
$$= (0,0,\quad 0,0,\quad 0,0,0,0,1,\quad 0,0,0,0,0)^T$$

(11) $$e(\tilde{k},\tilde{m};(2,1,3),0;\quad (2,1,5),0;(1,2,2),1)$$
$$= (0,0,\quad 0,1,\quad 0,0,0,0,0,\quad 0,0,0,0,0)^T$$

$$e(\tilde{k},\tilde{m};(2,1,3),1;\quad (2,1,5),0;(1,2,2),1)$$
$$= (0,0,\quad 0,1,\quad 0,0,1,0,0,\quad 0,0,0,0,0)^T$$

$$e(\tilde{k},\tilde{m};(2,1,3),0;\quad (2,1,5),1;(1,2,2),0)$$
$$= (0,0,\quad 0,0,\quad 0,0,0,0,1,\quad 0,0,0,0,0)^T$$

(12) $$e(\tilde{k},\tilde{m};(2,1,3),0;\quad (2,1,5),0;(2,2,5),1)$$
$$= (0,0,\quad 0,0,\quad 0,0,0,0,0,\quad 0,0,0,0,1)^T$$

$$e(\tilde{k},\tilde{m};(2,1,3),0;\quad(2,1,5),1;(2,2,5),0)$$
$$=(0,0,\quad 0,0,\quad 0,0,0,0,1,\quad 0,0,0,0,0)^T$$

(13) $$e(\tilde{k},\tilde{m};(2,1,4),0;\quad(2,1,5),0;(1,2,2),1)$$
$$=(0,0,\quad 0,1,\quad 0,0,0,0,0,\quad 0,0,0,0,0)^T$$

$$e(\tilde{k},\tilde{m};(2,1,4),1;\quad(2,1,5),0;(1,2,2),1)$$
$$=(0,0,\quad 0,1,\quad 0,0,0,1,0,\quad 0,0,0,0,0)^T$$

$$e(\tilde{k},\tilde{m};(2,1,4),0;\quad(2,1,5),1;(1,2,2),0)$$
$$=(0,0,\quad 0,0,\quad 0,0,0,0,1,\quad 0,0,0,0,0)^T$$

(14) $$e(\tilde{k},\tilde{m};(2,1,4),0;\quad(2,1,5),0;(2,2,5),1)$$
$$=(0,0,\quad 0,0,\quad 0,0,0,0,0,\quad 0,0,0,0,1)^T$$

$$e(\tilde{k},\tilde{m};(2,1,4),0;\quad(2,1,5),1;(2,2,5),0)$$
$$=(0,0,\quad 0,0,\quad 0,0,0,0,1,\quad 0,0,0,0,0)^T.$$

2.3.4 Normal Forms of Two-Region Piecewise-Linear Vector Fields

Definition 2.3.35. Recall a two-region piecewise-linear map $f(x;(A,q_A),(B,q_B),(\alpha,\beta))$ in Definition 2.2.3:

$$\begin{aligned} f(x) &= f(x;(A,q_A),(B,q_B),(\alpha,\beta)) \\ &= \begin{cases} Ax+q_A & (x\in R_0) \\ Bx+q_B & (x\in R_1) \end{cases} \qquad (2.3.7) \\ R_0 &= \{x\in \mathbb{R}^n \mid \langle\alpha,x\rangle-\beta\le 0\} \\ R_1 &= \{x\in \mathbb{R}^n \mid \langle\alpha,x\rangle-\beta> 0\}, \end{aligned}$$

where $A,B\in M(n), q_A,q_B\in\mathbb{R}^n, \alpha\in\mathbb{R}^n-\{0\}$ and $\beta\in\mathbb{R}$. For this f, a piecewise-linear map $\bar{f}:\mathbb{R}^n\to\mathbb{R}^n$ defined by

$$\begin{aligned} \bar{f}(x) &= f(x;(B,q_B),(A,q_A),(\alpha,\beta)) \\ &= \begin{cases} Bx+q_B & (x\in R_0) \\ Ax+q_A & (x\in R_1) \end{cases} \qquad (2.3.8) \end{aligned}$$

is called the*complement* of f. For a piecewise-linear vector field X_f, defined by f in (2.3.7), the piecewise-linear vector field $X_{\bar{f}}$ defined by $\bar{f}$ in (2.3.8) is called *the* *complement* of X_f. For $f(x;(A,q_A),(B,q_B),(\alpha,\beta))$, *an exchange of the signs of regions* is defined by

$$
\begin{aligned}
& f(x;(B,q_B),(A,q_A),(-\alpha,-\beta)) \\
& = \begin{cases} Bx+q_B & (x \in R_0(-\alpha,-\beta)) \\ Ax+q_A & (x \in R_1(-\alpha,-\beta)) \end{cases} \qquad (2.3.9) \\
& R_0(-\alpha,-\beta) = \{\langle -\alpha, x\rangle - (-\beta) \le 0\} = \{\langle \alpha, x\rangle - \beta \ge 0\} = \bar{R}_1, \\
& R_1(-\alpha,-\beta) = \{\langle -\alpha, x\rangle - (-\beta) > 0\} = \{\langle \alpha, x\rangle - \beta < 0\} = \mathrm{int}(R_0).
\end{aligned}
$$

A two-region piecewise-linear map $f(x) = f(x;(A,q_A),(B,q_B),(\alpha,\beta))$ is *degenerate* if $f(x) \neq 0$ and $\bar{f}(x) \neq 0$ for all $x \in \mathbb{R}^n$. A two-region piecewise-linear map f is *non-degenerate* if it is not degenerate. A non-degenerate f is *regular* if, for some $x_0 \notin U(\alpha,\beta) = \{\langle\alpha,x\rangle = \beta\}, f(x_0) = 0$ or $\bar{f}(x_0) = 0$. A non-degenerate f is *singular* if it is not regular. If f is singular, there is an $x_0 \in U(\alpha,\beta)$ such that $f(x_0) = 0$ and $\bar{f}(x_0) = 0$. If f is degenerate, non-degenerate, regular, or singular, the vector field X_f is *degenerate, non-degenerate, regular*, or *singular*, respectively.

First we consider the normal forms of non-degenerate two-region piecewise-linear vector fields, which will be derived in Theorem 2.3.41. Suppose X_f is non-degenerate. Then, taking a parallel translation $h(x) = x - x_0$ (where $x_0 \in \mathbb{R}^n$ such that $f(x_0) = 0$ or $\bar{f}(x_0) = 0$), and if necessary, taking its complement and its exchange of signs of regions, X_f reduces to the following form:

$$
X_f(x) = f(x) = \begin{cases} Ax+q_A & (x \in R_1) \\ Bx & (x \in R_0) \end{cases} \qquad (2.3.10)
$$

$$
R_\omega = \{x \in \mathbb{R}^n \mid \mathrm{sgn}(\langle\alpha,x\rangle - \beta) = \omega\}, \quad \omega = 0,1.
$$

Applying Theorem 2.3.21, we can assume that B is a real Jordan matrix J, α is a canonical normal vector for J, and $\beta = 0,1$:

$$
X_f(x) = f(x) = \begin{cases} Ax+q_A & (x \in R_1) \\ Jx & (x \in R_0) \end{cases} \qquad (2.3.11)
$$

$$
R_\omega = \{x \in \mathbb{R}^n \mid \mathrm{sgn}(\langle\alpha,x\rangle - \varepsilon) = \omega\}, \quad \omega = 0,1.
$$

$$
\alpha = \text{a canonical normal vector for } J,
$$

$$
\varepsilon = \begin{cases} 1 & (\text{if } f \text{ is regular}) \\ 0 & (\text{if } f \text{ is singular}). \end{cases}
$$

If X_f in (2.3.11) is continuous, by Theorem 2.2.4, we have

$$
A = J + (A-J)h\alpha^T, q_A = -\varepsilon(A-J)h,
$$

where $h = \alpha/\langle\alpha,\alpha\rangle$. Set $p = (A-J)h \in \mathbb{R}^n$ then

$$A = J + p\alpha^T, \quad q_A = -\varepsilon p.$$

Hence, X_f is written as

$$X_f(x) = f(x) = \begin{cases} (J + p\alpha^T)x - \varepsilon p & (x \in R_1) \\ Jx & (x \in R_0) \end{cases} \tag{2.3.12}$$

or, by Theorem 2.2.4(5), as

$$X_f(x) = f(x) = Jx + (1/2)p\{|\langle \alpha, x\rangle - \varepsilon| + \langle \alpha, x\rangle - \varepsilon\}.$$

Our goal is to simplify $p \in \mathbb{R}^n$ as much as possible, under an affine transformation by which J and $U(\alpha, \varepsilon)$ are invariant. If $h(x) = Hx + b \quad (H \in GL(n, \mathbb{R}), b \in \mathbb{R}^n)$, then

$$h_*X_f(x) = \begin{cases} H(J + p\alpha^T)H^{-1}x - H(J + p\alpha^T)H^{-1}b - \varepsilon Hp, & (x \in R_1) \\ HJH^{-1}x - HJH^{-1}b, & (x \in R_0) \end{cases}$$

$$h(U) = \{\langle (H^{-1})^T\alpha, x\rangle = \langle (H^{-1})^T\alpha, b\rangle + \varepsilon\}.$$

Assume the following conditions, which guarantee that J and $U(\alpha, \varepsilon)$ are invariant:

$$\text{(i)} HJH^{-1} = J, \quad \text{(ii)} Jb = 0, \quad \text{(iii)} \alpha^T H = k\alpha^T, \quad \text{(iv)} \langle (H^{-1})^T\alpha, b\rangle = 0,$$

where $k \neq 0$ (if $\varepsilon = 0$), and $k = 1$ (if $\varepsilon = 1$) (If $\det J \neq 0$, Condition (ii) implies Condition (iv). If $\det J = 0$, we may be able to take b such that $\langle (H^{-1})^T\alpha, b\rangle + \varepsilon = 0$. However, when $\det J = 0$ and $\varepsilon = 1$, X_f should be regular by Definition 2.3.35, because there exists $x_0 \notin U$ such that $X_f(x_0) = 0$. Hence we need to assume Condition (iv)). Setting $H' = (1/k)H$ and $h'(x) = H'x + b$, we have

$$h'_*X_f(x) = \begin{cases} (J + H'p\alpha^T)x - \varepsilon H'p, & (x \in R_1) \\ Jx, & (x \in R_0) \end{cases}$$

$$h'(U) = \{\langle \alpha, x\rangle = \varepsilon\}.$$

Definition 2.3.36. Assume $\tilde{k} = (k_1, \cdots, k_r) \in \mathbb{N}^r_<$, $\tilde{m} = (m_1, \cdots, m_r) \in \mathbb{N}^r$, $n = \tilde{k} \cdot \tilde{m}$ and $d = 1$ or 2. Set $\alpha = e_d(\tilde{k}, \tilde{m}; (i_*, 1, u_*))$.

(1) $U(\tilde{k}, \tilde{m}; \alpha)$ is a subset of $\mathbb{Z}^r_+$ defined by

$$\{u_*(i)\}_{i=1}^r \in U(\tilde{k}, \tilde{m}; \alpha) \subset \mathbb{Z}^r_+$$

$$\Longleftrightarrow$$

(i) $0 \leq u_*(i) \leq k_i$ for $1 \leq i \leq r$

(ii) $m_{i_*} = 1 \Rightarrow u_*(i_*) = 0$ and

(iii) $i < j$ and $u_*(i)u_*(j) \neq 0 \Rightarrow 0 < u_*(j) - u_*(i) < k_j - k_i$.

(2) For $\{u_*(i)\} \in U(\tilde{k}, \tilde{m}; \alpha)$, a subset $\tilde{P}_d(\tilde{k}, \tilde{m}; \alpha; \{u_*(i)\})$ of $M(dn, 1)$ is defined by

$$\tilde{p} = (\tilde{p}(i,s,u)) \in \tilde{P}_d(\tilde{k}, \tilde{m}; \alpha; \{u_*(i)\}) \subset M(dn, 1)$$
$$(\tilde{p}(i,s,u) \in \mathbb{R}^d, 1 \le i \le r, 1 \le s \le m_i, 1 \le u \le k_i)$$

$$\Longleftrightarrow$$

(i) $\tilde{p}(i,s,u) = 0$ unless $i = i_*$ and $s = 1$ and $u_l \le u \le u_h$

(ii) $\tilde{p}(i_*, 1, u_h) \neq 0 \quad \text{if} \quad u_h \neq 0$

(iii) $0 \le u_h < \min_{i:u_*(i)\neq 0} \{u_*(i) + \max\{0, k_{i_*} - k_i\}, k_{i_*} + 1\}$ and

(iv) $u_l = \max_{i:u_*(i)\neq 0} \{u_h - k_{i_*} + u_*, u_*(i) + \min\{u_* - k_i, 1\}, 1\}.$

(3) For $\{u_*(i)\} \in U(\tilde{k}, \tilde{m}; \alpha)$, a subset $P_d(\tilde{k}, \tilde{m}; \alpha; \{u_*(i)\})$ of $M(dn, 1)$ is defined by

$$p \in P_d(\tilde{k}, \tilde{m}; \alpha; \{u_*(i)\})$$

$$\Longleftrightarrow$$

(i) $p = p_* + \tilde{p}$

(ii) $p_* = \sum_{i=1}^{r} e_d(\tilde{k}, \tilde{m}; (i, m_i, u_*(i)))$

(iii) $\tilde{p} \in \tilde{P}_d(\tilde{k}, \tilde{m}; \alpha; \{u_*(i)\})$.

This p_* is called the*discrete part* of p, and $\tilde{p}$ is called the*continuous part* of p (see Example 2.3.44 at the end of this subsection).

(4) $P_d(\tilde{k}, \tilde{m}; \alpha)$ is a subset of $M(dn, 1)$ such that

$$p \in P_d(\tilde{k}, \tilde{m}; \alpha) \subset M(dn, 1)$$

$$\Longleftrightarrow$$

there exist $\quad \{u_*(i)\}_{i=1}^{r} \in U(\tilde{k}, \tilde{m}; \alpha) \quad$ such that

$$p \in P_d(\tilde{k}, \tilde{m}; \alpha; \{u_*(i)\}),$$

i.e.,

$$P_d(\tilde{k}, \tilde{m}; \alpha) = \cup\{P_d(\tilde{k}, \tilde{m}; \alpha; \{u_*(i)\}) \mid \{u_*(i)\} \in U(\tilde{k}, \tilde{m}; \alpha)\}.$$

Standing Assumption. In this subsection we assume

$$\tilde{k} = (k_1, \cdots, k_r) \in \mathbb{N}^r_<, \quad \tilde{m} = (m_1, \cdots, m_r) \in \mathbb{N}^r,$$

$$n = \tilde{k} \cdot \tilde{m}, \quad d = 1 \quad \text{or} \quad 2,$$

$$\alpha = e_d(\tilde{k}, \tilde{m}; (i_*, 1, u_*)), \quad 1 \le i_* \le r, 1 \le u_* \le k_{i_*},$$

and

$$C_d^*(\tilde{k}, \tilde{m}; \alpha) = \{H \in C_d(\tilde{k}, \tilde{m}; \alpha) \mid \alpha^T = \alpha^T H\},$$

i.e., the case of $a = 1$ in Lemma 2.3.14), unless otherwise stated.

Lemma 2.3.37. *For any $p \in \mathbb{R}^{dn}$, there exist*

$$H \in C_d^*(\tilde{k}, \tilde{m}; \alpha) \cap GL(dn, \mathbb{R}) \quad \text{and} \quad \{u_*(i)\}_{i=1}^r \in U(\tilde{k}, \tilde{m}; \alpha)$$

such that

$$Hp \in P_d(\tilde{k}, \tilde{m}; \alpha; \{u_*(i)\}_{i=1}^r).$$

Proof. Case of $d = 1$ (see Example 2.3.45). Let

$$p = (p(i, s, u)) \in \mathbb{R}^n \quad (1 \le i \le r, \quad 1 \le s \le m_i, \quad 1 \le u \le k_i).$$

For each $1 \le i \le r$ and $1 \le s \le m_i$, set

$$u'_*(i, s) = \begin{cases} 0 & \text{if } p(i, s, u) = 0 \quad \text{for all } 1 \le u \le k_i \\ \max\{u \mid p(i, s, u) \ne 0\} & \text{otherwise.} \end{cases}$$

Step 1. For each $1 \le i \le r$ and $1 \le s \le m_i$, define

$$F_{is} \in C_1(k_i) \cap GL(k_i, \mathbb{R})$$

as follows:

(i) If $i = i_*$ and $s = 1$, then, using $p(u) = p(i_*, 1, u)$ and $u'_* = u'_*(i_*, 1)$, define

$$F_{i_*1} = \Delta_1(1, \underbrace{0, \cdots, 0}_{k_{i_*} - u_*}, a_1, \cdots, a_{u_*-1}),$$

where

$$a_1 = -p(u'_* - k_{i_*} + u_* - 1)/p(u'_*)$$

$$a_k = -\{p(u'_* - k_{i_*} + u_* - k) + a_{k-1} p(u'_* - 1) + \cdots$$
$$+ a_1 p(u'_* - k + 1)\}/p(u'_*)$$
$$\text{for} \quad 1 < k \le u'_* - k_{i_*} + u_* - 1, \quad \text{and}$$

$$a_k = 0 \qquad \text{for} \quad k > u'_* - k_{i_*} + u_* - 1.$$

(ii) If $i = i_*$ and $s \neq 1$, then, using $p(u) = p(i_*, s, u)$ and $u'_* = u'_*(i_*, s)$, define

$$F_{i_* s} = \Delta_1(a_1, \cdots, a_{k_{i_*}}),$$

where

$$\begin{aligned} a_1 &= 1/p(u'_*) \\ a_k &= -\{a_{k-1}p(u'_* - 1) + \cdots + a_1 p(u'_* - k + 1)\}/p(u'_*) \\ &\qquad \text{for} \quad 1 < k \leq u'_*, \quad \text{and} \\ a_k &= 0 \qquad \text{for} \quad k > u'_*. \end{aligned}$$

(iii) If $i \neq i_*$, then, using $p(u) = p(i, s, u)$ and $u'_* = u'_*(i, s)$, define

$$F_{is} = \Delta_1(a_1, \cdots, a_{k_i}),$$

where

$$\begin{aligned} a_1 &= 1/p(u'_*) \\ a_k &= -\{a_{k-1}p(u'_* - 1) + \cdots + a_1 p(u'_* - k + 1)\}/p(u'_*) \\ &\qquad \text{for} \quad 1 < k \leq u'_* \quad \text{and} \\ a_k &= 0 \qquad \text{for} \quad k > u'_*. \end{aligned}$$

Define $F \in C_1(\tilde{k}, \tilde{m}) \cap GL(n, \mathbb{R})$ by

$$F = \dot{\sum}_{i=1}^{r} \dot{\sum}_{s=1}^{m_i} F_{is}.$$

Then it follows from the definition of F_{is} that, for $p' = Fp = (p'(i, s, u)) \in \mathbb{R}^n \quad (1 \leq i \leq r, 1 \leq s \leq m_i, 1 \leq u \leq k_i)$,

$$p'(i, s, u) = \begin{cases} 1 & \text{if } (i, s, u) = (i, s, u'_*(i, s)) \text{ and } u'_*(i, s) \neq 0 \\ p(i, s, u) & \text{if } (i, s) = (i_*, 1) \text{ and} \\ & u'_*(i_*, 1) - k_{i_*} + u_* - 1 < u \leq u'_*(i_*, 1) \\ 0 & \text{otherwise} \end{cases}$$

$$u'_*(i, s) = \begin{cases} 0 & \text{if } p'(i, s, u) = 0 \quad \text{for all } 1 \leq u \leq k_i \\ \max\{u \mid p'(i, s, u) \neq 0\} & \text{otherwise.} \end{cases}$$

Also, from the definition of F_{is}, we have, for $\alpha = e(\tilde{k}, \tilde{m}; (i_*, 1, u_*))$,

$$\alpha^T F = \alpha^T.$$

Step 2. For each $1 \leq i \leq r$, define $u'_*(i) \in \mathbb{Z}_+$ with $0 \leq u'_*(i) \leq k_i$ as follows:

(i) If $i = i_*$, then

$$u'_*(i_*) = \begin{cases} 0 & \text{if } u'_*(i_*, s) = 0 \quad \text{for all } 2 \le s \le m_{i_*} \\ 0 & \text{if } m_{i_*} = 1 \\ \max\{u'_*(i_*, s) \mid 2 \le s \le m_{i_*}\} & \text{otherwise.} \end{cases}$$

(ii) If $i \neq i_*$, then

$$u'_*(i_*) = \begin{cases} 0 & \text{if } u'_*(i, s) = 0 \quad \text{for all } 1 \le s \le m_i \\ \max\{u'_*(i, s) \mid 1 \le s \le m_i\} & \text{otherwise.} \end{cases}$$

Changing the coordinate of p' by a suitable linear transformation which belongs to $C_1^*(\tilde{k}, \tilde{m}; \alpha) \cap GL(n, \mathbb{R})$, we can assume that, for each $1 \le i \le r$,

$$u'_*(i) = u'_*(i, m_i) \quad \text{if} \quad u'_*(i) \neq 0.$$

For each $1 \le i \le r$, define $F_i = (F_{i,st}) \in C_1(k_i m_i) \cap GL(k_i m_i, \mathbb{R})$ $(F_{i,st} \in M(k_i, k_i), 1 \le s, t \le m_i)$ as follows:

(iii) If $i = i_*$, then

$$F_{i_*,st} = \begin{cases} I(k_{i_*}) & \text{if } s = t \\ -N_1(k_{i_*}; u'_*(i_*) - u'_*(i_*, s)) & \text{if } t = m_{i_*} \ge 2 \text{ and } 2 \le s \le m_{i_*} \\ & \text{with } u'_*(i_*, s) \neq 0 \\ O & \text{otherwise.} \end{cases}$$

(iv) If $i \neq i_*$, then

$$F_{i,st} = \begin{cases} I(k_i) & \text{if } s = t \\ -N_1(k_i; u'_*(i) - u'_*(i, s)) & \text{if } t = m_i \text{ and } 1 \le s \le m_i \\ & \text{with } u'_*(i, s) \neq 0 \\ O & \text{otherwise.} \end{cases}$$

Set

$$F = \dot{\sum}_{i=1}^{r} F_i$$

and $p'' = Fp'$. For simplicity, rewrite p'' as p', then $p' = (p'(i, s, u)) \in \mathbb{R}^n$ satisfies

$$p'(i, s, u) = \begin{cases} 1 & \text{if } (i, s, u) = (i, m_i, u'_*(i)) \text{ and } u'_*(i) \neq 0 \\ p(i, s, u) & \text{if } (i, s) = (i_*, 1) \text{ and} \\ & u'_*(i_*, 1) - k_{i_*} + u_* - 1 < u \le u'_*(i_*, 1) \\ 0 & \text{otherwise.} \end{cases}$$

However, $\{u'_*(i)\}_{i=1}^r$ may not belong to $U(\tilde{k}, \tilde{m}; \alpha)$.

Step 3. For $p' = Fp = (p'(i, s, u)) \in \mathbb{R}^n$ and $\{u'_*(i)\}_{i=1}^r$, remake them by the following algorithm:

(i) Set

$$
\begin{aligned}
i_0 &= \min\{1 \le i \le r \mid u'_*(i) \ne 0\} \quad \text{and} \\
j_0 &= \min\{i_0 + 1 \le j \le r \mid u'_*(j) \ne 0\}.
\end{aligned}
$$

If there are no i_0 or j_0, stop.

(ii) Set

$$
\begin{aligned}
i_1 &= \min\{i_0 \le i \le r \mid u'_*(i) \ne 0\} \quad \text{and} \\
j_1 &= \min\{j_0 \le j \le r \mid u'_*(j) \ne 0\}.
\end{aligned}
$$

If there are no i_1 or j_1, stop.

(iii) If

$$k_{i_1} - u'_*(i_1) < k_{j_1} - u'_*(j_1),$$

then go to (v); otherwise, i.e.,

$$k_{i_1} - u'_*(i_1) \ge k_{j_1} - u'_*(j_1),$$

go to (iv).

(iv) Define $G = (G_{ij,st}) \in C_1^*(\tilde{k}, \tilde{m}; \alpha) \cap GL(n, \mathbb{R})(G_{ij,st} \in M(k_i, k_j), \quad 1 \le i, j \le r, 1 \le s \le m_i, \quad 1 \le t \le m_j)$ by

$$G_{ij,st} = \begin{cases} I(k_i) & \text{if } (i,s) = (j,t) \\ [ON] & \text{if } (i,s) = (i_1, m_{i_1}) \text{ and } (j,t) = (j_1, m_{j_1}) \\ O & \text{otherwise,} \end{cases}$$

where $N = -N_1(k_{i_1}; u'_*(j_1) - k_{j_1} - u'_*(i_1) + k_{i_1})$. Set

$$p'' = Gp' = (p''(i, s, u)) \in \mathbb{R}^n,$$

then we can easily verify that

$$p''(i_1, m_{i_1}, u'_*(i_1)) = 0 \quad \text{and} \quad p''(j_1, m_{j_1}, u'_*(j_1)) = 1.$$

Define $\{u''_*(i)\}_{i=1}^r$ by

$$u''_*(i) = \begin{cases} 0 & \text{if } i = i_1 \\ u'_*(i) & \text{otherwise.} \end{cases}$$

Rewrite p'' as p', and $\{u''_*(i)\}_{i=1}^r$ as $\{u'_*(i)\}_{i=1}^r$. Set

$$i_0 := j_1 \quad \text{and} \quad j_0 := j_1 + 1$$

(where $A := B$ means that the variable A takes the value of B), and go to (ii).

(v) If $u'_*(i_1) < u'_*(j_1)$, then go to (vii); otherwise, go to (vi).

(vi) Define $G = (G_{ij,st}) \in C_1^*(\tilde{k}, \tilde{m}; \alpha) \cap GL(n, \mathbb{R})(G_{ij,st} \in M(k_i, k_j), \quad 1 \le i, j \le r, 1 \le s \le m_i, \quad 1 \le t \le m_j)$ by

$$G_{ij,st} = \begin{cases} I(k_i) & \text{if } (i,s) = (j,t) \\ N\#O & \text{if } (i,s) = (j_1, m_{j_1}) \text{ and } (j,t) = (i_1, m_{i_1}) \\ O & \text{otherwise,} \end{cases}$$

where $N = -N_1(k_{i_1}; u'_*(i_1) - u'_*(j_1))$. Set

$$p'' = Gp' = (p''(i,s,u)) \in \mathbb{R}^n.$$

Then we can easily verify that

$$p''(i_1, m_{i_1}, u'_*(i_1)) = 1 \quad \text{and} \quad p''(j_1, m_{j_1}, u'_*(j_1)) = 0.$$

Define $\{u''_*(i)\}_{i=1}^r$ by

$$u''_*(i) = \begin{cases} 0 & \text{if } i = j_1 \\ u'_*(i) & \text{otherwise.} \end{cases}$$

Rewrite p'' as p', and $\{u''_*(i)\}_{i=1}^r$ as $\{u'_*(i)\}_{i=1}^r$. Set

$$i_0 := i_1 \quad \text{and} \quad j_0 := j_1 + 1,$$

and go to (ii).

(vii) Set

$$i_0 := j_1 \quad \text{and} \quad j_0 := j_1 + 1,$$

and go to (ii).

Since i_1 and j_1 are finite, the above algorithm will stop in finite steps. Then we have $p' = (p'(i,s,u)) \in \mathbb{R}^n$ and $\{u'_*(i)\}_{i=1}^r$ such that

$$p'(i,s,u) = \begin{cases} 1 & \text{if } (i,s,u) = (i, m_i, u'_*(i)) \text{ and } u'_*(i) \neq 0 \\ p(i,s,u) & \text{if } (i,s) = (i_*, 1) \text{ and} \\ & u'_*(i_*, 1) - k_{i_*} + u_* - 1 < u \leq u'_*(i_*, 1) \\ 0 & \text{otherwise} \end{cases}$$

and $\{u'_*(i)\}_{i=1}^r \in U(\tilde{k}, \tilde{m}; \alpha)$.

Step 4. For $p' \in \mathbb{R}^n$ and $\{u'_*(i)\}_{i=1}^r$, remake them by the following algorithm.

(i) Set $i_0 = 1$.

(ii) Set $i_1 = \min\{i_0 \leq i \leq r \mid u'_*(i) \neq 0\}$. If i_1 does not exist, stop.

(iii) If

$$u'_*(i_1) > u'_*(i_*, 1) + \min\{0, k_{i_1} - k_{i_*}\},$$

then go to (v); otherwise, i.e.,

$$u'_*(i_1) \leq u'_*(i_*, 1) + \min\{0, k_{i_1} - k_{i_*}\},$$

go to (iv).

(iv) Define $G = (G_{ij,st}) \in C_1(\tilde{k}, \tilde{m}; \alpha) \cap GL(n, \mathbb{R}) (G_{ij,st} \in M(k_i, k_j),\quad 1 \leq i,j \leq r, 1 \leq s \leq m_i,\quad 1 \leq t \leq m_j)$ as follows:

(a) If $i_1 < i_*$, then

$$G_{ij,st} = \begin{cases} I(k_i) & \text{if } (i,s) = (j,t) \\ [ON] & \text{if } (i,s) = (i_1, m_{i_1}) \text{ and } (j,t) = (i_*, 1) \\ O & \text{otherwise,} \end{cases}$$

where, using $p'(u) = p'(i_*, 1, u)$ and $u'_* = u'_*(i_*, 1)$,

$$\begin{aligned} N &= N_1(k_{i_1}; u'_* - k_{i_*} - u'_*(i_1) + k_{i_1})\Delta_1(a_1, \cdots, a_{k_{i_1}}) \\ a_1 &= -1/p'(u'_*), \\ a_k &= -\{a_{k-1}p'(u'_* - 1) + \cdots + a_1 p'(u'_* - k + 1)\}/p'(u'_*) \quad \text{for} \quad 2 \le k \le k_{i_1}. \end{aligned}$$

(b) If $i_1 \ge i_*$, then

$$G_{ij,st} = \begin{cases} I(k_i) & \text{if } (i,s) = (j,t) \\ N\#O & \text{if } (i,s) = (i_1, m_{i_1}) \text{ and } (j,t) = (i_*, 1) \\ O & \text{otherwise,} \end{cases}$$

where, using $p'(u) = p'(i_*, 1, u)$ and $u'_* = u'_*(i_*, 1)$,

$$\begin{aligned} N &= N_1(k_{i_*}; u'_* - u'_*(i_1))\Delta_1(a_1, \cdots, a_{k_{i_*}}) \\ a_1 &= -1/p'(u'_*), \\ a_k &= -\{a_{k-1}p'(u'_* - 1) + \cdots + a_1 p'(u'_* - k + 1)\}/p'(u'_*) \quad \text{for} \quad 2 \le k \le k_{i_*}. \end{aligned}$$

Set $p'' = Gp' = (p''(i,s,u)) \in \mathbb{R}^n$. Then we can verify

$$p''(i_1, m_{i_1}, u) = 0 \quad \text{for all} \quad 1 \le u \le k_{i_1}.$$

Define $\{u''_*(i)\}_{i=1}^r$ by

$$u''_*(i) = \begin{cases} 0 & \text{if } i = i_1 \\ u'_*(i) & \text{otherwise.} \end{cases}$$

Rewrite p'' as p', and $\{u''_*(i)\}_{i=1}^r$ as $\{u'_*(i)\}_{i=1}^r$. Set $i_0 := i_1 + 1$ and go to (ii).

(v) Define $G = (G_{ij,st}) \in C_1^*(\tilde{k}, \tilde{m}; \alpha) \cap GL(n, \mathbb{R})(G_{ij,st} \in M(k_i, k_j),\ 1 \le i, j \le r, 1 \le s \le m_i,\ 1 \le t \le m_j)$ as follows:

(a) If $i_1 < i_*$, then

$$G_{ij,st} = \begin{cases} I(k_i) & \text{if } (i,s) = (j,t) \\ N\#O & \text{if } (i,s) = (i_*, 1) \text{ and } (j,t) = (i_1, m_{i_1}) \\ O & \text{otherwise,} \end{cases}$$

where

$$\begin{aligned} N &= \Delta_1(a_1, \cdots, a_{k_{i_1}}) \\ a_k &= 0 \quad \text{for} \quad 1 \le k \le k_{i_1} - u_* + 1 \quad \text{or} \quad u'_*(i_1) < k \le k_{i_1}, \\ a_k &= -p'(i_*, 1, u'_*(i_1) - k + 1) \quad \text{for} \quad k_{i_1} - u_* + 1 < k \le u'_*(i_1). \end{aligned}$$

(b) If $i_1 \ge i_*$, then

$$G_{ij,st} = \begin{cases} I(k_i) & \text{if } (i,s) = (j,t) \\ [ON] & \text{if } (i,s) = (i_*, 1) \text{ and } (j,t) = (i_1, m_{i_1}) \\ O & \text{otherwise,} \end{cases}$$

where

$$\begin{aligned} N &= \Delta_1(a_1,\cdots,a_{k_{i_*}}), \quad \text{and} \\ a_k &= 0 \quad \text{for} \quad 1 \le k \le k_{i_*} - u_* + 1 \\ &\qquad \text{or} \quad u'_*(i_1) - (k_{i_1} - k_{i_*}) < k \le k_{i_*}, \\ a_k &= -p'(i_*, 1, u'_*(i_1) - (k_{i_1} - k_{i_*}) - k + 1) \\ &\qquad \text{for} \quad k_{i_*} - u_* + 1 < k \le u'_*(i_1) - (k_{i_1} - k_{i_*}). \end{aligned}$$

Set $p'' = Gp'$ as p'. Set $i_0 := i_1 + 1$ and go to (ii).

Since i_1 is finite, the algorithm will stop in finite steps. Then we obtain $p' \in \mathbb{R}^n$ and $\{u'_*(i)\}_{i=1}^r$ such that

$$\{u'_*(i)\}_{i=1}^r \in U(\tilde{k}, \tilde{m}; \alpha) \quad \text{and} \quad p' \in P_1(\tilde{k}, \tilde{m}; \alpha; \{u'_*(i)\}_{i=1}^r).$$

Case of $d = 2$. Proof follows *mutatis mutandis.* □

Lemma 2.3.38. *Let $\{u_*(i)\}_{i=1}^r, \{u'_*(i)\}_{i=1}^r \in U(\tilde{k}, \tilde{m}; \alpha)$ be given. Set*

$$p = \sum\nolimits_{i=1}^r e_d(\tilde{k}, \tilde{m}; (i, m_i, u_*(i))), \quad p' = \sum\nolimits_{i=1}^r e_d(\tilde{k}, \tilde{m}; (i, m_i, u'_*(i)))$$

and $H \in C_d(\tilde{k}, \tilde{m}; \alpha)$ (notice that if $u_*(i) = 0$, $e_d(\tilde{k}, \tilde{m}; (i, m_i, u_*(i))) = O$). Then

(1) for any $1 \le i \le r$ with $u_*(i) \ne 0$ and for any $u_*(i) < u \le k_i$,

$$\text{“the } (i, m_i, u)\text{-component of } Hp\text{ ”} = 0.$$

(2) If $H \in GL(dn, \mathbb{R})$, then

$$Hp = p' \Rightarrow p = p'.$$

Proof. Case of $d = 1$.

(1) Suppose that there are an i_0, $1 \le i_0 \le r$ with $u_*(i_0) \ne 0$, and a u_0, $u_*(i_0) < u_0 \le k_{i_0}$, such that

$$\text{“the } (i_0, m_{i_0}, u_0)\text{-component of } Hp\text{ ”} \ne 0.$$

Then there exists an $i_1 (\ne i_0)$ such that $u_*(i_1) \ne 0$ and

$$\text{“the } (i_0, m_{i_0}, u_0)\text{-component of } He\text{ ”} \ne 0,$$

where $e = e(\tilde{k}, \tilde{m}; (i_1, m_{i_1}, u_*(i_1)))$. Since $H \in C_1(\tilde{k}, \tilde{m})$, it follows from Definition 2.3.4(5) that

$$\begin{aligned} i_1 > i_0 &\Rightarrow k_{i_0} - u_0 \ge k_{i_1} - u_*(i_1); \\ i_1 < i_0 &\Rightarrow u_*(i_1) \ge u_0. \end{aligned}$$

Since $u_*(i_0) < u_0$, we have

$$k_{i_0} - u_*(i_0) \geq k_{i_1} - u_*(i_1) \quad \text{or} \quad u_*(i_1) \geq u_*(i_0).$$

This contradicts $\{u'_*(i)\}_{i=1}^r \in U(\tilde{k}, \tilde{m}; \alpha)$ (see Definition 2.3.36(1)).

(2) Let $1 \leq i_0 \leq r$ with $u_*(i_0) \neq 0$ be given. For any $u_*(i_0) < u \leq k_{i_0}$,

$$\text{“the } (i_0, m_{i_0}, u)\text{-component of } Hp\text{ ''} = 0.$$

Hence $0 \leq u'_*(i_0) \leq u_*(i_0)$. If $u'_*(i_0) \neq 0$, since

$$\text{“the } (i_0, m_{i_0}, u)\text{-component of } H^{-1}p'\text{ ''} = 0,$$

for any $u'_*(i_0) < u \leq k_{i_0}$, we have $u_*(i_0) \leq u'_*(i_0)$, hence $u_*(i_0) = u'_*(i_0)$. If $u'_*(i_0) = 0$, there exists an $i_1 (\neq i_0)$ such that $u_*(i_0) \neq 0$ and

$$\text{“the } (i_0, m_{i_0}, u_*(i_0))\text{-component of } He\text{ ''} \neq 0,$$

where $e = e(\tilde{k}, \tilde{m}; (i_1, m_{i_1}, u_*(i_1))$. Since $H \in C_1(\tilde{k}, \tilde{m})$, we have

$$k_{i_0} - u_*(i_0) \geq k_{i_1} - u_*(i_1) \quad \text{or} \quad u_*(i_0) \leq u_*(i_1).$$

This contradicts $\{u_*(i)\}_{i=1}^r \in U(\tilde{k}, \tilde{m}; \alpha)$. Next, let $1 \leq i_0 \leq r$ with $u_*(i_0) = 0$, be given. If $u'_*(i_0) \neq 0$, there exists an $i_1 (\neq i_0)$ such that $u_*(i_1) \neq 0$ and

$$\text{“the } (i_0, m_{i_0}, u_*(i_0))\text{-component of } He\text{ ''} \neq 0,$$

where $e = e(\tilde{k}, \tilde{m}; (i_1, m_{i_1}, u_*(i_1))$. Since $H \in C_1(\tilde{k}, \tilde{m})$, we have

$$k_{i_0} - u_*(i_0) \geq k_{i_1} - u_*(i_1) \quad \text{or} \quad u_*(i_0) \leq u_*(i_1).$$

This contradicts $\{u_*(i)\}_{i=1}^r \in U(\tilde{k}, \tilde{m}; \alpha)$. Therefore, we have $u_*(i_0) = u'_*(i_0) = 0$.

Case of $d = 2$. Proof follows *mutatis mutandis.* □

Lemma 2.3.39. *Assume*

$$\{u_*(i)\}_{i=1}^r, \{u'_*(i)\}_{i=1}^r \in U(\tilde{k}, \tilde{m}; \alpha),$$

$$p \in P_d(\tilde{k}, \tilde{m}; \alpha; \{u_*(i)\}_{i=1}^r), \quad p' \in P_d(\tilde{k}, \tilde{m}; \alpha; \{u'_*(i)\}_{i=1}^r),$$

and

$$H \in C_d(\tilde{k}, \tilde{m}; \alpha) \cap GL(n, \mathbb{R}).$$

(1) For any $1 \leq i \leq r$ with $u_*(i) \neq 0$ and for any u with $u_*(i) < u \leq k_i$,

$$\text{“the } (i, m_i, u)\text{-component of } Hp\text{ ''} = 0, \quad \text{and}$$

$$\text{“the } (i, m_i, u)\text{-component of } H\tilde{p}\text{ ''} = 0,$$

where $\tilde{p}$ denotes the continuous part of p.

(2) $Hp = p'$ implies $p = p'$.

Proof. Case of $d = 1$. Set

$$p_* = \sum_{i=1}^{r} e(\tilde{k}, \tilde{m}; (i, m_i, u_*(i))), \quad \tilde{p} = p - p_*,$$

$$p'_* = \sum_{i=1}^{r} e(\tilde{k}, \tilde{m}; (i, m_i, u'_*(i))), \quad \tilde{p}' = p' - p'_*.$$

From Definition 2.3.36(3), it follows that all components of $\tilde{p}$ and $\tilde{p}'$ are zero except the $(i_*, 1)$-blocks, which are denoted by

$$\tilde{p}(i_*, 1) = (\tilde{p}(1), \cdots, \tilde{p}(k_{i_*})), \quad \tilde{p}'(i_*, 1) = (\tilde{p}'(1), \cdots, \tilde{p}'(k_{i_*})).$$

(1) Let $1 \le i_0 \le r$ with $u_*(i_0) \ne 0$ be given. By Definition 2.3.36(2), we have

$$u_*(i_0) > u_*(i_*, 1) + \min\{0, k_{i_0} - k_{i_*}\},$$

where $u_*(i_*, 1) = \max\{u \mid \tilde{p}(u) \ne 0, 1 \le u \le k_{i_*}\}$. Hence

$$i_0 \ge i_* \Rightarrow u_*(i_0) > u_*(i_*, 1)$$

$$i_0 < i_* \Rightarrow k_{i_0} - u_*(i_0) < k_{i_*} - u_*(i_*, 1).$$

Since $H \in C_1(\tilde{k}, \tilde{m}; \alpha)$, from Lemma 2.3.16 it follows that

$$\text{“the } (i_0, m_{i_0}, u)\text{-component of } H\tilde{p}\text{ ”} = 0$$

for any $u_*(i_0) < u \le k_{i_0}$. By Lemma 2.3.38(1), for any $u_*(i_0) < u \le k_{i_0}$,

$$\text{“the } (i_0, m_{i_0}, u)\text{-component of } Hp_*\text{ ”} = 0.$$

Since $Hp = Hp_* + H\tilde{p}$,

$$\text{“the } (i_0, m_{i_0}, u)\text{-component of } Hp\text{ ”} = 0.$$

for any $u_*(i_0) < u \le k_{i_0}$.

(2) Let $1 \le i_0 \le r$ with $u_*(i_0) \ne 0$ be given. Since the (i_0, m_{i_0}, u)-components of Hp and $\tilde{p}'$ are 0 for any u with $u_*(i_0) < u \le k_{i_0}$, and since $p'_* = Hp - \tilde{p}'$, we have $0 \le u'_*(i_0) \le u_*(i_0)$.

If $u'_*(i_0) \ne 0$, since the (i_0, m_{i_0}, u)-components of $H^{-1}p'$ and $\tilde{p}$ are 0 for any u with $u'_*(i_0) < u \le k_{i_0}$, we have $u_*(i_0) \le u'_*(i_0)$. Hence $u_*(i_0) = u'_*(i_0)$.

If $u'_*(i_0) = 0$, since the $(i_0, m_{i_0}, u_*(i_0))$-components of $H\tilde{p}$ and $\tilde{p}'$ are 0,

$$\text{“the } (i_0, m_{i_0}, u_*(i_0))\text{-component of } Hp_*\text{ ”} = 0.$$

Hence, there is an $i_1 (\ne i_0)$ such that $u_*(i_1) \ne 0$ and

$$\text{“the } (i_0, m_{i_0}, u_*(i_0))\text{-component of } He\text{ ”} \ne 0,$$

where $e = e(\tilde{k}, \tilde{m}; (i_1, m_{i_1}, u_*(i_1)))$. Therefore, by $H \in C_1(\tilde{k}, \tilde{m})$,

$$k_{i_0} - u_*(i_0) \geq k_{i_1} - u_*(i_1) \quad \text{or} \quad u_*(i_0) \leq u_*(i_1),$$

this contradicts with $\{u_*(i)\}_{i=1}^r \in U(\tilde{k}, \tilde{m}; \alpha)$. Consequently, $u_*(i_0) = u'_*(i_0)$ for all $1 \leq i_0 \leq r$ with $u_*(i_0) \neq 0$.

Next, let $1 \leq i_0 \leq r$ with $u_*(i_0) = 0$ be given. If $u'_*(i_0) \neq 0$, since the $(i_0, m_{i_0}, u'_*(i_0))$-components of $H^{-1}\tilde{p}'$ and $\tilde{p}$ are 0, and since $p_* = H^{-1}\tilde{p}' + H^{-1}p'_* - \tilde{p}$, there exists an $i_1 (\neq i_0)$ such that $u'_*(i_1) \neq 0$ and the $(i_0, m_{i_0}, u'_*(i_0))$-component of $H^{-1}e(\tilde{k}, \tilde{m}; (i_1, m_{i_1}, u'_*(i_1)))$ is non-zero. Hence, by $H^{-1} \in C_1(\tilde{k}, \tilde{m})$, we have

$$k_{i_0} - u'_*(i_0) \geq k_{i_1} - u'_*(i_1) \quad \text{or} \quad u'_*(i_0) \leq u'_*(i_1).$$

This contradicts $\{u'_*(i)\}_{i=1}^r \in U(\tilde{k}, \tilde{m}; \alpha)$. Hence, we have $u_*(i) = u'_*(i)$ for all $1 \leq i \leq r$, that is, $p_* = p'*$.

To show $\tilde{p} = \tilde{p}'$, set

$$\begin{aligned} u_*(i_*, 1) &= \max\{1 \leq u \leq k_{i_*} \mid \tilde{p}(u) \neq 0\} \\ u'_*(i_*, 1) &= \max\{1 \leq u \leq k_{i_*} \mid \tilde{p}'(u) \neq 0\} \\ u_0(i_*, 1) &= \min\{1 \leq u \leq k_{i_*} \mid \tilde{p}(u) \neq 0\} \\ u'_0(i_*, 1) &= \min\{1 \leq u \leq k_{i_*} \mid \tilde{p}'(u) \neq 0\}. \end{aligned}$$

Notice $u_*(i_*, 1) = u_h$ and $u_0(i_*, 1) \geq u_l$ where u_h and u_l are integers as in Definition 2.3.36(2). Since $H \in C_1(\tilde{k}, \tilde{m}; \alpha)$,

$$\text{"the } (i_*, 1, u)\text{-component of } Hp_* \text{"} = 0,$$

for all u with $u_0(i_*, 1) \leq u \leq u_*(i_*, 1)$. Since

$$u_*(i_*, 1) - u_0(i_*, 1) + 1 \leq k_{i_*} - u_* + 1,$$

and since the $((i_*, 1), (i_*, 1))$-block of H has the form

$$H_{i_*i_*,11} = \Delta(a_1, \cdots, a_{k_{i_*}}),$$

$$a_1 = 1, \quad a_2 = \cdots = a_{k_{i_*} - u_* + 1} = 0,$$

we obtain, for each $u_0(i_*, 1) \leq u \leq u_*(i_*, 1)$,

$$\text{"the } (i_*, 1, u)\text{-component of } H\tilde{p} \text{"} = \tilde{p}(u).$$

Since $Hp = Hp_* + H\tilde{p}$,

$$\text{"the } (i_*, 1, u)\text{-component of } Hp \text{"} = \tilde{p}(u)$$

for each $u_0(i_*, 1) \leq u \leq u_*(i_*, 1)$. Since the $(i_*, 1, u)$-component of p'_* is equal to 0 for $1 \leq u \leq k_{i_*}$, and since $\tilde{p}' = Hp - p'_*$, we have $\tilde{p}(u) = \tilde{p}'(u)$ for $u_0(i_*, 1) \leq$

$u \leq u_*(i_*, 1)$. Clearly, for u such that $1 \leq u \leq u_0(i_*, 1)$ or $u_*(i_*, 1) < u \leq k_{i_*}$, we have $\tilde{p}(u) = 0$, and also $\tilde{p}(u_0(i_*, 1)) \neq 0$ and $\tilde{p}(u_*(i_*, 1)) \neq 0$. Applying the same argument to $\tilde{p}'$, we obtain

$$\tilde{p}'(u) = \tilde{p}(u) \quad \text{for} \quad u_0'(i_*, 1) \leq u \leq u_*'(i_*, 1)$$

$$\tilde{p}'(u) = 0 \quad \text{for} \quad 1 \leq u \leq u_0'(i_*, 1) \quad \text{or} \quad u_*'(i_*, 1) < u \leq k_{i_*}$$

$$\tilde{p}'(u_0'(i_*, 1)) \neq 0 \quad \text{and} \quad \tilde{p}'(u_*'(i_*, 1)) \neq 0.$$

Comparing these results, we obtain $u_0'(i_*, 1) = u_0(i_*, 1)$, $u_*'(i_*, 1) = u_*(i_*, 1)$, and

$$\tilde{p}'(u) = \tilde{p}(u) \quad \text{for all} \quad 1 \leq u \leq k_{i_*}.$$

Hence, $\tilde{p} = \tilde{p}'$. Therefore,

$$p = p_* + \tilde{p} = p_*' + \tilde{p}' = p'.$$

Case of $d = 2$. Proof follows *mutatis mutandis.* □

Definition 2.3.40. Denote a real Jordan matrix by

$$J = \dot{\sum}_{i=1}^{r} J_1(\lambda_i; \tilde{k}_i, \tilde{m}_i) \dot{+} \dot{\sum}_{i=1}^{s} J_2(\omega_i; \tilde{h}_i, \tilde{n}_i),$$

where

$$\begin{aligned}
\tilde{k}_i &= (k_{i1}, \cdots, k_{ir(i)}) \in \mathbb{N}_<(r(i)), \quad 1 \leq i \leq r \\
\tilde{h}_i &= (h_{i1}, \cdots, h_{is(i)}) \in \mathbb{N}_<(s(i)), \quad 1 \leq i \leq s \\
\tilde{m}_i &= (m_{i1}, \cdots, m_{ir(i)}) \in \mathbb{N}(r(i)), \quad 1 \leq i \leq r \\
\tilde{n}_i &= (n_{i1}, \cdots, n_{is(i)}) \in \mathbb{N}(s(i)), \quad 1 \leq i \leq s
\end{aligned}$$

and denote a canonical normal vector for J by

$$\alpha = \dot{\prod}_{i=1}^{r} e_1(\tilde{k}_i, \tilde{m}_i; (k(i), 1, m(i))) \dot{\times} \dot{\prod}_{i=1}^{s} e_2(\tilde{h}_i, \tilde{n}_i; (h(i), 1, n(i))).$$

(1) Set $n = \sum \tilde{k}_i \cdot \tilde{m}_i + 2 \sum \tilde{h}_i \cdot \tilde{n}_i$,

$$\begin{aligned}
\tilde{k}_{di} &= \tilde{k}_i(d = 1); = \tilde{h}_i(d = 2) \\
\tilde{m}_{di} &= \tilde{m}_i(d = 1); = \tilde{n}_i(d = 2) \\
\alpha_{di} &= \begin{cases} e_1(\tilde{k}_i, \tilde{m}_i; (k(i), 1, m(i))) & (d = 1) \\ e_2(\tilde{h}_i, \tilde{n}_i; (h(i), 1, n(i))) & (d = 2). \end{cases}
\end{aligned}$$

A vector $p \in \mathbb{R}^n$ is called a *canonical parameter* for (J, α) if there exist $p_{di} \in P_d(\tilde{k}_{di}, \tilde{m}_{di}; \alpha_{di})$ such that

$$p = \dot{\prod}_{i=1}^{r} p_{1i} \dot{\times} \dot{\prod}_{j=1}^{s} p_{2j}.$$

(2) By p_{*di}, denote the discrete part of p_{di}, by $\tilde{p}_{di}$ the continuous part of p_{di}

$$p_{di} = p_{*di} + \tilde{p}_{di}.$$

Set

$$\begin{aligned} p_* &= \dot{\prod}_{i=1}^{r} p_{*1i} \dot{\times} \dot{\prod}_{j=1}^{s} p_{*2j} \\ \tilde{p} &= \dot{\prod}_{i=1}^{r} \tilde{p}_{1i} \dot{\times} \dot{\prod}_{j=1}^{s} \tilde{p}_{2j} \end{aligned}$$

then $p = p_* + \tilde{p}$. This p_* is called the *discrete part* of p, and the $\tilde{p}$ is called the *continuous part* of p.

Theorem 2.3.41. (Normal Forms of Non-Degenerate Two-Region Systems)

(1) If necessary, taking the complement and exchanging the signs of regions, any continuous n-dimensional non-degenerate two-region piecewise-linear vector field is affine conjugate to the vector field with the following form:

$$\begin{aligned} X_f(x) &= f(x) = \begin{cases} (J + p\alpha^T)x - \varepsilon p, & (x \in R_1) \\ Jx, & (x \in R_0) \end{cases} \quad (2.3.13) \\ &= Jx + (1/2)p\{|\langle \alpha, x\rangle - \varepsilon| + \langle \alpha, x\rangle - \varepsilon\}. \\ R_\omega &= \{x \in \mathbb{R}^n \mid \mathrm{sgn}(\langle \alpha, x\rangle - \varepsilon) = \omega\}, \quad \omega = 0, 1, \end{aligned}$$

where J is a real Jordan matrix, $\alpha \in \mathbb{R}^n$ is a canonical normal vector for J, p is a canonical parameter for (J, α), $\varepsilon = 1$ if X_f is regular, and $\varepsilon = 0$ if X_f is singular.

(2) The complement vector field of X_f is given by

$$\begin{aligned} X_{\bar{f}}(x) &= \bar{f}(x) = \begin{cases} (J + p\alpha^T)x - \varepsilon p, & (x \in R_0) \\ Jx, & (x \in R_1) \end{cases} \\ &= Jx + (1/2)p\{|\langle \alpha, x\rangle - \varepsilon| - \langle \alpha, x\rangle - \varepsilon\}. \end{aligned}$$

(3) Assume

$$X_{f_1}(x) = Jx + (1/2)p_1\{|\langle \alpha, x\rangle - \varepsilon| + \langle \alpha, x\rangle - \varepsilon\}$$

$$X_{f_2}(x) = Jx + (1/2)p_2\{|\langle \alpha, x\rangle - \varepsilon| + \langle \alpha, x\rangle - \varepsilon\}$$

both p_1 and p_2 are canonical parameters for (J, α).

Then there exists a linear transformation $G \in GL(n, \mathbb{R})$ such that $GX_{f_1} = X_{f_2}G$ if and only if $p_1 = p_2$. In this sense, the canonical parameter p is uniquely determined for a linear conjugate class of vector fields with fixed (J, α).

Proof. (1) Without loss of generality, we can assume that X_f has the same form as in Equation (2.3.11). Set $h = \alpha/\langle\alpha, \alpha\rangle$. Then, since $\langle h, \alpha\rangle = 1$, and since f is continuous, Theorem 2.2.4 implies

$$A = J + (A - J)h\alpha^T, \quad q_A = -\varepsilon(A - J)h.$$

Setting $p' = (A - J)h$, we have

$$A = J + p'\alpha^T, \quad q_A = -\varepsilon p'.$$

Applying Lemma 2.3.37 to the real Jordan matrix corresponding to each eigenvalue, we have an $H \in GL(n, \mathbb{R})$ such that

$$\text{(i)} HJH^{-1} = J, \quad \text{(ii)} \alpha^T H = \alpha^T,$$

$$\text{(iii)} p = Hp' \quad \text{is a canonical parameter for } (J, \alpha).$$

Since $Hp'\alpha^T H^{-1} = p\alpha^T$, the X_f is transformed into a vector field in (2.3.13) under the linear transformation H.

(2) Follows easily from the definition of the complement.

(3) The "if" statement is clear. We will prove the "only if" statement. First, let us consider the regular case, i.e., $\varepsilon = 1$. From the assumption, it follows that

$$\text{(i)} GJ = JG, \quad \text{(ii)} \alpha^T G = \alpha^T, \quad \text{(iii)} Gp_1 = p_2.$$

Divide J and α into the forms in Definition 2.3.40. Then, by (i) and (ii), it follows from Lemma 2.3.11 and Lemma 2.3.15 that G is split into the direct sum

$$G = \dot{\sum}_{i=1}^{r} G_{1i} \dot{+} \dot{\sum}_{i=1}^{s} G_{2i}, \quad G_{di} \in C_d(\tilde{k}_{di}, \tilde{m}_{di}; \alpha_{di}).$$

Splitting p_1 and p_2 into the form as in Definition 2.3.40, and applying Lemma 2.3.39(2) to it, we obtain $p_1 = p_2$.

Next, let us consider the singular case, i.e., $\varepsilon = 0$. By the assumption, there is a $k \neq 0$ such that

$$\text{(i)} GJ = JG, \quad \text{(ii)} \alpha^T G = k\alpha^T, \quad \text{(iii)} Gp_1\alpha^T = p_2\alpha^T G.$$

It follows from (ii) and (iii) that

$$\begin{aligned} p_2 &= p_2\alpha^T\alpha/\langle\alpha, \alpha\rangle = Gp_1\alpha^T G^{-1}\alpha/\langle\alpha, \alpha\rangle \\ &= (1/k)Gp_1\alpha^T\alpha/\langle\alpha, \alpha\rangle = (1/k)Gp_1, \end{aligned}$$

that is, $Gp_1 = kp_2$. Setting $G' = (1/k)G$, we have

$$\text{(i')}G'J = JG', \quad \text{(ii')}\alpha^T G' = \alpha^T, \quad \text{(iii')}G'p_1 = p_2.$$

Therefore, similar to the regular case, we obtain $p_1 = p_2$. □

Next we will consider the case of degenerate two-region piecewise-linear vector fields.

Definition 2.3.42. Let a real Jordan matrix

$$J = J_* \overset{\bullet}{+} J_1(0; \tilde{k}, \tilde{m}),$$

and a canonical normal vector for J

$$\alpha = \alpha_* \overset{\bullet}{\times} \alpha_0, \quad \alpha_0 = e(\tilde{k}, \tilde{m}; (i_*, 1, u_*))$$

be given. Set $J_0 = J_1(0; \tilde{k}, \tilde{m})$ and $n_2 = \tilde{k} \cdot \tilde{m}$.

(1) Let q_0 be a canonical constant vector for (J_0, α_0):

$$q_0 = e(\tilde{k}, \tilde{m}; (i_*, 1, u_* - 1), \tau_3; (i_*, 1, k_{i_*}), \tau_2; (l_*, m_{l_*} k_{l_*}), \tau_1).$$

The triplet $(\tau_1, \tau_2, \tau_3) \in \{(1,0,0), (1,0,1), (1,1,0), (0,1,0), (1,1,1)\}$ is determined from Rule (τ) in Theorem 2.3.32 (where we set $(\tau_1, \tau_2, \tau_3) = (1,1,1)$ and $i_* = 0$ if $\alpha_0 = 0$). $U(\tilde{k}, \tilde{m}; \alpha_0, q_0)$ is a subset of $U(\tilde{k}, \tilde{m}; \alpha_0) \subset \mathbb{Z}_+^r$ consisting of all elements $\{u_*(i)\}_{i=1}^r$ satisfying conditions (i)–(iii) of Definition 2.3.36(1) and, in addition,

(iv) $\tau_1 = 1 \quad \text{and} \quad m_{l_*} = 1 \Rightarrow u_*(l_*) = 0.$

(2) For $\{u_*(i)\} \in U(\tilde{k}, \tilde{m}; \alpha_0, q_0)$, subset $\tilde{P}(\tilde{k}, \tilde{m}; \alpha_0, q_0; \{u_*(i)\})$ of $M(n_2, 1)$ consists of all $\tilde{p} = (\tilde{p}(i, s, u))(1 \le i \le r, 1 \le s \le m_i, 1 \le u \le k_i)$ satisfying the following four conditions:

(i) $\tilde{p}(i, s, u) = 0$ unless $(i, s) = (i_*, 1)$ and $u_l(i_*) \le u \le u_h(i_*)$, or $(i, s) = (l_*, m_{l_*})$ and $u_l(l_*) \le u \le u_h(l_*)$

(ii) $p(i_*, 1, u_h(i_*)) \ne 0$ if $u_h(i_*) \ne 0$ and $i_* \ne 0$, and $p(l_*, m_{l_*}, u_h(l_*)) \ne 0$ if $u_h(l_*) \ne 0$

(iii) If $(\tau_1, \tau_2, \tau_3) = (1,1,1)$, then $u_h(i_*) = 0$ and $u_l(i_*) = 1$. Otherwise,

$$0 \le u_h(i_*) < \min_{i \,|\, u_*(i) \ne 0} \{[u_*(i) + \max\{k_{i_*} - k_i, \tau_3(u_* - 1)\}]^{\tau_2},$$

$$[u_h(l_*) + \max\{k_{i_*} - k_{l_*}, u_* - 1\}]^{1-\tau_3}, k_{i_*} + 1\},$$

$$u_l(i_*) = \max_{i \,|\, u_*(i) \ne 0} \{[u_h(i_*) - k_{i_*} + u_*]^{\tau_2 \tau_3}, [u_h(i_*) - u_* + 2]^{\tau_2(1-\tau_3)},$$

$$u_*(i) + \min\{u_* - k_i, 1\}, 1\}$$

(iv) If $(\tau_1, \tau_2, \tau_3) = (0, 1, 0)$, then $u_h(l_*) = 0$ and $u_l(l_*) = 1$. Otherwise,

$$0 \le u_h(l_*) < \min_{i \mid u_*(i) \neq 0} \{[u_*(i) + \max\{k_{l_*} - k_i, 0\},$$

$$u_h(i_*) + \max\{k_{l_*} - k_{i_*}, 0\}]^{\tau_2 \tau_3}, k_{l*} + 1\},$$

$$u_l(l_*) = \max_{i \mid u_*(i) \neq 0} \{[u_*(i_*) - k_{i_*} + k_{l*} + 1]^{\tau_2 \tau_3},$$

$$u_*(i) + \min\{k_{l_*} - k_i + 1, 1\}, 1\},$$

where $[E]^\tau = E$ if $\tau = 0$, while nothing if $\tau = 1$ for any expression E. For example,

$$\max\{[a]^\tau, b, c\} = \begin{cases} \max\{a, b, c\} & \text{if} \quad \tau = 0 \\ \max\{b, c\} & \text{if} \quad \tau = 1 \end{cases}$$

(3) For $\{u_*(i)\} \in U(\tilde{k}, \tilde{m}; \alpha_0, q_0)$, subset $P(\tilde{k}, \tilde{m}; \alpha_0, q_0; \{u_*(i)\})$ of $M(n_2, 1)$ consists of all $p = \tilde{p} + \sum_{i=1}^{r} e(\tilde{k}, \tilde{m}; (i, m_i, u_*(i)))$, $\tilde{p} \in \tilde{P}(\tilde{k}, \tilde{m}; \alpha_0, q_0; \{u_*(i)\})$. $P(\tilde{k}, \tilde{m}; \alpha_0, q_0)$ is union of $P(\tilde{k}, \tilde{m}; \alpha_0, q_0; \{u_*(i)\})$ for all $\{u_*(i)\} \in U(\tilde{k}, \tilde{m}; \alpha_0, q_0)$.

(4) This $p = p_* \overset{\bullet}{\times} p_0$ is a *canonical parameter* for (J, α, q) if p_* is a canonical parameter for (J_*, α_*) and $p_0 \in P(\tilde{k}, \tilde{m}; \alpha_0, q_0)$.

Theorem 2.3.43. (Normal Forms of Degenerate Two-Region Systems)

(1) If necessary, taking the complement and changing the signs of regions, any continuous n-dimensional degenerate two-region piecewise-linear vector field is linearly conjugate to the vector field with the following form:

$$\begin{aligned} X_f(x) &= f(x) = \begin{cases} (J + p\alpha^T)x + q - \varepsilon p, & (x \in R_1) \\ Jx + q, & (x \in R_0) \end{cases} \\ &= Jx + q + (1/2)p\{|\langle \alpha, x\rangle - \varepsilon| + \langle \alpha, x\rangle - \varepsilon\}, \\ R_\omega &= \{x \in \mathbb{R}^n \mid \operatorname{sgn}(\langle \alpha, x\rangle - \varepsilon) = \omega\}, \quad \omega = 0, 1, \end{aligned}$$

where

J is a real Jordan matrix; $J = J_* \overset{\bullet}{\times} J_1(0; \tilde{k}, \tilde{m})$,

$\alpha \in \mathbb{R}^n$ is a canonical normal vector for J; $\alpha = \alpha_* \overset{\bullet}{\times} \alpha_0$,

$q \in \mathbb{R}^n$ is a canonical constant vector for (J, α); $q = O \overset{\bullet}{\times} q_0$,

$p \in \mathbb{R}^n$ is a canonical parameter for (J, α), and

$\varepsilon = 0$ if $\alpha_0 \neq 0$, while $\varepsilon = 0$ or 1 if $\alpha_0 = 0$.

(2) The complement vector field of X_f *is given by*

$$X_{\bar{f}}(x) = \bar{f}(x) = \begin{cases} (J + p\hat{\alpha}^T)x + q - \varepsilon p, & (x \in R_0) \\ Jx + q, & (x \in R_1) \end{cases}$$
$$= Jx + q + (1/2)p\{|\langle\alpha, x\rangle - \varepsilon| - \langle\alpha, x\rangle - \varepsilon\}.$$

(3) Assume

$$X_{f_1}(x) = Jx + q + (1/2)p_1\{|\langle\alpha, x\rangle - \varepsilon| + \langle\alpha, x\rangle - \varepsilon\},$$

$$X_{f_2}(x) = Jx + q + (1/2)p_2\{|\langle\alpha, x\rangle - \varepsilon| + \langle\alpha, x\rangle - \varepsilon\},$$

where both p_1 *and* p_2 *are canonical parameters for* (J, α, q). *Then there exists a linear transformation* $G \in GL(n, \mathbb{R})$ *such that* $GX_{f_1} = X_{f_2}G$ *if and only if* $p_1 = p_2$. *In this sense, the canonical parameter* p *is uniquely determined for a linear conjugate class of vector fields with fixed* (J, α, q).

Proof. (1) Without loss of generality, we can assume that X_f has the following form (See Theorems 2.2.4 and 2.3.32):

$$X_f(x) = f(x) = \begin{cases} (J + p'\alpha^T)x + q - \varepsilon p', & (x \in R_1) \\ Jx + q, & (x \in R_0) \end{cases}$$
$$R_\omega = \{x \in \mathbb{R}^n \mid \mathrm{sgn}(\langle\alpha, x\rangle \dot{-} \varepsilon) = \omega\}, \quad \omega = 0, 1,$$

where J, q, α and ε are canonical. It remains to transform p' into a canonical parameter p for (J, α, q). Split $p' = p'_* \dot{\times} p'_0$ corresponding to $J = J_* \dot{+} J_1(0; \tilde{k}, \tilde{m})$. Then p'_* is transformed into a canonical parameter p_* for (J_*, α_*) by applying Lemma 2.3.37. To see that p'_0 can be transformed to a canonical parameter p_0 for (J_0, α_0, q_0), we use an algorithm similar to that in Lemma 2.3.37 for $\{u_*(i)\} \in U(\tilde{k}, \tilde{m}; \alpha_0, q_0)$ and $H \in C^*(\tilde{k}, \tilde{m}; \alpha_0) \cap GL(\tilde{k} \cdot \tilde{m}, \mathbb{R})$ with $Hq_0 = q_0$ in each case of (τ_1, τ_2, τ_3). Then we obtain a canonical parameter $p = p_* \dot{\times} p_0$.

(2) Clear.

(3) The "if" statement is clear. We will prove the "only if" statement. First, let us consider the case of $\varepsilon = 0$. By assumption, there are $G \in GL(n, \mathbb{R})$ and $k \neq 0$ such that

$$\text{(i)} GJ = JG, \quad \text{(ii)} \alpha^T G = k\alpha^T, \quad \text{(iii)} Gp_1\alpha^T = p_2\alpha^T G.$$

Setting $G' = (1/k)G$, we have

$$\text{(i')} G'J = JG', \quad \text{(ii')} \alpha^T G' = \alpha^T, \quad \text{(iii')} G'p_1\alpha^T = p_2\alpha^T G'.$$

From this, it follows that $p_1 = p_2$ by an argument similar to that in Lemma 2.3.39(2).

The case of $\varepsilon = 1$. By assumption, there exists $G \in GL(n, \mathbb{R})$ such that

$$\text{(i)} GJ = JG, \quad \text{(ii)} \alpha^T G = \alpha^T, \quad \text{(iii)} Gp_1\alpha^T = p_2\alpha^T G.$$

From this, it follows $p_1 = p_2$ by a similar argument to Lemma 2.3.39(2). □

Example 2.3.44. (Canonical Parameter for (J, α)).
Assume

$$J = J(\lambda; \tilde{k}, \tilde{m}), \quad \tilde{k} = (2,5), \quad \tilde{m} = (2,2),$$

$$\alpha = e(\tilde{k}, \tilde{m}; (i_*, 1, u_*)), \quad i_* = 2, \quad u_* = 3, \quad k_{i_*} = 5.$$

(1) $U(\tilde{k}, \tilde{m}; \alpha)$ consists of the following sequences:

$$\{0,0\}, \quad \{0,1\}, \quad \{0,2\}, \quad \{0,3\}, \quad \{0,4\}, \quad \{0,5\}, \quad \{1,0\},$$

$$\{2,0\}, \quad \{1,2\}, \quad \{1,3\}, \quad \{2,3\}, \quad \{2,4\}.$$

(2) Since

$$0 \le u_h < \min\{5+1\} = 6 \quad \text{and} \quad u_l = \max\{u_h - 5 + 3\},$$

$P(\tilde{k}, \tilde{m}; \alpha; \{0,0\})$ consists of the following vectors:

$$(0,0, \quad 0,0, \quad 0,0,0,0,0, \quad 0,0,0,0,0)^T$$

$$(0,0, \quad 0,0, \quad p_1,0,0,0,0, \quad 0,0,0,0,0)^T$$

$$(0,0, \quad 0,0, \quad p_2,p_1,0,0,0, \quad 0,0,0,0,0)^T$$

$$(0,0, \quad 0,0, \quad p_3,p_2,p_1,0,0, \quad 0,0,0,0,0)^T$$

$$(0,0, \quad 0,0, \quad 0,p_3,p_2,p_1,0, \quad 0,0,0,0,0)^T$$

$$(0,0, \quad 0,0, \quad 0,0,p_3,p_2,p_1, \quad 0,0,0,0,0)^T,$$

where $p_1 \neq 0$.

(3) Since

$$0 \le u_h < \min\{3 + \max\{0, 5-5\}, 5+1\} = 3 \quad \text{and}$$

$$u_l = \max\{u_h - 5 + 3, 3 - 5 + 3\},$$

$P(\tilde{k}, \tilde{m}; \alpha; \{0,3\})$ consists of the following vectors:

$$(0,0, \quad 0,0, \quad 0,0,0,0,0, \quad 0,0,1,0,0)^T$$

$$(0,0, \quad 0,0, \quad p_1,0,0,0,0, \quad 0,0,1,0,0)^T$$

$$(0,0, \quad 0,0, \quad p_2,p_1,0,0,0, \quad 0,0,1,0,0)^T,$$

where $p_1 \neq 0$.

(4) Since

$$0 \le u_h < \min\{1+\max\{0,5-2\},3+\max\{0,5-5\},5+1\} = 3 \quad \text{and}$$

$$u_l = \max\{u_h - 5 + 3, 1-2+3, 3-5+3\} = 2,$$

$P(\tilde{k},\tilde{m};\alpha;\{1,3\})$ consists of the following vectors:

$$(0,0,\quad 1,0,\quad 0,0,0,0,0,\quad 0,0,1,0,0)^T$$

$$(0,0,\quad 1,0,\quad 0,p_1,0,0,0,\quad 0,0,1,0,0)^T,$$

where $p_1 \neq 0$.

(5) Since

$$0 \le u_h < \min\{2+\max\{0,5-2\},3+\max\{0,5-5\},5+1\} = 3$$

and

$$u_l = \max\{u_h - 5 + 3, 2-2+3, 3-5+3\} = 3,$$

$P(\tilde{k},\tilde{m};\alpha;\{2,3\})$ consists of the following vectors:

$$(0,0,\quad 0,1,\quad 0,0,0,0,0,\quad 0,0,1,0,0)^T.$$

Example 2.3.45. (Derivation of Canonical Parameter; see Lemma 2.3.37) Suppose $J = J(\lambda;\tilde{k},\tilde{m})$, $\alpha = e(\tilde{k},\tilde{m};(i_*,1,u_*))$, $\tilde{k}=(2,3), \tilde{m}=(2,2), i_*=2$, and $u_*=2$. All elements of $C^*(\tilde{k},\tilde{m};\alpha)$ have the following form:

$$\left(\begin{array}{cc|cc|ccc|ccc}
\sigma & \phi & \sigma' & \phi' & 0 & \eta & \kappa & 0 & \omega & \varsigma \\
0 & \sigma & 0 & \sigma' & 0 & 0 & \eta & 0 & 0 & \omega \\
\hline
\gamma & \delta & \gamma' & \delta' & 0 & \chi & \psi & 0 & \nu & \mu \\
0 & \gamma & 0 & \gamma' & 0 & 0 & \chi & 0 & 0 & \nu \\
\hline
0 & \iota & 0 & \rho & 1 & 0 & \upsilon & 0 & 0 & \xi \\
0 & 0 & 0 & 0 & 0 & 1 & 0 & 0 & 0 & 0 \\
0 & 0 & 0 & 0 & 0 & 0 & 1 & 0 & 0 & 0 \\
\hline
\omega' & \zeta' & \theta' & \tau' & \upsilon' & \xi' & \sigma' & \varepsilon & \theta & \tau \\
0 & \omega' & 0 & \theta' & 0 & \upsilon' & \xi' & 0 & \varepsilon & \theta
\end{array}\right).$$

Step 1. Assume a, c, f and k are non-zero.

$$\left(\begin{array}{cc|cc|ccc|ccc}
a' & 0 & 0 & 0 & 0 & 0 & 0 & 0 & 0 & 0 \\
0 & a' & 0 & 0 & 0 & 0 & 0 & 0 & 0 & 0 \\
\hline
0 & 0 & c' & b' & 0 & 0 & 0 & 0 & 0 & 0 \\
0 & 0 & 0 & c' & 0 & 0 & 0 & 0 & 0 & 0 \\
\hline
0 & 0 & 0 & 0 & 1 & 0 & d' & 0 & 0 & 0 \\
0 & 0 & 0 & 0 & 0 & 1 & 0 & 0 & 0 & 0 \\
0 & 0 & 0 & 0 & 0 & 0 & 1 & 0 & 0 & 0 \\
\hline
0 & 0 & 0 & 0 & 0 & 0 & 0 & k' & h' & g' \\
0 & 0 & 0 & 0 & 0 & 0 & 0 & 0 & k' & h' \\
0 & 0 & 0 & 0 & 0 & 0 & 0 & 0 & 0 & k'
\end{array}\right)
\begin{pmatrix} a \\ 0 \\ b \\ c \\ d \\ e \\ f \\ g \\ h \\ k \end{pmatrix}
=
\begin{pmatrix} 1 \\ 0 \\ 0 \\ 1 \\ 0 \\ e \\ f \\ 0 \\ 0 \\ 1 \end{pmatrix},$$

where $a' = 1/a, c' = 1/c, b' = -b/c^2, k' = 1/k, h' = -h/k^2, d' = -d/f$ and $g' = -g/k^2 + h^2/k^3$.

Step 2. $u'_*(1) = 2, u'_*(2) = 3$.

$$\left(\begin{array}{cc|cc|ccc|ccc} 1 & 0 & 0 & -1 & 0 & 0 & 0 & 0 & 0 & 0 \\ 0 & 1 & 0 & 0 & 0 & 0 & 0 & 0 & 0 & 0 \\ \hline 0 & 0 & 1 & 0 & 0 & 0 & 0 & 0 & 0 & 0 \\ 0 & 0 & 0 & 1 & 0 & 0 & 0 & 0 & 0 & 0 \\ \hline 0 & 0 & 0 & 0 & 1 & 0 & 0 & 0 & 0 & 0 \\ 0 & 0 & 0 & 0 & 0 & 1 & 0 & 0 & 0 & 0 \\ 0 & 0 & 0 & 0 & 0 & 0 & 1 & 0 & 0 & 0 \\ \hline 0 & 0 & 0 & 0 & 0 & 0 & 0 & 1 & 0 & 0 \\ 0 & 0 & 0 & 0 & 0 & 0 & 0 & 0 & 1 & 0 \\ 0 & 0 & 0 & 0 & 0 & 0 & 0 & 0 & 0 & 1 \end{array}\right) \begin{pmatrix} 1 \\ 0 \\ 0 \\ 1 \\ 0 \\ e \\ f \\ 0 \\ 0 \\ 1 \end{pmatrix} = \begin{pmatrix} 0 \\ 0 \\ 0 \\ 1 \\ 0 \\ e \\ f \\ 0 \\ 0 \\ 1 \end{pmatrix}$$

Step 3.

$$\left(\begin{array}{cc|cc|ccc|ccc} 1 & 0 & 0 & 0 & 0 & 0 & 0 & 0 & 0 & 0 \\ 0 & 1 & 0 & 0 & 0 & 0 & 0 & 0 & 0 & 0 \\ \hline 0 & 0 & 1 & 0 & 0 & 0 & 0 & 0 & -1 & 0 \\ 0 & 0 & 0 & 1 & 0 & 0 & 0 & 0 & 0 & -1 \\ \hline 0 & 0 & 0 & 0 & 1 & 0 & 0 & 0 & 0 & 0 \\ 0 & 0 & 0 & 0 & 0 & 1 & 0 & 0 & 0 & 0 \\ 0 & 0 & 0 & 0 & 0 & 0 & 1 & 0 & 0 & 0 \\ \hline 0 & 0 & 0 & 0 & 0 & 0 & 0 & 1 & 0 & 0 \\ 0 & 0 & 0 & 0 & 0 & 0 & 0 & 0 & 1 & 0 \\ 0 & 0 & 0 & 0 & 0 & 0 & 0 & 0 & 0 & 1 \end{array}\right) \begin{pmatrix} 0 \\ 0 \\ 0 \\ 1 \\ 0 \\ e \\ f \\ 0 \\ 0 \\ 1 \end{pmatrix} = \begin{pmatrix} 0 \\ 0 \\ 0 \\ 0 \\ 0 \\ e \\ f \\ 0 \\ 0 \\ 1 \end{pmatrix}$$

Step 4.

$$\left(\begin{array}{cc|cc|ccc|ccc} 1 & 0 & 0 & 0 & 0 & 0 & 0 & 0 & 0 & 0 \\ 0 & 1 & 0 & 0 & 0 & 0 & 0 & 0 & 0 & 0 \\ \hline 0 & 0 & 1 & 0 & 0 & 0 & 0 & 0 & 0 & 0 \\ 0 & 0 & 0 & 1 & 0 & 0 & 0 & 0 & 0 & 0 \\ \hline 0 & 0 & 0 & 0 & 1 & 0 & 0 & 0 & 0 & 0 \\ 0 & 0 & 0 & 0 & 0 & 1 & 0 & 0 & 0 & 0 \\ 0 & 0 & 0 & 0 & 0 & 0 & 1 & 0 & 0 & 0 \\ \hline 0 & 0 & 0 & 0 & f' & g' & h' & 1 & 0 & 0 \\ 0 & 0 & 0 & 0 & 0 & f' & g' & 0 & 1 & 0 \\ 0 & 0 & 0 & 0 & 0 & 0 & f' & 0 & 0 & 1 \end{array}\right) \begin{pmatrix} 0 \\ 0 \\ 0 \\ 0 \\ 0 \\ e \\ f \\ 0 \\ 0 \\ 1 \end{pmatrix} = \begin{pmatrix} 0 \\ 0 \\ 0 \\ 0 \\ 0 \\ e \\ f \\ 0 \\ 0 \\ 0 \end{pmatrix},$$

where $f' = -1/f, g' = -ef'/f$ and $h' = -eg'/f$.

Example 2.3.46. (Normal Forms of Three-Dimensional Non-Degenerate Two-Region Systems)
For each Jordan matrix J, α is a canonical normal vector, and p is a canonical parameter for (J, α).

$$(1)\begin{pmatrix} \lambda & & \\ & \mu & \\ & & \nu \end{pmatrix}$$

α	p
$(1,1,1)$	(q,r,s)
$(1,1,0)$	$(q,r,1)$,$(q,r,0)$
$(1,0,0)$	$(q,1,1)$,$(q,1,0)$,$(q,0,0)$

$$(2)\begin{pmatrix} \lambda & & \\ & \mu & \\ & & \mu \end{pmatrix}$$

α	p
$(1,1,0)$	$(q,r,0)$,$(q,0,1)$
$(1,0,0)$	$(q,0,1)$,$(q,0,0)$
$(0,1,0)$	$(1,q,0)$,$(1,0,1)$,$(0,q,0)$,$(0,0,1)$

$$(3)\begin{pmatrix} \lambda & & \\ & \lambda & \\ & & \lambda \end{pmatrix}$$

α	p
$(1,0,0)$	$(q,0,0)$,$(0,0,1)$

$$(4)\begin{pmatrix} \lambda & & \\ & \mu & 1 \\ & & \mu \end{pmatrix}$$

α	p
$(1,1,0)$	(q,r,s)
$(1,0,1)$	$(q,0,s)$,$(q,r,0)$
$(1,0,0)$	$(q,0,1)$,$(q,1,0)$,$(q,0,0)$
$(0,1,0)$	$(1,r,s)$,$(0,r,s)$
$(0,0,1)$	$(1,0,s)$,$(1,r,0)$,$(0,0,s)$,$(0,r,0)$

$$(5)\begin{pmatrix} \lambda & & \\ & \lambda & 1 \\ & & \lambda \end{pmatrix}$$

α	p
$(1,0,0)$	$(q,0,1)$,$(q,0,0)$,$(0,1,0)$
$(0,1,0)$	$(0,r,s)$,$(1,r,0)$
$(0,0,1)$	$(0,0,s)$,$(0,r,0)$,$(1,0,0)$

$$(6)\begin{pmatrix} \lambda & 1 & \\ & \lambda & 1 \\ & & \lambda \end{pmatrix}$$

α	p
$(1,0,0)$	(q,r,s)
$(0,1,0)$	$(0,r,s)$,$(q,r,0)$
$(0,0,1)$	$(0,0,s)$,$(0,r,0)$,$(q,0,0)$

$$(7)\begin{pmatrix} \lambda & & \\ & a & -b \\ & b & a \end{pmatrix}$$

α	p
$(1,1,0)$	(q,r,r')
$(1,0,0)$	$(q,1,0)$,$(q,0,0)$
$(0,1,0)$	$(1,r,r')$,$(0,r,r')$

Example 2.3.47. (Normal Forms of Three-Dimensional Degenerate Two-Region Systems)
For a Jordan matrix J, α is a canonical normal vector, q is a canonical constant vector for (J,α), and p is a canonical parameter for (J,α,q).

$$(1)\begin{pmatrix} \lambda & & \\ & \mu & \\ & & 0 \end{pmatrix}$$

α	q	p
$(1,1,0)$	$(0,0,1)$	(r,s,t)
$(1,0,0)$	$(0,0,1)$	$(r,1,t)$,$(r,0,t)$
$(1,1,1)$	$(0,0,1)$	(r,s,t)
$(1,0,1)$	$(0,0,1)$	$(r,0,t)$,$(r,1,t)$
$(0,0,1)$	$(0,0,1)$	$(1,0,t)$,$(0,0,t)$

$$(2)\begin{pmatrix} \lambda & & \\ & \lambda & \\ & & 0 \end{pmatrix}$$

α	q	p
$(1,0,0)$	$(0,0,1)$	$(r,0,t)$,$(0,1,t)$
$(1,0,1)$	$(0,0,1)$	$(r,0,t)$,$(0,1,t)$
$(0,0,1)$	$(0,0,1)$	$(0,1,t)$,$(0,0,t)$

$$(3)\begin{pmatrix} \lambda & 1 & \\ & \lambda & \\ & & 0 \end{pmatrix}$$

α	q	p
$(1,0,0)$	$(0,0,1)$	(r,s,t)
$(0,1,0)$	$(0,0,1)$	$(0,s,t)$,$(r,0,t)$
$(1,0,1)$	$(0,0,1)$	(r,s,t)
$(0,1,1)$	$(0,0,1)$	$(0,s,t)$,$(r,0,t)$
$(0,0,1)$	$(0,0,1)$	$(0,1,t)$,$(1,0,t)$,$(0,0,t)$

$$(4)\begin{pmatrix} a & -b & \\ b & a & \\ & & 0 \end{pmatrix}$$

α	q	p
$(1,0,0)$	$(0,0,1)$	(r,r',t)
$(1,0,1)$	$(0,0,1)$	(r,r',t)
$(0,0,1)$	$(0,0,1)$	$(1,0,t)$,$(0,0,t)$

$$(5)\begin{pmatrix} \lambda & & \\ & 0 & 1 \\ & & 0 \end{pmatrix}$$

α	q	p
$(1,0,0)$	$(0,0,1)$	(r,s,t)
$(1,1,0)$	$(0,0,1)$	(r,s,t)
$(1,0,1)$	$(0,0,1)$	(r,s,t)
$(0,1,0)$	$(0,0,1)$	$(1,s,t)$,$(0,s,t)$
$(0,0,1)$	$(0,0,1)$	$(1,s,t)$,$(0,s,t)$

$$(6)\begin{pmatrix} \lambda & & \\ & 0 & \\ & & 0 \end{pmatrix}$$

α	q	p
$(1,0,0)$	$(0,0,1)$	$(r,1,0),(r,0,t)$
$(1,1,0)$	$(0,0,1)$	$(r,s,0),(r,0,t)$
	$(0,1,0)$	$(r,0,1)$
$(0,1,0)$	$(0,0,1)$	$(1,s,0),(0,s,0),(1,0,t),(0,0,t)$
	$(0,1,0)$	$(1,s,t),(0,s,t)$

$$(7)\begin{pmatrix} 0 & 1 & \\ & 0 & 1 \\ & & 0 \end{pmatrix}$$

α	q	p
$(1,0,0)$	$(0,0,1)$	(r,s,t)
$(0,1,0)$	$(0,0,1)$	(r,s,t)
$(0,0,1)$	$(0,0,1)$	(r,s,t)

$$(8)\begin{pmatrix} 0 & & \\ & 0 & 1 \\ & & 0 \end{pmatrix}$$

α	q	p
$(1,0,0)$	$(0,0,1)$	$(r,0,t),(0,s,t)$
	$(1,0,0)$	$(r,0,1),(r,1,0)$
	$(1,0,1)$	(r,s,t)
$(0,1,0)$	$(1,0,0)$	$(0,s,t),(1,s,0)$
	$(0,0,1)$	$(1,s,t)$
$(0,0,1)$	$(1,0,0)$	$(0,0,t),(r,s,0)$
	$(0,0,1)$	$(0,s,t),(1,0,t)$

$$(9)\begin{pmatrix} 0 & & \\ & 0 & \\ & & 0 \end{pmatrix}$$

α	r	p
$(1,0,0)$	$(0,0,1)$	$(r,0,t),(r,1,t)$
	$(1,0,0)$	$(r,0,1),(r,1,0),(r,0,0)$

2.3.5 Normal Forms of Proper Two-Region Piecewise-Linear Vector Fields

As stated in Subsection 2.3.4, any non-degenerate two-region piecewise-linear vector field is transformed by a parallel translation into the following vector field:

$$X_f(x) = f(x) = \begin{cases} Ax + q_A & (x \in R^+) \\ Bx & (x \in R^-) \end{cases} \tag{2.3.14}$$

$$R^{\pm} = \{x \in \mathbb{R}^n \mid \pm(\langle \alpha, x\rangle - \beta) > 0\}.$$

A linear subspace E in $\mathbb{R}^n$ is *B-invariant* if $B(E) \subset E$. A non-degenerate two-region piecewise-linear vector field is *proper* if there exists no linear subspace $E \subset \mathbb{R}^n (0 < \dim E < n)$ which is parallel to $U = \{x \in \mathbb{R}^n \mid \langle \alpha, x\rangle = \beta\}$ and is B-invariant. In this section, we will consider normal forms of continuous proper two-region piecewise-linear vector fields.

By operating a linear transformation, we may assume that (2.3.14) has the following form:

$$X_f(x) = f(x) = \begin{cases} Ax + q_A & (x \in R^+) \\ Jx & (x \in R^-) \end{cases} \tag{2.3.15}$$

$$R^{\pm} = \{x \in \mathbb{R}^n \mid \pm(\langle \alpha, x\rangle - \varepsilon) > 0\}, (\varepsilon = 0, \pm 1)$$

$$J = \dot{\sum}_{i=1}^{r} J(\lambda_i; k_i) \dot{+} \dot{\sum}_{j=1}^{s} J(a_j, b_j; l_j),$$

α is a canonical normal vector corresponding to J.

Notice that the vector field with $\varepsilon = -1$ is the complement of the one with $\varepsilon = 1$, up to the point symmetry with respect to the origin.

Proposition 2.3.48. *Suppose X_f in (2.3.15) is proper, then the following hold:*

(1) the eigenvalues of J, $\lambda_i (1 \le i \le r), a_j \pm \sqrt{-1} b_j (1 \le j \le s)$ are distinct.

(2)

$$\begin{aligned} \alpha &= (u_1, \cdots, u_r, v_1, \cdots, v_s) \\ u_i &= (1, 0, \cdots, 0) \in M(1, k_i) \\ v_j &= (\tilde{1}, \tilde{0}, \cdots, \tilde{0}) \in M(1, 2l_j), \\ \tilde{1} &= (1, 0), \quad \tilde{0} = (0, 0). \end{aligned} \tag{2.3.16}$$

Proof. (1) Suppose that $\lambda_i (1 \le i \le r)$ is repeated. Without loss of generality, put $\lambda_1 = \lambda_2 = \lambda$, then the dimension of the real eigenspace $E \subset \mathbb{R}^n$ belonging to λ is greater than 2. Since E is J-invariant, E must intersect U, and $\dim(U \cap E) \geq 1$. Take $w_1, w_2 \in U \cap E$ with $w_1 \neq w_2$, then the span$\{w_1 - w_2\}$ is a J-invariant one-dimensional linear subspace which is parallel to U. This contradicts the fact that X_f is proper.

Next suppose that $a_j \pm \sqrt{-1}b_j (1 \leq j \leq s)$ is repeated. Without loss of generality, put $a_1 + \sqrt{-1}b_1 = a_2 + \sqrt{-1}b_2 = \omega$, then the complex dimension of complex eigenspace belonging to ω, $\tilde{E} = \{z \in \mathbb{C}^n \mid (J - \omega I)z = 0\}$, is greater than 2 (i.e. $\dim_{\mathbb{C}} \tilde{E} \geq 2$). Take two complex vectors $z_1, z_2 \in \tilde{E}$ which are linearly independent, and put

$$\xi_j = Re(z_j), \quad \eta_j = Im(z_j) \in \mathbb{R}^n \quad (j = 1, 2).$$

Then, since

$$J\xi_j = \frac{1}{2}(\omega + \bar{\omega})\xi_j - \frac{1}{2\sqrt{-1}}(\omega - \bar{\omega})\eta_j \tag{2.3.17}$$

$$J\eta_j = \frac{1}{2\sqrt{-1}}(\omega + \bar{\omega})\xi_j + \frac{1}{2}(\omega - \bar{\omega})\eta_j \quad (j = 1, 2) \tag{2.3.18}$$

the two-dimensional linear subspace $E_j = \text{span}\{\xi_j, \eta_j\}$ is J-invariant $(j = 1, 2)$. Since each E_j must intersect U, we can take $a_j, b_j \in \mathbb{R} (j = 1, 2)$ such that

(i) $a_j\xi_j + b_j\eta_j$ is parallel to U, i.e.$\langle a_j\xi_j + b_j\eta_j, \alpha\rangle = 0$,

(ii) $a_j\xi_j - b_j\eta_j \in U \cap E_j$.

Define

$$\hat{\xi} = (a_1\xi_1 + b_1\eta_1) - (a_2\xi_2 + b_2\eta_2),$$
$$\hat{\eta} = (a_1\xi_1 - b_1\eta_1) - (a_2\xi_2 - b_2\eta_2),$$

then by (i) and (ii), we have $\langle \hat{\xi}, \alpha\rangle = \langle \hat{\eta}, \alpha\rangle = 0$. Thus $E = \text{span}\{\hat{\xi}, \hat{\eta}\}$ is parallel to U. It follows from (2.3.17) and (2.3.18) that

$$J\hat{\xi} = \frac{1}{2}(\omega + \bar{\omega})\hat{\xi} - \frac{1}{2\sqrt{-1}}(\omega - \bar{\omega})\hat{\eta}$$

$$J\hat{\eta} = \frac{1}{2\sqrt{-1}}(\omega + \bar{\omega})\hat{\xi} + \frac{1}{2}(\omega - \bar{\omega})\hat{\eta},$$

and hence E is J-invariant. This contradicts the fact that X_f is proper.

(2) Suppose $u_i \neq (1, 0, \cdots, 0)$. By Definition 2.3.22 of canonical normal vectors, the first component of u_i is equal to 0. This contradicts the fact that the real eigenspace $E = \{x \in \mathbb{R}^n \mid (J - \lambda_i I)x = 0\}$ belonging to λ_i is not parallel to U. Next, suppose $v_j \neq (\hat{1}, \hat{0}, \cdots, \hat{0})$, then by Definition 2.3.22, the first and second components are equal to $\hat{0} = (0, 0)$. Thus, the real eigenspace $E = \text{span}\{Re(z), Im(z) \mid (J - \omega_j I)z = 0\}$ belonging to $\omega_j = a_j \pm \sqrt{-1}b_j$ is parallel to U. This is a contradiction. □

Proposition 2.3.49. (Sylvester Matrix) *For an $n \times n$ matrix defined by*

$$S = \begin{pmatrix} 0 & 1 & & \mathbf{0} & \\ 0 & 0 & 1 & & \\ \vdots & & \ddots & \ddots & \\ \vdots & & & \ddots & 1 \\ a_n & a_{n-1} & \cdots & \cdots & a_1 \end{pmatrix} \qquad (a_1, \cdots, a_n \in \mathbb{R})$$

(called a Sylvester matrix), the following hold:

(1) If λ is a real eigenvalue of S, the real eigenspace belonging to λ, $E(\lambda) = \{x \in \mathbb{R}^n \mid (S - \lambda I)x = 0\}$, is a one-dimensional linear subspace spanned by

$$u_\lambda^1 = (1, \lambda, \cdots, \lambda^{n-1})^T \in \mathbb{R}^n, \quad i.e., \quad E(\lambda) = \mathrm{span}\{u_\lambda^1\}.$$

(2) Moreover, if λ has a multiplicity k, the generalized real eigenspace belonging to λ, $E_w(\lambda) = \{x \in \mathbb{R}^n \mid (S - \lambda I)^k x = 0\}$, is a k-dimensional linear subspace spanned by

$$u_\lambda^j = (u_{j1}, u_{j2}, \cdots, u_{jn})^T \in \mathbb{R}^n, \quad (1 \le j \le k),$$

where

$$u_{ji} = \binom{i-1}{j-1} \lambda^{i-j} \quad (1 \le i \le n, 1 \le j \le k)$$

$$\binom{i}{j} = \begin{cases} i!/\{j!(i-j)!\} & \text{if } i \ge j \\ 0 & \text{if } i < j \end{cases},$$

i.e., $E_w(\lambda) = \mathrm{span}\{u_\lambda^1, \cdots, u_\lambda^k\}$.

(3) If ω and $\bar{\omega}$ are complex eigenvalues of S, the real eigenspace belonging to ω and $\bar{\omega}$, $E(\omega, \bar{\omega}) = \mathrm{span}\{Re(z), Im(z) \in \mathbb{R}^n \mid (S - \omega I)z = 0, z \in \mathbb{C}^n\}$ is a two-dimensional linear subspace spanned by $Re(v_\omega^1)$ and $Im(v_\omega^1)$, where

$$v_\omega^1 = (1, \omega, \cdots, \omega^{n-1})^T \in \mathbb{C}^n,$$

i.e., $E(\omega, \bar{\omega}) = \mathrm{span}\{Re(v_\omega^1), Im(v_\omega^1)\}$.

(4) Moreover, if ω and $\bar{\omega}$ have a multiplicity l, the generalized real eigenspace belonging to ω and $\bar{\omega}$, $E_w(\omega, \bar{\omega}) = \mathrm{span}\{Re(z), Im(z) \mid (S - \omega I)^l z = 0, z \in \mathbb{C}^n\}$, is a $2l$-dimensional linear subspace spanned by $Re(v_\omega^j)$ and $Im(v_\omega^j) \in \mathbb{R}^n (1 \le j \le l)$, where

$$v_\omega^j = (v_{j1}, v_{j2}, \cdots, v_{jn})^T \in \mathbb{C}^n, \quad (1 \le j \le l)$$

$$v_{ji} = \binom{i-1}{j-1} \omega^{i-j} \quad (1 \le i \le n, 1 \le j \le l)$$

$$\binom{i}{j} = \begin{cases} i!/\{j!(i-j)!\} & \text{if } i \ge j \\ 0 & \text{if } i < j \end{cases},$$

i.e., $E_w(\omega,\bar{\omega}) = \text{span}\{Re(v_\omega^j), Im(v_\omega^j) \mid 1 \le j \le l\}$.

(5) Let $\mu_1, \cdots, \mu_n \in \mathbb{C}$ be all eigenvalues of S including multiplicities, then

$$a_k = (-1)^{k-1} \sum \mu_{i_1}\mu_{i_2} \cdots \mu_{i_k} \quad (1 \le k \le n),$$

where $\sum$ takes all $i_1, i_2, \cdots, i_k$ such that $i_1 < i_2 < \cdots < i_k$. In particular,

$$a_1 = \text{tr}S \quad \textit{and} \quad a_n = (-1)^{n-1} \det S.$$

Proof. (1) Let us denote $x = (x^1, \cdots, x^n)^T \in \mathbb{R}^n$. In order to obtain an eigenvector of λ, from the equation $(S - \lambda I)(x^1, \cdots, x^n)^T = 0$, we have

$$-\lambda x^i + x^{i+1} = 0 \quad (1 \le i \le n).$$

Since $x^1 = 0$ implies $x^1 = \cdots = x^n = 0$, x^1 is not 0. Putting $x^1 = 1$, we obtain $(x^1, \cdots, x^n) = (1, \lambda, \cdots, \lambda^{n-1})$.

To see (2), it is sufficient to prove that $u_\lambda^1, \cdots, u_\lambda^k$ are linearly independent and satisfy

$$(S - \lambda I) u_\lambda^{j+1} = u_\lambda^j \quad (1 \le j < k) \tag{2.3.19}$$

From (2.3.19) and (2.3.19), we have $u_{ji} = 0 (1 \le i \le j, 2 \le j \le k)$ and $u_{jj} = 1 (1 \le j \le k)$. Thus, $\text{rank}[u_\lambda^1, \cdots, u_\lambda^k] = k$, hence $u_\lambda^1, \cdots, u_\lambda^k$ are linearly independent. Notice that (2.3.19) is equivalent to

$$-\lambda u_{j+1,i} + u_{j+1,i+1} = u_{j,i} \quad (1 \le i \le j, 1 \le j \le k) \tag{2.3.20}$$

and

$$\sum_{i=1}^{n} a_{n-i+1} u_{j+1,i} - \lambda u_{j+1,n} = u_{jn} \quad (1 \le j < k). \tag{2.3.21}$$

It is easy to see that (2.3.20) follows from the relation $\binom{i}{j} = \binom{i-1}{j} + \binom{i-1}{j-1}$. Since λ is a root of $\det(tI - S) = 0$ with multiplicity k,

$$\begin{aligned}
\frac{d^j}{d\lambda^j}\{\det(\lambda I - S)\} &= \frac{d^j}{d\lambda^j}\{\lambda^n - \sum_{i=1}^{n} a_{n-i+1}\lambda^{i-1}\} \\
&= j!\{\binom{n}{j}\lambda^{n-j} - \sum_{i=1}^{n} a_{n-i+1}\binom{i-1}{j}\lambda^{i-j-1}\} \\
&= j!\{\binom{n-1}{j-1}\lambda^{n-j} + \lambda\binom{n-1}{j}\lambda^{n-j-1} - \sum_{i=1}^{n} a_{n-i+1}u_{j+1,i}\} \\
&= j!\{u_{jn} + \lambda u_{j+1,n} - \sum_{i=1}^{n} a_{n-i+1}\binom{i-1}{j}\lambda^{i-j-1}\} = 0
\end{aligned}$$

for $1 \le j < k$, hence (2.3.21) holds.

(3) and (4) follow from the same argument as in (1) and (2). (5) follows from

$$\det(tI - S) = t^n - \sum_{i=1}^{n} a_{n-i+1} t^{i-1}$$

$$= \prod_{i=1}^{n} (t - \mu_i)$$

by comparing the coefficients. □

If the piecewise-linear vector field X_f in (2.3.15) is proper, the number of real Jordan blocks belonging to one eigenvalue is only one by Proposition 2.3.48. Thus, J is transformed into a Sylvester matrix S under linear transformation by Proposition 2.3.49. Indeed, using column vectors $u_\lambda^i \in \mathbb{R}^n$ and $v_\omega^j \in \mathbb{C}^n$ in Proposition 2.3.49(2) and (4), define a matrix G by

$$G = [u_1^1, \cdots, u_1^{k_1}, u_2^1, \cdots, u_r^{k_r}, Rev_1^1, Imv_1^1, \cdots,$$
$$Rev_1^{l_1}, Imv_1^{l_1}, Rev_2^1, Imv_2^1, \cdots, Rev_s^{l_s}, Imv_s^{l_s}],$$

where $u_i^j = u_{\lambda_i}^j, v_i^j = v_{\omega_i}^j$ and $\omega_i = a_i + \sqrt{-1} b_i$. Then we can verify that

$$S = GJG^{-1}$$
$$G(U) = \{x \in \mathbb{R}^n \mid \langle \alpha', x \rangle = 1\}$$
$$\alpha' = (1, 0, \cdots, 0)^T,$$

where $U = \{x \in \mathbb{R}^n \mid \langle \alpha, x \rangle = 1\}$ and α is a normal vector in (2.3.16). Thus, we have the following theorem.

Theorem 2.3.50. (Normal Forms of Proper Two-Region Systems)

If necessary, taking the complement and exchanging the signs of regions, any continuous n-dimensional proper two-region piecewise-linear vector field is linearly conjugate to the vector field with the following form:

$$X_f(x) = f(x) = \begin{cases} Cx + q, & (x \in R^+) \\ Sx, & (x \in R^-) \end{cases}, \tag{2.3.22}$$

where

$$R^{\pm} = \{x \in \mathbb{R}^n \mid \pm (\langle \alpha, x \rangle - \varepsilon) > 0\}, \quad (\varepsilon = 0, \pm 1) \tag{2.3.23}$$

$$\alpha = (1, 0, \cdots, 0)^T, \quad q = -(c_1, \cdots, c_n)^T \tag{2.3.24}$$

$$S = \begin{pmatrix} 0 & 1 & & & \mathbf{0} \\ 0 & 0 & 1 & & \\ \vdots & & \ddots & \ddots & \\ \vdots & & & \ddots & 1 \\ a_n & a_{n-1} & \cdots & \cdots & a_1 \end{pmatrix} \tag{2.3.25}$$

$$C = \begin{pmatrix} c_1 & 1 & & & \mathbf{0} \\ c_2 & 0 & 1 & & \\ \vdots & & \ddots & \ddots & \\ c_{n-1} & & & \ddots & 1 \\ c_n + a_n & a_{n-1} & \cdots & \cdots & a_1 \end{pmatrix}. \tag{2.3.26}$$

Moreover, the following hold:

(1) Put $p = (c_1, \cdots, c_n)^T \in \mathbb{R}^n$*, then (2.3.22) is written as*

$$X_f(x) = Sx + (1/2)p\{|\langle \alpha, x\rangle - \varepsilon| - \langle \alpha, x\rangle - \varepsilon\}(\varepsilon = 0, \pm 1). \tag{2.3.27}$$

(2) Let $\mu_1, \cdots, \mu_n \in \mathbb{C}$ *be eigenvalues of* S *including multiplicities, and define*

$$a_k = (-1)^{k-1} \sum \mu_{i_1}\mu_{i_2} \cdots \mu_{i_k} \quad (1 \leq k \leq n), \tag{2.3.28}$$

where $\sum$ *takes all* $i_1, i_2, \cdots, i_k$ *such that* $i_1 < i_2 < \cdots < i_k$.

(3) Let $\nu_1, \cdots, \nu_n \in \mathbb{C}$ *be eigenvalues of* C *including multiplicities, and define*

$$b_k = (-1)^{k-1} \sum \nu_{i_1}\nu_{i_2} \cdots \nu_{i_k} \quad (1 \leq k \leq n), \tag{2.3.29}$$

where $\sum$ *takes all* $i_1, i_2, \cdots, i_k$ *such that* $i_1 < i_2 < \cdots < i_k$*. Then*

$$c_k = b_k - a_k + \sum_{i=1}^{k-1} a_i c_{k-i} \quad (1 \leq k \leq n). \tag{2.3.30}$$

(4) Assume C^{-1} *exists, and put* $Q = -C^{-1}q \in \mathbb{R}^n$*. Then (2.3.22) is written as*

$$X_f(x) = \begin{cases} C(x - Q), & (x \in R^+) \\ Sx, & (x \in R^-) \end{cases} \tag{2.3.31}$$

and

$$Q = (1 - a_n/b_n, c_1 a_n/b_n, c_2 a_n/b_n, \cdots, c_{n-1} a_n/b_n). \tag{2.3.32}$$

(5) Assume $p = (c_1, \cdots, c_n)^T, p' = (c'_1, \cdots, c'_n)^T, \varepsilon = 0, \pm 1$ *and*

$$X(x) = Sx + (1/2)p\{|\langle \alpha, x\rangle - \varepsilon| - \langle \alpha, x\rangle - \varepsilon\}$$

$$X'(x) = Sx + (1/2)p'\{|\langle \alpha, x\rangle - \varepsilon| - \langle \alpha, x\rangle - \varepsilon\}.$$

Then X *and* X' *are linearly conjugate if and only if* $p = p'$.

Proof. It has been shown that X_f is given by (2.3.22) and that $R^{\pm}$, α and S can be taken as (2.3.23)-(2.3.26). Since X_f is continuous, by Theorem 2.2.4,

$$C = S + (C - S)m\alpha^T, \quad q = -(C - S)m \tag{2.3.33}$$

holds for $m = (1, 0, \cdots, 0)^T \in \mathbb{R}^n$. Define $p = (C - S)m = (c_1, \cdots, c_n)^T$, then we have q in (2.3.24) and C in (2.3.26).

Conversely, for any $p = (c_1, \cdots, c_n)^T$, define C by (2.3.26) and q by (2.3.24), then (2.3.33) holds, therefore X_f is continuous.

To see (1), substitute $C = S + p\alpha^T$ and $q = -p$ into (2.3.22). Then (1) follows from Theorem 2.2.4 (5). (2) has already been proven in Proposition 2.3.49(5). To see (3), using the expression of C in (2.3.26), expand $\det(tI - C)$ as a polynomial of t:

$$\det(tI - C) = t^n - \sum_{i=1}^{n} t^{n-i}\{a_i + c_i - \sum_{j=1}^{i-1} c_j a_{i-j}\}. \tag{2.3.34}$$

By the definition of b_k in (2.3.29),

$$\det(tI - C) = t^n - \sum_{i=1}^{n} t^{n-i} b_i. \tag{2.3.35}$$

Comparing the coefficients in (2.3.34) and (2.3.35), we obtain the conclusion.

We will prove (4). Equation (2.3.31) easily follows from the definition of Q. Using (2.3.26) and (2.3.32), we have

$$CQ = (c_1, \cdots, c_{n-1}, d)^T$$

$$d = (c_n + a_n)(1 - a_n/b_n) + (c_1 a_{n-1} + \cdots + c_{n-1} a_1) a_n/b_n. \tag{2.3.36}$$

By (2.3.30), we have

$$c_1 a_{n-1} + \cdots + c_{n-1} a_1 = c_n + a_n - b_n. \tag{2.3.37}$$

Thus, substituting (2.3.37) into (2.3.36), we obtain $CQ = p = -q$. Therefore, $Q = -C^{-1}q$ is given by (2.3.32).

(5) The "if " statement is clear. To see the "only if " statement, let b_i (respectively b_i') ($1 \le i \le n$) be the elementary symmetric function of eigenvalues for $C = S + p\alpha^T$ (respectively $C' = S + p'\alpha^T$) as in (2.3.29). Since $GC = C'G$ for some $G \in GL(n, \mathbb{R})$, we have $b_i = b_i' (1 \le i \le n)$. Thus, by (2.3.30), $c_i = c_i' (1 \le i \le n)$, i.e. $p = p'$. □

Example 2.3.51. (Normal Form of 3-Dimensional Proper Two-Region Piecewise Linear Vector Fields)

$$\begin{aligned}
\dot{x}^1 &= x^2 + (1/2)c_1\{|x^1 - \varepsilon| + x^1 - \varepsilon\} \\
\dot{x}^2 &= x^3 + (1/2)c_2\{|x^1 - \varepsilon| + x^1 - \varepsilon\} \\
\dot{x}^3 &= a_3 x^1 + a_2 x^2 + a_1 x^3 + (1/2)c_3\{|x^1 - \varepsilon| + x^1 - \varepsilon\}
\end{aligned}$$

where $(x^1, x^2, x^3)^T \in \mathbb{R}^3$, and $\varepsilon = 0, \pm 1$.

Remark 2.3.52. Theorem 2.3.50 means that the linearly conjugate classes of proper two-region systems are determined by the fundamental symmetric expression of eigenvalues for linear vector fields in each region, up to the complement vector fields. Since the condition for properness (i.e., invariant subspaces are not parallel to the boundary of regions) is generic, proper piecewise-linear vector fields are important. Theorem 2.3.50 is very useful when we study global bifurcations of piecewise-linear vector fields because they are completely characterized by the values of the elementary symmetric function of eigenvalues in each region, and hence these latter can be thought of as universal bifurcation parameters for piecewise-linear vector fields.

2.4 Multiregion Systems and Chaotic Attractors

In this section, as an application of Theorem 2.3.50, we will derive normal forms of three-dimensional three-region systems with point symmetry, three-dimensional four-region systems with axial symmetry, and four-dimensional three-region systems with point symmetry. We will deal with only three- and four-dimensional systems, while the method of derivation of normal forms is valid for n-dimensional systems with many regions.

2.4.1 Attractors in Three-Dimensional Three-Region System

Let X_f be a piecewise-linear vector field having many regions. If each two-region system derived from the adjacent two regions of X_f is regular (respectively proper), X_f is *regular* (respectively *proper*).

A vector field $X_f : \mathbb{R}^n \to \mathbb{R}^n$ is symmetric with respect to the origin if $X_f(-x) = -X_f(x)$ for all $x \in \mathbb{R}^n$. Then a three-region piecewise-linear vector field with the symmetry with respect to the origin is given by

$$X_f(x) = f(x) = \begin{cases} Ax + q & (x \in R^+) \\ Bx & (x \in R^0) \\ Ax - q & (x \in R^-) \end{cases} \tag{2.4.1}$$

$$R^{\pm} = \{x \in \mathbb{R}^n \mid \pm(\langle \alpha, x \rangle \mp \beta) \geq 0\}$$
$$R^0 = \{x \in \mathbb{R}^n \mid |\langle \alpha, x \rangle| < \beta\}, \quad \beta > 0.$$

Theorem 2.4.1. (Normal Forms of a Three-Dimensional Three-Region System with Point Symmetry)

Any continuous three-dimensional proper three-region system with symmetry with respect to the origin is linearly conjugate to the vector field with the following form:

$$\begin{aligned} \dot{x} &= c_1 x + y + (1/2)c_1\{|x-1| - |x+1|\} \\ \dot{y} &= c_2 x + z + (1/2)c_2\{|x-1| - |x+1|\} \\ \dot{z} &= (c_3 + a_3)x + a_2 y + a_1 z + (1/2)c_3\{|x-1| - |x+1|\}, \end{aligned} \quad (2.4.2)$$

where $(x,y,z)^T \in \mathrm{IR}^3$. *Moreover the following hold:*

(1) Equation (2.4.2) is written as

$$X_f(x,y,z) = \begin{cases} M_1(x,y,z)^T + q & x \geq 1 \\ M_0(x,y,z)^T & |x| < 1 \\ M_1(x,y,z)^T - q & x \leq -1 \end{cases} \quad (2.4.3)$$

$$M_0 = \begin{pmatrix} 0 & 1 & 0 \\ 0 & 0 & 1 \\ a_3 & a_2 & a_1 \end{pmatrix}, \quad M_1 = \begin{pmatrix} c_1 & 1 & 0 \\ c_2 & 0 & 1 \\ c_3 + a_3 & a_2 & a_1 \end{pmatrix}, \quad q = \begin{pmatrix} -c_1 \\ -c_2 \\ -c_3 \end{pmatrix}. \quad (2.4.4)$$

(2) Let μ_i *and* $\nu_i (i = 1,2,3)$ *be the eigenvalues of* M_0 *and* M_1*, respectively. Then*

$$\begin{aligned} &a_1 = \mu_1 + \mu_2 + \mu_3, \quad a_2 = -(\mu_1\mu_2 + \mu_2\mu_3 + \mu_3\mu_1), \quad a_3 = \mu_1\mu_2\mu_3, \\ &b_1 = \nu_1 + \nu_2 + \nu_3, \quad b_2 = -(\nu_1\nu_2 + \nu_2\nu_3 + \nu_3\nu_1), \quad b_3 = \nu_1\nu_2\nu_3, \\ &c_1 = b_1 - a_1, \quad c_2 = b_2 - a_2 + a_1 c_1, c_3 = b_3 - a_3 + a_2 c_1 + a_1 c_2. \end{aligned} \quad (2.4.5)$$

(3) Linearly conjugate classes are uniquely determined by the values of a_i *and* $b_i (i = 1,2,3)$.

(4) If $b_3 \neq 0$ *and* $a_3/b_3 < 0$*, singular points of* X_f *are the three points* P^+, P^- *and* O*, which are given by*

$$O = (0,0,0)^T, \quad P^{\pm} = \pm(1 - a_3/b_3, c_1 a_3/b_3, c_2 a_3/b_3)^T. \quad (2.4.6)$$

Proof. Regard the case of $n = 3$ in (2.3.22)-(2.3.26) of Theorem 2.3.50. From the symmetry of X_f with respect to the origin, we obtain (2.4.3)-(2.4.4). It is easily verified that (2.4.2) and (2.4.3)-(2.4.4) are equivalent. (2) follows from Theorem 2.3.50(2) and (3), and (4) follows from Theorem 2.3.50(4), In symmetric system with respect to the origin, the origin is necessarily a singular point, therefore, the complement vector field cannot be defined. Therefore, (3) follows from Theorem 2.3.50(3) and (5). □

Figs. 2.4.1-2.4.6 show typical attractors which are observed in the normal form (2.4.2) of three-region system with symmetry with respect to the origin. All figures are obtained by the Runge-Kutta method, and drawn after running several thousand steps starting at $(x,y,z) = (0.01, 0.01, 0.01)$. The size h of step

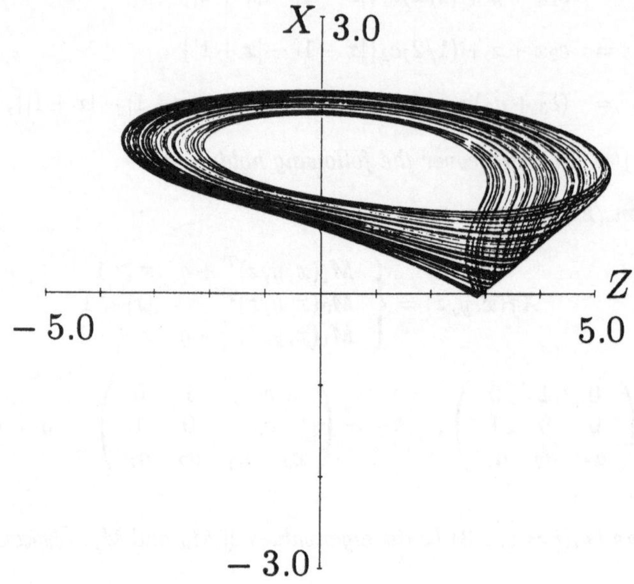

Fig. 2.4.1. Spiral-type attractor

in the Runge-Kutta method is $h = 0.005$ for Figs. 2.4.1, 2.4.2 and 2.4.4, whereas $h = 0.001$ for Figs. 2.4.3, 2.4.5 and 2.4.6.

Fig. 2.4.1 shows an attractor projected onto the (z,x)-plane where

$$(a_1, a_2, a_3) = (0.19, -1.9644, 6.31222),$$

$$(c_1, c_2, c_3) = (-2.77, -2.61, -13.91).$$

Eigenvalues are 1.55 and $-0.68 \pm 1.90\sqrt{-1}$ for M_0, -2.76 and $0.09 \pm 2.13\sqrt{-1}$ for M_1. This attractor has the structure of Rössler's spiral-type attractor [Rössler 1979b].

Fig. 2.4.2 shows an attractor projected onto the (z,x)-plane where

$$(a_1, a_2, a_3) = (0.19, -1.964, 6.312),$$

$$(c_1, c_2, c_3) = (-2.69, -2.383, -14.049).$$

Eigenvalues are 1.55 and $-0.68 \pm 1.90\sqrt{-1}$ for M_0, -2.76 and $0.13 \pm 2.13\sqrt{-1}$ for M_1. This attractor has the structure of Double Scroll attractor. Its geometric structure and bifurcation are studied in [Matsumoto, Chua and Komuro 1985, Chua, Komuro and Matsumoto 1986].

Fig. 2.4.3 shows an attractor projected onto the (y,z)-plane where

$$(a_1, a_2, a_3) = (0.604, -6.3047, 3.1264),$$

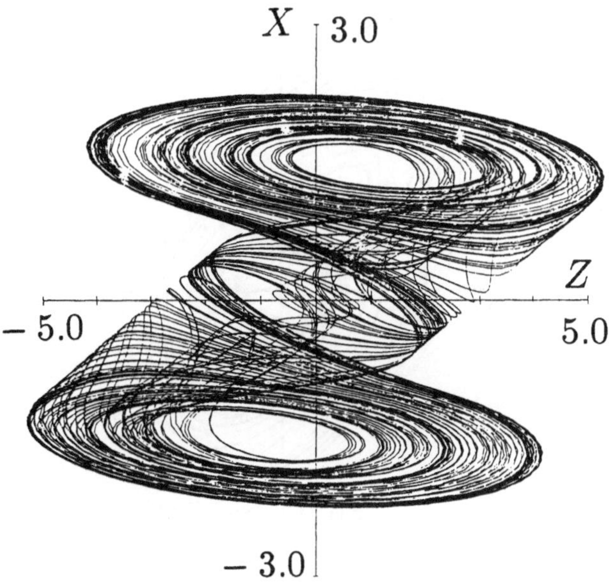

Fig. 2.4.2. Double Scroll-type attractor

$$(c_1, c_2, c_3) = (-1.284, 5.2756, 7.1022).$$

Eigenvalues are 0.5 and $0.052 \pm 2.5\sqrt{-1}$ for M_0, -1.2 and $0.26 \pm 0.9\sqrt{-1}$ for M_1. This attractor has the structure of Rössler's double screw-type attractor [Rössler 1979b].

Fig. 2.4.4 shows an attractor projected onto the (y,z)-plane where

$$(a_1, a_2, a_3) = (-1.164, -0.827824, -1.306011),$$

$$(c_1, c_2, c_3) = (1.2482, -8.532633, 16.98886).$$

Eigenvalues are -1.3 and $0.068 \pm 1.0\sqrt{-1}$ for M_0, 0.8 and $-0.3579 \pm 2.89\sqrt{-1}$ for M_1. This attractor has the structure of Sparrow's attractor [Sparrow 1981].

Fig. 2.4.5 shows an attractor projected onto the (y,z)-plane where

$$(a_1, a_2, a_3) = (0.1, -1.0, 0.2),$$

$$(c_1, c_2, c_3) = (-0.17, -0.01699, -0.1717).$$

Eigenvalues are 0.1963 and $-0.04815 \pm 1.0083\sqrt{-1}$ for M_0, -0.13868 and $0.03434 \pm 1.00416\sqrt{-1}$ for M_1. This attractor has the structure of Rössler's toroidal-type attractor [Rössler 1979b].

Fig. 2.4.6 shows an attractor projected onto the (x,y)-plane where

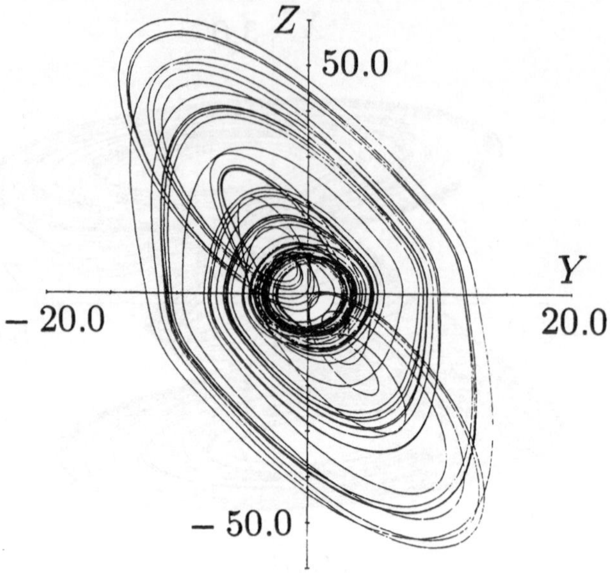

Fig. 2.4.3. Double Screw-type attractor

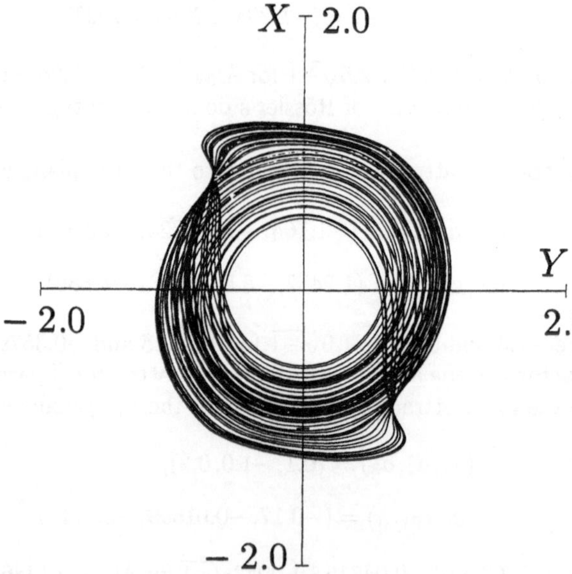

Fig. 2.4.4. Sparrow-type attractor

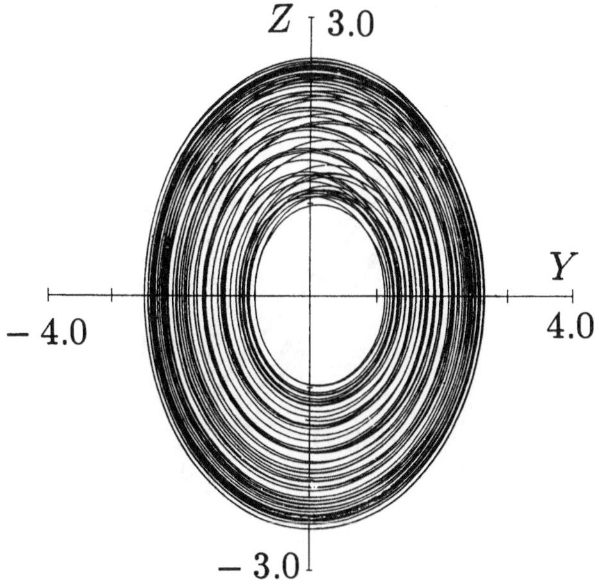

Fig. 2.4.5. Toroidal-type attractor

$$(a_1, a_2, a_3) = (-4.86, 4.158739, 5.995615),$$

$$(c_1, c_2, c_3) = (4.8, -29.518749, 156.227451).$$

Eigenvalues are $-5.423, -0.807$ and 1.37 for M_0, -0.526799 and $0.2334\pm$ 1.49112 $\sqrt{-1}$ for M_1. This attractor is called point symmetric Lorenz-type attractor. The original Lorenz attractor [Lorenz 1963] has axial symmetry, and its singular points are the origin O, which is a saddle with the three real eigenvalues λ_1, λ_2 and λ_3 such that $\lambda_1 < \lambda_2 < 0 < \lambda_3$, and the symmetric points P^+ and P^- with respect to the $z-$axis which are saddle focuses with a real eigenvalue ν and complex conjugate eigenvalues $\sigma \pm \sqrt{-1}\omega$ such that $\nu < 0$ and $\sigma > 0$. The point symmetric Lorenz-type attractor has three singular points P^+, P^- and O which have eigenvalues of the same type as the original Lorenz attractor, while the vector field itself, and so P^+ and P^-, are symmetric with respect to the origin O.

2.4.2 The Piecewise-Linear Lorenz Attractor

A three-dimensional vector field $X_f : \mathbb{R}^3 \to \mathbb{R}^3$ defined by

$$X_f(x, y, z) = (f_1(x, y, z), f_2(x, y, z), f_3(x, y, z))$$

is *symmetric with respect to the z-axis* if

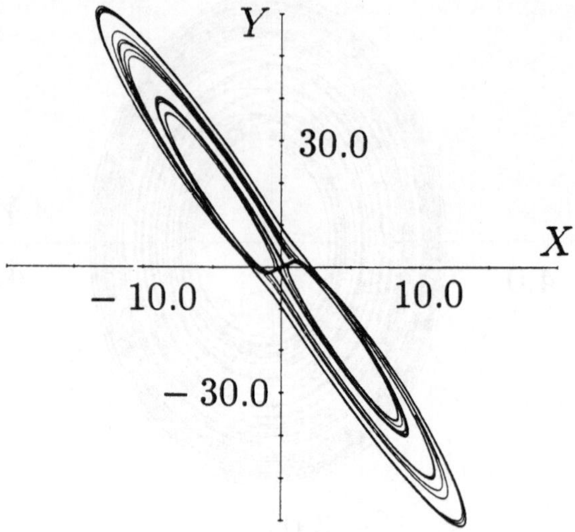

Fig. 2.4.6. Point symmetric Lorenz-type attractor

$$X_f(-x,-y,z) = (-f_1(x,y,z), -f_2(x,y,z), f_3(x,y,z))$$

is satisfied for all $(x, y, z) \in \mathbb{R}^3$. Two planes in $\mathbb{R}^3$ with general position divide $\mathbb{R}^3$ into four regions. We will consider three-dimensional four-region systems with axial symmetry.

Theorem 2.4.2. (Normal Forms of Three-Dimensional Four-Region System with Axial Symmetry)

If a continuous three-dimensional four-region piecewise-linear vector field is regular, proper and has the z-axial symmetry such that the boundaries are not parallel to the z-axis, then it is linearly conjugate with the following vector fields:

$$\begin{aligned}
\dot{x} &= ((\alpha_1+\alpha_0)/(2\alpha_0))y + ((\alpha_1-\alpha_0)/(4\alpha_0))\{|y+z| - |y-z|\} \\
\dot{y} &= \alpha_0 x + ((\beta_1+\beta_0)/2)y + ((\beta_1-\beta_0)/4)\{|y+z| - |y-z|\} \\
\dot{z} &= ((\gamma_1+\gamma_0)/2)z + ((\gamma_1-\gamma_0)/4)\{|y+z| + |y-z|\} - 1.
\end{aligned} \tag{2.4.7}$$

Moreover the following hold:

(1) *Equation (2.4.7) is equivalent to the following:*

$$X_f(x,y,z)=\begin{cases} M_0(x,y,z)^T+q & (x,y,z)\in R_0\\ M_1(x,y,z)^T+q & (x,y,z)\in R_1\\ M_2^+(x,y,z)^T+q & (x,y,z)\in R_2^+\\ M_2^-(x,y,z)^T+q & (x,y,z)\in R_2^-\end{cases}, \tag{2.4.8}$$

$$M_0=\begin{pmatrix} 0 & 1 & 0\\ \alpha_0 & \beta_0 & 0\\ 0 & 0 & \gamma_0\end{pmatrix},\quad M_1=\begin{pmatrix} 0 & \alpha_1/\alpha_0 & 0\\ \alpha_0 & \beta_1 & 0\\ 0 & 0 & \gamma_1\end{pmatrix}, \tag{2.4.9}$$

$$M_2^\pm=\begin{pmatrix} 0 & (\alpha_1+\alpha_0)/2\alpha_0 & \pm(\alpha_1-\alpha_0)/2\alpha_0\\ \alpha_0 & (\beta_1+\beta_0)/2 & \pm(\beta_1-\beta_0)/2\\ 0 & \pm(\gamma_1-\gamma_0)/2 & (\gamma_1+\gamma_0)/2\end{pmatrix},\quad q=\begin{pmatrix}0\\0\\-1\end{pmatrix} \tag{2.4.10}$$

$$\begin{aligned} R_0&=\{(x,y,z)\in\mathbb{R}^3\mid y+z\le 0, y-z\ge 0\},\\ R_1&=\{(x,y,z)\in\mathbb{R}^3\mid y+z\ge 0, y-z\le 0\},\\ R_2^+&=\{(x,y,z)\in\mathbb{R}^3\mid y+z\ge 0, y-z\ge 0\},\\ R_2^-&=\{(x,y,z)\in\mathbb{R}^3\mid y+z\le 0, y-z\le 0\}.\end{aligned} \tag{2.4.11}$$

(2) Define

$$P_i=-M_i^{-1}q=\begin{pmatrix}0\\0\\1/\gamma_i\end{pmatrix}\quad (i=0,1) \tag{2.4.12}$$

$$Q^\pm=\frac{1}{\alpha_0\gamma_1+\alpha_1\gamma_0}\begin{pmatrix}\pm(\alpha_1\beta_0-\alpha_0\beta_1)/\alpha_0\\ \pm(\alpha_0-\alpha_1)\\ (\alpha_0+\alpha_1)\end{pmatrix}. \tag{2.4.13}$$

If $P_i\in R_i$ $(i=0,1)$ (respectively $Q^\pm\in R_2^\pm$), then P_i (respectively $Q^\pm$) are singular points of X_f.

(3) If

$$\alpha_i+\beta_i\gamma_i-\gamma_i^2\neq 0\quad (i=0,1),$$

then (2.4.8) is proper. The linearly conjugate classes of (2.4.8) are uniquely determined by the values of α_i,β_i and γ_i $(i=0,1)$.

Proof. Without loss of generality, we can take the four regions as in (2.4.11). By the z-axial symmetry, matrices M_0 and M_1 must have the following form:

$$M_i=\begin{pmatrix} A_i & 0\\ 0 & \gamma_i\end{pmatrix}\quad (i=0,1), \tag{2.4.14}$$

where A_i is a 2×2 matrix. We take a Sylvester matrix

$$\begin{pmatrix} 0 & 1 \\ \alpha_1 & \beta_0 \end{pmatrix}$$

as A_0. Since X_f is continuous, applying Theorem 2.3.50 to two-regions between R_0 and $R_2^{\pm}$, and between R_2^+ and R_1, we have

$$M_2^+ = M_0 + \begin{pmatrix} c_1 \\ c_2 \\ c_3 \end{pmatrix} (0,1,1) = \begin{pmatrix} 0 & c_1+1 & c_1 \\ \alpha_0 & c_2+\beta_0 & c_2 \\ 0 & c_3 & c_3+\gamma_0 \end{pmatrix} \tag{2.4.15}$$

$$M_1 = M_2^+ + \begin{pmatrix} d_1 \\ d_2 \\ d_3 \end{pmatrix} (0,1,-1) = \begin{pmatrix} 0 & c_1+d_1+1 & c_1-d_1 \\ \alpha_0 & c_2+d_2+\beta_0 & c_2-d_2 \\ 0 & c_3+d_3 & c_3-d_3+\gamma_0 \end{pmatrix}. \tag{2.4.16}$$

Comparing (2.4.16) and (2.4.14), we have $c_1 = d_1, c_2 = d_2$ and $c_3 = -d_3$. Also, setting $\alpha_1 = -\det A_1$ and $\beta_1 = \text{trace}A_1$, we have $c_1 = (\alpha_1 - \alpha_0)/2\alpha_0, c_2 = (\beta_1 - \beta_0)/2$ and $c_3 = (\gamma_1 - \gamma_0)/2$. Substitute these into (2.4.15) and (2.4.16). Then we have M_1 and M_2^+ in (2.4.9) and (2.4.10). The formula for M_2^- follows from M_2^+ by the z-axial symmetry. Since the origin $O = (0,0,0)$ is on the intersection of four regions, defining $q = X_f(0,0,0)$, we have (2.4.8). Since X_f is regular, $q \neq 0$. Since X_f is z-axial symmetry, we can define $q = (0,0,-1)^T$, without loss of generality. Comparing the vector fields in each region, we conclude that (2.4.8) and (2.4.7) are equivalent. (2) follows immediately from (2.4.9) and (2.4.10).

To see (3), notice that if $\alpha_i + \beta_i\gamma_i - \gamma_i^2 = 0$, then M_i has γ_i and $\beta_i - \gamma_i$ as eigenvalues, and the dimension of eigenspace belonging to γ_i is not less than 2, for $i = 0, 1$. Then we can verify that $M_2^{\pm}$ has eigenspaces parallel to the boundary. Thus, if $\alpha_i + \beta_i\gamma_i - \gamma_i^2 = 0$ for either $i = 0$ or $i = 1$, then X_f is not proper. Conversely, if $\alpha_i + \beta_i\gamma_i - \gamma_i^2 \neq 0$ for both $i = 0$ and $i = 1$, then both M_0 and M_1 do not have eigenspaces parallel to the boundary. Define R_0 as the region to which $X_f(0,0,0) = q$ points. Then, since the values of the fundamental symmetric expression of eigenvalues in each region are uniquely determined by α_i, β_i and $\gamma_i (i = 0,1)$, we obtain the conclusion of the latter half of (3) by Theorem 2.3.50 (2),(3) and (5). □

Figs. 2.4.7 and 2.4.8 show attractors observed in the normal forms (2.4.7). The parameter values are $(\alpha_0, \beta_0, \gamma_0) = (2.0, -1.0, -0.5)$ and $(\alpha_1, \beta_1, \gamma_1) = (-2.8, -1.6, 0.3)$, and eigenvalues are $-2.0, -0.5$ and 1.0 for M_0; 0.3 and $-0.8 \pm 1.47\sqrt{-1}$ for M_1; -1.43328 and $0.01664 \pm 0.83512\sqrt{-1}$ for $M_2^{\pm}$. The step size of the Runge-Kutta method is equal to 0.01. Fig. 2.4.7 shows the projection onto the (y,z)-plane, while Fig. 2.4.8 the projection onto the (y,x)-plane of the attractor. Fig. 2.4.9 shows the Lorenz plot of local maximal values of z, that is, if $\phi(t) = (\phi^1(t), \phi^2(t), \phi^3(t))$ is the trajectory originating (x_0, y_0, z_0) at $t = 0$, and if $\phi^3(t)$ has local maximal values at $t = t_i (i = 1, 2, \cdots)$ with $0 < t_1 < t_2 < \cdots$, then the Lorenz plot is a plot of points $(\phi^3(t_i), \phi^3(t_{i+1}))$ for $i = 1, 2, \cdots$. The shape of the attractor in Figs. 2.4.7 and 2.4.8, and the shape of the Lorenz plot resemble the

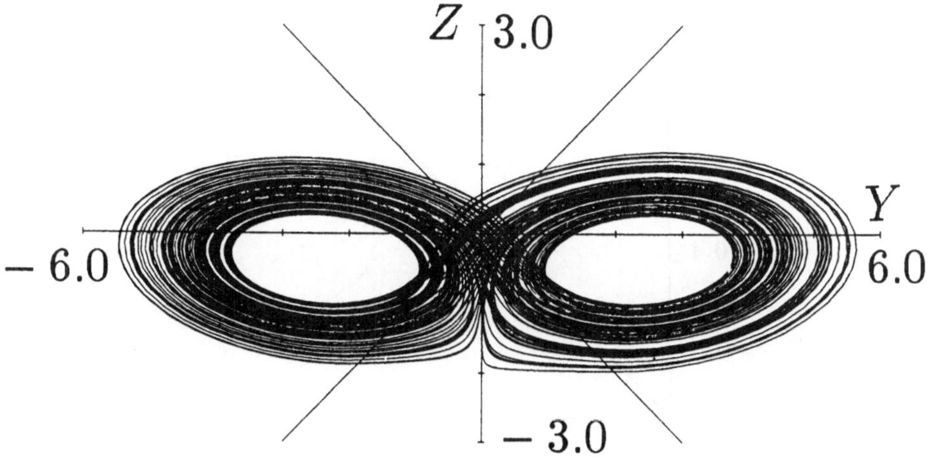

Fig. 2.4.7. Piecewise-linear Lorenz attractor (1)

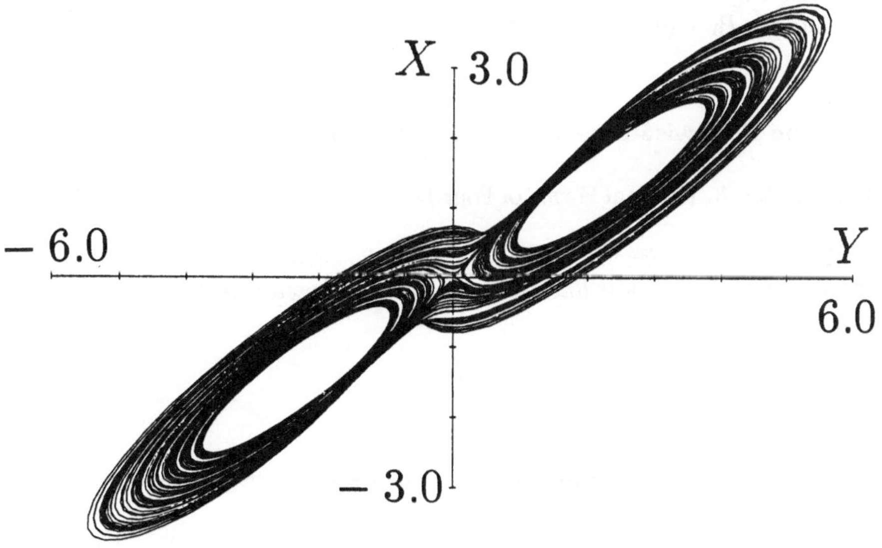

Fig. 2.4.8. Piecewise-linear Lorenz attractor (2)

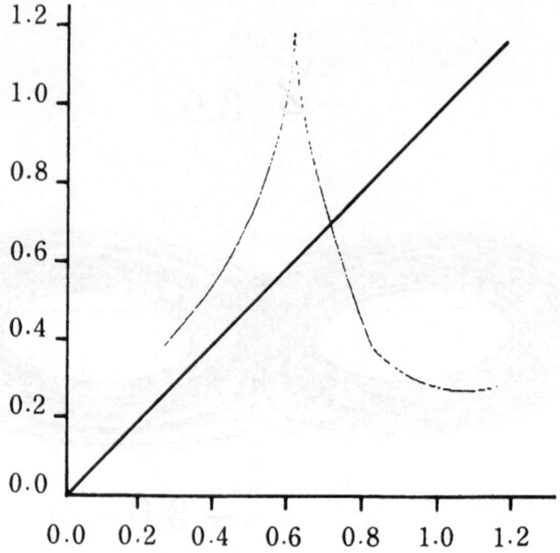

Fig. 2.4.9. Lorenz plot

ones of the original Lorenz attractors [Lorenz 1963] very closely. Locations of singular points are $P_0 = (0, 0, -2.0)^T, P_1 = (0, 0, 3.33)^T$ and $Q^{\pm} = (\pm 1.0, \pm 1.5, -0.4)$.

2.4.3 The Piecewise-Linear Duffing Attractor

Theorem 2.4.3. (Normal Forms of Four-Dimensional Three-Region System with Point Symmetry)

Any continuous four-dimensional proper three-region system with symmetry with respect to the origin is linearly conjugate to the vector field with the following form:

$$\begin{aligned}
\dot{x} &= c_1 x + y + (1/2)c_1\{|x-1| - |x+1|\} \\
\dot{y} &= c_2 x + z + (1/2)c_2\{|x-1| - |x+1|\} \\
\dot{z} &= c_3 x + w + (1/2)c_3\{|x-1| - |x+1|\} \\
\dot{w} &= (c_4 + a_4)x + a_3 y + a_2 z + a_1 w + (1/2)c_4\{|x-1| - |x+1|\}.
\end{aligned} \tag{2.4.17}$$

Moreover the following hold:

(1) *Equation (2.4.17) is written as*

$$X_f(x,y,z,w) = \begin{cases} M_1(x,y,z,w)^T + q, & x \geq 1 \\ M_0(x,y,z,w)^T, & |x| < 1 \\ M_1(x,y,z,w)^T - q, & x \leq -1 \end{cases} \tag{2.4.18}$$

$$M_0 = \begin{pmatrix} 0 & 1 & 0 & 0 \\ 0 & 0 & 1 & 0 \\ 0 & 0 & 0 & 1 \\ a_4 & a_3 & a_2 & a_1 \end{pmatrix}, \quad M_1 = \begin{pmatrix} c_1 & 1 & 0 & 0 \\ c_2 & 0 & 1 & 0 \\ c_3. & 0 & 0 & 1 \\ c_4 + a_4 & a_3 & a_2 & a_1 \end{pmatrix},$$

$$q = \begin{pmatrix} -c_1 \\ -c_2 \\ -c_3 \\ -c_4 \end{pmatrix}. \tag{2.4.19}$$

(2) Let μ_i and $\nu_i (i = 1,2,3,4)$ be the eigenvalues of M_0 and M_1 respectively, then

$$\begin{aligned}
a_1 &= \mu_1 + \mu_2 + \mu_3 + \mu_4, \\
a_2 &= -(\mu_1\mu_2 + \mu_1\mu_3 + \mu_1\mu_4 + \mu_2\mu_3 + \mu_2\mu_4 + \mu_3\mu_4), \\
a_3 &= \mu_1\mu_2\mu_3 + \mu_1\mu_2\mu_4 + \mu_1\mu_3\mu_4 + \mu_2\mu_3\mu_4, \\
a_4 &= \mu_1\mu_2\mu_3\mu_4, \\
b_1 &= \nu_1 + \nu_2 + \nu_3 + \nu_4, \\
b_2 &= -(\nu_1\nu_2 + \nu_1\nu_3 + \nu_1\nu_4 + \nu_2\nu_3 + \nu_2\nu_4 + \nu_3\nu_4), \\
b_3 &= \nu_1\nu_2\nu_3 + \nu_1\nu_2\nu_4 + \nu_1\nu_3\nu_4 + \nu_2\nu_3\nu_4, \\
b_4 &= \nu_1\nu_2\nu_3\nu_4, \\
c_1 &= b_1 - a_1,\ c_2 = b_2 - a_2 + a_1c_1,\ c_3 = b_3 - a_3 + a_2c_1 + a_1c_2. \\
c_4 &= b_4 - a_4 + a_3c_1 + a_2c_2 + a_1c_3.
\end{aligned} \tag{2.4.20}$$

(3) Linearly conjugate classes are uniquely determined by the values of a_i and b_i $(i = 1,2,3,4)$.

(iv) If $b_4 \neq 0$ and $a_4/b_4 < 0$, singular points of X_f are the three points P^+, P^- and O which are given by

$$O = (0,0,0,0)^T, \quad P^{\pm} = \pm(1 - a_4/b_4, c_1a_4/b_4, c_2a_4/b_4, c_3a_4/b_4)^T. \tag{2.4.21}$$

Proof. Regard the case of $n = 4$ in (2.3.22) and (2.3.26) of Theorem 2.3.50. The symmetry of X_f with respect to the origin implies (2.4.18). Other statements follow similarly to the proof of Theorem 2.4.1. □

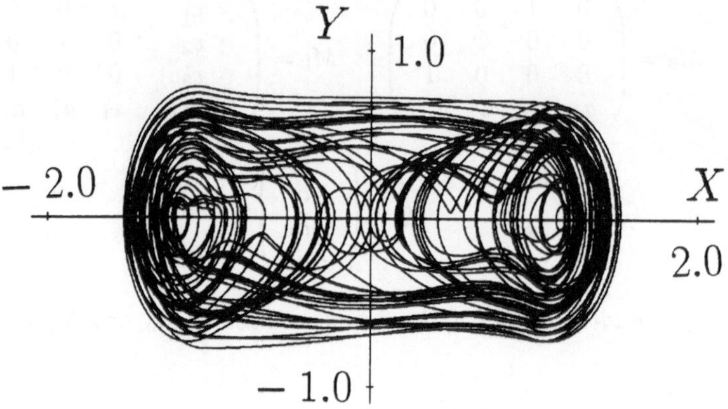

Fig. 2.4.10. Piecewise-linear Duffing flow

Consider a piecewise-linear version of Duffing equation (cf. [Ueda 1980]):

$$\begin{aligned}\dot{x} &= y \\ \dot{y} &= -f(x) - ky + B\cos t,\end{aligned} \tag{2.4.22}$$

where

$$\begin{aligned} f(x) &= \begin{cases} m_1(x-1) + m_0 & \text{if } x \geq 1 \\ m_0 x & \text{if } |x| < 1 \\ m_1(x+1) - m_0 & \text{if } x \leq -1 \end{cases} \\ &= m_0 x + (1/2)(m_1 - m_0)\{|x-1| - |x+1| + 2x\}. \end{aligned} \tag{2.4.23}$$

Equation (2.4.22) has a chaotic attractor, for example, when $(k, b, m_0, m_1) = (0.2, 0.3, -0.3, 3.0)$ with initial condition $(x, y, t) = (0, 0, 0)$. Fig. 2.4.10 shows the attractor of the flow in the (x,y)-plane, and Fig. 2.4.11 shows the attractor of the time 2π-map (the step size of Runge-Kutta method is 0.01).

This non-autonomous equation is embedded in the following four-dimensional three-region system with symmetry with respect to the origin:

$$\begin{aligned} \dot{x} = y, &\quad \dot{y} = -f(x) - ky + z \\ \dot{z} = -w, &\quad \dot{w} = z, \end{aligned} \tag{2.4.24}$$

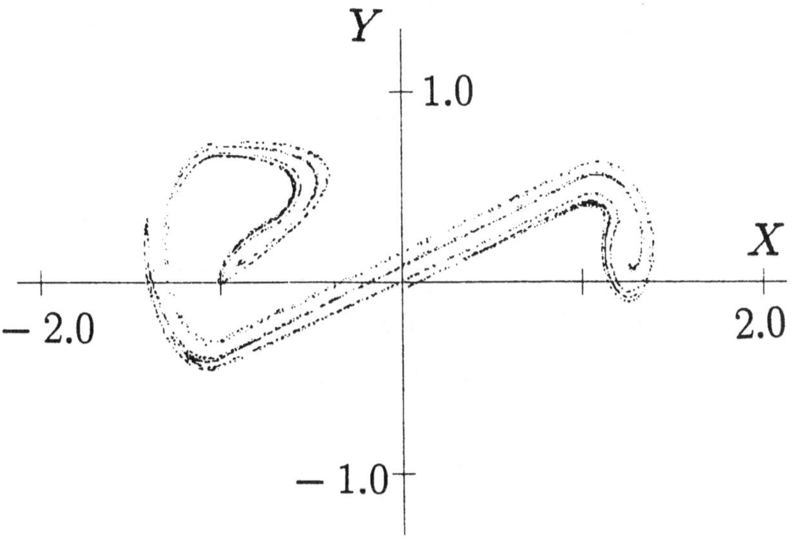

Fig. 2.4.11. Piecewise-linear Duffing attractor

where we take an initial condition $(x, y, z, w) = (0, 0, B, 0)$. Since (2.4.24) has the eigenvalues $(-k \pm \sqrt{k^2 - 4m_0})/2$ and $\pm\sqrt{-1}$ in the center region, and $(-k \pm \sqrt{k^2 - 4m_1})/2$ and $\pm\sqrt{-1}$ in the outer region, (2.4.24) is linearly conjugate to (2.4.17) with

$$a_1 = a_3 = b_1 = b_3 = -k, \quad a_2 = -m_0 - 1, \quad a_4 = -m_0,$$
$$b_2 = -m_1 - 1, \quad b_4 = -m_1.$$

2.5 Bifurcation Equations of Piecewise-Linear Vector Fields

In this section we will derive global equations of bifurcation sets (bifurcation equations) for continuous piecewise-linear vector fields. We will study homoclinic/heteroclinic bifurcation for singular points, and saddle-node, period-doubling and Hopf bifurcations for periodic orbits.

Essential tools to derive these bifurcation equations are (1) our normal forms of piecewise-linear vector fields, (2) the tangent maps of Poincaré full return maps, and (3) the return time coordinate on boundary. We will state the normal forms for three-dimensional two-region continuous proper piecewise-linear vector fields in Subsection 2.5.1, the tangent maps of a Poincaré full return maps in Subsection 2.5.2, and the return time coordinate in Subsection 2.5.3.

In Subsections 2.5.4 and 2.5.5, the bifurcation equations are derived. We will derive the equations of homoclinic bifurcation sets in A of Subsection 2.5.4, the equations of heteroclinic bifurcation sets in B of Subsection 2.5.4, and the equations of bifurcation sets for periodic orbits (saddle-node, period-doubling, and Hopf) in Subsection 2.5.5.

We discuss mainly the three-dimensional two-region system, however, the arguments can be easily extended to the case of general n-dimensional multi-region systems.

2.5.1 Normal Forms of Three-Dimensional Two-Region Systems

Given a non-zero vector $\alpha \in \mathbb{R}^3$, define a plane

$$V = \{x \in \mathbb{R}^3 \mid \langle \alpha, x \rangle = 1\}$$

(where $\langle \cdot, \cdot \rangle$ denotes the usual inner product), and half spaces

$$R^{\pm} = \{x \in \mathbb{R}^3 \mid \pm (\langle \alpha, x \rangle - 1) > 0\}.$$

Consider a vector field defined by an ordinary differential equation

$$\frac{dx}{dt} = f(x) = \begin{cases} \text{Ax}, & (x \in R^-) \\ \text{Bx-p}, & (x \in R^+), \end{cases} \tag{2.5.1}$$

where A and B are 3×3 matrices, and $p \in \mathbb{R}^3$(all elements of $\mathbb{R}^3$ are column vectors, unless otherwise stated). We call the vector field a *three-dimensional two-region piecewise-linear vector field*, and the plane V the *boundary* of the vector field. This vector field may be discontinuous on boundary V.

The tangent space of $\mathbb{R}^3$ is identified with $\mathbb{R}^3$ itself via natural basis. Then the vector field is written as

$$f : \mathbb{R}^3 \to \mathbb{R}^3, \tag{2.5.2}$$

and the corresponding flow is written as

$$\varphi^t : \mathbb{R}^3 \to \mathbb{R}^3 \ (t \in \mathbb{R}). \tag{2.5.3}$$

When we use the exponential matrix

$$e^{At} = \sum_{n=0}^{\infty} \frac{t^n}{n!} A^n, \tag{2.5.4}$$

the flow $\varphi^t(x)$ is partially represented as follows:

(1) For $x \in R^- \cup V$ and $t_0 > 0$, if $e^{At}x \in R^- \cup V$ for all t with $0 \le t \le t_0$, we have

$$\varphi^t(x) = e^{At}x \qquad \text{for} \qquad 0 \le t \le t_0. \tag{2.5.5}$$

(2) Assume there exists a $P \in \mathbb{R}^3$ such that $BP = p$. For $x \in R^+ \cup V$ and $t_0 > 0$, if $e^{Bt}(x - P) + P \in R^+ \cup V$ for all t with $0 \le t \le t_0$, we have

$$\varphi^t(x) = e^{Bt}(x - P) + P \qquad for \quad 0 \le t \le t_0. \tag{2.5.6}$$

Lemma 2.5.1. *The vector field defined by (2.5.1) is continuous on the boundary V, if and only if*

$$B = A + p\alpha^T. \tag{2.5.7}$$

Moreover, the vector field $f(x)$ is represented by

$$f(x) = Ax + \frac{1}{2}p\{|\langle \alpha, x\rangle - 1| + (\langle \alpha, x\rangle - 1)\}. \tag{2.5.8}$$

Proof. Assume the vector field (2.5.1) is continuous on V. Let $h_1 \in V$ be given. Since $Ah_1 = Bh_1 - p$, we have

$$p = (B - A)h_1. \tag{2.5.9}$$

Take $h_2, h_3 \in \mathbb{R}^3$ such that h_1, h_2, h_3 are linearly independent, and such that $\langle \alpha, h_i\rangle = 0 \quad (i = 2, 3)$. Define a nonsingular matrix $H \in GL(3, \mathbb{R})$ by $H = [h_1, h_2, h_3]$. Since $h_1 + h_i \in V$ $(i = 2, 3)$, we have $A(h_1 + h_i) = B(h_1 + h_i) - p$. By (2.5.9),

$$(A - B)h_i = (B - A)h_1 - p = 0.$$

Since $\alpha^T H = \alpha^T[h_1, h_2, h_3] = (1, 0, 0)$, we have

$$(A - B)H = (A - B)[h_1, h_2, h_3] = [-p, 0, 0] = -p(1, 0, 0) = -p\alpha^T H.$$

Multiplying H^{-1} from the right hand side, we have (2.5.7).

Conversely, assume (2.5.7) holds. For any $x \in V$, since $\alpha^T x = 1$, we have

$$Ax = (B - p\alpha^T)x = Bx - p\alpha^T x = Bx - p.$$

Thus the vector field (2.5.1) is continuous on the boundary V. Equation (2.5.8) is easily derived from (2.5.1) and (2.5.7).

Definition 2.5.2. Two vector fields f and f' on $\mathbb{R}^3$ are *linearly conjugate* if there is a nonsingular matrix $H \in GL(3, \mathbb{R})$ such that

$$Hf(x) = f'(Hx) \quad \text{for all} \quad x \in \mathbb{R}^3. \tag{2.5.10}$$

Definition 2.5.3. A vector field f defined by (2.5.8) is *proper* if any A-invariant proper linear subspace $E \in \mathbb{R}^3$ intersects with the boundary V, i.e.,

$$A(E) \subset E \quad \text{and} \quad 0 < \dim(E) < 3 \implies E \cap V \neq \emptyset. \tag{2.5.11}$$

Theorem 2.5.4. (Normal Forms of Proper Three-dimensional Two-Region Systems)

Any proper continuous three-dimensional two-region piecewise-linear vector field defined by

$$f'(x) = \begin{cases} A'x, & (\langle \alpha', x\rangle - 1 \le 0) \\ B'x - p', & (\langle \alpha', x\rangle - 1 \ge 0) \end{cases} \tag{2.5.12}$$

$$= A'x + \frac{1}{2}p'\{|\langle \alpha', x\rangle - 1| + (\langle \alpha', x\rangle - 1)\} \tag{2.5.13}$$

is linearly conjugate to the vector field defined by

$$f(x) = Ax + \frac{1}{2}p\{|\langle \alpha, x\rangle - 1| + (\langle \alpha, x\rangle - 1)\}, \tag{2.5.14}$$

$$= \begin{cases} Ax, & (x \in R^-) \\ Bx\text{-}p, & (x \in R^+), \end{cases} \tag{2.5.15}$$

where

$$R^{\pm} = \{x \in \mathbb{R}^3 \mid \pm(\langle \alpha, x\rangle - 1) > 0\}$$

$$\alpha = (1,0,0)^T, \quad p = (c_1, c_2, c_3)^T,$$

$$A = \begin{pmatrix} 0 & 1 & 0 \\ 0 & 0 & 1 \\ a_3 & a_2 & a_1 \end{pmatrix}, \quad B = \begin{pmatrix} c_1 & 1 & 0 \\ c_2 & 0 & 1 \\ c_3 + a_3 & a_2 & a_1 \end{pmatrix} = A + p\alpha^T$$

$$a_1 = \mu_1 + \mu_2 + \mu_3, a_2 = -(\mu_1\mu_2 + \mu_2\mu_3 + \mu_3\mu_1), a_3 = \mu_1\mu_2\mu_3,$$

$$b_1 = \nu_1 + \nu_2 + \nu_3, b_2 = -(\nu_1\nu_2 + \nu_2\nu_3 + \nu_3\nu_1), b_3 = \nu_1\nu_2\nu_3,$$

$$c_1 = b_1 - a_1, c_2 = b_2 - a_2 + c_1a_1, c_3 = b_3 - a_3 + a_2c_1 + a_1c_2,$$

μ_1, μ_2, μ_3 are eigenvalues of A, and ν_1, ν_2, ν_3 are eigenvalues of B.

Moreover, when $\det(B) = b_3 \neq 0$, *if we define*

$$P = (1 - a_3/b_3, c_1a_3/b_3, c_2a_3/b_3), \tag{2.5.16}$$

then

$$f(x) = \begin{cases} Ax, & (x \in R^-) \\ Bx\text{-}p, & (x \in R^+). \end{cases} \tag{2.5.17}$$

Proof. Notice that a three-dimensional two-region continuous piecewise-linear vector field

$$f(x) = Ax + \frac{1}{2}p\{|\langle \alpha, x\rangle - 1| + (\langle \alpha, x\rangle - 1)\},$$

is transformed by a nonsingular matrix H to

$$Hf(H^{-1}x) = HAH^{-1}x + \frac{1}{2}Hp\{|\langle (h^{-1})^T\alpha, x\rangle - 1| + (\langle (h^{-1})^T\alpha, x\rangle - 1)\},$$

Without loss of generality, we can assume matrix A is a Jordan matrix. We divide the proof into three steps.

Step 1. If there are two eigenvectors v_1, v_2 of A with respect to an eigenvalue μ, they are linearly dependent. Because, if v_1 and v_2 are linearly independent, there are $k_1, k_2 \in \mathbb{R}$ $(k_1^2 + k_2^2 \neq 0)$ such that

$$\langle \alpha, k_1v_1 + k_2v_2\rangle = 0.$$

Define a proper linear subspace E by

$$E = \{r(k_1v_1 + k_2v_2) \mid r \in \mathbb{R}\},$$

then E satisfies

$$AE \subset E \quad \text{and} \quad E \cap V = \emptyset.$$

This contradicts the properness condition.

Step 2. By Step 1, matrix A should be a Jordan matrix with one of the following four types:

$$(1)\begin{pmatrix} a & 1 & 0 \\ 0 & a & 1 \\ 0 & 0 & a \end{pmatrix} \quad (2)\begin{pmatrix} a & 0 & 0 \\ 0 & b & 1 \\ 0 & 0 & b \end{pmatrix} \quad (3)\begin{pmatrix} a & 0 & 0 \\ 0 & b & 0 \\ 0 & 0 & c \end{pmatrix} \quad (4)\begin{pmatrix} a & 0 & 0 \\ 0 & \sigma & -\omega \\ 0 & \omega & \sigma \end{pmatrix}$$

where a,b and c are distinct, and $\omega \neq 0$.

Case 1. By the assumption, $\alpha = (\alpha^1, \alpha^2, \alpha^3)^T$ satisfies $\alpha^1 \neq 0$. So α is transformed by matrix $\alpha^1 I$ to

$$\alpha = (1, y, z)^T.$$

It is further transformed by matrix

$$H = \begin{pmatrix} 1 & y & z \\ 0 & 1 & y \\ 0 & 0 & 1 \end{pmatrix}$$

into $\alpha = (1, 0, 0)^T$. Under these transformations, matrix

$$A = \begin{pmatrix} a & 1 & 0 \\ 0 & a & 1 \\ 0 & 0 & a \end{pmatrix}$$

is invariant. By a transformation

$$K = \begin{pmatrix} 1 & 0 & 0 \\ a & 1 & 0 \\ a^2 & 2a & 1 \end{pmatrix},$$

we have

$$KAK^{-1} = \begin{pmatrix} 0 & 1 & 0 \\ 0 & 0 & 1 \\ a_3 & a_2 & a_1 \end{pmatrix} = A', a_3 = a^3, a_2 = -3a^2, a_1 = 3a,$$

$$(K^{-1})^T\alpha = (1,0,0)^T = \alpha'.$$

Case 2. By the assumption, $\alpha = (\alpha^1, \alpha^2, \alpha^3)^T$ satisfies $\alpha^1 \neq 0$ and $\alpha^2 \neq 0$. Therefore α is transformed by matrix

$$\begin{pmatrix} \alpha^1 & 0 & 0 \\ 0 & \alpha^2 & 0 \\ 0 & 0 & \alpha^2 \end{pmatrix}$$

into $\alpha = (1,1,z)^T$. And this is further transformed by matrix

$$H = \begin{pmatrix} 1 & 0 & 0 \\ 0 & 1 & z \\ 0 & 0 & 1 \end{pmatrix}$$

into $\alpha = (1,1,0)^T$. Under these transformations, matrix

$$A = \begin{pmatrix} a & 0 & 0 \\ 0 & b & 1 \\ 0 & 0 & b \end{pmatrix}$$

is invariant. By transformation

$$K = \begin{pmatrix} 1 & 1 & 0 \\ a & b & 1 \\ a^2 & b^2 & 2b \end{pmatrix},$$

we have

$$KAK^{-1} = \begin{pmatrix} 0 & 1 & 0 \\ 0 & 0 & 1 \\ a_3 & a_2 & a_1 \end{pmatrix} = A', a_3 = ab^2, a_2 = -(2a+b)b, a_1 = a + 2b,$$

$$(K^{-1})^T\alpha = (1,0,0)^T = \alpha'.$$

Case 3. By the assumption, $\alpha = (\alpha^1, \alpha^2, \alpha^3)^T$ satisfies $\alpha^1 \neq 0$, $\alpha^2 \neq 0$ and $\alpha^3 \neq 0$. So α is transformed by matrix

$$\begin{pmatrix} \alpha^1 & 0 & 0 \\ 0 & \alpha^2 & 0 \\ 0 & 0 & \alpha^3 \end{pmatrix}$$

into $\alpha = (1,1,1)^T$. Under this transformation, matrix

$$A = \begin{pmatrix} a & 0 & 0 \\ 0 & b & 0 \\ 0 & 0 & c \end{pmatrix}$$

is invariant. By transformation

$$K = \begin{pmatrix} 1 & 1 & 1 \\ a & b & c \\ a^2 & b^2 & c^2 \end{pmatrix},$$

we have

$$KAK^{-1} = \begin{pmatrix} 0 & 1 & 0 \\ 0 & 0 & 1 \\ a_3 & a_2 & a_1 \end{pmatrix} = A', a_3 = abc, a_2 = -(ab+bc+ca), a_1 = a+b+c,$$

$$(K^{-1})^T\alpha = (1,0,0)^T = \alpha'.$$

Case 4. By the assumption, $\alpha = (\alpha^1, \alpha^2, \alpha^3)^T$ satisfies $\alpha^1 \neq 0$ and $(\alpha^2)^2 + (\alpha^3)^2 \neq 0$. So α is transformed by matrix

$$\begin{pmatrix} \alpha^1 & 0 & 0 \\ 0 & \alpha^2 & \alpha^3 \\ 0 & -\alpha^3 & \alpha^2 \end{pmatrix}$$

into $\alpha = (1,1,0)^T$. Under this transformation, matrix

$$A = \begin{pmatrix} a & 0 & 0 \\ 0 & \sigma & -\omega \\ 0 & \omega & \sigma \end{pmatrix}$$

is invariant. By transformation

$$K = \begin{pmatrix} 1 & 1 & 1 \\ a & \sigma & -\omega \\ a^2 & \sigma^2+\omega^2 & -2\sigma\omega \end{pmatrix},$$

we have

$$KAK^{-1} = \begin{pmatrix} 0 & 1 & 0 \\ 0 & 0 & 1 \\ a_3 & a_2 & a_1 \end{pmatrix} = A',$$

$$a_3 = a(\sigma^2+\omega^2), \quad a_2 = -(2a\sigma + \sigma^2 + \omega^2), \quad a_1 = a + 2\sigma,$$

$$(K^{-1})^T\alpha = (1,0,0)^T = \alpha'.$$

Step 3. In any case of Step 2, if we set

$$p' = Kp = (c_1, c_2, c_3)^T,$$

then matrix

$$p'(\alpha')^T + A' = \begin{pmatrix} c_1 & 1 & 0 \\ c_2 & 0 & 1 \\ c_3 + a_3 & a_2 & a_1 \end{pmatrix}$$

has μ_1, μ_2, μ_3 as its eigenvalues, because it is similar to matrix B. Since

$$|p'(\alpha')^T + A' - tI| = -t^3 + (a_1 + c_1)t^2 - (c_1 a_1 - a_2 - c_2)t + (c_3 + a_3 - a_2 c_1 - a_1 c_2)$$
$$= -t^3 + b_1 t^2 + b_2 t + b_3,$$

we have

$$c_1 = b_1 - a_1, \quad c_2 = b_2 - a_2 + c_1 a_1, \quad c_3 = b_3 - a_3 + a_2 c_1 + a_1 c_2.$$

The equation (2.5.16) follows from $BP = p$. □

2.5.2 The Tangent Map of Poincaré Full Return Maps

Consider a vector field

$$f(x) = Ax + \frac{1}{2}p\{|\langle\alpha, x\rangle - 1| + (\langle\alpha, x\rangle - 1)\} \tag{2.5.18}$$

$$= \begin{cases} \text{Ax}, & (x \in R^-) \\ \text{Bx-p}, & (x \in R^+), \end{cases} \tag{2.5.19}$$

where $B = A + p\alpha^T$ and $BP = p$, and define

$$\begin{aligned} V &= \{x \in \mathrm{I\!R}^3 \mid \alpha^T x - 1 = 0\}, \\ V_- &= \{x \in V \mid \alpha^T Ax < 0\}, \\ V_+ &= \{x \in V \mid \alpha^T Ax > 0\}. \end{aligned}$$

(1) Let $x_0 \in V_-$ be given. Assume there exists a $t_0 > 0$ such that

$$y_0 = e^{At_0} x_0 \in V_+$$

and

$$\alpha^T e^{At} x_0 - 1 \neq 0 \quad \text{for all} \quad t \in (0, t_0).$$

Then define a function

$$G(x, t) = \alpha^T e^{At} x - 1. \tag{2.5.20}$$

Since $G(x_0, t_0)) = \alpha^T y_0 - 1 = 0$ and

$$\frac{\partial G}{\partial t}(x_0, t_0) = \alpha^T A e^{At_0} x_0 = \alpha^T A y_0 \neq 0, \tag{2.5.21}$$

there exist a neighborhood $V_-(x_0)$ of x_0 on V_- and a function (called a *return time function*)

$$t : V_-(x_0) \longrightarrow \mathrm{I\!R},$$

such that $G(x, t(x)) = 0$ and $t(x_0) = t_0$. Then

$$Dt(x_0) = -[\frac{\partial G}{\partial t}(x_0, t_0)]^{-1}\frac{\partial G}{\partial x}(x_0, t_0) \quad (2.5.22)$$

$$= -[\alpha^T A e^{At_0} x_0]^{-1}\alpha^T e^{At_0} \quad (2.5.23)$$

$$= -[\alpha^T A y_0]^{-1}\alpha^T e^{At_0}. \quad (2.5.24)$$

Define a map $g : V_-(x_0) \to V_+$ by $g(x) = e^{At(x)}x$, which is called a *return map* from V_- to V_+. From (2.5.24) we have

$$Dg(x_0) = Ae^{At_0}x_0 Dt(x_0) + e^{At_0} \quad (2.5.25)$$

$$= -Ay_0[\alpha^T A y_0]^{-1}\alpha^T e^{At_0} + e^{At_0} \quad (2.5.26)$$

$$= \{I - \frac{Ay_0\alpha^T}{\alpha^T A y_0}\}e^{At_0}. \quad (2.5.27)$$

(2) Let $y_0 \in V_+$ be given. Assume there exists an $s_0 > 0$ such that

$$z_0 = e^{Bs_0}(y_0 - P) + P \in V_-,$$

and

$$\alpha^T\{e^{Bs}(y_0 - P) + P\} - 1 \neq 0 \quad \text{for all} \quad s \in (0, s_0).$$

Then define a function

$$H(y, s) = \alpha^T\{e^{Bs}(y - P) + P\} - 1. \quad (2.5.28)$$

Since $H(y_0, s_0)) = \alpha^T z_0 - 1 = 0$ and

$$\frac{\partial H}{\partial t}(y_0, s_0) = \alpha^T B e^{Bs_0}(y_0 - P) = \alpha^T A z_0 \neq 0,$$

there exist a neighborhood $V_+(y_0)$ of y_0 on V_+ and a function (called a *return time function*)

$$s : V_+(y_0) \longrightarrow \mathbb{R},$$

such that $H(y, s(y)) = 0$ and $s(y_0) = s_0$. Then

$$Ds(y_0) = -[\frac{\partial H}{\partial t}(y_0, s_0)]^{-1}\frac{\partial H}{\partial y}(y_0, s_0)$$

$$= -[\alpha^T A z_0]^{-1}\alpha^T e^{Bs_0}. \quad (2.5.29)$$

Define a map $h : V_+(y_0) \to V_-$ by $h(y) = e^{Bs(y)}(y - P) + P$, which is called a *return map* from V_+ to V_-. From (2.5.29) we have

$$Dh(y_0) = Be^{Bs_0}(y_0 - P)Ds(y_0) + e^{Bs_0}$$

$$= B(z_0 - P)Ds(y_0) + e^{Bs_0}$$

$$= -Az_0[\alpha^T A z_0]^{-1}\alpha^T e^{Bs_0} + e^{Bs_0}$$

$$= \{I - \frac{Az_0\alpha^T}{\alpha^T A z_0}\}e^{Bs_0}. \quad (2.5.30)$$

(3) The composition map $h \circ g : V_-(x_0) \to V_-$ is called a *return map* from V_- to V_-. Since

$$e^{Bs_0}Ay_0 = e^{Bs_0}B(y_0 - P) = Be^{Bs_0}(y_0 - P) = B(z_0 - P) = Az_0,$$

we have

$$\begin{aligned} D(h \circ g)(x_0) &= Dh(y_0)Dg(x_0) \\ &= \{I - \frac{Az_0\alpha^T}{\alpha^T Az_0}\}e^{Bs_0}\{I - \frac{Ay_0\alpha^T}{\alpha^T Ay_0}\}e^{At_0} \\ &= \{I - \frac{Az_0\alpha^T}{\alpha^T Az_0}\}\{e^{Bs_0} - \frac{Az_0\alpha^T}{\alpha^T Ay_0}\}e^{At_0} \\ &= \{I - \frac{Az_0\alpha^T}{\alpha^T Az_0}\}e^{Bs_0}e^{At_0} - \{\frac{Az_0\alpha^T}{\alpha^T Ay_0} - \frac{Az_0(\alpha^T Az_0)\alpha^T}{(\alpha^T Az_0)\alpha^T Ay_0}\}e^{At_0} \\ &= \{I - \frac{Az_0\alpha^T}{\alpha^T Az_0}\}e^{Bs_0}e^{At_0} - \{\frac{Az_0\alpha^T}{\alpha^T Ay_0} - \frac{Az_0\alpha^T}{\alpha^T Ay_0}\}e^{At_0} \\ &= \{I - \frac{Az_0\alpha^T}{\alpha^T Az_0}\}e^{Bs_0}e^{At_0}. \end{aligned}$$

(4) Assume $x_0 \in V_-$ is a *periodic point* and the orbit of x_0 is *transverse* to V, that is, there exist an integer $m > 0$ and

$$x_i \in V_-,\ y_i \in V_+,\ t_i > 0,\ s_i > 0 \quad (0 \le i \le m-1),$$

such that

$$y_i = e^{At_i}x_i, x_{i+1} = e^{Bs_i}(y_i - P) + P \quad (0 \le i \le m-1)$$

$$\alpha^T e^{At}x_i - 1 \neq 0 \quad \text{for all } t \in (0, t_0) \quad (0 \le i \le m-1)$$

and $x_m = x_0$. By (1) above, for each i, there exist a neighborhood $V_-(x_i)$ of x_i in V_-, a return time function

$$t^i : V_-(x_i) \longrightarrow \mathbb{R},$$

and a return map

$$g_i : V_-(x_i) \longrightarrow V_+; g_i(x) = e^{At^i(x)}x.$$

Also, by (2) above, for each i, there exist a neighborhood $V_-(y_i)$ of y_i in V_+, a return time function

$$s^i : V_+(y_i) \longrightarrow \mathbb{R},$$

and a return map

$$h_i : V_+(y_i) \longrightarrow V_-; h_i(y) = e^{Bs^i(y)}(y - P) + P.$$

Define *Poincaré full return map* g of the periodic point $x_0 \in V_-$ by

$$g : V_-(x_0) \to V_-; g(x) = h_{m-1} \circ g_{m-1} \circ \cdots \circ h_0 \circ g_0(x). \tag{2.5.31}$$

Since

$$e^{At_i} Ax_i = Ae^{At_i} x_i = Ay_i$$

and

$$e^{Bs_i} Ay_i = e^{Bs_i} B(y_i - P) = Be^{Bs_i}(y_i - P) = B(x_{i+1} - P) = Ax_{i+1},$$

we have

$$\begin{aligned}
&\{I - \frac{Ax_{i+1}\alpha^T}{\alpha^T Ax_{i+1}}\} e^{Bs_i} e^{At_i} \{I - \frac{Ax_i\alpha^T}{\alpha^T Ax_i}\} \\
&= \{I - \frac{Ax_{i+1}\alpha^T}{\alpha^T Ax_{i+1}}\}\{e^{Bs_i} e^{At_i} - \frac{Ax_{i+1}\alpha^T}{\alpha^T Ax_i}\} \\
&= \{I - \frac{Ax_{i+1}\alpha^T}{\alpha^T Ax_{i+1}}\} e^{Bs_i} e^{At_i} - \{\frac{Ax_{i+1}\alpha^T}{\alpha^T Ax_i} - \frac{Ax_{i+1}(\alpha^T Ax_{i+1})\alpha^T}{(\alpha^T Ax_{i+1})\alpha^T Ax_i}\} \\
&= \{I - \frac{Ax_{i+1}\alpha^T}{\alpha^T Ax_{i+1}}\} e^{Bs_i} e^{At_i}.
\end{aligned}$$

Hence, by (3) above, the tangent map of the Poincaré full return map is given by

$$Dg(x_0) = Dh_{m-1}(y_{m-1}) Dg_{m-1}(x_{m-1}) \cdots Dh_0(y_0) Dg_0(x_0) =$$

$$\{I - \frac{Ax_m\alpha^T}{\alpha^T Ax_m}\} e^{Bs_{m-1}} e^{At_{m-1}} \{I - \frac{Ax_{m-1}\alpha^T}{\alpha^T Ax_{m-1}}\} e^{Bs_{m-2}} e^{At_{m-2}} \cdots \{I - \frac{Ax_1\alpha^T}{\alpha^T Ax_1}\} e^{Bs_0} e^{At_0}$$

$$= \{I - \frac{Ax_m\alpha^T}{\alpha^T Ax_m}\} e^{Bs_{m-1}} e^{At_{m-1}} e^{Bs_{m-2}} e^{At_{m-2}} \cdots e^{Bs_0} e^{At_0}. \tag{2.5.32}$$

2.5.3 The Return Time Coordinates

Consider the vector fields defined by (2.5.18)-(2.5.19). For $x \in V_+$, assume there exist $y, z \in V_-$ and $t, s > 0$ such that

$$y = e^{Bs}(x - P) + P \quad \text{where } s = \inf\{s' > 0 \mid e^{Bs'}(x - P) + P \in V_-\},$$

$$z = e^{-At}x \quad \text{where } t = \inf\{t' > 0 \mid e^{-At'}x \in V_-\}.$$

See Fig. 2.5.1. Also assume A and B are nonsingular, and set $C = A^{-1}BA$.

Lemma 2.5.5. *Let $x \in V$ and $s > 0$ be given.*

(1) If $e^{Bs}(x - P) + P \in V$, then $e^{Cs}x = e^{Bs}(x - P) + P$.

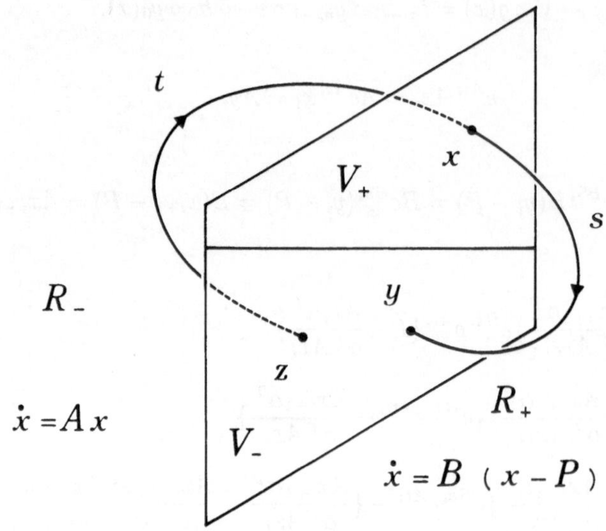

Fig. 2.5.1. Return time coordinate.

(2) If $e^{Cs}x \in V$, then $e^{Cs}x = e^{Bs}(x-P) + P$.

Proof. Since the vector field is continuous, $Aw = B(w-P)$ for all $w \in V$. Since A and B are non-singular, $w = A^{-1}B(w-P)$ and $w = B^{-1}Aw + P$ for all $w \in V$.

(1) Since $x \in V$ and $e^{Bs}(x-P) + P \in V$,

$$\begin{aligned} e^{Cs}x &= A^{-1}e^{Bs}Ax = A^{-1}e^{Bs}B(x-P) \\ &= A^{-1}B\{e^{Bs}(x-P) + P - P\} = e^{Bs}(x-P) + P. \end{aligned}$$

(2) Since $x \in V$ and $e^{Cs}x \in V$,

$$e^{Cs}x = B^{-1}Ae^{Cs}x + P = e^{Bs}B^{-1}Ax + P = e^{Bs}(x-P) + P.$$

□

Definition 2.5.6. By Lemma 2.5.5, $y = e^{Cs}x$. Since $x, y, z \in V$, we have

$$\alpha^T e^{-At}x = 1, \quad \alpha^T x = 1, \quad \alpha^T e^{Cs}x = 1. \tag{2.5.33}$$

Set

$$e_1 = (1,0,0)^T, e_2 = (0,1,0)^T, e_3 = (0,0,1)^T, h = (1,1,1)^T,$$

then (2.5.33) is equivalent to

$$[e_1\alpha^T e^{-At} + e_2\alpha^T + e_3\alpha^T e^{Cs}]x = h. \tag{2.5.34}$$

If the 3×3 matrix $[e_1\alpha^T e^{-At} + e_2\alpha^T + e_3\alpha^T e^{Cs}]$ is nonsingular, we set the inverse matrix as

$$K(s,t) = [e_1\alpha^T e^{-At} + e_2\alpha^T + e_3\alpha^T e^{Cs}]^{-1}. \tag{2.5.35}$$

Then we have

$$x = K(s,t)h, \tag{2.5.36}$$

and the pair (s,t) is called a *return time coordinate* of $x \in V_+$.

2.5.4 Bifurcation Equations of Three-Dimensional Two-Region Systems

In this subsection, we derive bifurcation equations. Since bifurcation phenomena are invariant under linear conjugacy, we can assume that the piecewise-linear vector fields are represented by the normal forms.

Consider a three-dimensional two-region continuous proper piecewise-linear vector field defined by the following normal form:

$$\begin{aligned} f(x) &= Ax + \frac{1}{2}p\{|\langle\alpha, x\rangle - 1| + (\langle\alpha, x\rangle - 1)\}, \\ &= \begin{cases} \text{Ax}, & (x \in R^-) \\ \text{Bx-p}, & (x \in R^+) \end{cases} \end{aligned} \tag{2.5.37}$$

where

$$R^{\pm} = \{x \in \mathbb{R}^3 \mid \pm(\langle\alpha, x\rangle - 1) > 0\}$$
$$\alpha = (1,0,0)^T, \quad p = (c_1, c_2, c_3)^T,$$
$$P = (1 - a_3/b_3, c_1 a_3/b_3, c_2 a_3/b_3)^T,$$
$$A = \begin{pmatrix} 0 & 1 & 0 \\ 0 & 0 & 1 \\ a_3 & a_2 & a_1 \end{pmatrix}, \quad B = \begin{pmatrix} c_1 & 1 & 0 \\ c_2 & 0 & 1 \\ c_3 + a_3 & a_2 & a_1 \end{pmatrix} = A + p\alpha^T$$
$$a_1 = \mu_1 + \mu_2 + \mu_3, a_2 = -(\mu_1\mu_2 + \mu_2\mu_3 + \mu_3\mu_1), a_3 = \mu_1\mu_2\mu_3,$$
$$b_1 = \nu_1 + \nu_2 + \nu_3, b_2 = -(\nu_1\nu_2 + \nu_2\nu_3 + \nu_3\nu_1), b_3 = \nu_1\nu_2\nu_3,$$
$$c_1 = b_1 - a_1, c_2 = b_2 - a_2 + c_1 a_1, c_3 = b_3 - a_3 + a_2 c_1 + a_1 c_2.$$

μ_1, μ_2, μ_3 are eigenvalues of A, and ν_1, ν_2, ν_3 are eigenvalues of B .

The vector field f is determined by a $\mu = (a_1, a_2, a_3, b_1, b_2, b_3) \in \mathbb{R}^6$, which will be called the *eigenvalue parameter*. The set of all μ is called the μ *-space*:

$$\mathbb{R}^6 = \{\mu = (a_1, a_2, a_3, b_1, b_2, b_3) \mid a_i, b_i \in \mathbb{R}, i = 1, 2, 3\}. \tag{2.5.38}$$

Assume A and B are non-singular, and set $C = A^{-1}BA$. Define the boundary V by

$$V = \{x \in \mathbb{R}^3 \mid \langle \alpha, x \rangle = 1\}.$$

If $\mu_i \quad (i = 1, 2, 3)$ is real, we define a point $C_i \in V$ by

$$C_i = (1, \mu_i, \mu_i^2)^T. \tag{2.5.39}$$

Then the vector $\overrightarrow{OC_i}$ is an eigenvector of A associated with μ_i. If $\nu_i \quad (i = 1, 2, 3)$ is real, we define a point $D_i \in V$ by

$$D_i = (1, \nu_i a_3/b_3, \nu_i(\nu_i - c_1)a_3/b_3)^T. \tag{2.5.40}$$

Then the vector $\overrightarrow{PD_i}$ is an eigenvector of B associated with ν_i.

A Homoclinic Bifurcations Assume μ_1 is positive real, and μ_2 and μ_3 are negative real or a pair of complex-conjugate with negative real part. Since an eigenvector for μ_i is given by (2.5.39), a one-dimensional unstable eigenspace $E^u(O)$ and a two-dimensional stable eigenspace $E^s(O)$ for $O = (0, 0, 0)^T$ are given by

$$\begin{aligned} E^u(O) &= \{x \in \mathbb{R}^3 \mid x = r(1, \mu_1, \mu_1^2)^T,\ \alpha^T x - 1 \le 0,\ r \in \mathbb{R}\}, \\ E^s(O) &= \{x \in \mathbb{R}^3 \mid u^T x = 0,\ \alpha^T x - 1 \le 0\}, \end{aligned}$$

where $u = (0, (\mu_2 + \mu_3)/(\mu_2\mu_3), -1/(\mu_2\mu_3))^T$. The intersection $E^s(O) \cap V$ is given by

$$E^s(O) \cap V = \{x = (x^1, x^2, x^3) \in \mathbb{R}^3 \mid u^T x - 1 = 0,\ x^1 = 1\}. \tag{2.5.41}$$

Definition 2.5.7. A point $x \in \mathbb{R}^3$ is called a *homoclinic point* for a singular point O if

$$\lim_{t \to \pm\infty} \varphi^t(x) = O. \tag{2.5.42}$$

The orbit of a homoclinic point x is called a *homoclinic orbit*:

$$O(x) = \{\varphi^t(x) \mid t \in \mathbb{R}\}.$$

Assume a point $C_1 = (1, \mu_1, \mu_1^2)^T \in V_+$ is a homoclinic point for O. The homoclinic point x or its orbit is *transverse* to the boundary V if there exist an integer $m > 0$, real numbers $s_1, t_i, s_i > 0$ and $x_i, y_i \in V\ (2 \le i \le m)$ such that

$$y_i = e^{Bs_i}(x_i - P) + P \in V_- \quad (1 \le i \le m) \tag{2.5.43}$$

$$x_{i+1} = e^{At_{i+1}} y_i \in V_+ \quad (1 \le i \le m-1) \tag{2.5.44}$$

$$\alpha^T\{e^{Bs}(x_i - P) + P\} - 1 \ne 0 \quad \text{for all } s \in (0, s_i) \quad (1 \le i \le m) \tag{2.5.45}$$

$$\alpha^T e^{At} y_i - 1 \ne 0 \quad \text{for all} \quad t \in (0, t_{i+1}) \quad (1 \le i \le m-1) \tag{2.5.46}$$

$$\alpha^T e^{At} y_m - 1 \neq 0 \quad \text{for all} \quad t > 0, \tag{2.5.47}$$

where $x_1 = C_1$. The integer $2m > 0$ is called the *intersection number* of the homoclinic point C_1 (or the homoclinic orbit) with the boundary V, i.e.,

$$2m = \sharp(O(C_1) \cap V). \tag{2.5.48}$$

The sequence $\{s_1, t_2, s_2, \cdots, t_m, s_m\}$ is called a *return time sequence* of the homoclinic point $C_1 \in V_+$.

Theorem 2.5.8. (Homoclinic Bifurcation Equations)
Set

$$h = (1,1,1)^T, e_1 = (1,0,0)^T, e_2 = (0,1,0)^T, e_3 = (0,0,1)^T,$$

$$C_1 = (1, \mu_1, \mu_1^2)^T, \tag{2.5.49}$$

$$u = (0, (\mu_2 + \mu_3)/(\mu_2\mu_3), -1/(\mu_2\mu_3))^T, \tag{2.5.50}$$

$$K(t,s) = [e_1\alpha^T e^{-At} + e_2\alpha^T + e_3\alpha^T e^{Cs}]^{-1} \tag{2.5.51}$$

and

$$N = \begin{pmatrix} 0 & 1 & 0 \\ 0 & 0 & 1 \end{pmatrix}. \tag{2.5.52}$$

(1) If there exists an $s_1 > 0$ such that

$$\alpha^T e^{Cs_1} C_1 - 1 = 0 \tag{2.5.53}$$

$$u^T e^{Cs_1} C_1 = 0 \tag{2.5.54}$$

$$\alpha^T e^{Cs} C_1 - 1 \neq 0 \quad \text{for all} \quad s \in (0, s_i) \tag{2.5.55}$$

$$\alpha^T e^{At} e^{Cs_1} C_1 - 1 \neq 0 \quad \text{for all} \quad t > 0, \tag{2.5.56}$$

then point C_1 is a transverse homoclinic point for O with the intersection number 2.

(2) If there exist $s_1, t_i, s_i > 0 \quad (2 \leq i \leq m)$ such that

$$\alpha^T e^{Cs_1} C_1 - 1 = 0 \tag{2.5.57}$$

$$N(e^{At_2} e^{Cs_1} C_1 - K(t_2, s_2)h) = O \tag{2.5.58}$$

$$N(e^{At_{i+1}} e^{Cs_i} K(t_i, s_i) - K(t_{i+1}, s_{i+1}))h = O \tag{2.5.59}$$

$$u^T e^{Cs_m} K(t_m, s_m)h = 0 \tag{2.5.60}$$

$$\alpha^T e^{Cs} C_1 - 1 \neq 0 \quad \text{for all} \quad s \in (0, s_1) \tag{2.5.61}$$

$$\alpha^T e^{-At} K(t_i, s_i)h - 1 \neq 0 \quad \text{for all} \quad t \in (0, t_i) \quad (2 \leq i \leq m) \tag{2.5.62}$$

$$\alpha^T e^{Cs} K(t_i, s_i)h - 1 \neq 0 \quad \text{for all} \quad s \in (0, s_i) \quad (2 \leq i \leq m) \tag{2.5.63}$$

$$\alpha^T e^{At} e^{Cs_m} K(t_m, s_m)h - 1 \neq 0 \quad \text{for all} \quad t > 0 \tag{2.5.64}$$

then point C_1 is a transverse homoclinic point for O with the intersection number $2m \quad (m \geq 2)$.

Proof. (1) Equations (2.5.53) and (2.5.55) guarantee that the point C_1 goes to the point $y_1 = e^{Cs_1}C_1 \in V$ in the first return time s_1. Equation (2.5.54) guarantees that the point y_1 lies on the stable eigenspace $E^s(O)$ of the origin O, and (2.5.56) guarantees that the point goes to the origin O as $t \to \infty$. Hence C_1 is a transverse homoclinic point for the origin O.

(2) Equations (2.5.57) and (2.5.61) guarantee that the point C_1 goes to the point $y_1 = e^{Cs_1}C_1 \in V$ in the first return time s_1. Equations (2.5.58), (2.5.59), (2.5.62) and (2.5.63) guarantee that the point y_i goes to the point $x_{i+1} = K(t_{i+1}, s_{i+1})h \in V$ in the first return time t_{i+1}, and that the point x_{i+1} goes to the point $y_{i+1} = e^{Cs_{i+1}}x_{i+1} \in V$ in the first return time s_{i+1} $(1 \leq i < m)$. Equation (2.5.60) guarantees that the point y_m lies on the stable eigenspace $E^s(O)$ of the origin O, and (2.5.64) guarantees that the point goes to the origin O as $t \to \infty$. Hence, C_1 is a transverse homoclinic point for the origin O. □

Definition 2.5.9. Equations (2.5.57)-(2.5.60) are regarded as $2m$ scalar equations between $(\mu, s_1, t_2, s_2, \cdots, t_m, s_m) \in \mathbb{R}^{6+2m-1}$. Equations (2.5.61)-(2.5.64) are open conditions, which are called *first return time conditions.* Therefore the solution set for (2.5.57)-(2.5.64) is a five-dimensional subset in $\mathbb{R}^{6+2m-1}$. This statement is also valid for $m = 1$. The projection of the solution set to the μ-space $\mathbb{R}^6$ which has the codimension one is called a *homoclinic bifurcation set.*

B Heteroclinic Bifurcations As for eigenvalues of A, assume μ_1 is positive real, and μ_2 and μ_3 are negative real or a pair of complex-conjugates with negative real parts. As for eigenvalues of B, assume ν_1 is negative real, and ν_2 and ν_3 are positive real or a pair of complex-conjugate with positive real parts. Since eigenvectors for μ_i and ν_i are given by (2.5.38)-(2.5.39), a one-dimensional unstable eigenspace $E^u(O)$ of O and a one-dimensional stable eigenspace $E^s(P)$ of P are given by

$$E^u(O) = \{x \in \mathbb{R}^3 \mid x = r\overrightarrow{OC_1}, \quad \alpha^T x - 1 \leq 0, \quad r \in \mathbb{R}\}, \tag{2.5.65}$$

$$E^s(P) = \{x \in \mathbb{R}^3 \mid x = r\overrightarrow{PD_1}, \quad \alpha^T x - 1 \geq 0, \quad r \in \mathbb{R}\}. \tag{2.5.66}$$

Definition 2.5.10. A point $x \in \mathbb{R}^3$ is called a *heteroclinic point* from O to P if

$$\lim_{t \to -\infty} \varphi^t(x) = O \quad \text{and} \quad \lim_{t \to \infty} \varphi^t(x) = P. \tag{2.5.67}$$

The orbit of a heteroclinic point x is called a *heteroclinic orbit.* Assume a point $C_1 = (1, \mu_1, \mu_1^2)^T \in V_+$ is a heteroclinic point from O to P. The heteroclinic point C_1 or its orbit is *transverse* to the boundary V if there exist an integer $m > 0$ and real numbers $s_1, t_i, s_i, t_m > 0$ $(2 \leq i \leq m-1)$ such that

$$y_i = e^{Bs_i}(x_i - P) + P \in V_- \quad (1 \leq i \leq m-1) \tag{2.5.68}$$

$$x_{i+1} = e^{At_{i+1}}y_i \in V_+ \quad (1 \leq i \leq m-1) \tag{2.5.69}$$

$$\alpha^T\{e^{Bs}(x_i - P) + P\} - 1 \neq 0 \quad \text{for all} \quad s \in (0, s_i) \quad (1 \leq i \leq m-1) \tag{2.5.70}$$

$$\alpha^T e^{At} y_i - 1 \neq 0 \quad \text{for all} \quad t \in (0, t_{i+1}) \quad (1 \leq i \leq m-1), \tag{2.5.71}$$

where $x_1 = C_1$ and $x_m = D_1$. The integer $2m - 1 > 0$ is called the *intersection number* of the heteroclinic point C_1 (or the heteroclinic orbit) with the boundary V, i.e.,

$$2m - 1 = \sharp(O(C_1) \cap V).$$

The sequence $\{s_1, t_2, s_2, \cdots, t_{m-1}, s_{m-1}, t_m\}$ is called a *return time sequence* of the homoclinic point $C_1 \in V_+$.

Theorem 2.5.11. (Heteroclinic Bifurcation Equations)
Set

$$h = (1,1,1)^T, \quad e_1 = (1,0,0)^T, \quad e_2 = (0,1,0)^T, \quad e_3 = (0,0,1)^T,$$

$$C_1 = (1, \mu_1, \mu_1^2)^T,$$

$$D_1 = (1, \nu_1 a_3/b_3, \nu_1(\nu_1 - c_1)a_3/b_3)^T, \tag{2.5.72}$$

$$K(t,s) = [e_1 \alpha^T e^{-At} + e_2 \alpha^T + e_3 \alpha^T e^{Cs}]^{-1}$$

$$N = \begin{pmatrix} 0 & 1 & 0 \\ 0 & 0 & 1 \end{pmatrix}.$$

(1) Point C_1 is a transverse heteroclinic point with the intersection number 1 if and only if $C_1 = D_1$, that is,

$$\mu_2 + \mu_3 = \nu_2 + \nu_3 \quad \text{and} \quad \mu_2\mu_3 = \nu_2\nu_3. \tag{2.5.73}$$

(2) If there exist $s_1, t_2 > 0$ such that

$$\alpha^T e^{Cs_1} C_1 - 1 = 0 \tag{2.5.74}$$

$$\alpha^T e^{-At_2} D_1 - 1 = 0 \tag{2.5.75}$$

$$N(D_1 - e^{At_2} e^{Cs_1} C_1) = O \tag{2.5.76}$$

$$\alpha^T e^{Cs} C_1 - 1 \neq 0 \quad \text{for all} \quad s \in (0, s_1) \tag{2.5.77}$$

$$\alpha^T e^{-At} D_1 - 1 \neq 0 \quad \text{for all} \quad t \in (0, t_2), \tag{2.5.78}$$

then point C_1 is a transverse heteroclinic point for O to P with the intersection number 3.

(3) If there exist $s_1, t_i, s_i, t_m > 0 \quad (2 \leq i \leq m-1)$ such that

$$\alpha^T e^{Cs_1} C_1 - 1 = 0 \tag{2.5.79}$$

$$N(e^{At_2} e^{Cs_1} C_1 - K(t_2, s_2)h) = O \tag{2.5.80}$$

$$N(e^{At_{i+1}} e^{Cs_i} K(t_i, s_i) - K(t_{i+1}, s_{i+1}))h = O \tag{2.5.81}$$

$$N(D_1 - e^{At_m} e^{Cs_{m-1}} K(t_{m-1}, s_{m-1}))h) = O \tag{2.5.82}$$

$$\alpha^T e^{-At_m} D_1 - 1 = 0 \tag{2.5.83}$$

$$\alpha^T e^{C s} C_1 - 1 \neq 0 \quad \textit{for all} \quad s \in (0, s_1) \tag{2.5.84}$$

$$\alpha^T e^{-At} K(t_i, s_i)h - 1 \neq 0 \quad \textit{for all} \quad t \in (0, t_i) \quad (2 \leq i \leq m-1) \tag{2.5.85}$$

$$\alpha^T e^{C s} K(t_i, s_i)h - 1 \neq 0 \quad \textit{for all} \quad s \in (0, s_i) \quad (2 \leq i \leq m-1) \tag{2.5.86}$$

$$\alpha^T e^{-At} D_1 - 1 \neq 0 \quad \textit{for all} \quad t \in (0, t_m), \tag{2.5.87}$$

then point C_1 is a transverse homoclinic point from O to P with the intersection number $2m-1$ $(m \geq 2)$.

Proof. (1) This statement immediately follows from (2.5.72).

(2) Equations (2.5.74) and (2.5.77) guarantee that point C_1 goes to point $y_1 = e^{Cs_1}C_1 \in V$ in the first return time s_1. Equations (2.5.75),(2.5.76) and (2.5.78) guarantee that the point y_1 goes to point $D_1 \in V$ in the first return time t_2. Since D_1 lies on the stable eigenspace $E^s(O)$ of P, $e^{Bs}(D_1 - P) + P$ goes to point P as $s \to \infty$. Hence point C_1 is a transverse heteroclinic point.

(3) Equations (2.5.79) and (2.5.84) guarantee that point C_1 goes to point $y_1 = e^{Cs_1}C_1 \in V$ in the first return time s_1. Equations (2.5.80), (2.5.81), (2.5.85) and (2.5.86) guarantee that point y_i goes to point $x_{i+1} = K(t_{i+1}, s_{i+1})h \in V$ in the first return time $t_{i+1}(1 \leq i \leq m-1)$, and that the point x_{i+1} goes to point $y_{i+1} = e^{Cs_{i+1}}x_{i+1} \in V$ in the first return time $s_{i+1}(1 \leq i \leq m-2)$. Equations (2.5.83) and (2.5.87) guarantee that point x_m goes to point $D_1 \in V$ in the first return time t_m. Since D_1 lies on the stable eigenspace $E^s(P)$ of P, $e^{Bs}(D_1 - P) + P$ goes to point P as $s \to \infty$. Hence point C_1 is a transverse heteroclinic point. □

Definition 2.5.12. Equations (2.5.79)-(2.5.83) are regarded as $2m$ scalar equations between $(\mu, s_1, t_2, s_2, \cdots, t_{m-1}, s_{m-1}, t_m) \in \mathbb{R}^{6+2m-2}$. Equations (2.5.84)-(2.5.87) are open conditions, which are called the *first return time conditions.* Thus, the solution set for (2.5.79)-(2.5.87) is a four-dimensional subset in $\mathbb{R}^{6+2m-2}$. This statement is also valid for $m = 1, 2$. The projection of the solution set to the μ-space $\mathbb{R}^6$ which is the codimension two subset is called a *heteroclinic bifurcation set.*

2.5.5 Bifurcation Equations of Periodic Orbits

In this subsection, we will derive bifurcation equations of periodic orbits. Recall the three-dimensional two-region continuous proper piecewise-linear vector field defined by (2.5.37) in Subsection 2.5.2.

Definition 2.5.13. Point $x \in \mathbb{R}^3$ is called a *periodic point* with period $T > 0$ if $\varphi^T(x) = x$ and

$$\varphi^t(x) - x \neq 0 \quad \text{for all} \quad 0 < t < T. \tag{2.5.88}$$

The orbit of a periodic point x is called a *periodic orbit*:

$$O(x) = \{\varphi^t(x) \mid t \in \mathbb{R}\}.$$

Assume point $x \in V_+$ is periodic. The periodic point x or its orbit $O(x)$ is *transverse* to the boundary V, if there exist an integer $m > 0$ and real numbers $s_i, t_i > 0 \quad (1 \le i \le m)$ such that

$$y_i = e^{Bs_i}(x_i - P) + P \in V_- \quad (1 \le i \le m) \tag{2.5.89}$$

$$x_{i+1} = e^{At_{i+1}} y_i \in V_+ \quad (1 \le i \le m) \tag{2.5.90}$$

$$\alpha^T\{e^{Bs}(x_i - P) + P\} - 1 \neq 0 \quad \text{for all} \quad s \in (0, s_i) \quad (1 \le i \le m) \tag{2.5.91}$$

$$\alpha^T e^{At} y_i - 1 \neq 0 \quad \text{for all} \quad t \in (0, t_{i+1}) \quad (1 \le i \le m), \tag{2.5.92}$$

where $x_1 = x_{m+1} = x$ and $t_{m+1} = t_1$. The integer $2m > 0$ is called the *intersection number* of the periodic point x (or the periodic orbit $O(x)$) with the boundary V, i.e.,

$$2m = \sharp(O(x) \cap V) \tag{2.5.93}$$

and the integer $m > 0$ is called the *intersection index*. The sequence $\{s_1, t_2, s_2, \cdots, t_m, s_m, t_1\}$ is called a *return time sequence* of the periodic point $x \in V_+$.

Theorem 2.5.14. (Periodic Point Condition)
Set

$$h = (1,1,1)^T, \quad e_1 = (1,0,0)^T, \quad e_2 = (0,1,0)^T, \quad e_3 = (0,0,1)^T,$$

$$K(t,s) = [e_1\alpha^T e^{-At} + e_2\alpha^T + e_3\alpha^T e^{Cs}]^{-1} \tag{2.5.94}$$

$$N = \begin{pmatrix} 0 & 1 & 0 \\ 0 & 0 & 1 \end{pmatrix}.$$

If there exist $t_i, s_i > 0$ and $K(t_i, s_i) \quad (1 \le i \le m)$ such that

$$N(e^{At_{i+1}} e^{Cs_i} K(t_i, s_i) - K(t_{i+1}, s_{i+1}))h = O \quad (1 \le i \le m) \tag{2.5.95}$$

$$\alpha^T e^{-At} K(t_i, s_i)h - 1 \neq 0 \quad \textit{for all} \quad t \in (0, t_i) \quad (1 \le i \le m) \tag{2.5.96}$$

$$\alpha^T e^{Cs} K(t_i, s_i)h - 1 \neq 0 \quad \textit{for all} \quad s \in (0, s_i) \quad (1 \le i \le m), \tag{2.5.97}$$

where $t_{m+1} = t_1$ and $s_{m+1} = s_1$, then point $x = K(t_1, s_1)h \in V_+$ is a transverse periodic point with the intersection number $2m$.

Proof. Equations (2.5.95), (2.5.96) and (2.5.97) guarantee that point $x_i = K(t_i, s_i)h$ goes to point $y_{i+1} = e^{Cs_i}x_i \in V$ in the first return time s_i, and that point y_{i+1} goes to the point $x_{i+1} = e^{At_{i+1}} y_{i+1} \in V$ in the first return time $t_{i+1} \quad (1 \le i \le m)$. Hence, x_1 is a transverse periodic point with the intersection number $2m$. □

Matrices A and $C = A^{-1}BA$ are determined by six-dimensional eigenvalue parameter

$$\mu = (a_1, a_2, a_3, b_1, b_2, b_3)^T \in \mathbb{R}^6,$$

and matrix $K(t,s)$ is determined by $\mu \in \mathbb{R}^6$ and $(t,s) \in \mathbb{R}^2$. Hence, equation (2.5.95) is regarded as $2m$ scalar equations between

$$(\mu, t_1, s_1, \cdots, t_m, s_m) \in \mathbb{R}^{6+2m}.$$

Equations (2.5.96) and (2.5.97) are open conditions which guarantee that t_i and s_i $(1 \le j \le m)$ are first return times (thus they are called the *first return time conditions*). Therefore, the solution set for (2.5.95)-(2.5.97) is six-dimensional subset in $\mathbb{R}^{6+2m}$.

Theorem 2.5.15. (Bifurcation Conditions for Periodic Points)
Let $x \in V_+$ be a transverse periodic point with a return time sequence $\{s_1, t_2, s_2, \cdots, t_m, s_m, t_1\}$. By (2.5.32) in Subsection 2.5.2, the tangent map of the Poincaré full return map for point x is given by

$$D\pi(x) = \{I - \frac{Ax\alpha^T}{\alpha^T Ax}\} e^{At_1} e^{Bs_m} e^{At_m} \cdots e^{Bs_2} e^{At_2} e^{Bs_1}. \tag{2.5.98}$$

Define

$$M = e^{Bs_m} e^{At_m} \cdots e^{Bs_1} e^{At_1}, \tag{2.5.99}$$

$$T = \mathrm{tr}(M) \quad \text{and} \quad D = \det(M). \tag{2.5.100}$$

(1) $D\pi(x)$ has an eigenvalue 1 if and only if

$$2 - T + D = 0. \tag{2.5.101}$$

(2) $D\pi(x)$ has an eigenvalue -1 if and only if

$$T + D = 0. \tag{2.5.102}$$

(3) $D\pi(x)$ has a pair of complex conjugate eigenvalues whose absolute values are equal to 1 if and only if

$$D - 1 = 0, \quad \text{and} \tag{2.5.103}$$

$$-1 < T < 3. \tag{2.5.104}$$

Definition 2.5.16. The following three bifurcations at periodic point x are defined in terms of the tangent map:

(1) *Saddle-node bifurcation:* $D\pi(x)$ has an eigenvalue 1.

(2) *Period-doubling bifurcation:* $D\pi(x)$ has an eigenvalue -1.

(3) *Hopf bifurcation*: $D\pi(x)$ has a pair of complex conjugate eigenvalues whose absolute values are equal to 1.

Remark 2.5.17. For smooth vector fields (see Chapter 3), saddle-node, period-doubling and Hopf bifurcations usually mean the following phenomena:

Saddle-node bifurcation: an attractive periodic orbit and a saddle-type periodic orbit approach each other, collide and disappear when a parameter changes continuously. When this bifurcation occurs, the tangent map of Poincaré full return map for the periodic orbit has an eigenvalue 1.

Period-doubling bifurcation: an attractive periodic orbit with the period T deforms into a saddle-type periodic orbit, and simultaneously an attractive periodic orbit with the period $2T$ is born. When this bifurcation occurs, the tangent map of Poincaré full return map for the original periodic orbit with the period T has an eigenvalue -1.

Hopf bifurcation: an attractive periodic orbit changes into a repellent periodic orbit, and simultaneously an attractive invariant torus is born. When this bifurcation occurs, the tangent map of Poincaré full return map for the periodic orbit has a pair of complex-conjugate eigenvalues whose absolute values are equal to 1.

If the time-reversed vector field exhibits the above bifurcations, the original vector field is also said to have the same bifurcations. For continuous piecewise-linear vector fields, paying attention to the eigenvalues, we define these bifurcations as in Definition 2.5.16.

Proof of Theorem. Set

$$M' = e^{At_1}e^{Bs_m}e^{At_m}\cdots e^{Bs_2}e^{At_2}e^{Bs_1}, \tag{2.5.105}$$

then M' has an eigenvector Ax corresponding to eigenvalue 1. Indeed, if we define

$$y_{i+1} = e^{Bs_i}(x_i - P) + P, x_{i+1} = e^{At_{i+1}}y_{i+1} \quad (1 \le i \le m), \tag{2.5.106}$$

where $x = x_{m+1} = x_1, y_{m+1} = y_1$ and $t_{m+1} = t_1$, then

$$\begin{aligned} e^{At_{i+1}}e^{Bs_i}Ax_i &= e^{At_{i+1}}e^{Bs_i}B(x_i - P) \\ &= e^{At_{i+1}}B(y_{i+1} - P) = e^{At_{i+1}}Ay_{i+1} = Ax_{i+1}\ (1 \le i \le m). \end{aligned} \tag{2.5.107}$$

Hence, we have

$$\begin{aligned} M'Ax &= e^{At_1}e^{Bs_m}e^{At_m}\cdots e^{Bs_2}e^{At_2}e^{Bs_1}Ax_1 \\ &= e^{At_1}e^{Bs_m}e^{At_m}\cdots e^{Bs_2}Ax_2 \\ &= e^{At_1}e^{Bs_m}Ax_m = e^{At_1}Ay_m = Ax. \end{aligned}$$

Matrix

$$L = \{I - \frac{Ax\alpha^T}{\alpha^T Ax}\}$$

is a projection from $\mathbb{R}^3$ to

$$TV = \{x \in \mathbb{R}^3 \mid \langle \alpha, x \rangle = 0\} \tag{2.5.108}$$

along Ax. Hence, if matrix M' has eigenvalues $\{1, \lambda_2, \lambda_3\}$ and corresponding eigenvectors $\{Ax, v_2, v_3\}$, matrix LM' has eigenvalues $\{0, \lambda_2, \lambda_3\}$ and corresponding eigenvectors $\{Ax, u_2, u_3\}$, where $u_i = Lv_i \quad (i = 2, 3)$.

Since M and M' are linearly conjugate, i.e.,

$$M = e^{-At_1} M' e^{-At_1}, \tag{2.5.109}$$

we have

$$\det(\lambda I - M') = \det(\lambda I - M) = (\lambda - 1)(\lambda^2 + (1 - T)\lambda + D). \tag{2.5.110}$$

Set $g(\lambda) = (\lambda^2 + (1 - T)\lambda + D)$. Then LM' has an eigenvalue 1 if and only if $g(1) = 0$. This shows Equation (2.5.101). And LM' has an eigenvalue -1 if and only if $g(-1) = 0$. This shows Equation (2.5.102). Also, LM' has a pair of complex-conjugate eigenvalues whose absolute values are equal to 1 if and only if $D = 1$ and $(1 - T)^2 - 4D < 0$. This shows (2.5.103) and (2.5.104). □

Definition 2.5.18. Equation (2.5.101) is a scalar equation between $(\mu, t_1, s_1, \cdots, t_m, s_m) \in \mathbb{R}^{6+2m}$ where $\mu = (a_1, a_2, a_3, b_1, b_2, b_3) \in \mathbb{R}^6$. Thus, the solution set for (2.5.95)-(2.5.97) and (2.5.101) is a five-dimensional subset in $\mathbb{R}^{6+2m}$. The projection of the solution set to the μ-space $\mathbb{R}^6$ which has codimension 1 is called a *saddle-node bifurcation set.* Similarly, the projection of the solution set for (2.5.95)-(2.5.97) and (2.5.102) (respectively (2.5.103) and(2.5.104)) to the μ-space $\mathbb{R}^6$ which has codimension 1 is called a *period-doubling bifurcation set* (respectively a *Hopf bifurcation set*).

2.6 Bifurcation Sets

In this section we will numerically solve the bifurcation equations of Theorems 2.5.8, 2.5.11, 2.5.14 and 2.5.15. In Subsection 2.6.3, we will discuss numerical method to obtain the bifurcation curves.

The linear conjugate classes for three-dimensional two-region systems are determined by six parameters (Theorem 2.5.4). We will fix three or four parameters, and describe the structure of the bifurcation sets in spaces of the remaining two or three parameters. Since the bifurcation equations include no approximations, we can solve the equations to any degree of precision by computer. The numerical results in this section may not be enough to capture a whole bifurcation set, but they will provide a useful guide for computing bifurcation sets for general piecewise-linear systems.

In Subsections 2.6.1 and 2.6.2, we will describe the global structure of the bifurcation sets. We assume that the piecewise-linear vector fields have eigenvalues

$\gamma_0 > 0$ and $\sigma_0 \pm i\omega_0 (\sigma_0 < 0)$ in one region, and $\gamma_1 < 0$ and $\sigma_1 \pm i\omega_1 (\sigma_1 > 0)$ in another region.

In A of Subsection 2.6.1, the bifurcation sets for principal homoclinic orbits are described in the following parameter spaces:

(i) the (σ_0, σ_1)-plane when (ω_0, ω_1) is fixed and (γ_0, γ_1) varies,

(ii) the (σ_0, ω_0)-plane when $(\omega_1, \gamma_0, \gamma_1)$ is fixed and σ_1 varies.

In B of Subsection 2.6.1, we describe the bifurcation sets for subsidiary homoclinic/heteroclinic orbits in the (σ_0, ω_0)-plane when$(\omega_1, \gamma_0, \gamma_1)$ is fixed and σ_1 varies.

In Subsection 2.6.2, we study the periodic point set and their bifurcation sets. We will describe saddle-node bifurcation sets in A of Subsection 2.6.2, and period-doubling bifurcation sets in B of Subsection 2.6.2, in the (σ_0, σ_1)-plane when $(\omega_0, \omega_1, \gamma_0, \gamma_1)$ is fixed. In C of Subsection 2.6.2, we will describe periodic point sets (periodic surface) in the (σ_0, σ_1, T)-space (T denotes the period of periodic point). In D of Subsection 2.6.2, we will observe periodic windows in the (σ_0, σ_1)-plane, which are parameter regions corresponding to attractive or repelling periodic orbits.

2.6.1 Homoclinic/Heteroclinic Bifurcation Sets

Recall the continuous three-dimensional two-region proper piecewise-linear vector field defined by Equation (2.5.37). For eigenvalues of matrices A and B, we assume

$$\mu_1 = \gamma_0 > 0, \quad \mu_2,\ \mu_3 = \sigma_0 \pm i\omega_0 \quad (\sigma_0 < 0)$$

$$\nu_1 = \gamma_1 < 0, \quad \nu_2,\ \nu_3 = \sigma_1 \pm i\omega_1 \quad (\sigma_1 > 0).$$

A Bifurcation Sets for Principal Homoclinic Orbits A homoclinic orbit with the intersection number 2 is called *principal.* In a parameter space, the bifurcation set which connects with a principal homoclinic orbit is called the *principal homoclinic bifurcation set.* We will consider the part of principal homoclinic bifurcation set whose intersection number $2m$ satisfies $2m \leq 10$.

Fig. 2.6.1(a) shows the bifurcation set of principal homoclinic orbits for the origin O in the (σ_0, σ_1)-space $(-1.0 \leq \sigma_0 \leq 0.0, \quad 0.0 \leq \sigma_1 \leq 1.0)$ when $(\omega_0, \omega_1) = (1.0, 1.0)$ and $(\gamma_0, \gamma_1) = (0.5, -0.5)$.
Figs. 2.6.1(b) and (c) show the homoclinic orbits corresponding to the parameter points (a) and (b) in Fig. 2.6.1(a). Dots on the orbits denote the intersection points with boundary V. Here we have taken the Jordan matrix as the matrix A:

$$A = \begin{pmatrix} \sigma_0 & -\omega_0 & 0 \\ \omega_0 & \sigma_0 & 0 \\ 0 & 0 & \gamma_0 \end{pmatrix}.$$

Therefore, the boundary V and homoclinic point C are defined by

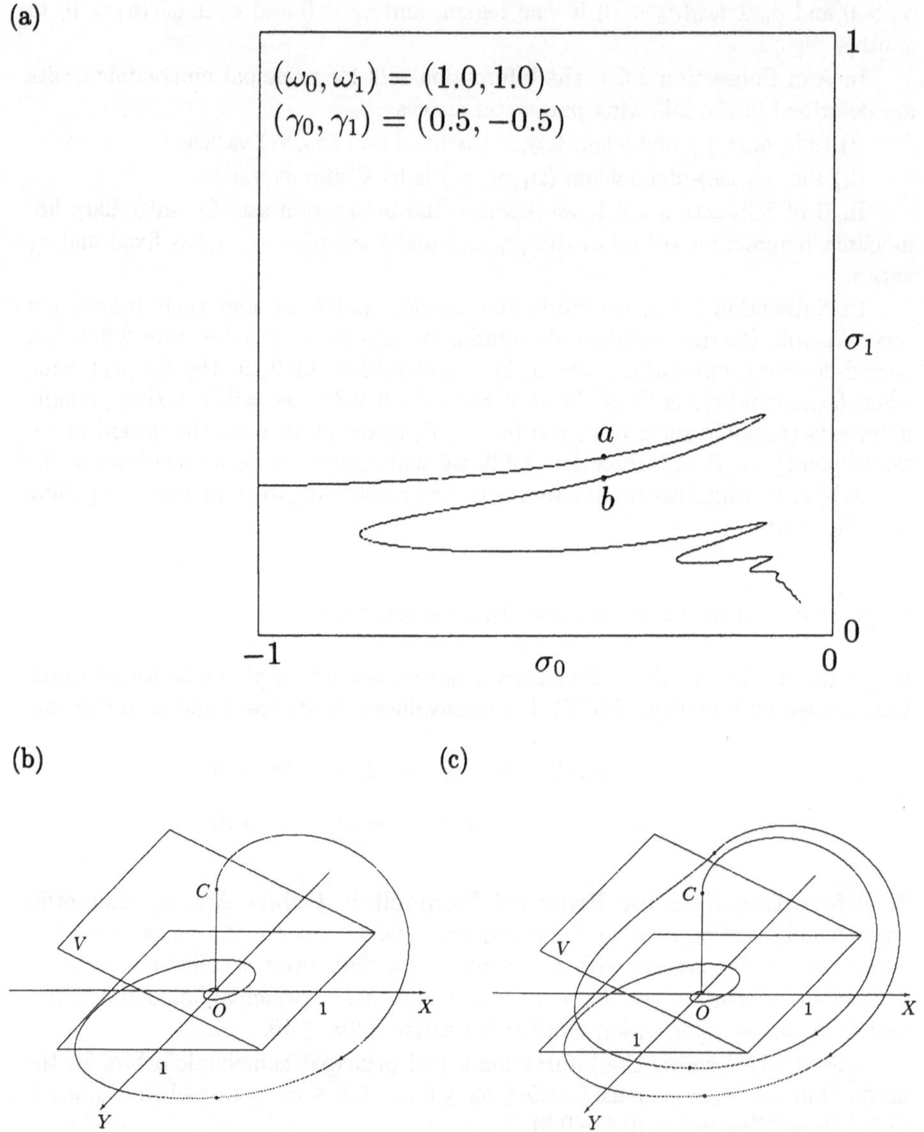

Fig. 2.6.1. Principal homoclinicity.
(a) Bifurcation set in the (σ_0, σ_1)-space.$(\omega_0, \omega_1, \gamma_0, \gamma_1)$=(1.0, 1.0, 0.5, -0.5).
(b) Homoclinic orbit at point (a):$(\sigma_0, \sigma_1, s_1)$ = (-0.400536, 0.300978, 4.556805).
(c) Homoclinic orbit at point (b):$(\sigma_0, \sigma_1, s_1, t_2, s_2)$ = (-0.401938, 0.266684, 4.692735, 3.301122, 3.922416).

$$V = \{(x, y, z) \in \mathbb{R}^3 | x + z = 1\}, \quad \text{and} \tag{2.6.1}$$

$$C = (1, 0, 0)^T. \tag{2.6.2}$$

Fig. 2.6.2 shows the principal homoclinic bifurcation set in the (σ_0, σ_1)-space $(-1.0 \leq \sigma_0 \leq 0.0, \quad 0.0 \leq \omega_0 \leq 4.0)$ when $(\omega_1, \gamma_0, \gamma_1) = (1.0, 0.5, -0.5)$ is fixed and σ_1 is varied as

$$\text{(a) } \sigma_1 = 0.10, \quad \text{(b) } \sigma_1 = 0.15, \quad \text{(c) } \sigma_1 = 0.20,$$

$$\text{(d) } \sigma_1 = 0.25, \quad \text{(e) } \sigma_1 = 0.30, \quad \text{(f) } \sigma_1 = 0.35.$$

Also, Fig. 2.6.3(a) shows a deformation of the principal homoclinic bifurcation set in the (σ_0, ω_0)-space $(-1.0 \leq \sigma_0 \leq 0.0, \quad 0.0 \leq \omega_0 \leq 3.0)$, when $(\omega_1, \gamma_0, \gamma_1) = (1.0, 0.5, -0.5)$ is fixed and σ_1 is varied from 0.10 to 0.37 with step size 0.01. Fig. 2.6.3(b) is an illustration of Fig. 2.6.3(a). We can observe that the $(\sigma_0, \sigma_1, \omega_0)$-space separates into two parts, $\mathcal{A}$ and $\mathcal{B}$, by the principal O-homoclinic bifurcation set $\mathcal{S}$. In $\mathcal{A}$, the orbit from point C returns to boundary V where $z < 0$, and it cannot return to V (it goes to $z = -\infty$, because $\gamma_0 > 0$.) Hence, inside of $\mathcal{A}$, O-homoclinic orbit does not exist. In $\mathcal{B}$, the orbit from point C returns to the boundary V where $z > 0$, and it may return to line $\{(x, y, z) \in V | z = 0\}$. Hence, inside of $\mathcal{B}$, O-homoclinic orbits may exist.

B Subsidiary Homoclinic Bifurcation Sets and Heteroclinic Bifurcation Sets An O-homoclinic orbit which does not belong to the principal homoclinic bifurcation set is called a *subsidiary O-homoclinic orbit*. We will observe a bifurcation set for the simplest subsidiary O-homoclinic orbit.

Fig. 2.6.4 shows a bifurcation set of a subsidiary homoclinic orbit for the origin O in the (σ_0, ω_0)-space $(-1.2 \leq \sigma_0 \leq 0.0, \quad 0.0 \leq \omega_0 \leq 3.0)$ when $(\omega_1, \gamma_0, \gamma_1, \sigma_1) = (1.0, 0.5, -0.5, 0.27)$ is fixed. Center (a) of the spiral in Fig. 2.6.4 corresponds to a heteroclinic bifurcation set.

Fig. 2.6.5 shows the subsidiary homoclinic orbit corresponding to point (b) in Fig. 2.6.4.

Fig. 2.6.6 shows a deformation of the principal homoclinic bifurcation set and the subsidiary homoclinic bifurcation set in the (σ_0, ω_0)-space $(-2.0 \leq \sigma_0 \leq 0.0, \quad 0.0 \leq \omega_0 \leq 6.0)$ when $(\omega_1, \gamma_0, \gamma_1) = (1.0, 0.5, -0.5)$ is fixed and σ_1 varies as

$$\text{(a) } 0.20 \sim 0.23, \quad \text{(b) } 0.24 \sim 0.27, \quad \text{(c) } 0.28 \sim 0.31.$$

In Figs. 2.6.6(a)-(c), the bifurcation set for a heteroclinic orbit from O to P with the intersection number 3 is also drawn.

2.6.2 Bifurcation Sets for Periodic Orbits

Next we will consider bifurcation sets for periodic orbits. We fix four eigenvalue parameters as

$$(\omega_0, \gamma_0, \omega_1, \gamma_1) = (1.0, 0.25, 1.0, -0.25).$$

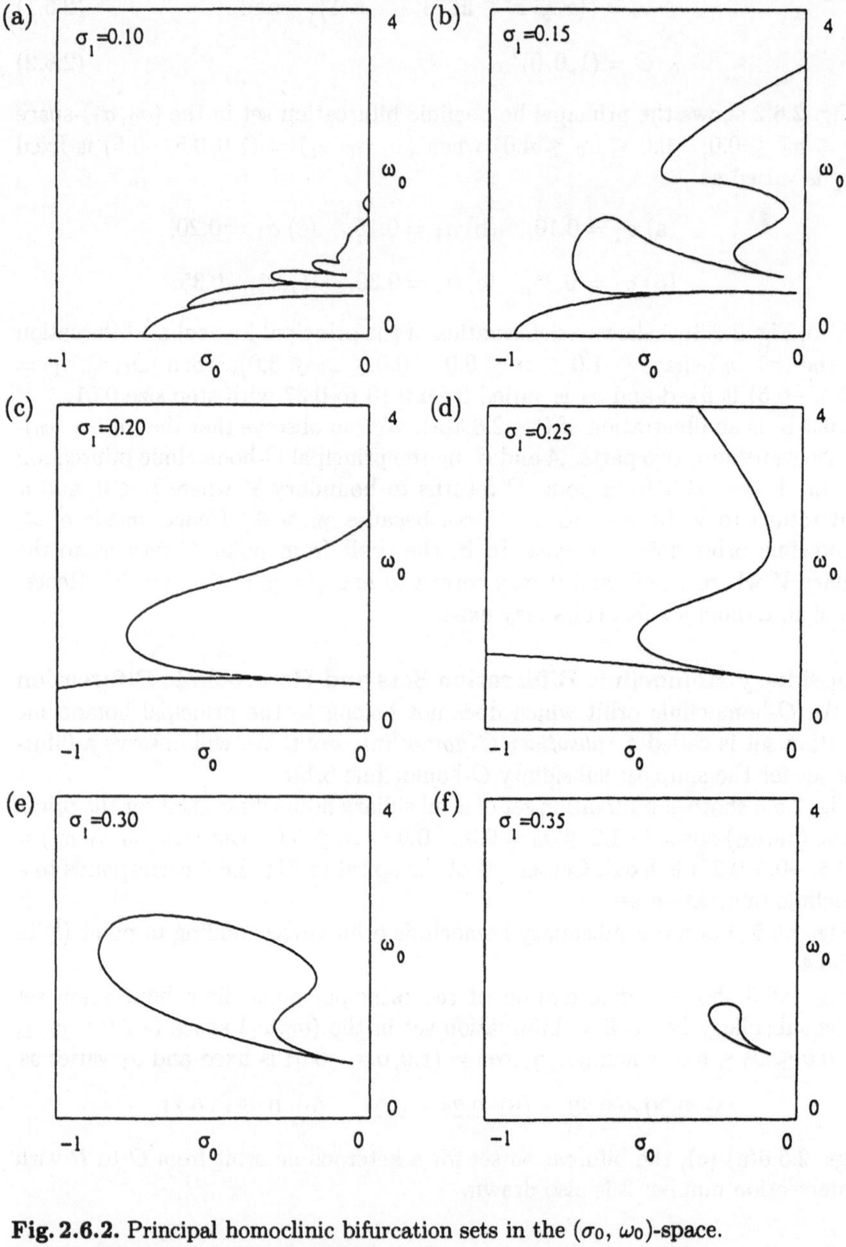

Fig. 2.6.2. Principal homoclinic bifurcation sets in the (σ_0, ω_0)-space. $(\omega_1, \gamma_0, \gamma_1)$=(1.0, 0.5, -0.5). Initial point:$(\sigma_0, \sigma_1, \omega_0, s_1)$= (-0.470065, 0.10, 0.403451,4.120000) for (a); (-0.712232, 0.15, 0.404871,4.412918) for (b); (-1.038328, 0.20, 0.491987, 4.690000) for (c); (-2.241998, 0.25, 0.500000, 5.137040) for (d); (-0.411290, 0.30, 1.005700, 4.570000) for (e); (-0.195345, 0.35, 1.002707, 4.410000) for (f).

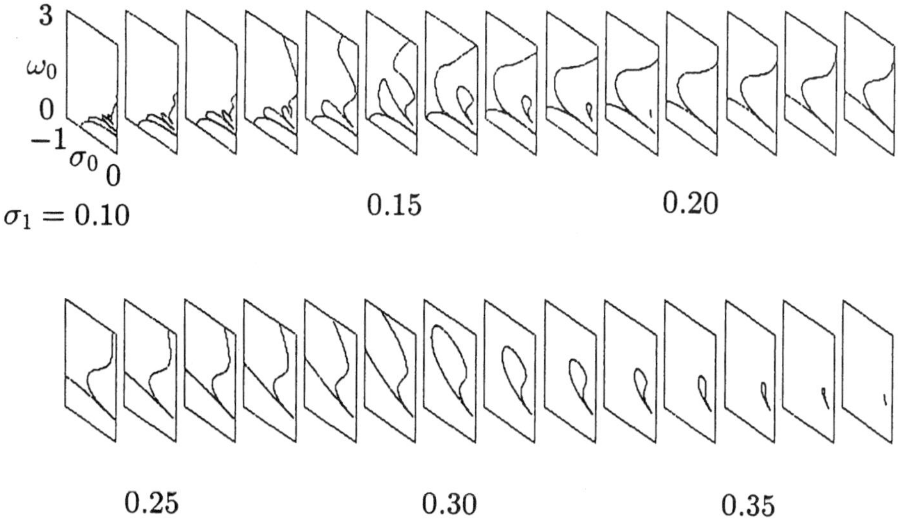

(a)

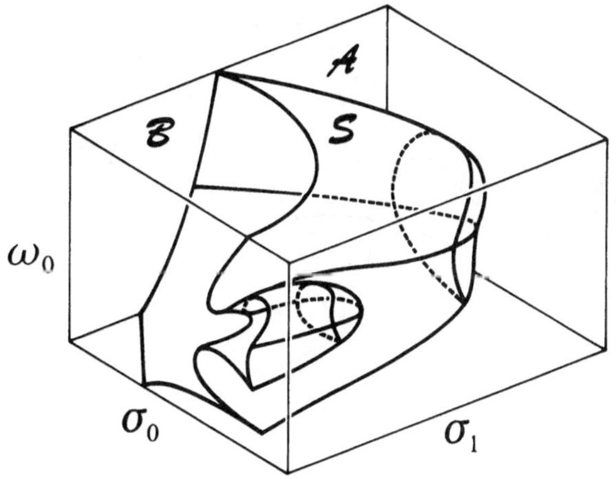

(b)

Fig. 2.6.3. Principal homoclinicity.
(a) Bifurcation set in the $(\sigma_0, \sigma_1, \omega_0)$-space. $(\omega_1, \gamma_0, \gamma_1) = (1.0, 0.5, \text{-}0.5)$.
(b) An illustration of the bifurcation set in (a).

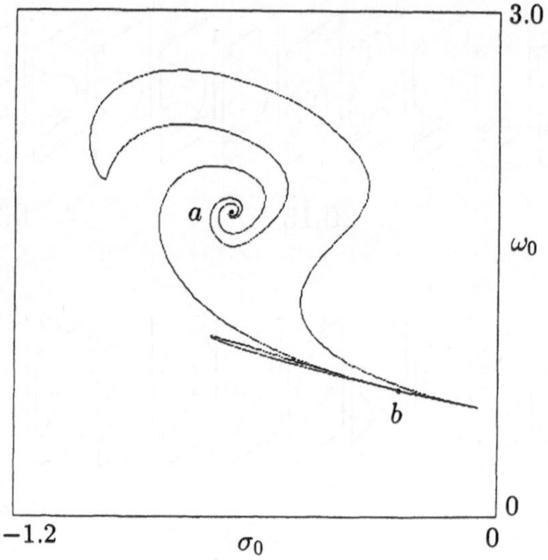

Fig. 2.6.4. A subsidiary homoclinic bifurcation set in the (σ_0, ω_0)-space. $(\omega_1, \gamma_0, \gamma_1, \sigma_1) = (1.0, 0.5, -0.5, 0.27)$.

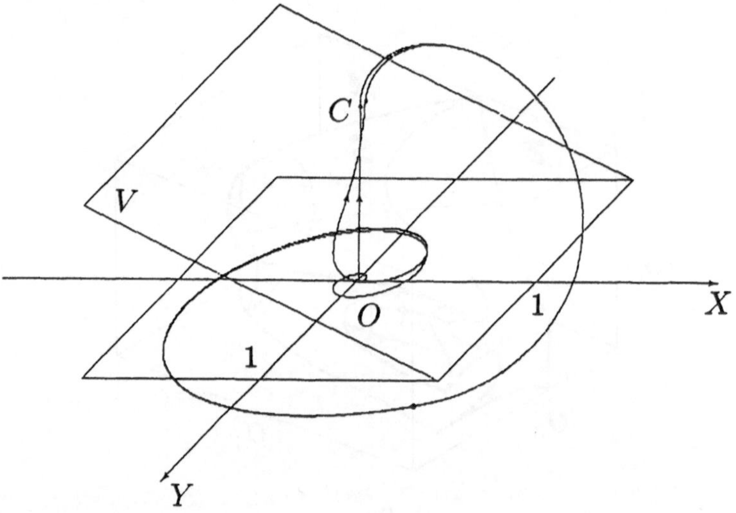

Fig. 2.6.5. A subsidiary homoclinic orbit at point (b) in Fig. 2.6.4. $(\sigma_0, \omega_0, s_1, t_2, s_2) = (-0.251856, 0.766962, 4.129208, 13.154304, 4.038031)$. $(\omega_1, \gamma_0, \gamma_1, \sigma_1) = (1.0, 0.5, -0.5, 0.27)$.

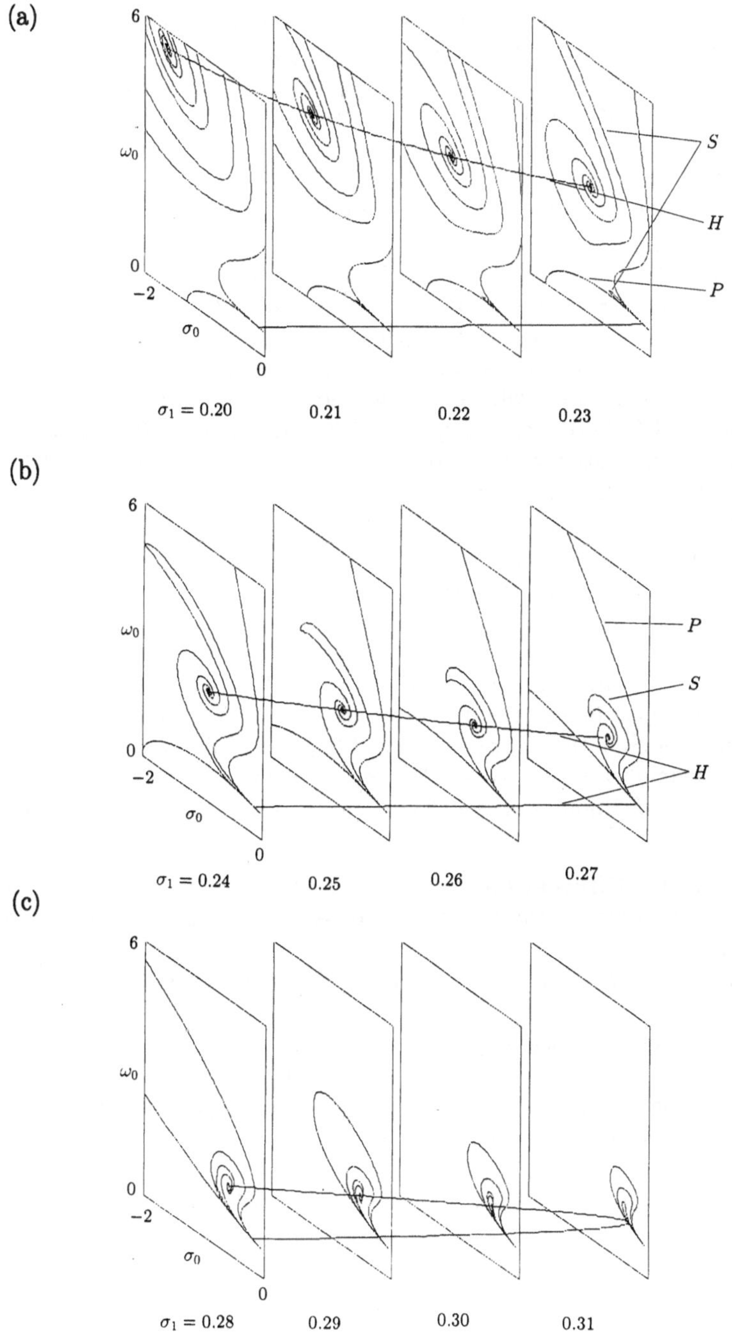

Fig. 2.6.6. A principal/subsidiary homoclinic bifurcation set (P/S) and heteroclinic bifurcation set (H) in the (σ_0, ω_0, σ_1)-space.
(a) $\sigma_1 = 0.20 \sim 0.23$. (b) $\sigma_1 = 0.24 \sim 0.27$. (c) $\sigma_1 = 0.28 \sim 0.31$.

A Saddle-Node Bifurcation Sets Fig. 2.6.7(a) shows the principal homoclinic bifurcation set H_o (respectively H_p) at O (respectively at P) in the (σ_0, σ_1)-space $(-1.0 \le \sigma_0 \le 0.0, \quad 0.0 \le \sigma_1 \le 1.0)$. Since $\omega_0 = \omega_1$ and $\gamma_0 = -\gamma_1$, two curves H_o and H_p are symmetric with respect to line $\{\sigma_0 + \sigma_1 = 0\}$. Figs. 2.6.7(b)-(l) show saddle-node bifurcation curves in the (σ_0, σ_1)-space. We denote them $S_1, S_2, \cdots$ in order of the period.
Fig. 2.6.7(b) shows set S_1 in the (σ_0, σ_1)-space $(-1.0 \le \sigma_0 \le 0.0, \quad 0.0 \le \sigma_1 \le 1.0)$.
Fig. 2.6.7(c) is a magnification of (b) $(-0.5 \le \sigma_0 \le -0.4, \quad 0.4 \le \sigma_1 \le 0.5)$, and a butterfly catastrophe (see p. 178 of [Poston and Stewart 1978]) is observed.
Fig. 2.6.7(d) shows set S_2 in the (σ_0, σ_1)-space $(-1.0 \le \sigma_0 \le 0.0, \quad 0.0 \le \sigma_1 \le 1.0)$. Two triangle-like closed curves are seen with three cusp points, which are located in a symmetric position with respect to line $\{\sigma_0 + \sigma_1 = 0\}$.
Fig. 2.6.7(e) shows set S_3 in the (σ_0, σ_1)-space $(-1.0 \le \sigma_0 \le 0.0, \quad 0.0 \le \sigma_1 \le 1.0)$.
Fig. 2.6.7(f) is a magnification of (e) $(-0.45 \le \sigma_0 \le -0.35, \quad 0.35 \le \sigma_1 \le 0.45)$, and a butterfly catastrophe is also observed.

Fig. 2.6.7(g) is a magnification of (e) $(-0.25 \le \sigma_0 \le -0.23, \quad 0.5 \le \sigma_1 \le 0.7)$. Set S_3 consists of two triangle-like closed curves with three cusp points, and a closed curve with a butterfly catastrophe.
Fig. 2.6.7(h) shows set S_4 in the (σ_0, σ_1)-space $(-1.0 \le \sigma_0 \le 0.0, \quad 0.0 \le \sigma_1 \le 1.0)$.
Fig. 2.6.7(i) is a magnification of (h) $(-0.25 \le \sigma_0 \le -0.23, \quad 0.5 \le \sigma_1 \le 0.7)$, and a swallowtail catastrophe (see p. 178 of [Poston and Stewart 1978]) is observed.
Fig. 2.6.7(j) shows set S_5 in the (σ_0, σ_1)-space $(-1.0 \le \sigma_0 \le 0.0, \quad 0.0 \le \sigma_1 \le 1.0)$.
Fig. 2.6.7(k) is a magnification of (j) $(-0.25 \le \sigma_0 \le -0.23, \quad 0.5 \le \sigma_1 \le 0.7)$, and Fig. 2.6.7(l) is a magnification of (k) $(-0.237 \le \sigma_0 \le -0.233, \quad 0.57 \le \sigma_1 \le 0.61)$.

B Period-Doubling Bifurcation Sets Fig. 2.6.8(a)-(f) show the period-doubling bifurcation curves in the (σ_0, σ_1)-space. We denote them $D_1, D_2, \cdots$ in order of the period.
Fig. 2.6.8(a) shows set D_1 in the (σ_0, σ_1)-space $(-1.0 \le \sigma_0 \le 0.0, \quad 0.0 \le \sigma_1 \le 1.0)$.
Fig. 2.6.8(b) shows set D_2 in the (σ_0, σ_1)-space $(-1.0 \le \sigma_0 \le 0.0, \quad 0.0 \le \sigma_1 \le 1.0)$.
Fig. 2.6.8(c) is a magnification of (b) $(-0.5 \le \sigma_0 \le -0.4, 0.4 \le \sigma_1 \le 0.5)$.
Fig. 2.6.8(d) shows set D_3 in the (σ_0, σ_1)-space $(-1.0 \le \sigma_0 \le 0.0, \quad 0.0 \le \sigma_1 \le 1.0)$.
Fig. 2.6.8(e) is a magnification of (d) $(-0.45 \le \sigma_0 \le -0.35, 0.35 \le \sigma_1 \le 0.45)$, and
Fig. 2.6.8(f) is also a magnification of (d) $(-0.3 \le \sigma_0 \le -0.2, 0.4 \le \sigma_1 \le 0.8)$. Observe that set D_3 consists of three components.

C Windows Denote by T the period of a periodic orbit, i.e., if a periodic orbit has a return time sequence $\{t_1, s_1, \cdots, t_m, s_m\}$, then

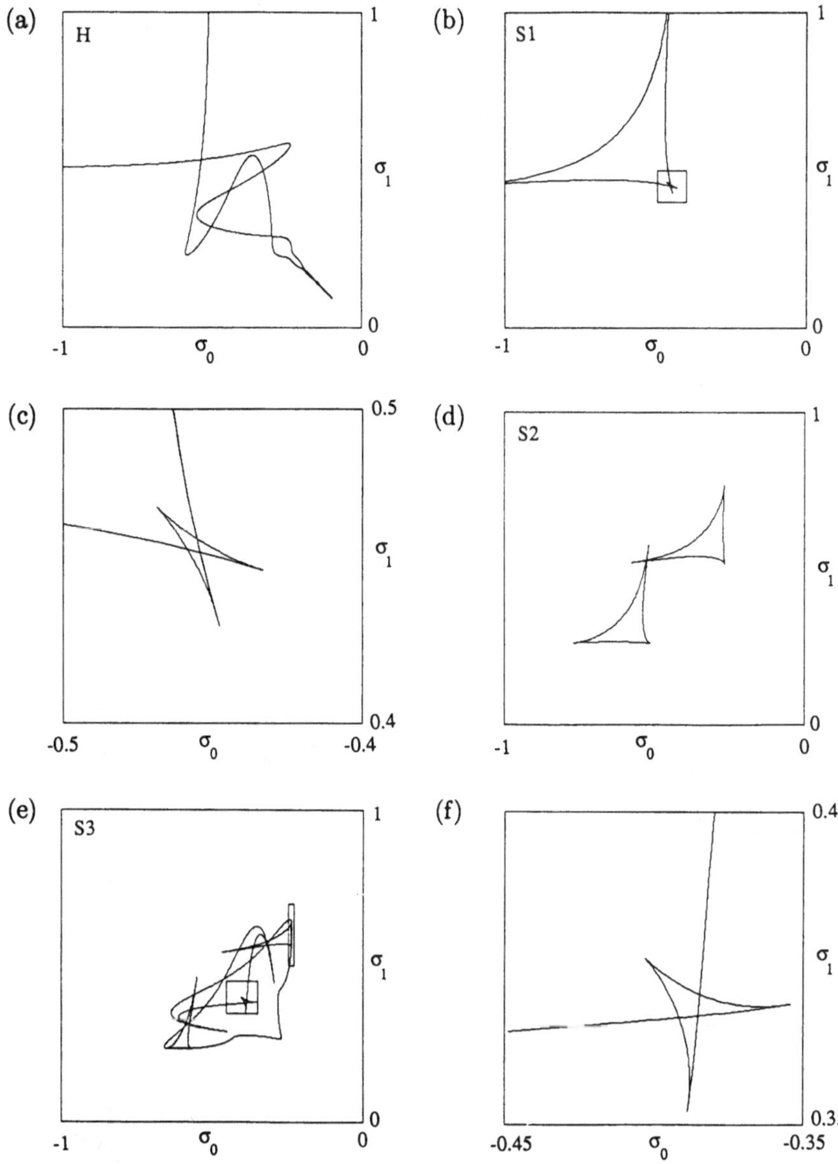

Fig. 2.6.7. Saddle-node bifurcation sets in the (σ_0, σ_1)-space.
(a) Principal homoclinic bifurcation sets. (b) S_1-line. (c) Magnification of (b). (d) S_2-line. (e) S_3-line. (f) Magnification of (e). Initial point:$(\sigma_0, \sigma_1, t_1, s_1)$= (-0.500001, 0.853310, 4.550078, 4.186404) for (b) ; $(\sigma_0, \sigma_1, t_1, s_1, t_2, s_2)$= (-0.277925, 0.718356, 4.692716, 0.210000, 6.180635, 3.991401) for (d); (-0.244997, 0.570363, 15.053611, 4.078229, 4.403001, 0.895610),and (-0.499935, 0.374849, 7.713634, 4.508817, 3.013453, 3.689954) for (e).

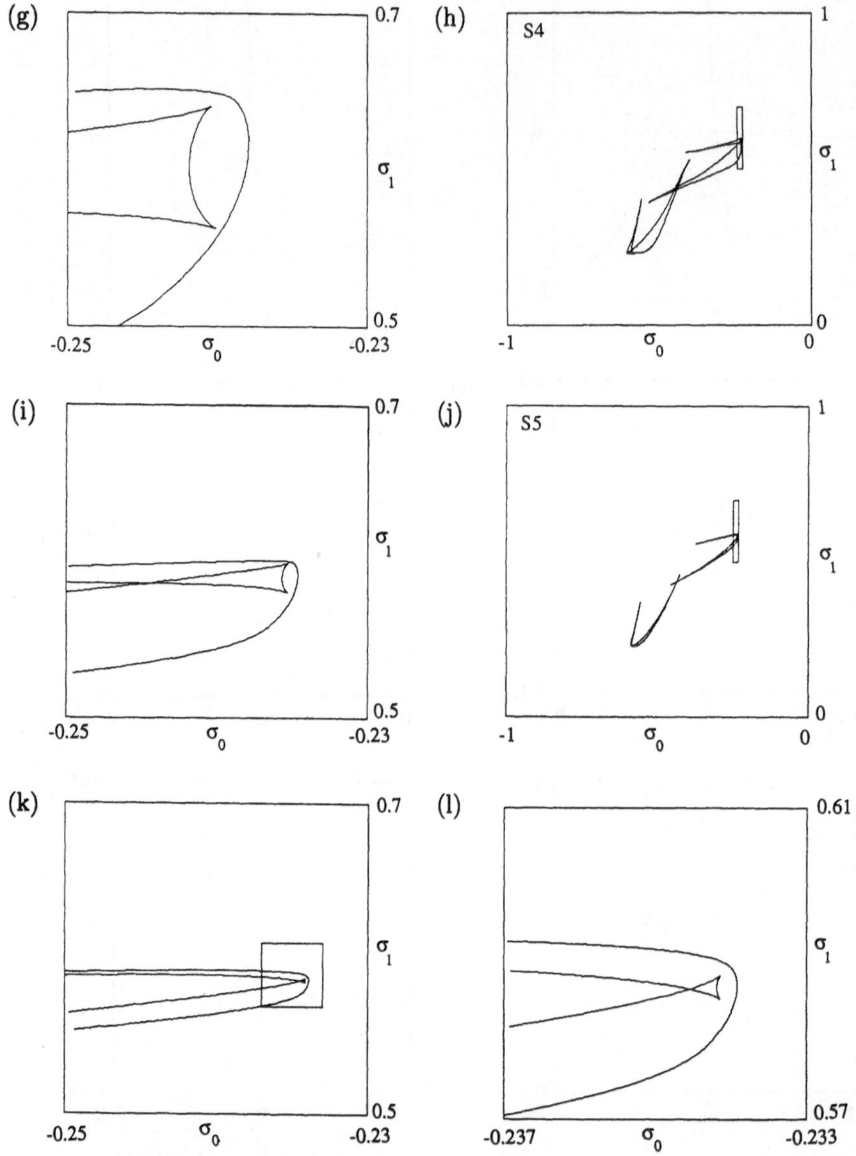

Fig. 2.6.7. continued.
(g) Magnification of (e). (h) S_4-line.
(i) Magnification of (h). (j) S_5-line.
(k) Magnification of (j). (l) Magnification of (k).

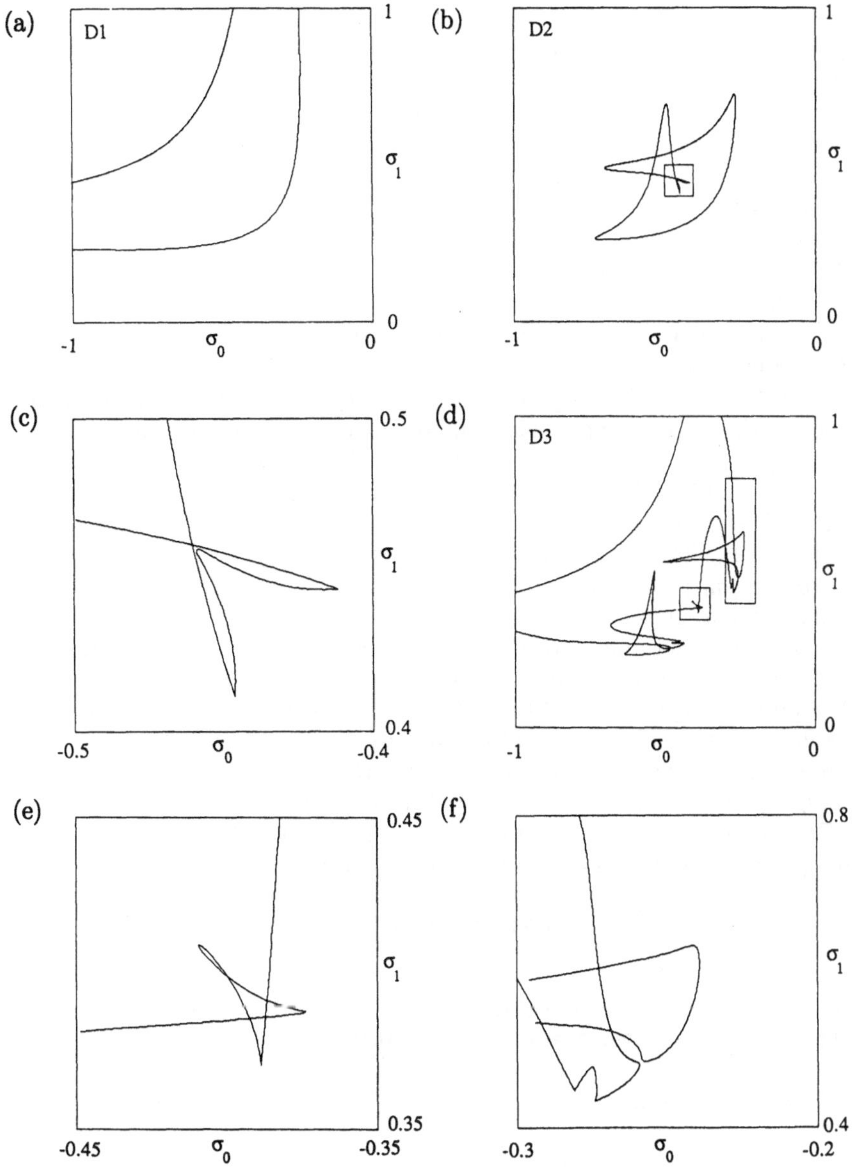

Fig. 2.6.8. Period-doubling bifurcation sets in the (σ_0,σ_1)-space.
(a) D_1-line. (b) D_2-line.
(c) Magnification of (b). (d) D_3-line.
(e) Magnification of (d). (f) Magnification of (d).

$$T = t_1 + s_1 + \cdots + t_m + s_m.$$

Also denote by Σ the solution set of the equation for a periodic point condition (see Theorem 2.5.14) in the (σ_0, σ_1, T)-space. The solution set Σ becomes a surface with codimension 1 in the (σ_0, σ_1, T)-space, hence it is called the *periodic surface.* Plate 2.1 shows a bifurcation topography of periodic surface. A region on the periodic surface where the corresponding periodic orbit is attractive (respectively repelling) is called an attractive (respectively a repelling) *window.* We will observe some windows in the (σ_0, σ_1)-space (i.e., projections of windows) in order of the period T.

Fig. 2.6.9(a) shows two windows, W_0^a and W_0^r $(-1.0 \le \sigma_0 \le 0.0,\quad 0.0 \le \sigma_1 \le 1.0)$. Window W_0^a is an attractive window whose boundary consists of curve D_1, Hopf bifurcation set $G_1 = \{(\sigma_0, \sigma_1, T) \in \Sigma : \sigma_0 + \sigma_1 = 0\}$ and edge $E_1 = \{(\sigma_0, \sigma_1, T) \in \Sigma : \sigma_1 = 0\}$. Edge E_1 is a Hopf bifurcation set for singular point P. Window W_0^r is a repelling window whose boundary consists of curves D_1, G_1 and edge $E_0 = \{(\sigma_0, \sigma_1, T) \in \Sigma : \sigma_0 = 0\}$. Edge E_0 is a Hopf bifurcation set for singular point O. Fig. 2.6.9(a) also shows two windows, W_1^a and W_1^r. Window W_1^a is an attractive window whose boundary consists of curves D_1, S_1 and homoclinic bifurcation set H_o. There is a repelling window W_1^r at the symmetric position with respect to line $\{\sigma_0 + \sigma_1 = 0\}$. The boundary of W_1^r consists of curves D_1, S_1 and homoclinic bifurcation set H_p.

Fig. 2.6.9(b) shows an attractive window W_2^a, whose boundary consists of curves D_1, S_1, D_2, and H_o $(-1.0 \le \sigma_0 \le 0.0,\quad 0.0 \le \sigma_1 \le 1.0)$.
Fig. 2.6.9(c) is a magnification of (b) $(-0.6 \le \sigma_0 \le -0.4, 0.4 \le \sigma_1 \le 0.6)$.
Fig. 2.6.9(d) is a magnification of (c) $(-0.5 \le \sigma_0 \le -0.4, 0.4 \le \sigma_1 \le 0.5)$.
Fig. 2.6.9(e) also shows two windows, W_3^a and W_3^r, whose boundary consists of curves $D_2, S_2, D_3, G_2, H_o, H_p$ and edge ED_1 $(-1.0 \le \sigma_0 \le 0.0,\quad 0.0 \le \sigma_1 \le 1.0)$. Edge ED_1 is set of all points in the (σ_0, σ_1, T)-space which correspond to the periodic points created by the period-doubling bifurcation D_1, i.e.,

$$ED_1 = \{(\sigma_0, \sigma_1, 2T) : (\sigma_0, \sigma_1, T) \in D_1\}.$$

Fig. 2.6.9(f) is a magnification of (e) $(-0.6 \le \sigma_0 \le -0.4, 0.5 \le \sigma_1 \le 0.55)$.

Fig. 2.6.9(g) is a magnification of (e) $(-0.35 \le \sigma_0 \le -0.3, 0.625 \le \sigma_1 \le 0.675)$.
Fig. 2.6.9(h) is a magnification of (g) $(-0.315 \le \sigma_0 \le -0.3225, 0.6425 \le \sigma_1 \le 0.6475)$.
Fig. 2.6.9(i) is a magnification of (e) $(-0.3 \le \sigma_0 \le -0.2, 0.7 \le \sigma_1 \le 0.8)$.
Fig. 2.6.9(j) is a magnification of (e) $(-0.3 \le \sigma_0 \le -0.2, 0.45 \le \sigma_1 \le 0.55)$.
Fig. 2.6.9(k) is a magnification of (j) $(-0.275 \le \sigma_0 \le -0.260, 0.515 \le \sigma_1 \le 0.530)$.
Fig. 2.6.9(l) shows two windows, W_4^a and W_4^r, whose boundaries consist of curves D_3, S_3, D_4, G_1, G_3 and H_o or H_p $(-1.0 \le \sigma_0 \le 0.0,\quad 0.0 \le \sigma_1 \le 1.0)$.

Fig. 2.6.9(m) is a magnification of (l) $(-0.6 \le \sigma_0 \le -0.2, 0.5 \le \sigma_1 \le 0.8)$.
Fig. 2.6.9(n) is a magnification of (m) $(-0.6 \le \sigma_0 \le -0.35, 0.52 \le \sigma_1 \le 0.58)$.
Fig. 2.6.9(o) is a magnification of (n) $(-0.48 \le \sigma_0 \le -0.4, 0.545 \le \sigma_1 \le 0.56)$.

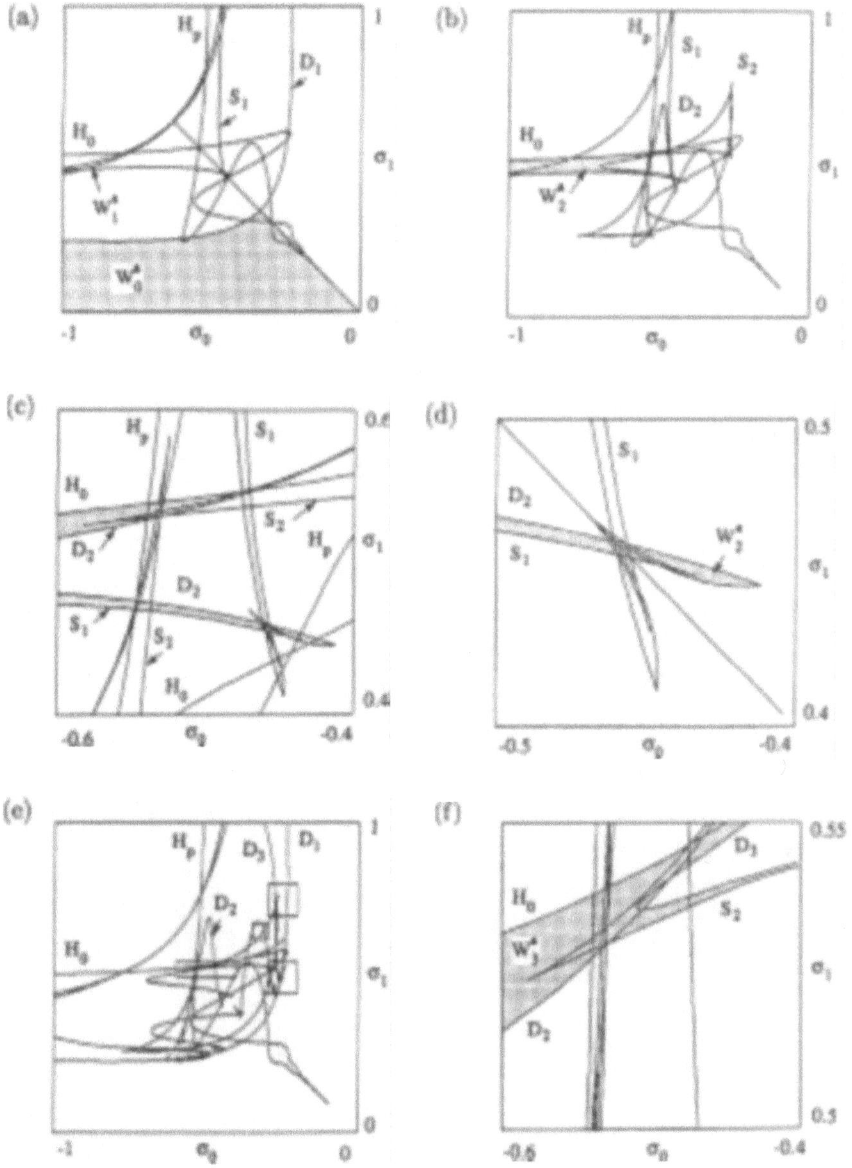

Fig. 2.6.9. Periodic windows in the (σ_0,σ_1)-space.
(a) W_0^a, W_1^a-windows and S_1,D_1, H_o, H_p-lines.
(b) W_2^a-window and S_1, S_2, D_2, H_o, H_p-lines.
(c) Magnification of (b). (d) Magnification of (b).
(e) S_2, D_1, D_2, D_3, H_o and H_p-lines.
(f) Magnification of (e).

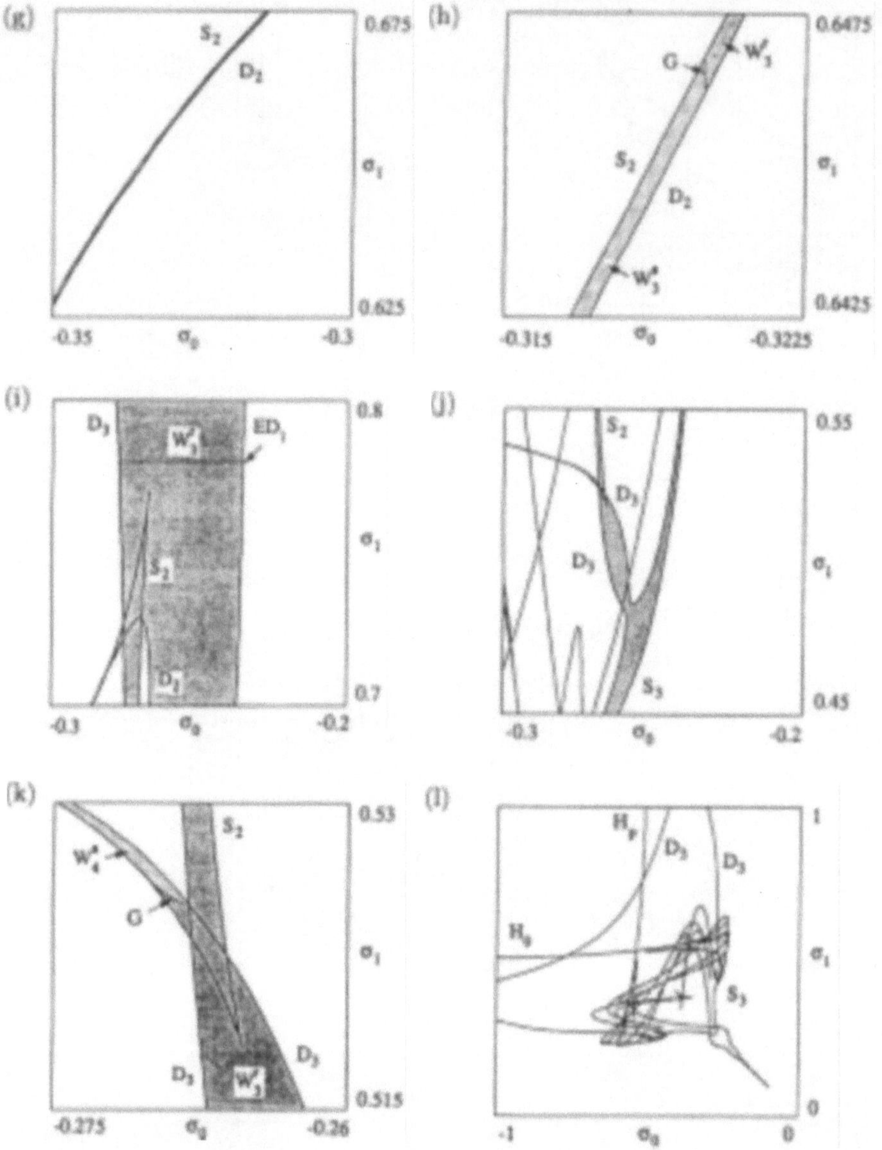

Fig. 2.6.9. continued.
(g) Magnification of (e).
(h) Magnification of (g) added G-line and W_3^a, W_3^r-windows.
(i) Magnification of (e).
(j) Magnification of (e).
(k) Magnification of (j) added G-line and W_4^a, W_3^r-windows.
(l) S_3, D_3, H_o and H_p-lines.

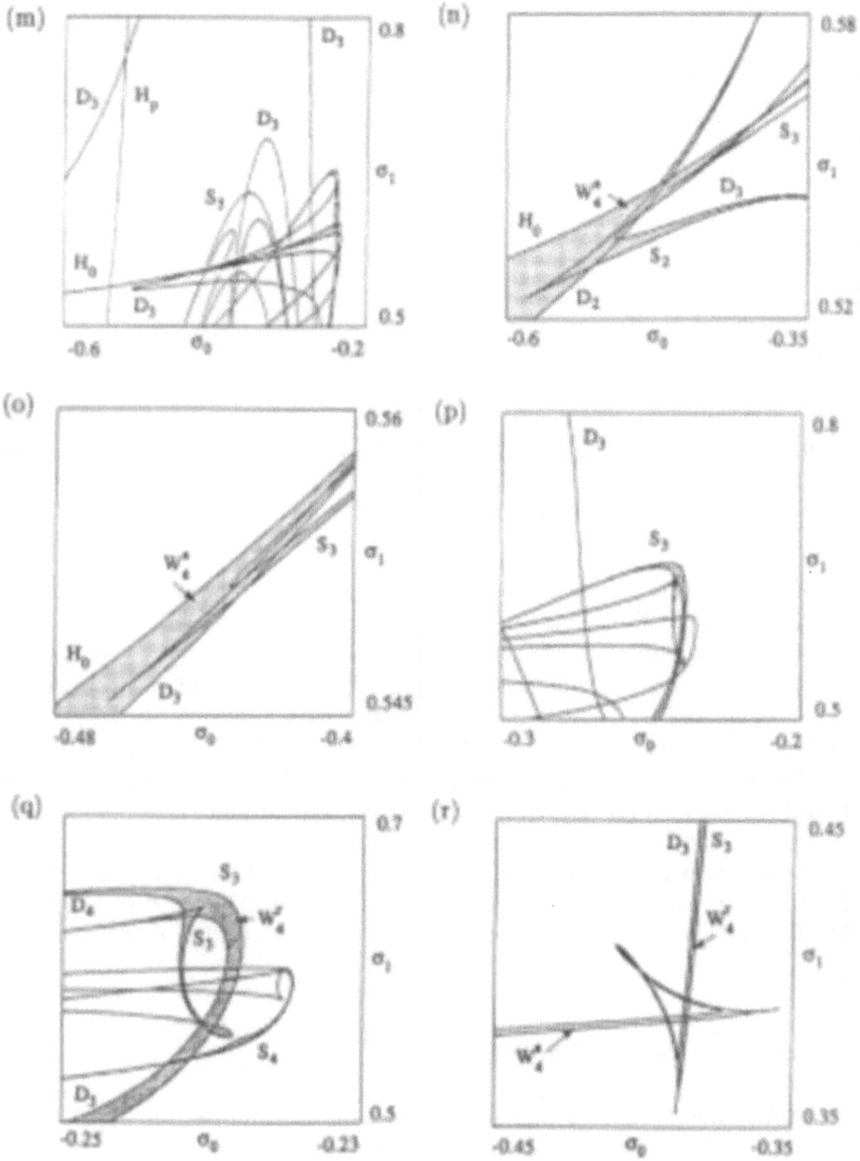

Fig. 2.6.9. continued.
(m) Magnification of (l).
(n) Magnification of (m) added S_2, D_2-lines and W_4^a-window.
(o) Magnification of (n).
(p) Magnification of (m) added D_4-line.
(q) Magnification of (p) added S_4-line and W_4^r-window.
(r) Magnification of (l) added W_4^a and W_4^r-windows.

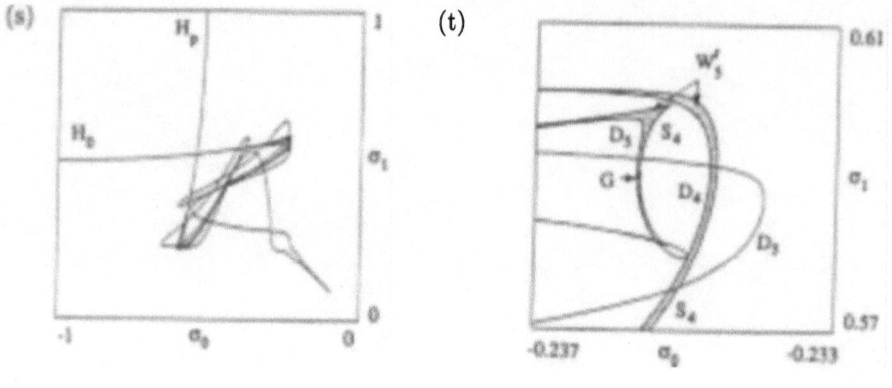

Fig. 2.6.9. continued. (s) S_4, D_4, D_5, H_o and H_p-lines. (t) Magnification of (s) added G-line.

Fig. 2.6.9(p) is a magnification of (m) $(-0.3 \leq \sigma_0 \leq -0.2, 0.5 \leq \sigma_1 \leq 0.8)$, and
Fig. 2.6.9(q) is a magnification of (p) $(-0.25 \leq \sigma_0 \leq -0.23, 0.5 \leq \sigma_1 \leq 0.7)$.
Fig. 2.6.9(r) is a magnification of (m) $(-0.45 \leq \sigma_0 \leq -0.35, 0.35 \leq \sigma_1 \leq 0.45)$.

Fig. 2.6.9(s) shows two windows, W_5^a and W_5^r, whose boundaries consist of curves D_4, S_4, D_5 and G_4 $(-1.0 \leq \sigma_0 \leq 0.0, \quad 0.0 \leq \sigma_1 \leq 1.0)$.
Fig. 2.6.9(t) is a magnification of (s) $(-0.237 \leq \sigma_0 \leq -0.233, 0.57 \leq \sigma_1 \leq 0.61)$.

Fig. 2.6.10(a)-(d) show the relative position of W_3, W_4, W_5 and W_6.
Fig. 2.6.10(a) shows curves D_2, S_2, D_3, S_3, D_4, S_4, D_5, S_5 and D_6 $(-0.3 \leq \sigma_0 \leq -0.2,\ 0.5 \leq \sigma_1 \leq 0.8)$.
Fig. 2.6.10(b) is a magnification of (a) $(-0.25 \leq \sigma_0 \leq -0.23,\ 0.5 \leq \sigma_1 \leq 0.7)$.
Fig. 2.6.10(c) is a magnification of (b) $(-0.237 \leq \sigma_0 \leq -0.233,\ 0.57 \leq \sigma_1 \leq 0.61)$.
Fig. 2.6.10(d) is a magnification of (c) $(-0.2344 \leq \sigma_0 \leq -0.2336,\ 0.584 \leq \sigma_1 \leq 0.592)$.

2.6.3 Computing Bifurcation Sets

We now explain how to compute a solution curve of, for example, the homoclinic bifurcation equation by the Newton method. First, we compute an orbit of point C of Equation (2.6.2) for various parameter values (σ_0, σ_1), and choose an approximate solution $(\sigma'_{01}, \sigma'_{11}, s'_{11})$ of the parameter (σ_0, σ_1) and return time (s_1) under which point C becomes the simplest homoclinic point (i.e., intersection number = 2). To compute the orbit, we can use any numerical integration method (e.g. the Runge-Kutta method) or formulas (2.5.5) and (2.5.6). For this approximate solution $(\sigma'_{01}, \sigma'_{11}, s'_{11})$, we carry out the Newton method by fixing σ'_{01}, for example, and obtain a solution $(\sigma''_{01}, \sigma''_{11}, s''_{11})$ (where $\sigma''_{01} = \sigma'_{01}$) for Equations (2.5.53) and (2.5.54). We then check if $(\sigma''_{01}, \sigma''_{11}, s''_{11})$ satisfies the first return time

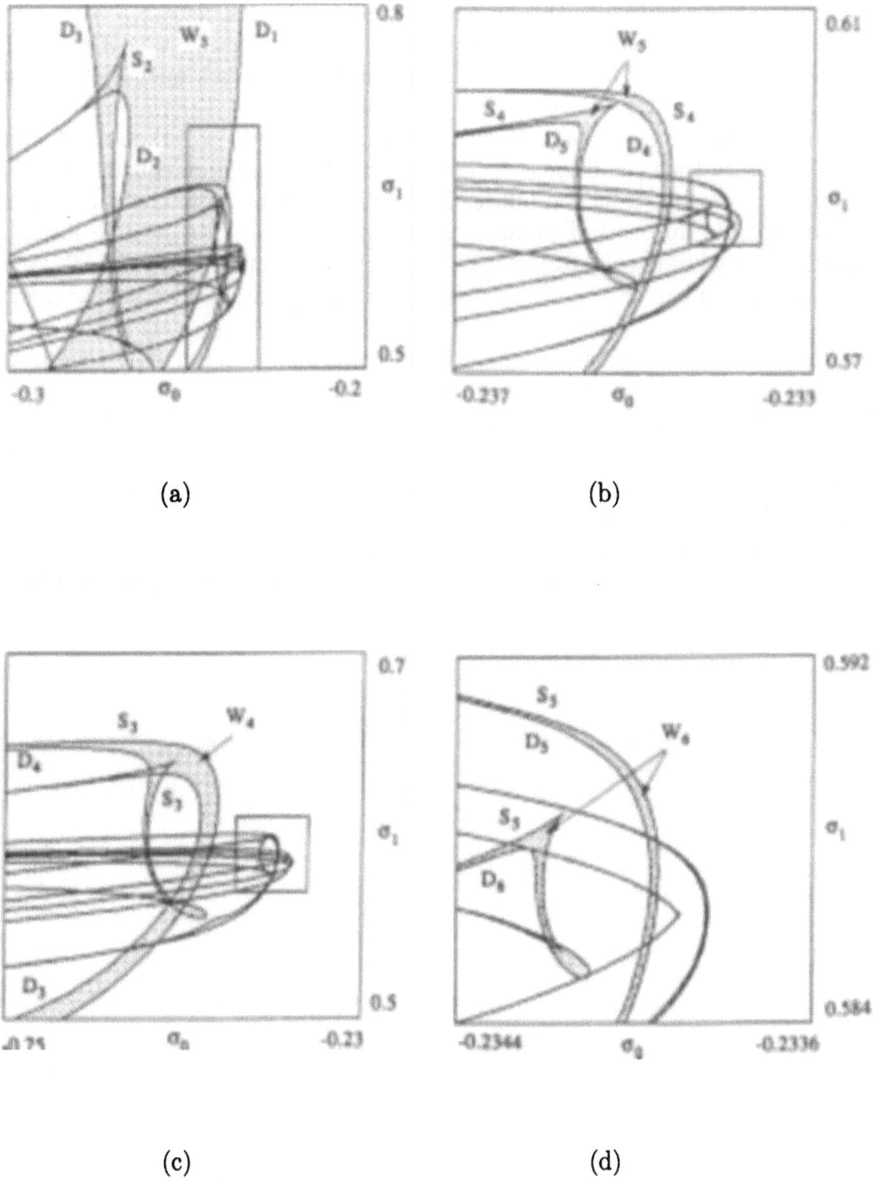

Fig. 2.6.10. Relative position of windows. (a) W_3-window. (b) W_4-window. (c) W_5-window. (d) W_6-window.

open conditions (2.5.55) and (2.5.56). If $(\sigma_{01}'', \sigma_{11}'', s_{11}'')$ does satisfy them, we set $(\sigma_{01}, \sigma_{11}, s_{11}) = (\sigma_{01}'', \sigma_{11}'', s_{11}'')$ as a solution for the homoclinic bifurcation equation (2.5.53)-(2.5.56) when intersection number $2m = 2$.

To get the next solution, we set an approximate solution $(\sigma_{02}', \sigma_{12}', s_{12}') = (\sigma_{01} + \varepsilon, \sigma_{11}, s_{11})$ (a small step ε should be taken), and carry out the Newton method by fixing σ_{02}' to obtain $(\sigma_{02}'', \sigma_{12}'', s_{12}'')$, and set $(\sigma_{02}, \sigma_{12}, s_{12}) = (\sigma_{02}'', \sigma_{12}'', s_{12}'')$ as a solution of the homoclinic bifurcation equation if it satisfies the first return time conditions. In this manner we obtain a sequence of solutions $(\sigma_{0i}, \sigma_{1i}, s_{1i}) \quad (i = 1, 2, \cdots, n-1)$.

If a solution $(\sigma_{0n}'', \sigma_{1n}'', s_{1n}'')$ for Equations (2.5.53) and (2.5.54) does not satisfy the first return time condition, we must replace the bifurcation equation with the intersection number 2 by the bifurcation equation with the intersection number 4. We compute an orbit of point C when $(\sigma_0, \sigma_1) = (\sigma_{0n}'', \sigma_{1n}'')$, and obtain an approximate solution $(\sigma_{0n}', \sigma_{1n}', s_{1n}', t_{2n}', s_{2n}')$ for Equations (2.5.57)-(2.5.60) for the intersection number $2m = 4$. From this approximate solution, by a similar manner as above, we obtain a sequence of solutions $(\sigma_{0i}, \sigma_{1i}, s_{1i}, t_{2i}, s_{2i}) \quad (i = n, n+1, \cdots)$ for the homoclinic bifurcation equation. When the first return time condition is violated for $2m \geq 4$, we may need to consider the case for $2m - 2$.

In order to obtain an approximate solution for the equation of the periodic point condition of Theorem 2.5.14, it is convenient to find an attractive periodic orbit by, for example, the Runge-Kutta method. From a sequence of solutions of the equation of the periodic point condition, by computing the value of the left-hand side of Equation (2.5.101), (2.5.102) or (2.5.103) in Theorem 2.5.15, we can obtain an approximate solution of the bifurcation equation for a periodic orbit.

3. Fundamental Concepts in Bifurcations

3.1 Introduction

The objective of this chapter is to give some fundamental notions and results in the theory of dynamical systems as well as their bifurcations, which are important in the qualitative study of dynamical systems and are used in the other two chapters, so that readers can become familiar with those important underlying ideas without referring to other textbooks or articles. Since the other chapters mainly deal with vector fields, that is, continuous dynamical systems, the description of this chapter also places more emphasis on continuous dynamical systems than on discrete dynamical systems, although one section is devoted to these discrete systems.

A dynamical system is a mathematical notion of a system evolving in time and it includes an autonomous ordinary differential equation or a vector field, and a self-mapping. These are used to model the behaviors of mechanical systems, dynamics of electronic circuits, population growth in ecological systems, oscillation of chemicals, etc. The main idea of the study of those systems is "qualitative analysis" instead of obtaining explicit solutions describing the phenomena. Therefore, we are interested in the qualitative structure of solution curves drawn in the phase space given by a dynamical system, called the orbits of the dynamical system, and the study of the picture of the orbits from a qualitative point of view. The orbit structure may be very simple in some cases, but often it can exhibit a highly complicated, irregular aspect, which is now usually called a "chaotic" behavior. Moreover, the structure of orbits in the phase space can change drastically under a perturbation by slightly changing the parameters involved in the dynamical system. If this occurs we say the dynamical system undergoes a bifurcation, a qualitative change due to a perturbation. All such things are the subjects of the present chapter and of this book as well.

In this chapter, we shall give a brief explanation of the following topics: Section 3.2 is a preliminary section where basic notions of dynamical systems are defined in the case of continuous dynamical systems, namely ordinary differential equations, together with some fundamental theorems and examples. In particular, a more precise explanation of the "qualitative" study of dynamical systems is given and the term "bifurcation" is defined. In Section 3.3, we will concentrate on the study of continuous dynamical systems in a neighborhood of a single point, which is an equilibrium point of the system. In addition, we will examine several techniques as well as results for such local dynamics and bifurcations. Notions used in the other chapters, such as the saddle-node bifurcation, the Hopf bifurcation, etc., are treated here in detail.

Before going on to a more global study of dynamics, we briefly describe discrete dynamical systems in Section 3.4 also for later use, since continuous and discrete dynamical systems are closely related to each other. One of the most important topics in Section 3.4 is what is called the "horseshoe," which is one prototype of a chaotic dynamical system, though we do not explain it in a full detail because it is already very well-described in several textbooks, for instance, [Palis and de Melo 1982], [Moser 1973]. In Section 3.5, we come back to continuous dynamical systems and study more global aspects of the bifurcations, namely, the homoclinic and heteroclinic bifurcations. The study of these bifurcations has in some sense a local nature, because these bifurcations occur in neighborhoods of homoclinic and/or heteroclinic orbits. However, as we shall see in Section 3.5, these bifurcations have an effect on global dynamics, in particular, on chaotic behavior of dynamical systems. The celebrated Shil'nikov theorem is one such result, which is also used in Chapter 1 in order to show the existence of chaotic dynamics in the Double Scroll circuit. We will also describe another recent result due to Rychlik on the existence of a Lorenz-like attractor related to certain homoclinic bifurcations. Finally at the end of Section 3.5, we shall make some remarks about a relation between local bifurcations near equilibrium points and more global bifurcations for homoclinic/heteroclinic orbits.

In this chapter, we have tried to avoid very technical detail and to make underlying ideas clear in the description. This does not necessarily mean that the explanation in this chapter is not mathematically rigorous. The most fundamental notions and theorems are stated in a sufficiently rigorous manner and the ideas of their proofs are clearly explained together with examples illustrating basic results as well as how they are applied to concrete dynamical systems. However, the proofs of most theorems are not given completely. The outline of a proof is indicated or even sometimes omitted. Instead, we will give references for the technical details of the proofs as well as related information so that interested readers can follow them and obtain a better understanding.

Since most of the materials in this chapter have been chosen in order to explain the basic notions and techniques used in the other two chapters, there are many important topics in the theory of dynamical systems and their bifurcation remaining untouched here: for instance, bifurcations of periodic and quasi-periodic solutions, as well as useful functional analysis techniques for these bifurcations. Also missing are conservative systems and their bifurcations, complex dynamical systems, the ergodic theory for dynamical systems, and infinite dimensional dynamical systems including evolution equations. For discrete dynamical systems, the well-established theory for hyperbolic diffeomorphisms is not explained, except for the horseshoe which is briefly described in Subsection 3.4.6. All such topics can be found in the references given at the end of this book, or at least in the bibliography therein.

3.2 Fundamental Notions for Dynamical Systems

In this section we shall give several fundamental notions for continuous dynamical systems which will be used below. Similar notions are provided for discrete dynamical systems in Section 3.4.

3.2.1 Definitions and Examples of Dynamical Systems

Consider an ordinary differential equation

$$\dot{x} = v(x) \qquad \left(\dot{} = \frac{d}{dt}\right) \tag{3.2.1}$$

on $\mathbb{R}^n$ where v is a smooth mapping from $\mathbb{R}^n$ into itself. Here "smooth" means C^r, namely, r-times continuously differentiable for sufficiently large r. The most basic fact for this is the fundamental theorem for ordinary differential equations as follows:

Theorem 3.2.1. *Let $f : \mathbb{R} \times \mathbb{R}^n \to \mathbb{R}^n$ be a smooth mapping, then the ordinary differential equation*

$$\dot{x} = f(t, x)$$

has a unique local solution. Namely, for each $\xi \in \mathbb{R}^n$, there exists a number $\varepsilon > 0$ and a function

$$\varphi : (-\varepsilon, \varepsilon) \to \mathbb{R}^n,$$

such that

$$\frac{d\varphi(t)}{dt} = f(t, \varphi(t)) \quad \text{and} \quad \varphi(0) = \xi.$$

The function $\varphi(t)$ is unique with respect to the initial condition ξ, and we denote it by $\varphi(t) = \varphi(t, \xi)$.

If $f(t, x)$ is of C^r in (t, x), then so is $\varphi(t, \xi)$ in (t, ξ).

Here we do not give a proof of this theorem since it is well-known, and only refer to a textbook of ordinary differential equations, such as [Hale 1969] and [Arnold 1973]. However, we remark that one way of proving this theorem is to convert the ordinary differential equation to an equivalent integral equation given by

$$\varphi(t) = \xi + \int_0^t f(s, \varphi(s))ds,$$

and regard it as a fixed point equation for a solution $\varphi(t)$. More precisely, one may consider the right-hand side as defining a function $\mathcal{F}(\varphi)$ given by

$$\mathcal{F}(\varphi)(t) = \xi + \int_0^t f(s, \varphi(s))ds$$

for each fixed ξ, and solve the equation

$$\mathcal{F}(\varphi) = \varphi,$$

the solution of which gives a desired solution to the ordinary differential equation with the prescribed initial condition. This idea is important and will be repeatedly used later in various different forms.

Since the equation (3.2.1) does not contain t explicitly in its right-hand side, it can be thought of as defining a vector field v on $\mathbb{R}^n$ and in this context the solution map $\varphi(t, \xi)$ indicates the position of the solution at time t which starts from ξ at time 0 and which moves along the flowline generated by the vector field v.

The solution map satisfies the semi-group property coming from the uniqueness of solutions on initial conditions:

(G1) $\varphi(0, \xi) = \xi$;

(G2) $\varphi(t, \varphi(s, \xi)) = \varphi(t + s, \xi) \qquad (t, s \in \mathbb{R})$,

as long as it exists, and it may be more clearly described in the following way:

(G1) $\varphi^0 = id$;

(G2) $\varphi^t \circ \varphi^s = \varphi^{t+s} \qquad (t, s \in \mathbb{R})$,

where the mapping φ^t is the time t map given by $\varphi^t(\xi) = \varphi(t, \xi)$. Of course the map φ^t may not in general be defined for all $t \in \mathbb{R}$ as observed below.

Example 3.2.2. The equation $\dot{x} = -x^2$ on $\mathbb{R}^1$ is solved to be

$$\varphi(t, \xi) = \frac{1}{t + \frac{1}{\xi}},$$

and hence $\varphi(t, \xi) \to +\infty$ as $t \to -\frac{1}{\xi} + 0$ and $\varphi(t, \xi) \to -\infty$ as $t \to -\frac{1}{\xi} - 0$. Therefore, φ^t is not defined for all $t \in \mathbb{R}$.

We call a vector field *complete* if the corresponding family of solution maps $\{\varphi^t\}$ is well-defined for any $t \in \mathbb{R}$ and is defined on the whole $\mathbb{R}^n$. In this case, the family of solution maps $\{\varphi^t\}$ forms a group and is called the *flow* generated by the vector field on the corresponding ordinary differential equation. It is not hard to show that a vector field on a compact manifold always generates the flow. A flow is one type of dynamical systems. To be more general,

Definition 3.2.3. A *continuous dynamical system* or a flow on a manifold M is a family of continuous maps $\{\varphi^t\}$ from M into itself defined for all $t \in \mathbb{R}$ satisfying the properties (G1) and (G2) above. In this case the family $\{\varphi^t\}$ forms a group under composition.

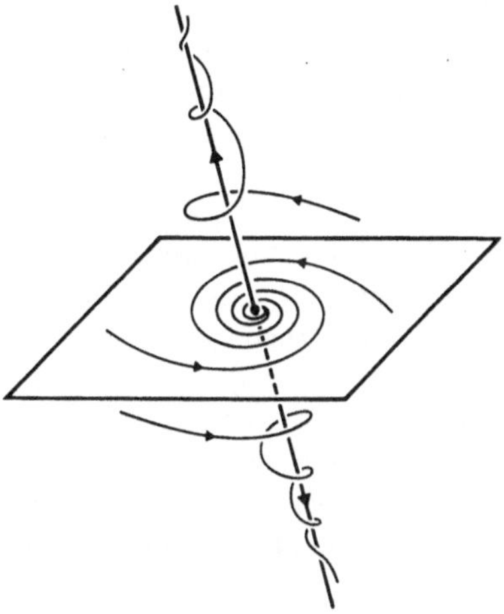

Fig. 3.2.1. Saddle-like behavior of a hyperbolic linear vector field.

Given a continuous dynamical system, we can consider the picture of oriented curves in M given by $t \mapsto \varphi^t(\xi)$ $(\xi \in M)$, called the *phase portrait* of the dynamical system. When we study a dynamical system, we are interested in the qualitative aspect of the phase portrait, which is sometimes referred to as the dynamics of the flow.

From the definition, any member of the flow φ^t has to be a bijective mapping, hence a homeomorphism. In this chapter, we mainly deal with smooth flows unless otherwise stated, that is, φ^t is continuously differentiable as many times as needed.

On the contrary, we can recover the vector field v from a smooth flow $\{\varphi^t\}$ as follows:

$$v(x) = \left.\frac{d}{dt}\right|_{t=0} \varphi^t(x).$$

This is well-defined because of the group property, and gives a smooth vector field on $\mathbb{R}^n$. Therefore a flow and a complete vector field are equivalent notions (as far as they are smooth).

Example 3.2.4. (Linear Vector Fields) A linear vector field on $\mathbb{R}^n$ is given by the equation

$$\dot{x} = Ax, \tag{3.2.2}$$

where A is an $n \times n$ real matrix. The solution map to this equation is given by $\varphi^t(\xi) = e^{At}\xi$ and hence it is linear again. The behavior of solutions to a linear ordinary differential equation, that is, the dynamics of the linear flow heavily depends on the eigenvalues of A and its Jordan normal form. In particular, if any eigenvalue of A has a negative real part, then $\varphi^t(\xi) = e^{At}\xi \to 0$ as $t \to +\infty$ and thus the solution to (3.2.2) goes to the origin O asymptotically. In this sense, the origin O is said to be (asymptotically) stable. On the other hand, if at least one eigenvalue of A has a positive real part, then $|\varphi^t(\xi)| \to +\infty$ as $t \to +\infty$ for an eigenvector $\xi \neq 0$ of the eigenvalue. In other words, there exists an arbitrarily small initial condition leaving from the origin, and hence the origin is unstable. See Definition 3.2.12 for the definition of stability.

Among linear vector fields, important are the ones whose coefficient matrices have no eigenvalues on the imaginary axis. Such a linear vector field is called a *hyperbolic linear vector field.* Since there are no eigenvalues with zero real parts in hyperbolic linear vector fields, their eigenvalues have either positive or negative real parts. If all the eigenvalues have the same sign in their real part, then the behavior of the solutions are as described above. Otherwise the linear vector fields have a saddle-like behavior as shown in Fig. 3.2.1, which illustrates the case of one positive real eigenvalue and one pair of complex conjugate eigenvalues with a negative real part. In general, for such linear vector fields, we define the eigenspace E^s (respectively, E^u) spanned by all the generalized eigenvectors corresponding to eigenvalues with negative (respectively, positive) real parts, which is called the *stable* (respectively, *unstable*) *eigenspace.* The solutions in E^s go to the origin O as the time goes to $+\infty$, whereas those in E^u go to O as the time goes to $-\infty$. Other solutions outside E^s and E^u always go away from O in a positive and negative finite time as depicted in Fig. 3.2.1, where the arrows indicate the direction of solutions as the time varies.

Example 3.2.5. (Lorenz Equation) Meteorologist E. N. Lorenz derived the following three-dimensional ordinary differential equation as a simplification of a partial differential equation modelling the dynamic movement of an atmospheric fluid ([Lorenz 1963]):

$$\begin{aligned} \dot{x} &= \sigma(y-x) \\ \dot{y} &= rx - y - xz \\ \dot{z} &= -bz + xy. \end{aligned} \tag{3.2.3}$$

Here x, y, z represent the amplitude of certain basic modes of the fluid motion and σ, r, b are the parameters determined by the physical characteristics of the fluid.

He then carried out a computer simulation of the ordinary differential equation with the parameter values $\sigma = 10$, $r = 28$, $b = \frac{8}{3}$ and found the picture of a solution as exhibited in Fig. 3.2.2. The initial condition is chosen very close to the origin, but only the long term behavior is depicted in Fig. 3.2.2 by omitting initial transient portion of the orbit. The solution seems making left or right turns around two holes in a very irregular way. Moreover, this computer-generated picture looks robust in the sense that the picture of the solution does not appear to change with a slight variation of the initial conditions or the parameter values. Therefore one may think this solution must give a robust attractive object.

Since then, this equation has been known as one of the typical examples of the complicated behavior in dynamical systems, now named "chaos" or a "strange at-

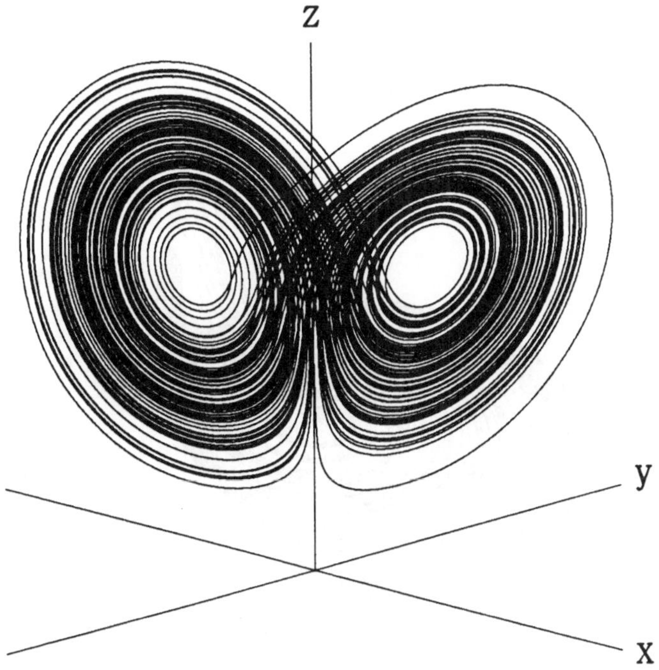

Fig. 3.2.2. The Lorenz attractor.

tractor". Equation (3.2.3) is called the Lorenz equation, and the solution depicted in Fig. 3.2.2 is called the Lorenz attractor.

It has been shown that the Lorenz equation defines a complete vector field (c.f. [Coomes 1989]) and hence it generates a smooth (even real analytic) flow on $\mathbb{R}^3$, though it is not explicitly solved analytically. Moreover, the equation itself does not change its expression if we replace the variables (x, y, z) with $(-x, -y, z)$, and hence the vector field admits a certain "symmetry". Namely, the vector field given by the Lorenz equation is symmetric under the rotation around the z-axis with the angle π. This fact explains the somewhat symmetric nature of Fig. 3.2.2, although the picture itself is not completely symmetric. See Subsection 3.3.5 for more information about the symmetry.

Example 3.2.6. (Rössler Equation) Another ordinary differential equation which we believe exhibits a complicated yet robust solution is the equation given by

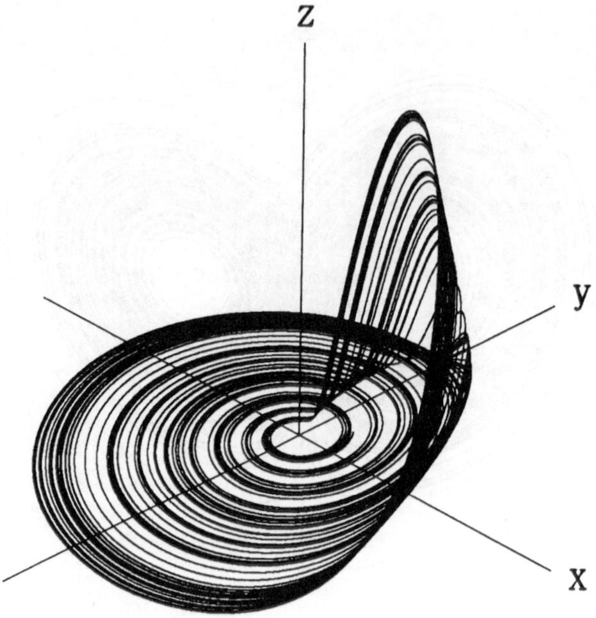

(a)

Fig. 3.2.3. The Rössler attractors. (a) A spiral attractor.

$$\begin{aligned}\dot{x} &= -y - z \\ \dot{y} &= x + ay \\ \dot{z} &= bx - cz + xz,\end{aligned} \tag{3.2.4}$$

with the parameters a, b, c. This equation was originally introduced by [Rössler 1979b]; it is hence called the Rössler equation.

For $(a, b, c) = (0.36, 0.4, 4.5)$, a computer simulation gives the picture in Fig. 3.2.3(a), while for $(a, b, c) = (0.5, 0.4, 4.5)$, it gives Fig. 3.2.3(b), both of which are thought of as "strange attractors", since, like the Lorenz attractor, the shape of the "strange attractors" does not seem to change with the slight variation of initial conditions, which means that it might "attract" nearby points. See Subsection 3.2.2 for the definition of an attractor. Furthermore, these attractors are seemingly persistent as well with a small perturbation of the parameters, although they are gradually deformed by largely changing the parameters from $(a, b, c) = (0.36, 0.4, 4.5)$ to $(a, b, c) = (0.5, 0.4, 4.5)$.

We remark that, unlike the Lorenz equation, the Rössler equation does not have any physical background from which the equation is derived.

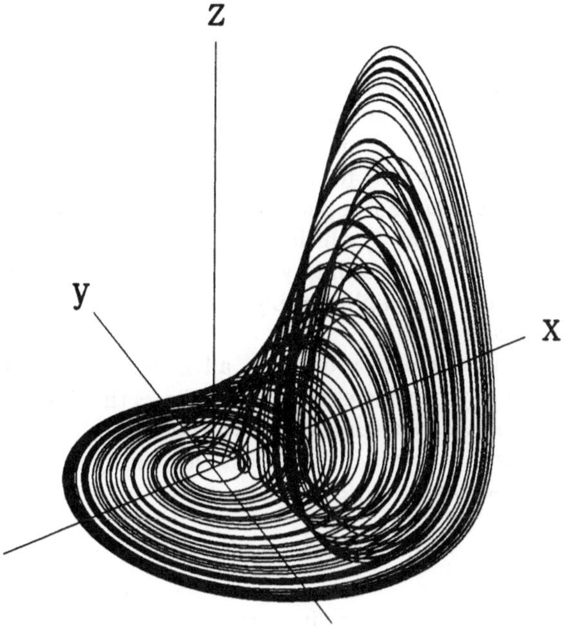

(b)

Fig. 3.2.3. continued (b) A screw attractor.

The Double Scroll circuit gives another example of continuous dynamical systems on $\mathbb{R}^3$, which is studied extensively in Chapters 1 and 2. There are, of course, many other interesting ordinary differential equations. In general, those equations do not always define flows on the whole space as we remarked before. Nevertheless, we are sometimes interested in just a part of the phase space where the vector field under consideration is defined. In such a case we can modify the vector field outside the interesting part so that it defines a flow on the whole space without changing the important part. In this way we can speak about a local dynamical system $\{\varphi^t\}$, which is only defined on an open subset in $\mathbb{R}^n$ and/or for t varying a part of the entire real line, although we can extend it suitably to a complete flow on $\mathbb{R}^n$. In what follows, we shall not distinguish such a local dynamical system and its extended (complete) dynamical system.

3.2.2 Orbits and Invariant Sets in Dynamical Systems

Let $\{\varphi^t\}$ be a flow on a manifold M. We call the set

$$\mathcal{O}(\xi) = \{\varphi^t(\xi) \mid t \in \mathbb{R}\}$$

the *orbit* of the flow through ξ. In particular, if an orbit $\mathcal{O}(\xi)$ is a point, then $\mathcal{O}(\xi) = \{\xi\}$, that is, $\varphi^t(\xi) = \xi$ for all $t \in \mathbb{R}$, and it is called an *equilibrium point* or a *singular point*. Equivalently, an equilibrium point ξ is a zero of the vector field generating the flow: $v(\xi) = 0$. If, on the other hand, an orbit $\mathcal{O}(\xi)$ is not a single point but compact, then it is easy to see that the orbit is homeomorphic to a circle and is called a *periodic orbit*. For a periodic orbit, there exists a minimal number $T > 0$ with $\varphi^T(\xi) = \xi$ for any ξ, which is called the *period* of the periodic orbit. The uniqueness of solutions to ordinary differential equations then implies that $\varphi^{t+T}(\xi) = \varphi^t(\xi)$ for any t. Equilibrium points and periodic orbits are sometimes called *elementary orbits.*

Orbits other than elementary ones are all non-compact, and among them, there can be an orbit approaching elementary orbits as the time goes to $\pm\infty$. Such an orbit is referred to as a *homoclinic* or *heteroclinic orbit.* To be more explicit, an orbit $\mathcal{O}(\xi)$ is a heteroclinic orbit if there exist elementary orbits $\mathcal{E}_+$ and $\mathcal{E}_-$ such that $\mathrm{dist}(\varphi^t(\xi), \mathcal{E}_\pm) \to 0$ as $t \to \pm\infty$, where

$$\mathrm{dist}(x, A) = \inf_{a \in A} \|x - a\|.$$

If $\mathcal{E}_+ = \mathcal{E}_-$, then it is called a homoclinic orbit.

Example 3.2.7.

(1) The origin O is an equilibrium point of a linear vector field. An equilibrium point of a low-dimensional linear vector field sometimes has a special name: for instance, a saddle refers to an equilibrium point where the coefficient matrix is hyperbolic and has eigenvalues with both positive and negative real parts. In two dimensions, a saddle has one positive and one negative eigenvalues. A node refers to an equilibrium point with real eigenvalues of the same sign, whereas a focus is the one with complex conjugate eigenvalues. In three dimensions, if a saddle has a complex eigenvalue, then it is called a saddle-focus. Otherwise it may be called a real saddle.

(2) The linear vector field

$$\begin{aligned} \dot{x} &= -y \\ \dot{y} &= x \end{aligned} \tag{3.2.5}$$

has infinitely many periodic orbits, since the equation can be put into

$$\begin{aligned} \dot{r} &= 0 \\ \dot{\theta} &= 1 \end{aligned}$$

by taking the polar coordinates. Hence every orbit is a concentric circle except the center O (see Fig. 3.2.4(a)). The period of the periodic orbits is the same, namely, 2π. A slightly different nonlinear vector field

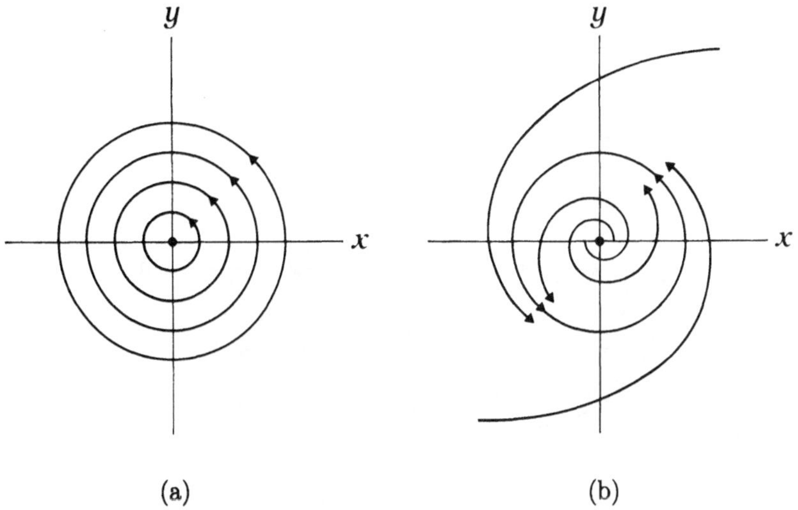

Fig. 3.2.4. The phase portraits of vector fields in Example 3.2.7(2). (a) Infinitely many periodic orbits. (b) One attracting periodic orbit.

$$\dot{r} = r - r^3$$
$$\dot{\theta} = 1,$$

or in Cartesian coordinates,

$$\dot{x} = x - y - x(x^2 + y^2)$$
$$\dot{y} = x + y - y(x^2 + y^2)$$

has, on the contrary, only one periodic orbit corresponding to the unit circle $r = 1$ and any other orbits except O converge to the periodic orbit as $t \to +\infty$ (see Fig. 3.2.4(b)). In particular, an orbit with the initial condition satisfying $0 < r < 1$ is a heteroclinic orbit connecting the equilibrium point O and the periodic orbit.

(3) The vector field

$$\dot{x} = y$$
$$\dot{y} = x^2 - 1$$

has an equilibrium point $E = (1, 0)$ and a homoclinic orbit to E. This can be seen by observing that the vector field is written as

$$\begin{aligned} \dot{x} &= \frac{\partial H}{\partial y}(x, y) \\ \dot{y} &= -\frac{\partial H}{\partial x}(x, y) \end{aligned} \qquad (3.2.6)$$

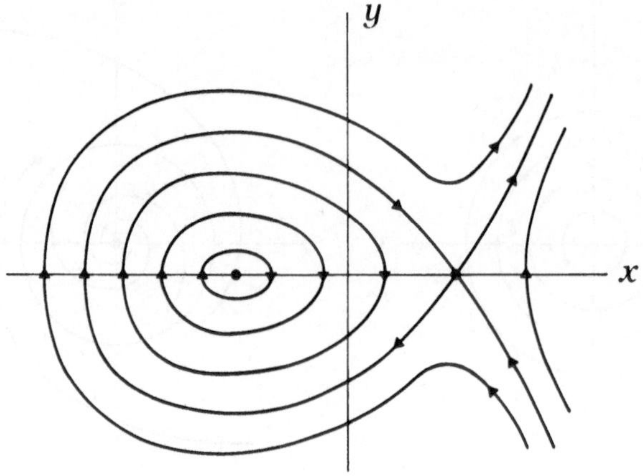

Fig. 3.2.5. A vector field with a homoclinic orbit.

with the function

$$H(x,y) = \frac{y^2}{2} - \frac{x^3}{3} + x,$$

which is invariant under the flow since

$$\frac{d}{dt}H(x,y) = \frac{\partial H}{\partial x}(x,y)\dot{x} + \frac{\partial H}{\partial y}(x,y)\dot{y} = 0$$

identically along orbits. Therefore, the set $H^{-1}(H(E)) = H^{-1}(\frac{2}{3})$ contains a homoclinic orbit converging to E as $t \to \pm\infty$, as depicted in Fig. 3.2.5. The interior of the region surrounded by the homoclinic orbit is filled with periodic orbits $H^{-1}(h)$ for $h < \frac{2}{3}$. A vector field with such an invariant function is sometimes called *conservative*. In particular, the vector field given by Equation (3.2.6) is called a Hamiltonian vector field with the Hamiltonian function H.

(4) According to a computer simulation (c.f. [Sparrow 1982]), the Lorenz equation possesses various kinds of periodic orbits, homoclinic orbits, and heteroclinic orbits for suitable parameter values. The existence of certain types of homoclinic orbits for the Double Scroll circuit is rigorously studied in Chapter 1 (Subsection 1.5.4).

A set S in M is an *invariant set* for a flow $\{\varphi^t\}$ if $\varphi^t(S) = S$ for all $t \in \mathbb{R}$. If $\varphi^t(S) \subset S$ for positive (respectively negative) t, it is called a *positively* (respectively *negatively*) *invariant set.* Orbits are invariant sets and hence so is a collection of orbits. Conversely, an invariant set is nothing but a collection of orbits in the set.

One of the most important invariant sets is an attractor: an invariant set $\mathcal{A}$ is an *attracting set* if there exists a neighborhood U of $\mathcal{A}$ such that any $\xi \in U$ satisfies $\text{dist}(\varphi^t(\xi), \mathcal{A}) \to 0$ as $t \to +\infty$. If, furthermore, it contains an orbit $\mathcal{O} \subset \mathcal{A}$ which is dense in $\mathcal{A}$, that is, $\overline{\mathcal{O}} = \mathcal{A}$, then it is called an *attractor*. The orbit $\mathcal{O}$ is called a dense orbit of the attractor. The above condition is sometimes referred to as the irreducibility of an attractor or the transitivity of the dynamics on the attractor. See also Subsection 3.4.5. The maximal open set U satisfying the above property is called the *basin* of the attractor $\mathcal{A}$, which is again an invariant set.

Example 3.2.8. For the Lorenz equation, consider the function

$$V(x, y, z) = rx^2 + \sigma y^2 + \sigma(z - 2r)^2$$

and compute its derivative along the orbit, namely,

$$\begin{aligned}\frac{dV}{dt} &= 2\{rx\dot{x} + \sigma y\dot{y} + \sigma(z - 2r)\dot{z}\} \\ &= 2\{rx \cdot \sigma(y - x) + \sigma y \cdot (rx - y - xz) + \sigma(z - 2r) \cdot (-bz + xy)\} \\ &= -2\sigma\{rx^2 + y^2 + b(z - r)^2 - br^2\}.\end{aligned}$$

This implies that the region given by $\left\{(x, y, z) | \frac{dV}{dt} \geq 0\right\}$ defines an ellipsoid. Let $M > 0$ be the maximal value of V on the ellipsoid. Since, by definition, $\frac{dV}{dt}$ takes a negative value outside the ellipsoid, the vector field given by the Lorenz equation points inward on the surface given by $V(x, y, z) = M + \varepsilon$ with $\varepsilon > 0$. This shows that the Lorenz equation admits a positively invariant bounded set

$$U = \{(x, y, z) \mid V(x, y, z) \leq M + \varepsilon\},$$

and therefore

$$\Lambda = \bigcap_{t \geq 0} \varphi^t(U)$$

defines an invariant attracting set where φ^t is the flow generated by the Lorenz equation. It may not be an attractor but definitely it contains, by its construction, all the bounded invariant sets, including the irregular motion exhibited by the computer generated picture of the Lorenz attractor in Fig. 3.2.2. See, for instance, [Sparrow 1982] for more detail. Furthermore, the set Λ is of zero volume in $\mathbb{R}^3$ since the divergence of the vector field given by the Lorenz equation equals $-(\sigma + b + 1)$, a negative constant. In fact, from the well-known general formula, we have

$$\frac{d}{dt}\text{vol}(\varphi^t(U)) = -(\sigma + b + 1) \cdot \text{vol}(\varphi^t(U)),$$

where $\text{vol}(U)$ stands for the volume of the set $U \subset \mathbb{R}^3$. This implies

$$\text{vol}(\Lambda) = \lim_{t \to +\infty} \text{vol}(\varphi^t(U)) = \lim_{t \to +\infty} e^{-(\sigma + b + 1)t} \cdot \text{vol}(U) = 0.$$

Any complicated behavior of the equation is thus contained in this very "thin" invariant set.

Another important invariant set is the limit set.

Definition 3.2.9. For a flow $\{\varphi^t\}$ on M, the *ω-limit set* for a point ξ is the set defined by

$$\omega(\xi) = \left\{ x \mid x = \lim_{n\to\infty} \varphi^{t_n}(\xi) \text{ for some } \{t_n\} \text{ with } t_n \to +\infty \right\}.$$

Similarly, we define the *α-limit set* for a point ξ as

$$\alpha(\xi) = \left\{ x \mid x = \lim_{n\to-\infty} \varphi^{t_n}(\xi) \text{ for some } \{t_n\} \text{ with } t_n \to -\infty \right\}.$$

The *limit set* for a point is just the union of the α- and ω-limit sets. Limit sets are also defined for M as follows:

$$\omega(M) = \overline{\bigcup_{\xi\in M} \omega(\xi)},$$

$$\alpha(M) = \overline{\bigcup_{\xi\in M} \alpha(\xi)}.$$

In particular, the *limit set* for the flow refers to the set $L(\{\varphi^t\}) = \alpha(M) \cup \omega(M)$.

Remark 3.2.10. The term "limit set" is used with a different meaning in literature, particularly for so-called dissipative systems. Interested readers can find the details in e.g. [Hale 1988].

An equilibrium point and a periodic orbit are the limit set of a point in itself. An attractor is the limit set of a point in its neighborhood. Another example of a limit set is the following:

Example 3.2.11. Consider a planar vector field which has a homoclinic orbit to a saddle equilibrium point. Assume that orbits inside the region D surrounded by the homoclinic orbit approach the homoclinic orbit except a focus equilibrium point inside, as shown in Fig. 3.2.6. Then clearly the homolinic orbit, together with the saddle point to which the homoclinic orbit is asymptotic, forms an ω-limit set of a point in D different from the focus, and the focus itself is an α-limit set of those points.

Vector fields on the plane possess a very remarkable property concerning their limit sets. Any bounded component of the limit set of a point for a vector field on $\mathbb{R}^2$ is either

- an isolated equilibrium point,
- a periodic orbit, or

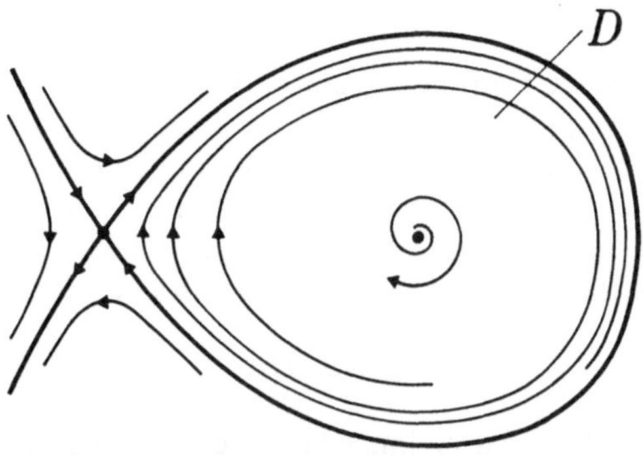

Fig. 3.2.6. A homoclinic orbit as a limit set.

- a set consisting of equilibrium points together with homoclinic or heteroclinic orbits asymptotic to these equilibrium points.

This result is known as the Poincaré-Bendixson theorem. This fact essentially follows from the Jordan curve theorem, which is only valid on the plane. See [Andronov, Gordon, Leontvich and Maier 1973a,b] and [Perko 1991] for the proof as well as more information on dynamical systems on the plane.

Before closing this subsection, we shall summarize some notions on the stability of equilibrium points.

Definition 3.2.12.

(1) An equilibrium point E of a vector field $\dot{x} = v(x)$ is said to be *Lyapunov stable* if, for any neighborhood V of the equilibrium point, there exists a (possibly smaller) neighborhood U such that any $x \in U$ satisfies $\varphi^t(x) \in V$ for any $t \geq 0$, where $\{\varphi^t\}$ is the flow generated by the vector field v.

(2) An equilibrium point E of a vector field $\dot{x} = v(x)$ is said to be (asymptotically) *stable* if it is Lyapunov stable and moreover, for any $x \in U$, $\varphi^t(x)$ converges to E as $t \to +\infty$.

(3) An equilibrium point E of a vector field $\dot{x} = v(x)$ is said to be *unstable* if it is not Lyapunov stable.

Lyapunov stable equilibrium point is not always (asymptotically) stable. For instance, the origin of the vector field (3.2.5) is Lyapunov stable but not (asymptotically) stable.

3.2.3 Linearization at Equilibrium Points and the Theorem of Hartman-Grobman

Let v be a smooth vector field on $\mathbb{R}^n$ and O be its equilibrium point. In other words, v is a smooth mapping from $\mathbb{R}^n$ into itself and O is a zero of the mapping; $v(O) = 0$. Then the derivative of v at O, denoted by $Dv(O)$, can be thought of as a linear vector field on the tangent space $T_O\mathbb{R}^n$, which is identified with $\mathbb{R}^n$ itself. This linear vector field is called the *linearized vector field* of v at the equilibrium point O. It is equivalently a linear ordinary differential equation

$$\dot{x} = Dv(O)x.$$

The notion of linearization can be immediately generalized to a vector field on a manifold M by taking a local chart around an equilibrium point.

Example 3.2.13. The Lorenz equation (3.2.3) has an equilibrium point at O where the linearization matrix is given by

$$\begin{pmatrix} -\sigma & \sigma & 0 \\ r & -1 & 0 \\ 0 & 0 & -b \end{pmatrix},$$

and hence its characteristic polynomial becomes

$$(\lambda + b)\{\lambda^2 + (\sigma + 1)\lambda + \sigma(1 - r)\} = 0.$$

Therefore the origin is a real saddle of the Lorenz equation if $\sigma > 0, r > 1, b > 0$. Similarly, we can show that the Rössler equation has a saddle-focus at the origin for the parameter values $(a, b, c) = (0.36, 0.4, 4.5)$ and $(a, b, c) = (0.5, 0.4, 4.5)$.

The following Proposition describes the relation between the stability of an equilibrium point in a vector field and that of its linearization, and provides us with an easy criterion for the stability of equilibrium points.

Proposition 3.2.14. *An equilibrium point O of a vector field $\dot{x} = v(x)$ on $\mathbb{R}^n$ is (asymptotically) stable if all the eigenvalues of the linearization $Dv(O)$ at the equilibrium point have negative real parts.*

Proof. Let $x(t)$ be a solution of $\dot{x} = v(x)$ with an initial condition $x(0)$ sufficiently close to the equilibrium point O and put $y(t) = x(t) - O$. Then it satisfies

$$\dot{y}(t) = Ay(t) + h(y(t)), \tag{3.2.7}$$

where $A = Dv(O)$ and $h(y) = v(y + O) - Ay$.

From the assumption on $A = Dv(O)$, there exists some constant $-\alpha < 0$ and $C > 0$ such that

$$|e^{At}| \leq Ce^{-\alpha t}.$$

Then since $h(y) = o(y)$ as $y \to 0$ from the definition, we can make

$$|h(y)| \leq \varepsilon |y|$$

for sufficiently small $|y|$ and ε. Here we take $\varepsilon < \alpha/C$ for later use.

The variation of constants formula applied to Equation (3.2.7) yields

$$y(t) = e^{At}y(0) + \int_0^t e^{A(t-s)}h(y(s))ds,$$

and hence, together with the above estimates, we have

$$\begin{aligned} |y(t)| &\leq |e^{At}y(0)| + \int_0^t |e^{A(t-s)}h(y(s))|ds \\ &\leq Ce^{-\alpha t}|y(0)| + \int_0^t Ce^{-\alpha(t-s)}\varepsilon|y(s)|ds, \end{aligned}$$

which is equivalent to

$$e^{\alpha t}|y(t)| \leq C|y(0)| + \int_0^t C\varepsilon \cdot e^{\alpha s}|y(s)|ds.$$

Then we apply the Gronwall technique ([Coddington and Levinson 1955]) to this inequality and obtain

$$e^{\alpha t}|y(t)| \leq C|y(0)|e^{C\varepsilon t},$$

and thus

$$|y(t)| \leq Ce^{(-\alpha+C\varepsilon)t}|y(0)|.$$

Since ε is chosen small so that the exponent $-\alpha + C\varepsilon < 0$, we have $|y(t)| = |x(t) - O| \to 0$ as $t \to +\infty$, which shows the desired asymptotic stability of the equilibrium point O. □

We note that asymptotic stability is not preserved under perturbation, though so is the assumption of Proposition 3.2.14. In particular, the converse of Proposition 3.2.14 is not true. A counterexample is given by a one-dimensional vector field $\dot{x} = -x^3$. This is perturbed to $\dot{x} = \mu x - x^3$ with small parameter μ, in which the origin is unstable for $\mu > 0$. See Fig. 3.2.7.

An equilibrium point of a (nonlinear) vector field is called *hyperbolic* if its linearized vector field at the equilibrium point is hyperbolic. The hyperbolicity is one of the most important notions in the theory of dynamical systems, and the next theorem ([Hartman 1960], [Grobman 1959]) gives us the first evidence of its importance.

Theorem 3.2.15. (Hartman-Grobman) *A vector field around a hyperbolic equilibrium point is locally topologically equivalent to the linearized vector field at the*

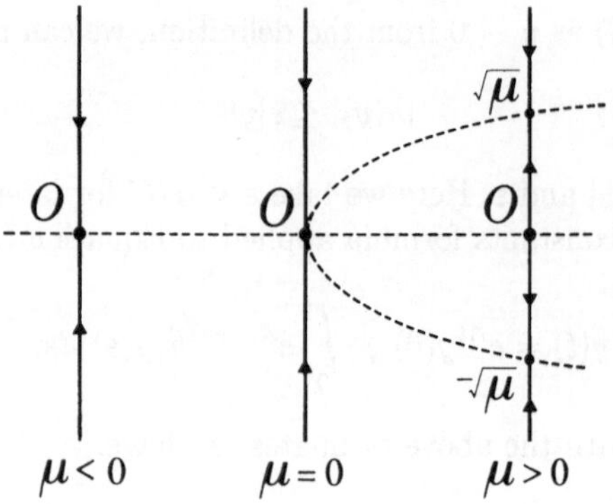

Fig. 3.2.7. The phase portrait of $\dot{x} = \mu x - x^3$.

equilibrium point. Namely, letting v be a vector field on $\mathbb{R}^n$ and O be its hyperbolic equilibrium point, there exists a neighborhood U of O and a homeomorphism H from U into $\mathbb{R}^n$ such that any orbit of v in U is sent to an orbit of the linearized vector field $Dv(O)$ in $H(U)$ with the direction of orbits preserved.

This means that the structure of orbits of the vector field in a neighborhood of the hyperbolic equilibrium point is qualitatively the same by means of homeomorphism as the structure of the orbits in its linearization.

Proof. The vector field $\dot{x} = v(x)$ in a neighborhood U of the origin O is decomposed into its linear part and essentially nonlinear part:

$$\dot{x} = Ax + h(x),$$

where $A = Dv(O)$ and $|h(x)| \le (\text{const}) \cdot |x|^2$ on U. Let $\{\varphi^t\}$ be the flow generated by the vector field, and $\{\psi^t\}$ be the flow generated by the linearized vector field $\dot{x} = Ax$, namely, $\psi^t(x) = e^{At}x$. Then it is easy to see that

$$D\varphi^t(O) = e^{At} = \psi^t.$$

In other words, $\{\psi^t\}$ is the linearization of the flow $\{\varphi^t\}$.

Fixing t_0 arbitrarily, we try to find a homeomorphism H, independent of t_0, satisfying

$$\varphi^{t_0} \circ H = H \circ \psi^{t_0}. \tag{3.2.8}$$

Let us give the name $G = e^{At_0} = \psi^{t_0}$ and put

$$\varphi^{t_0} = G + u$$

with $\| u \| = \sup_{x \in U} |u(x)|$ being arbitrarily small by making the neighborhood U small enough. We can assume the homeomorphism H to be of the form

$$H = Id + \eta$$

with $\| \eta \| < +\infty$, and from this, we can rewrite Equation (3.2.8) as

$$(G + u) \circ (Id + \eta) = (Id + \eta) \circ G,$$

which is equivalent to

$$G \circ \eta \circ G^{-1} + u \circ (G^{-1} + \eta \circ G^{-1}) = \eta.$$

We regard this expression as a fixed point equation, namely, we define the mapping

$$\mathcal{F}(\eta) = T\eta + \delta(\eta),$$

where

$$\begin{aligned} T\eta &= G \circ \eta \circ G^{-1} \\ \delta(\eta) &= u \circ (G^{-1} + \eta \circ G^{-1}), \end{aligned}$$

and solve the equation $\mathcal{F}(\eta) = \eta$. The map $\mathcal{F}$ is well-defined on the Banach space $B(U, \mathbb{R}^n)$ of all continuous bounded functions from U into $\mathbb{R}^n$ endowed with the supremum norm $\| \cdot \|$. Furthermore, from the assumption that the matrix A is hyperbolic, the linear map T expands $B(U, E^u)$ and contracts $B(U, E^s)$ with respect to the decomposition

$$B(U, \mathbb{R}^n) = B(U, E^u) \oplus B(U, E^s),$$

where E^u (respectively E^s) is the unstable (respectively stable) eigenspace of the matrix A. To be more precise, for any $\eta^u \in B(U, E^u)$ and $\eta^s \in B(U, E^s)$, it holds that

$$\| T\eta^u \| \geq \rho \| \eta^u \|, \quad \| T^{-1}\eta^s \| \geq \rho \| \eta^s \|$$

for some $\rho > 1$. Such a linear map T is also called hyperbolic. The remainder term $\delta(\eta)$, on the other hand, is a Lipschitz continuous map of η with an arbitrarily small Lipschitz constant if we make the neighborhood U small enough.

Lemma 3.2.16. *Let $T : B \to B$ be a hyperbolic linear map on a Banach space B and $\delta : B \to B$ be a Lipschitz mapping with a sufficiently small Lipschitz constant. Then the mapping $T + \delta : B \to B$ has a unique fixed point η_*, namely, $(T + \delta)(\eta_*) = \eta_*$.*

From this lemma, we show that the equation $\mathcal{F}(\eta) = T\eta + \delta(\eta) = \eta$ has a unique fixed point η_* in the space $B(U, \mathbb{R}^n)$, and hence $H = Id + \eta_*$ satisfies (3.2.8). It is not hard to show that H is a homeomorphism, since the fixed point η_* has a small Lipschitz constant. The proof of this lemma can be found in [Irwin 1980].

Finally we need to show that the homeomorphism $H = Id + \eta_*$ does not depend on the choice of t_0. We may first choose $t_0 = 1$. Then the above argument shows that there exists a homeomorphism H_1 such that

$$\varphi^1 \circ H_1 = H_1 \circ \psi^1$$

holds. Now let

$$H = \int_0^1 \varphi^{-s} \circ H_1 \circ \psi^s ds$$

and we claim $\varphi^t \circ H = H \circ \psi^t$ for any t. In fact

$$\begin{aligned}
\varphi^t \circ H &= \varphi^t \circ \left(\int_0^1 \varphi^{-s} \circ H_1 \circ \psi^s ds \right) \\
&= \left(\int_0^1 \varphi^{t-s} \circ H_1 \circ \psi^{s-t} ds \right) \circ \psi^t \\
&= \left(\int_{-t}^{1-t} \varphi^{-u} \circ H_1 \circ \psi^u du \right) \circ \psi^t \\
&= \left(\int_{-t}^{0} \varphi^{-u} \circ H_1 \circ \psi^u du + \int_0^{1-t} \varphi^{-u} \circ H_1 \circ \psi^u du \right) \circ \psi^t,
\end{aligned}$$

where the first integral term can be changed as

$$\begin{aligned}
\int_{-t}^{0} \varphi^{-u} \circ H_1 \circ \psi^u du &= \int_{-t}^{0} \varphi^{-u} \circ (\varphi^{-1} \circ H_1 \circ \psi^1) \circ \psi^u du \\
&= \int_{-t}^{0} \varphi^{-u-1} \circ H_1 \circ \psi^{u+1} du \\
&= \int_{1-t}^{1} \varphi^{-v} \circ H_1 \circ \psi^v dv.
\end{aligned}$$

Therefore, we have

$$\varphi^t \circ H = \left(\int_0^1 \varphi^{-v} \circ H_1 \circ \psi^v dv \right) \circ \psi^t = H \circ \psi^t.$$

The mapping H is clearly a homeomorphism, and thus the proof of the Hartman-Grobman theorem is completed. □

In general there is no hope for the mapping H to be a diffeomorphism, as shown in the following example.

Example 3.2.17. The following ordinary differential equation cannot be changed to its linearization by a smooth coordinate transformation:

$$\begin{aligned}
\dot{x} &= \alpha x \\
\dot{y} &= (\alpha - \gamma) y + \varepsilon x z \\
\dot{z} &= -\gamma z,
\end{aligned} \tag{3.2.9}$$

where $\alpha > \gamma > 0$ and $\varepsilon \neq 0$. The proof of non-existence of the smooth linearization of this equation is a little involved and we refer to [Hartman 1960] for the proof. However, this can be topologically linearized due to the Hartman-Grobman Theorem, and in this case, we can even give the explicit form of the homeomorphism $(x, y, z) \mapsto (x', y', z')$ which linearizes this example:

$$\begin{aligned} x &= x' \\ y &= y' - \tfrac{\varepsilon}{\gamma} x' z' \log |z'| \\ z &= z'. \end{aligned}$$

Observe that this transformation is extended to a homeomorphism on $\mathbb{R}^3$ but is not differentiable at $z' = 0$.

Despite the general non-existence of smooth linearization, it is sometimes possible to obtain an equivalence by a smooth coordinate transformation between a vector field near a hyperbolic equilibrium point and its linearization, provided we assume certain extra conditions on the coefficient matrix of the linearization. We give below three theorems of this kind.

Theorem 3.2.18. (Siegel) *Let*

$$\dot{x} = v(x) \tag{3.2.10}$$

be a real-analytic vector field defined on a neighborhood of the origin O with $v(O) = 0$. Suppose the linearization matrix $Dv(O)$ has eigenvalues λ_i $(i = 1, \cdots, n)$, which satisfy the following condition:

$$\exists \gamma > 0 \ \forall k \in \mathbb{Z}^n - \{0\}, \quad |(k, \lambda)| > \gamma |k|^{-n}, \tag{3.2.11}$$

where $k = (k_1, \cdots, k_n)$, $\lambda = (\lambda_1, \cdots, \lambda_n)$, $(k, \lambda) = k_1\lambda_1 + \cdots + k_n\lambda_n$, and $|k| = |k_1| + \cdots + |k_n|$. Then there exists a real-analytic transformation φ defined on a neighborhood of O such that it transforms the vector field (3.2.10) into its linearization $\dot{x} = Dv(O)x$. In other words, the vector field (3.2.10) is analytically linearizable under the assumption (3.2.11).

Theorem 3.2.19. (Sternberg) *Let*

$$\dot{x} = v(x) \tag{3.2.12}$$

be a C^∞ vector field defined on a neighborhood of the origin O with $v(O) = 0$. Suppose the linearization matrix $Dv(O)$ has eigenvalues λ_i $(i = 1, \cdots, n)$, which satisfy the following condition:

$$\forall i \forall k \in \mathbb{Z}^n - \{0\} \text{ with } |k| \geq 2, \quad \lambda_i - (k, \lambda) \neq 0. \tag{3.2.13}$$

Then there exists a C^∞ transformation φ defined on a neighborhood of O such that it transforms the vector field (3.2.12) into its linearization $\dot{x} = Dv(O)x$. In *other words, the vector field (3.2.12) is smoothly linearizable under the assumption (3.2.13).*

Theorem 3.2.20. (Belitskii) *Let*

$$\dot{x} = v(x) \tag{3.2.14}$$

be a C^2 vector field defined on a neighborhood of the origin O with $v(O) = 0$. Suppose the linearization matrix $Dv(O)$ has eigenvalues λ_i $(i = 1, \cdots, n)$, which satisfy the following condition:

$$\lambda_i \neq \lambda_j + \lambda_k, \tag{3.2.15}$$

where $1 \leq i, j, k \leq n$. Then there exists a C^1 transformation φ defined on a neighborhood of O such that it transforms the vector field (3.2.14) into its linearization $\dot{x} = Dv(O)x$. In other words, the vector field (3.2.14) is C^1 linearizable under the assumption (3.2.15).

In particular, since any $\alpha, \beta \in \mathbb{R}$ with $\beta < 0 < \alpha$, or $\alpha \in \mathbb{R}$, $\beta \in \mathbb{C} \setminus \mathbb{R}$ and $\bar{\beta}$ with $Re\beta < 0 < \alpha$ satisfy the condition (3.2.15), we can conclude from the Belitskii's theorem that a saddle in a two-dimensional vector field and a saddle-focus in a three-dimensional vector field are always C^1 linearizable. Note that the C^1 linearizability of two-dimensional vector fields near a saddle has already been given by [Hartman 1960]. See also [Stowe 1986].

Here we do not give any proofs for these three theorems, since they are very involved. Note that the proof of Siegel's theorem is analogous to the very deep perturbation theory for completely integrable Hamiltonian systems which is called the Kolmogorov-Arnold-Moser theory. Note also that Sternberg's condition (3.2.13) is nothing but the condition for the formal linearization of a formal vector field given by its Taylor expansion. For more information as well as proofs, see [Arnold 1983], [Sternberg 1958], and [Belitskii 1977].

3.2.4 Stable and Unstable Manifolds

According to the Hartman-Grobman theorem explained in the previous subsection, the qualitative nature of the orbits of a vector field near a hyperbolic equilibrium point is the same as that of its linearized vector field, and hence some characteristic behaviors of hyperbolic linear vector fields are inherited by nonlinear vector fields. The stable and unstable eigenspaces are such characteristics of hyperbolic linear vector fields: recall that the stable eigenspace E^s (respectively unstable eigenspace E^u) for a linear vector field $\dot{x} = Ax$ is a subspace spanned by (generalized) eigenvectors associated with eigenvalues with negative (respectively positive) real parts. By definition, the whole space $\mathbb{R}^n$ coincides with the direct product $E^s \oplus E^u$, and orbits starting at points in E^s (respectively E^u) converge to O as the time goes to $+\infty$ (respectively $-\infty$). Conversely, the stable (respectively unstable) subspace

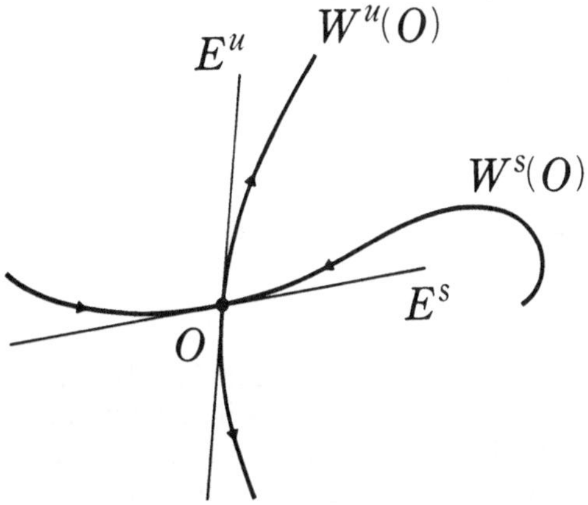

Fig. 3.2.8. The stable and unstable manifolds.

is characterized as the subspace consisting of the initial conditions of orbits which converge to O as $t \to +\infty$ (respectively $-\infty$).

These two subspaces describe typical saddle-like behavior of hyperbolic linear vector fields, which is inherited by nonlinear vector fields near hyperbolic equilibrium points. For a hyperbolic equilibrium point O of a vector field on M, there exist the stable and unstable sets defined by

$$\begin{aligned} W^s(O) &= \{x \in M \mid \lim_{t \to \infty} \varphi^t(x) = O\} \\ W^u(O) &= \{x \in M \mid \lim_{t \to -\infty} \varphi^t(x) = O\}, \end{aligned}$$

where $\{\varphi^t\}$ is the flow generated by the vector field. These sets are in correspondence with E^s and E^u, respectively, at least locally by means of the homeomorphism due to the Hartman-Grobman theorem. Surprisingly, however, these sets are not only the homeomorphic images of E^s and E^u but are nicely immersed global smooth submanifolds as shown in the next theorem.

Theorem 3.2.21. (Stable Manifold Theorem) *For a hyperbolic equilibrium point O of a C^r flow $\{\varphi^t\}$ on M, the set $W^s(O)$ is a C^r injectively immersed submanifold in M with $T_O W^s(O) = E^s$, where E^s is the stable eigenspace of the linearization.*

The same holds for $W^u(O)$ as well. The set $W^s(O)$ (respectively $W^u(O)$) is hence called the *stable* (respectively *unstable*) *manifold* of the hyperbolic equilibrium point O (with respect to the flow). See Fig. 3.2.8.

The essential part of this theorem is in the following local version of the stable manifold theorem.

Theorem 3.2.22. *Let $\{\varphi^t\}$ be a C^r flow on M and O be its hyperbolic equilibrium point. Let ε be a positive number and define*

$$W^s_{loc}(O) = \{x \in M \mid \mathrm{dist}(\varphi^t(x), O) < \varepsilon \ \ for\ \ \forall t > 0\}.$$

If ε is sufficiently small, then $W^s_{loc}(O)$ is a C^r embedded disc and is tangent to E^s at O. Furthermore it has the following characterization:

$$W^s_{loc}(O) = \{x \in M \mid \mathrm{dist}(\varphi^t(x), O) < Ce^{-\alpha t} \ \ for\ \ \forall t > 0\}$$

for some positive constants α and C.

The characterization of a local stable manifold tells us that an orbit staying in a sufficiently small neighborhood of a hyperbolic equilibrium point for all positive time in fact converges to the equilibrium point at an exponential manner. This is a very important and useful observation for the study of dynamics near hyperbolic equilibrium points.

Proof. Let us take the vector field of the form

$$\dot{x} = Ax + f(x), \qquad f(x) = O(|x|^2), \tag{3.2.16}$$

which generates the flow $\{\varphi^t\}$. Note that this is equivalent to taking the integral equation

$$x(t) = e^{At}x(0) + \int_0^t e^{A(t-s)} f(x(s))ds \tag{3.2.17}$$

due to the variation of constants formula.

The assumption that A is hyperbolic implies the decomposition of the phase space into its unstable and stable eigenspaces:

$$\mathbb{R}^n = E^u \oplus E^s$$

due to the splitting of the spectra of the linearization matrix A. Accordingly, we can take the projections

$$P^u : \mathbb{R}^n \to E^u, \qquad P^s : \mathbb{R}^n \to E^s,$$

with

$$Ker(P^u) = E^s, \qquad Ker(P^s) = E^u,$$

where the kernel $Ker(L)$ is the vector subspace consisting of vectors whose image by the linear map L goes to zero.

We modify the nonlinear term $f(x)$ in (3.2.16) outside a sufficiently small neighborhood U of the origin O in such a way that the modified nonlinear term, which we denote by the same notation $f(x)$, becomes bounded together with all the partial derivatives:

$$\sup_{x \in \mathbb{R}^n} |D^k f(x)| < +\infty \qquad (k = 0, 1, 2, \cdots, r),$$

and moreover, its C^1 norm $\sup_{x\in\mathbb{R}^n}(|f(x)|+|Df(x)|)$ becomes sufficiently small. This can be done because $f(x)=O(|x|^2)$. Since $f(x)$ is not changed in a fixed neighborhood U of O, the existence of the local stable manifold follows from the corresponding result with the modified nonlinear terms, and in fact, after this modification, we can take the global stable manifold as the graph of a function from E^s to $\mathbb{R}^n$, as we will see below.

Take an orbit $x(t)$ lying in the local stable set $W^s_{loc}(O)$ (not yet shown to be a manifold), namely,

$$|x(t)|<\varepsilon \quad \text{for } \forall t>0.$$

From (3.2.17), we have

$$e^{-At}x(t)=x(0)+\int_0^t e^{-As}f(x(s))ds,$$

to which we apply the projection P^u and get

$$e^{-At}P^u x(t)=P^u x(0)+\int_0^t e^{-As}P^u f(x(s))ds,$$

since P^u commutes with A, hence with e^{At}. Therefore, for $x(t)$ being in $W^s_{loc}(O)$, we obtain

$$P^u x(0)+\int_0^\infty e^{-As}P^u f(x(s))ds=0$$

by taking the limit $t\to+\infty$, which is used to rewrite (3.2.17) as

$$x(t)=e^{At}P^s x(0)+\int_0^t e^{A(t-s)}P^s f(x(s))ds-\int_t^\infty e^{A(t-s)}P^u f(x(s))ds. \quad (3.2.18)$$

Conversely, if a bounded function $x(t)$ $(t\geq 0)$ satisfies the above integral equation (3.2.18), it is easy to see that it is a solution of (3.2.16) with

$$\lim_{t\to\infty} x(t)=O,$$

and hence it lies in $W^s_{loc}(O)$. Thus we have shown the following:

Lemma 3.2.23. *A function $x(t)$ $(t\geq 0)$ is in $W^s_{loc}(O)$ if and only if it is a bounded function on $t\geq 0$ satisfying the integral equation (3.2.18).*

Now we consider the right-hand side of (3.2.18) as a mapping of $x(t)$, namely we define the mapping $\mathcal{F}$ as follows:

$$\mathcal{F}: B(\mathbb{R}_+,\mathbb{R}^n)\to B(\mathbb{R}_+,\mathbb{R}^n);$$
$$\mathcal{F}(x;\xi^s)(t)=e^{At}\xi^s+\int_0^t e^{A(t-s)}P^s f(x(s))ds-\int_t^\infty e^{A(t-s)}P^u f(x(s))ds$$

for each fixed $\xi^s \in E^s$, where $B(\mathbb{R}_+, \mathbb{R}^n)$ is the Banach space of all continuous bounded functions on $\mathbb{R}_+ = [0, +\infty)$ with the supremum norm $\| x \| = \sup_{t\geq 0}|x(t)|$. Our purpose is to find a fixed point $x = x_*(t;\xi^s)$ of the equation

$$\mathcal{F}(x_*(\bullet;\xi^s);\xi^s) = x_*(\bullet;\xi^s)$$

in the space $B(\mathbb{R}_+, \mathbb{R}^n)$. An elementary calculation using the hyperbolicity of A shows:

Lemma 3.2.24. *The mapping $\mathcal{F}$ is a contraction on a sufficiently small ball in the Banach space $B(\mathbb{R}_+, \mathbb{R}^n)$.*

Therefore, from the contraction mapping theorem (see [Berger 1977] for instance), we have a unique fixed point $x_*(t;\xi^s) \in B(\mathbb{R}_+, \mathbb{R}^n)$ for each $\xi^s \in E^s$. The fixed point defines a (C^0) submanifold W in $\mathbb{R}^n$ as the graph of the function $\xi^s \mapsto x_*(0;\xi^s)$, namely

$$W = \{x_*(0;\xi^s) \mid \xi^s \in E^s\}.$$

This submanifold is invariant under the flow and it must be the local stable manifold due to Lemma 3.2.23.

We furthermore need to show that the submanifold is of C^r and is tangent to E^s at O. Both of the assertions follow from the next lemma:

Lemma 3.2.25. *The function $x_*(0;\xi^s)$ is of C^r in $\xi^s \in E^s$. Moreover, its derivative vanishes at $\xi^s = 0$:*

$$\frac{\partial x_*}{\partial \xi^s}(0;0) = 0.$$

Proof. Recall that $x_*(t;\xi^s)$ satisfies the fixed point equation

$$x_*(\bullet;\xi^s) = \mathcal{F}(x_*(\bullet;\xi^s);\xi^s),$$

and observe that the functional $\mathcal{F}(x;\xi^s)$ is continuously differentiable with respect to $x \in B(\mathbb{R}_+, \mathbb{R}^n)$ and $\xi^s \in E^s$. Therefore we obtain

$$\frac{\partial x_*}{\partial \xi^s}(t;\xi^s) = \left(Id - \frac{\partial \mathcal{F}}{\partial x}(x_*;\xi^s)\right)^{-1} \frac{\partial \mathcal{F}}{\partial \xi^s}(x_*;\xi^s), \tag{3.2.19}$$

where the inverse of $Id - \frac{\partial \mathcal{F}}{\partial x}(x;\xi^s)$ exists because of $\frac{\partial \mathcal{F}}{\partial x}(0;0) = 0$ from $f(x) = O(|x|^2)$. This shows that $x_*(0;\xi^s)$ is continuously differentiable. Using the same argument repeatedly, we also show that $x_*(0;\xi^s)$ is as smooth as the original vector field. Moreover, the equality (3.2.19) implies

$$\frac{\partial x_*}{\partial \xi^s}(0;0) = 0,$$

which ends the proof of the lemma. □

Completion of the proof of Theorem 3.2.22. It remains to prove the characterization of the local stable manifold by an exponential decay. This simply follows from the estimate of the fixed point $x_*(t;\xi^s)$:

$$|x_*(t;\xi^s)| \leq \text{constant} \cdot e^{-\alpha t},$$

for some $\alpha > 0$, which is proved by replacing $x(t) = e^{-\alpha t}y(t)$ and by applying the same argument to $y(t) = e^{\alpha t}x(t)$, concluding that the fixed point $y_*(t;\xi^s)$ is a bounded function on $t \geq 0$, or equivalently $x_*(t;\xi^s)$ has the desired exponential decay of order $e^{-\alpha t}$. This completes the proof of Theorem 3.2.22. □

The (global) stable manifold is simply given from the local stable manifold as follows:

$$W^s(O) = \bigcup_{t>0} \varphi^{-t}(W^s_{loc}(O)).$$

Therefore the global stable manifold theorem is proved from this expression and the local stable manifold theorem.

3.2.5 Topological Equivalence and Structural Stability

Two ordinary differential equations are certainly thought of as the same when one is converted to the other by a change of variables. In other words, two vector fields v_1 and v_2 are *smoothly equivalent* if there exists a diffeomorphism h such that $h_*v_1 = v_2$ holds, where h_*v_1 is a transformed vector field from v_1 by h, that is,

$$h_*v_1(x) = Dh(h^{-1}(x)) \cdot v(h^{-1}(x))$$

in a coordinate expression. This is an equivalence relation among vector fields and one may try to classify them by this equivalence. It, however, immediately turns out that such classification is too fine: for instance, eigenvalues of the linearization at an equilibrium point are invariants for the smooth equivalence. On the other hand, we have already seen that local structure of orbits near a hyperbolic equilibrium point is well described by a coarser equivalence in terms of homeomorphisms preserving the orbit structure. Therefore it is natural to adopt the following notion of topological equivalence for the qualitative study of dynamical systems.

Definition 3.2.26. Two flows $\{\varphi_1^t\}$ and $\{\varphi_2^t\}$ are called *topologically equivalent* or C^0 *equivalent* if there exists a homeomorphism h such that it sends each orbit of $\{\varphi_1^t\}$ to an orbit of $\{\varphi_2^t\}$ with the direction of orbits preserved.

Example 3.2.27. Two linear vector fields v_1 and v_2 with the coefficient matrices

$$A_1 = \begin{pmatrix} 2 & 1 \\ 0 & 2 \end{pmatrix} \quad \text{and} \quad A_2 = \begin{pmatrix} 1 & -1 \\ 1 & 1 \end{pmatrix},$$

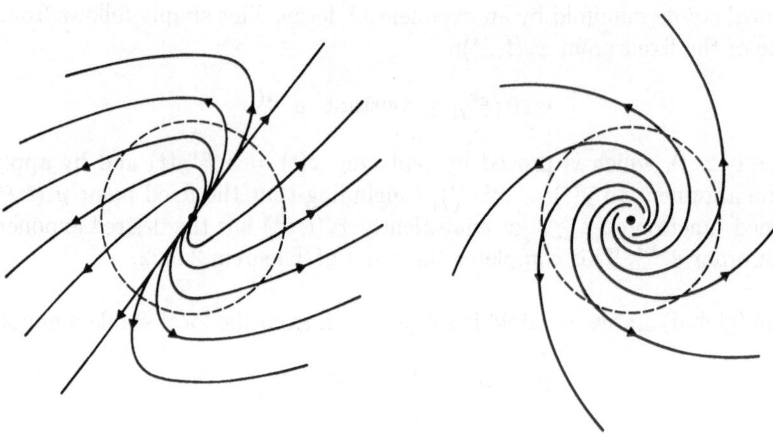

Fig. 3.2.9. The phase portraits of vector fields in Example 3.2.27. (a) $\dot{x} = A_1'x$. (b) $\dot{x} = A_2x$.

respectively, are topologically equivalent but not smoothly equivalent since the eigenvalues are different.

In order to construct a topological equivalence, first we notice that A_1 is linearly conjugate to

$$A_1' = \begin{pmatrix} 2 & \varepsilon \\ 0 & 2 \end{pmatrix},$$

where ε is arbitrarily small but positive, by which the topological equivalence is of course preserved. Then we take the unit circle in the plane $\mathbb{R}^2$ and show that every orbit of both $\dot{x} = A_1'x$ and $\dot{x} = A_2x$ except the origin intersects transversely with the circle precisely once. Indeed it is clear for $\dot{x} = A_2x$ by taking the polar coordinates. For $\dot{x} = A_1'x$, one sees the transverse intersection of orbits with the unit circle when $\varepsilon = 0$, and hence it persists for sufficiently small ε. See Fig. 3.2.9. Finally, the desired homeomorphism is obtained by extending the correspondence of the orbits on the unit circle to the whole plane together with the origin.

The above argument can be generalized to arbitrary hyperbolic linear vector fields as shown in the following

Proposition 3.2.28. *Two hyperbolic linear vector fields on* $\mathbb{R}^n$ *are topologically equivalent if and only if their stable (and hence unstable) eigenspaces are of equal dimension.*

Proof.
($\Longrightarrow$) If there exists a homeomorphism which gives a topological equivalence for two hyperbolic linear vector fields, then the homeomorphism preserves the stable and unstable subspaces, respectively, and hence the corresponding stable (unstable) subspaces for the vector fields must be of equal dimension.
($\Longleftarrow$) For two hyperbolic linear vector fields, if the corresponding stable (respectively unstable) eigenspaces are of the same dimension, then we can construct a topological equivalence h^s (respectively h^u) between the restrictions of the vector fields to the stable (respectively unstable) eigenspaces. This is easy to see if we take the Jordan normal form for the coefficient matrices whose off-diagonal entries are small enough (this is similar to the argument in the above example). Finally, the mapping $h = (h^u, h^s)$ defines a homeomorphism for the topological equivalence of the whole vector fields. □

There is a local version of the topological equivalence as seen in the Hartman-Grobman theorem when we are dealing with the local dynamics in a neighborhood of the specific orbit in which we are interested. Namely, two vector fields are *locally topologically equivalent* near equilibrium points if there exist neighborhoods of these equilibrium points and a homeomorphism from one neighborhood to another such that any orbit in the neighborhood is mapped to the orbit of the other with the direction preserved.

Remark 3.2.29.

(1) Although the topological equivalence is a coarser notion than the smooth equivalence, it is sometimes the case that there still exists an invariant quantity, called a *modulus of stability*, which is preserved even under topological equivalence. For example, a homoclinic orbit of a vector field on $\mathbb{R}^3$ to a hyperbolic equilibrium point whose eigenvalues are one real positive $\lambda > 0$ and a pair of complex conjugate numbers $-\mu \pm i\omega$ with $\mu > 0$, has a modulus of stability given by the ratio λ/μ, provided the ratio is bigger than the unity. In other words, two vector fields with such homoclinic orbits never be locally topologically equivalent unless they have the same ratio. This is a result due to [Togawa 1987], the proof of which uses the knot theory and the Shil'nikov's theorem (Theorem 3.5.5). The existence of moduli is far from being trivial and many works have been devoted to this subject. See, e.g. [Palis 1978], [van Strien 1982], [Beloqui 1986], and references therein.

(2) Even if the notion of smooth equivalence is too fine, it is still useful for the study of dynamical systems. For instance, we can make smooth coordinate transformations to a vector field on $\mathbb{R}^n$ in order to simplify its expression and make the study easier. This is the basic idea of normal forms which will be explained in the next section. Another importance of the smooth equivalence is that it does not lose information related to its differentiability. This point becomes more apparent when we study dynamics around a homoclinic or a heteroclinic orbit, which is the subject of Section 3.5.

If we are trying to classify dynamical systems under topological equivalence, then we first need to think of the set of all dynamical systems. This set is endowed with an appropriate topology. Here we mainly use the following smooth topology. The set of all smooth vector fields on $\mathbb{R}^n$ coincides with the set of all smooth mappings from $\mathbb{R}^n$ into itself, for which we can give the topology of uniform convergence on compact sets together with partial derivatives of any order. In the space of dynamical systems with smooth topology, we can consider topological equivalence classes. Each class consists of vector fields which are mutually topologically equivalent: some equivalence classes could be big, others could be small, and, among others, there can be an equivalence class which forms an open subset. Each member of such an equivalence class has the property that it does not change the qualitative nature of its orbit structure under a small perturbation (with respect to the topology), because any nearby members are topologically equivalent to each other. This property is called the structural stability.

Definition 3.2.30. A flow $\{\varphi^t\}$ is *structurally stable* if any flow in a neighborhood (with respect to the specified topology) of $\{\varphi^t\}$ is topologically equivalent to $\{\varphi^t\}$ itself. Otherwise it is called *structurally unstable.*

The notion of structural stability can also be considered for local dynamics. An orbit is said to be locally structurally stable if the dynamical system in consideration is so on a sufficiently small neighborhood of the orbit. Therefore the Hartman-Grobman theorem together with Proposition 3.2.28 implies

Corollary 3.2.31. *Hyperbolic equilibrium points are locally structurally stable.*

3.2.6 Bifurcation

In any small neighborhood of a structurally unstable dynamical system, there exist at least two dynamical systems which are not topologically equivalent. In other words, a structurally unstable dynamical system changes its qualitative nature under any small perturbation. In this case we say that the dynamical system undergoes *bifurcation.* This is the main subject of this book, and we will see various aspects of the bifurcations of dynamical systems in the proceeding sections. Here we only give two basic examples of such bifurcations.

Example 3.2.32. (Hopf Bifurcation) Consider the dynamical system given by the following ordinary differential equation:

$$\dot{x} = \mu x - y - x(x^2 + y^2)$$
$$\dot{y} = x + \mu y - y(x^2 + y^2)$$

defined in a neighborhood of the origin O of $\mathbb{R}^2$ with a real parameter μ. The eigenvalues of the linearization matrix at O is given by $\mu \pm \sqrt{-1}$, and hence it is not hyperbolic only when $\mu = 0$. This equation can be written as

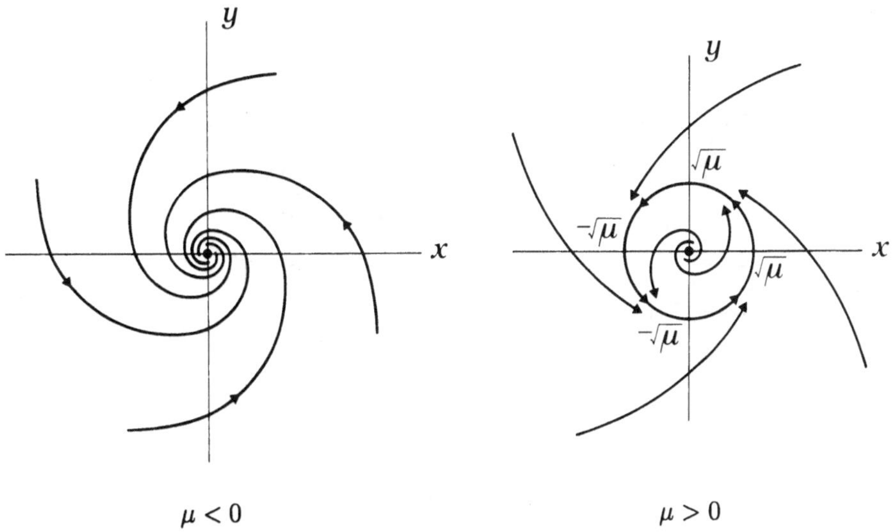

Fig. 3.2.10. The phase portrait of Equation (3.2.20).

$$\begin{aligned} \dot{r} &= r(\mu - r^2) \\ \dot{\theta} &= 1 \end{aligned} \tag{3.2.20}$$

using the polar coordinates. In particular, the $\dot{r}$-equation itself determines a one-dimensional vector field and the $\dot{\theta}$-equation simply gives a rotation with a constant speed, and hence we can easily describe the change of the entire dynamics as follows:

(1) When $\mu \leq 0$, the equilibrium point O is stable and the other orbits converge to O as $t \to +\infty$;

(2) When $\mu > 0$, O loses its stability and a circle of radius $\sqrt{\mu}$ appears as a stable periodic orbit. The other orbits converge to the periodic orbit as $t \to +\infty$.

See Fig. 3.2.10.

Such a change of the phase portraits is called the *Hopf bifurcation*, where a pair of complex conjugate eigenvalues for the linearization at an equilibrium point crosses the imaginary axis as the parameter changes and at the same time a periodic orbit appears in its neighborhood.

There are many physical, mechanical and even biological systems which create oscillatory motions by changing system parameters. For instance, a fluid starts showing a thermal convection if it is heated from below, which then changes to an oscillatory convective motion with increased heat. A chemical oscillation called the

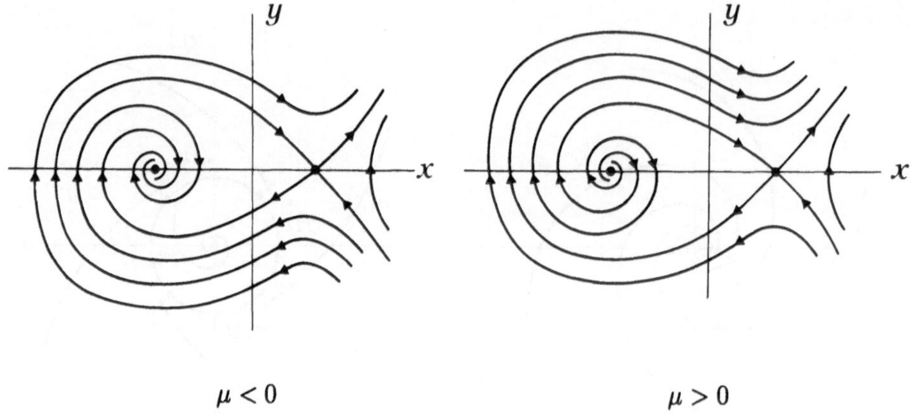

Fig. 3.2.11. The phase portrait of Equation (3.2.21).

Belouzov-Zhabotinsky reaction is another such system, in which the concentration of certain chemicals exhibits a periodic movement under some conditions. One can easily observe various kinds of oscillations in electric circuits, some of which are shown in Chapter 1. Moreover, the population of certain biological species changes periodically as is seen from long term statistical data, which may be explained as an oscillation of the biological system. Such examples of concrete systems with oscillations can be found in, for instance [Marsden and McCracken 1976] and references therein. As described there, the Hopf bifurcation can sometimes (not always) be thought of as explaining the mechanism of the generation of such oscillations.

The Hopf bifurcation is one of the most frequently observed bifurcations around an equilibrium point. See Subsection 3.3.3 for the detail.

Example 3.2.33. Consider the dynamical system given by the following ordinary differential equation on $\mathbb{R}^2$:

$$\begin{aligned} \dot{x} &= y \\ \dot{y} &= x^2 - 1 + \mu y \end{aligned} \tag{3.2.21}$$

with a real parameter μ. The equilibrium point $E = (1, 0)$ of this equation is always of the saddle type for small μ and both the stable manifold $W^s(E)$ and the unstable manifold $W^u(E)$ are of one dimension. In particular, when $\mu = 0$, Equation (3.2.21) is a Hamiltonian system with the integral $H(x, y) = \frac{y^2}{2} - \frac{x^3}{3} + x$ and the stable and unstable manifolds coincide in $\{x < 1\}$ since they are contained in the level set $H^{-1}(\frac{2}{3})$. The intersection of these two manifolds is an orbit homoclinic to the equilibrium point E. See Example 3.2.7(3).

A homoclinic orbit never be locally structurally stable. Indeed, there is no homoclinic orbit in (3.2.21) except when $\mu = 0$. This can be seen by observing that the function H is monotonically increasing or decreasing when $\mu \neq 0$ along the vector field given by (3.2.21) since

$$\frac{dH}{dt} = y\dot{y} - (x^2 - 1)\dot{x} = \mu y^2 \neq 0,$$

and hence there is no recurrent orbit other than the equilibrium point E. See Fig. 3.2.11. Therefore, in general, a vector field with a homoclinic orbit undergoes a bifurcation in its perturbation. *Homoclinic bifurcations* refer to various kinds of bifurcations from homoclinic orbits, which will be the main subject of Section 3.5.

It is important to note that, in vector fields, the stable and unstable manifolds of an equilibrium point, if they intersect, cannot be transverse along the homoclinic orbit. This is because the intersection necessarily contains at least one orbit, which is of one dimension, and hence the tangent spaces of the stable manifold and unstable manifold cannot span the whole space at any point of the intersection. This lack of transversality explains the above structural instability of homoclinic orbits to equilibrium points. This is a special property of vector fields, and it is not the case for generic diffeomorphisms as we shall see in C of Subsection 3.4.6. See also Subsection 3.5.1.

3.2.7 Framework for the Bifurcation Theory

For a k-parameter family of vector fields X_μ, a *singularity* or a bifurcation point of this family is a parameter value μ, or the vector field X_μ itself, for which X_μ is not structurally stable. Or conversely, a family of vector fields containing a structurally unstable vector field X as its member is called an *unfolding* of X. The purpose of the bifurcation theory for dynamical systems is to classify singularities, to study possible dynamics in an unfolding of each singularity, and to give its "standard model" of the dynamics. Here the standard model means the versal family as defined below.

Definition 3.2.34.

(1) Two k-parameter families of vector fields X_μ and Y_μ are *fiber C^0 equivalent* if there exists a homeomorphism h_μ for each μ which sends every orbit of X_μ to an orbit of Y_μ with the direction of orbits preserved. If, moreover, h_μ is continuous in μ, then these families are called *C^0 equivalent.*

(2) A family Y_μ ($\mu \in \mathbb{R}^k$) is a *versal family* of X_0, if, for any unfolding X_λ ($\lambda \in \mathbb{R}^l$) of X_0, there exists a C^∞ mapping $\alpha : (\mathbb{R}^l, 0) \to (\mathbb{R}^k, 0)$ such that X_λ and $Y_{\alpha(\lambda)}$ are fiber C^0 equivalent.

(3) The *codimension* of X_0 is the minimal number of parameters for versal families of X_0.

In words, the versal family is a family which involves any nearby dynamical systems appearing in a perturbation from the singularity. More precisely, for any vector field X close to X_0, there exists a member Y_μ of the versal family for X_0 which is topologically equivalent to the prescribed vector field X. The codimension, then, measures the degeneracy of the bifurcation point by counting the number of parameters needed in order to find all the possible perturbations.

There are several versions of the definition of "versal families," depending on the choice of equivalences. The versal family defined above was first introduced as the (fiber C^0, C^∞) versal family in [Dumortier 1978]. Sometimes we can speak about a stronger definition imposing the continuous dependence of the equivalence homeomorphism on the unfolding parameters, which is called a (C^0, C^∞) versal family in the above literature. The definition of versal families adopted here is suitable for our purpose, but readers should pay attention to the definition of versality in literatures, since there is no unanimously accepted definition of versal families at present.

Needless to say, there are infinitely many kinds of singularities and it is not clear whether every singularity admits a versal family. Therefore, and for a practical reason as well, the study of bifurcations should begin with the simplest singularities with low codimension. In Section 3.3, we are mainly concerned with codimension one and two bifurcation points near equilibrium points, which are the most well-understood bifurcations in vector fields. More global bifurcations are treated in Section 3.5, although it is difficult to give a versal family for global bifurcations even with the lowest codimensions. Therefore, we must admit that the description in Section 3.5 is far from complete.

3.3 Local Bifurcations around Equilibrium Points in Vector Fields

In this section we shall explain some aspects of local bifurcations around equilibrium points in vector fields, especially elementary bifurcations of codimension one and the Bogdanov-Takens bifurcation which occurs when the linearization of a vector field possesses a zero eigenvalue of multiplicity two at criticality. As a preliminary to the exposition of these bifurcations, we shall give a brief description of center manifolds and normal forms.

3.3.1 Center Manifolds

As we have seen in Section 3.2, hyperbolic equilibrium points are locally structurally stable and hence no bifurcation occurs from them. Therefore we turn to the study of non-hyperbolic equilibrium points where the linearization contains at least one eigenvalue with zero real part. In order for such study, it is convenient to split the dynamics into essentially non-hyperbolic part in which all the eigenvalues have zero real parts, and a complementary hyperbolic part. This idea is provided from the center manifold theorem as explained below.

Eigenvalues of the linearization $Dv(O)$ at an equilibrium point O in a C^r vector field $\dot{x} = v(x)$ are classified into three types: those with negative real parts, with positive real parts, and with zero real parts. Let E^s, E^u, E^c be the corresponding stable, unstable and center eigenspaces, respectively. Then we have the following:

Theorem 3.3.1.

(1) There exists a C^r submanifold $W^c(O)$ in a neighborhood of the equilibrium point O at which $W^c(O)$ is tangent to E^c, and it is invariant under the local flow generated by the vector field v. Thus it induces a vector field on the submanifold which we denote by v^c.

(2) The vector field $\dot{x} = v(x)$ is locally topologically equivalent to

$$\begin{aligned} \dot{x}_c &= v^c(x_c) \\ \dot{x}_s &= -x_s \\ \dot{x}_u &= x_u \end{aligned}$$

in a neighborhood of O, where

$$(x_c, x_s, x_u) \in W^c(O) \times W^s(O) \times W^u(O).$$

We call $W^c(O)$ a *center manifold* of O.

Proof. The idea of the proof of Theorem 3.3.1(1) is similar to that of the local stable manifold theorem (Theorem 3.2.22), namely, characterizing the center manifold as a graph of a function from the center subspace into its complementary one. The function is obtained by a fixed point argument applied to an integral equation, although the proof is more involved than that for the local stable manifold theorem. In what follows we give an outline of the proof based on [Vanderbauwhede 1989] where interested readers can find its complete detail.

We begin with a general C^r vector field with an equilibrium point at the origin O:

$$\dot{x} = v(x) = Ax + f(x), \qquad x \in \mathbb{R}^n, \tag{3.3.1}$$

where $f(x) = O(|x|^2)$ in a neighborhood of the origin. Analogous to the proof of the local stable manifold theorem, we decompose the phase space into the hyperbolic eigenspace $E^h = E^u \oplus E^s$ and the center eigenspace E^c, namely,

$$\mathbb{R}^n = E^h \oplus E^c$$

according to the splitting of the spectra of the linearization matrix A into eigenvalues with non-zero real parts and those with zero real parts. Accordingly, let

$$P^h : \mathbb{R}^n \to E^h, \qquad P^c : \mathbb{R}^n \to E^c$$

be the corresponding projections with

$$Ker(P^h) = E^c, \qquad Ker(P^c) = E^h.$$

We then modify the nonlinear term $f(x)$ outside a neighborhood of the origin O so that it is a globally defined C^r function with bounded partial derivatives up to the order r together with a small C^1 norm. We keep the same notation $f(x)$ after the modification. Similar to the local stable manifold theorem, if we could show the existence of a global center manifold for the modified $f(x)$, an (at least local) center manifold exists for the original vector field (3.3.1), since the vector field does not change in a neighborhood of O. See Remark 3.3.3(1).

Next we prepare the following:

Lemma 3.3.2. *Let $x(t;\xi)$ be a solution of the equation (3.3.1) with the initial condition $x(0;\xi) = \xi$. Then*

$$\sup_{t\in\mathbb{R}} |P^h x(t;\xi)| < +\infty,$$

if and only if $x(t;\xi)$ satisfies the following integral equation:

$$x(t;\xi) = e^{At}P^c\xi + \int_0^t e^{A(t-s)}P^c f(x(s;\xi))ds + \int_{-\infty}^{\infty} B(t-s)f(x(s;\xi))ds, \quad (3.3.2)$$

where

$$B(t) = \begin{cases} -e^{At}P^u & (t < 0) \\ e^{At}P^s & (t \geq 0) \end{cases}$$

and $P^h = P^u \oplus P^s$ is the decomposition of the projection with respect to decomposition $E^h = E^u \oplus E^s$.

From this Lemma, we introduce the functional equation

$$\mathcal{F}(x;\xi^c)(t) = x(t) \qquad (\xi^c \in E^c),$$

where

$$\mathcal{F}(x;\xi^c)(t) = e^{At}\xi^c + \int_0^t e^{A(t-s)}P^c f(x(s))ds + \int_{-\infty}^{\infty} B(t-s)f(x(s))ds$$

for a bounded function $x(t)$. If this equation has a solution $x(t) = x_*(t;\xi^c)$ for each $\xi^c \in E^c$, then we define the center manifold as follows:

$$W^c(O) = \{x_*(0;\xi^c) \mid \xi^c \in E^c\}.$$

Note that $P^c x_*(0;\xi^c) = \xi^c$ from the definition.

Technically, the solution $x_*(t;\xi^c)$ is shown to exist in the function space with small exponential growth:

$$Y_\eta = \{y(\bullet) \in B(\mathbb{R}, \mathbb{R}^n) \mid \sup_{t\in\mathbb{R}} e^{-\eta|t|}|y(t)| < +\infty\}$$

for some $\eta > 0$ smaller than the spectral gap given by

$$gap = \min\{Re\lambda^u, |Re\lambda^s|\},$$

where λ^u and λ^s are the unstable and stable eigenvalues of A, respectively.

After some estimates, one can show the existence of a unique fixed point $x_*(t;\xi^c) \in Y_\eta$ of the above functional equation for each $\xi^c \in E^c$ by using the standard contraction mapping theorem, which, as explained above, implies the existence of a (continuous) center manifold $W^c(O)$.

Furthermore, one can also show that the function $x_*(0;\xi^c)$ is of C^r with respect to ξ^c and that its first order derivative vanishes at $\xi^c = 0$, hence proving that the center manifold $W^c(O)$ is really a C^r submanifold tangent to E^c at the origin O. The assertion looks completely the same as that for the stable manifold, but in fact its proof is more delicate because of certain loss of differentiability, which is explained in the remark in [Vanderbauwhede 1989] following Theorem 3.1 in that article. In the case of center manifolds, the order of the exponential growth for the function space Y_η plays an important role in order to overcome such a difficulty. We do not explain it here but only refer to [Vanderbauwhede 1989] for the smoothness proof of the center manifold. See also [Vanderbauwhede and van Gils 1987].

The proof of the second part of Theorem 3.3.1 is technically more involved. This result can be considered as a generalization of the Hartman-Grobman theorem (Theorem 3.2.15), but the proof ([Palis and Takens 1977] and [Kirchgraber and Palmer 1990] as well as references therein) is different and more geometric: one uses so-called the stable and unstable foliations for constructing the equivalence homeomorphism.

Here the stable (respectively unstable) foliation means a decomposition of a neighborhood of O into a continuous family of submanifolds of equal dimension containing the stable (respectively unstable) manifold at O in it, and this family is also required to be invariant under the flow generated by the vector field. To be more precise, a family of manifolds $\{M_\xi\}$ is called a stable (respectively unstable) foliation on a neighborhood of O, provided that

(1) $\dim M_\xi = \dim E^s(O)$ (respectively $\dim M_\xi = \dim E^u(O)$) for $\forall\xi$;

(2) A neighborhood U of O is the union $\bigcup_\xi M_\xi$;

(3) $\exists\xi_0$ s.t. $M_{\xi_0} = W^s(O) \cap U$ (respectively $M_{\xi_0} = W^u(O) \cap U$);

(4) For each ξ and t, there exists ξ' such that if a point $x_0 \in M_\xi$ satisfies $\varphi^t(x_0) \in M_{\xi'}$, then every point $x \in M_\xi$ satisfies $\varphi^t(x) \in M_{\xi'}$ as long as $\varphi^t(x) \in U$.

If we could show the existence of such foliations, we can make use of them as a new coordinate near the equilibrium point O in which the dynamics is nicely separated into the stable, unstable and center parts. Then we recall the Hartman-Grobman theorem for further simplification of the expression of the vector field on its stable and unstable manifolds, hence obtaining the desired form. The detailed construction of the stable and unstable foliations can be found in the references cited above. We also remark that an alternative proof of the Hartman-Grobman theorem using the idea of foliations is given in [Palis and de Melo 1982]. □

If an eigenvalue of the linearization at an equilibrium point has a zero real part, then it is either zero or a pure imaginary number. Linear vector fields with such eigenvalues form an algebraic set Σ^1 of codimension one in the space of all linear vector fields whose generic member satisfies one of the following two conditions:

(1) 0 is the only eigenvalue with a zero real part and it is simple;

(2) A pair of pure imaginary numbers are the only eigenvalues with zero real part and they are simple.

More complicated cases such as those with zero eigenvalues of multiplicity two belong to the set of singularities of the algebraic set Σ^1.

From Theorem 3.3.1, the qualitative nature of the orbits around an equilibrium point is determined by its restriction to the center manifold, and hence for the study of vector fields near equilibrium points, we may only consider those vector fields whose linearization at the equilibrium point has no eigenvalues off the imaginary axis. For instance, the above two cases of eigenvalues with zero real part reduce to vector fields of dimension one and two, respectively, whose linearization only possesses these central eigenvalues, and the study of the dynamics and their bifurcations associated to these cases can be carried out for such low-dimensional vector fields, as we will see in Subsection 3.3.3.

Remark 3.3.3.

(1) The center manifold is not unique in general. In the proof of Theorem 3.3.1, we needed to modify the nonlinear term $f(x)$ in Equation (3.3.1) outside a neighborhood of the equilibrium point, in order to obtain a unique fixed point of the functional equation. In general, a different modification gives a different fixed point, and hence a different center manifold. In the case of the local stable manifold, however, the modification does not give the non-uniqueness since the integral equation (3.2.18) is considered only for $t \geq 0$. Of course the uniqueness of the local stable manifold is clear from the Hartman-Grobman theorem. We shall give an example of a non-hyperbolic

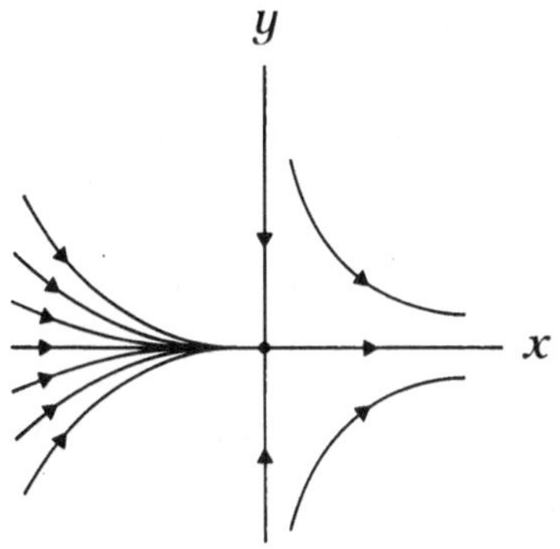

Fig. 3.3.1. Non-uniqueness of center manifolds.

equilibrium point which has more than one center manifolds following. The vector field

$$\dot{x} = x^2$$
$$\dot{y} = -y$$

has a non-hyperbolic equilibrium point at the origin O where there exists infinitely many center manifolds given by the graph of functions

$$y = \begin{cases} Ce^{1/x} & (x < 0) \\ 0 & (x \geq 0) \end{cases}$$

as shown in Fig. 3.3.1. Indeed the curve given by this function is tangent to the x-axis and is invariant under the flow since

$$\frac{dy}{dx} = Ce^{1/x} \cdot \left(-\frac{1}{x^2}\right) = \frac{y}{-x^2}.$$

Despite the non-uniqueness of center manifolds, any orbit staying in a neighborhood of O for all $t \in (-\infty, \infty)$ must be contained in every center manifold in general. See [Carr 1981] for details.

(2) Even if the vector field v is of C^∞, its center manifold is in general not necessarily C^∞, but it is C^r for any $r > 0$. This fact was first pointed out by [van Strien 1979]. For instance, it can be shown that the center manifold for the vector field

$$\dot{x} = -xz - x^3$$
$$\dot{y} = -y + x^2$$
$$\dot{z} = 0$$

at the origin is a two-dimensional submanifold tangent to the (x, z)-plane, which turns out to be not C^∞. This example is taken from [Carr 1981]. See also [Sijbrand 1985], [Vanderbauwhede 1989] for more details.

(3) The second statement of Theorem 3.3.1 tells us that any solution which is close to a center manifold is traced by a solution on the center manifold itself as long as it stays in a neighborhood of the equilibrium point. This follows from the fact that a sufficiently small neighborhood of the center manifold is fibered by the stable and unstable foliations and hence a solution is traced by the one projected down on the center manifold along these fibrations. This type of observation is often useful for the study of dynamics near center manifolds. Here we only refer to [Carr 1981], [Vanderbauwhede 1989] for related results of this kind, and to [Chow and Lin 1990] as an example showing the usefulness of such an observation.

(4) Although the center manifold is not unique, for a C^∞ vector field any two of the center manifolds are tangent at the equilibrium point of order higher than any power. Furthermore, it has recently been proved in [Burchard, Deng and Lu 1992] that vector fields on different center manifolds at the same equilibrium point are smoothly equivalent with each other. Therefore the choice of non-unique center manifolds is not important at all. It is also possible to give an approximation of a center manifold by using the power series expansion, as illustrated in the following example.

Example 3.3.4. Let us take the Rössler equation (Example 3.2.6) as an example:

$$\begin{aligned}\dot{x} &= -y - z\\ \dot{y} &= x + ay\\ \dot{z} &= bx - cz + xz.\end{aligned}$$

This equation has an equilibrium point at the origin $O = (0,0,0)$ and its characteristic polynomial of the linearization matrix at O is given by

$$\lambda^3 - (a-c)\lambda^2 - (ac - b - 1)\lambda - (ab - c) = 0,$$

and hence it has a zero eigenvalue of multiplicity two if

$$ab - c = 0, \quad ac - b - 1 = 0, \quad \text{and} \quad a - c \neq 0,$$

which is equivalent to

$$a \neq 0, \pm 1, \quad b = \frac{1}{a^2 - 1}, \quad c = \frac{a}{a^2 - 1}.$$

The other eigenvalue is $\alpha = a - c (\neq 0)$.

We shall compute the second order approximation of the center manifold $W^c(O)$ corresponding to the double zero eigenvalue. For that purpose, it is convenient to take coordinates for which the linearization matrix becomes the Jordan normal form. An eigenvector to the zero eigenvalue is given by

$$e_1 = (a, -1, 1)$$

and its generalized eigenvector is

$$e_2 = (-1, 0, -a),$$

while an eigenvector to the other eigenvalue α is

$$e_3 = (-c, 1, \alpha c - 1).$$

Therefore, we shall take the new coordinates (u, v, w) as

$$\begin{pmatrix} x \\ y \\ z \end{pmatrix} = \begin{pmatrix} a & -1 & -c \\ -1 & 0 & 1 \\ 1 & -a & \alpha c - 1 \end{pmatrix} \begin{pmatrix} u \\ v \\ w \end{pmatrix}.$$

Then we rewrite the Rössler equation using the new coordinates and obtain

$$\begin{aligned} \dot{u} &= v && -\tfrac{1}{\alpha^2}\varphi(u,v,w) \\ \dot{v} &= && -\tfrac{1}{\alpha}\varphi(u,v,w) \\ \dot{w} &= \alpha w && -\tfrac{1}{\alpha^2}\varphi(u,v,w), \end{aligned} \tag{3.3.3}$$

where

$$\varphi(u, v, w) = (au - v - cw)(u - av + (\alpha c - 1)w).$$

The center manifold $W^c(O)$ has to be tangent to the (u, v)-plane at O, and thus we may assume that it is given by the graph of a function of the form

$$w = h(u, v) = pu^2 + quv + rv^2 + O(3), \tag{3.3.4}$$

where $O(3)$ stands for terms of order three or higher with respect to u and v. Since $W^c(O)$ is invariant under the flow, it must satisfy

$$\dot{w} = \frac{\partial h}{\partial u}\dot{u} + \frac{\partial h}{\partial v}\dot{v}$$

along the orbits, which implies the equality

$$\begin{aligned} \alpha h(u,v) - &\frac{1}{\alpha^2}\varphi(u,v,h(u,v)) \\ &= \frac{\partial h}{\partial u}(u,v)\left(v - \frac{1}{\alpha^2}\varphi(u,v,h(u,v)\right) \\ &\quad + \frac{\partial h}{\partial v}(u,v)\left(-\frac{1}{\alpha}\varphi(u,v,h(u,v)\right). \end{aligned} \tag{3.3.5}$$

In order to obtain the coefficients p, q, r in (3.3.4), we may neglect terms of order three or higher in (3.3.5) and hence obtain

$$\alpha(pu^2 + quv + rv^2) - \frac{1}{\alpha^2}(au - v)(u - av) = (2pu + qv)v,$$

or equivalently,

$$\left(\alpha p-\frac{a}{\alpha^2}\right)u^2+\left(\alpha q+\frac{a^2+1}{\alpha^2}-2p\right)uv+\left(\alpha r-\frac{a}{\alpha^2}-q\right)=0.$$

This equality has to be identically satisfied, thus yielding

$$p=\frac{a}{\alpha^3},\quad q=\frac{2a-\alpha(a^2+1)}{\alpha^4},\quad r=\frac{2a-\alpha(a^2+1)+\alpha^2 a}{\alpha^5}.$$

This gives the second order approximation of the center manifold with the coordinates (u, v, w), and the original center manifold approximation can be obtained by returning to the original coordinates (x, y, z), which, of course, takes a messy form and hence we do not give it explicitly here.

A higher order approximation can be obtained as well by employing the same procedure. To be more precise, the third order approximation is given first by putting

$$h(u,v)=pu^2+quv+rv^2+ku^3+lu^2v+muv^2+nv^3+O(4),$$

where p, q, r are given above and k, l, m, n are to be determined. By substituting this into the equation (3.3.5) and comparing third order terms, we then obtain the equation for the coefficients k, l, m, n. We can proceed with such computation as high as we want in principle, although it becomes more and more cumbersome as the order goes higher. A symbolic manipulation system on computers would be useful for such calculation for concrete examples. See, for instance, [Rand and Armbruster 1987]. We also remark that the computation of a center manifold for the Lorenz equation is carried out in [Guckenheimer and Holmes 1983].

We refer to [Carr 1981], [Sijbrand 1985] and [Vanderbauwhede 1989] for general references on the center manifold.

3.3.2 Normal Forms

The center manifold theorem reduces the study of local dynamics of a vector field near an equilibrium point to one with purely nonhyperbolic linear part. In order to study such reduced vector fields on the center manifold, it is convenient to bring it into a simpler form by a coordinate transformation. Such a simplified expression is called a *normal form*. Here "simplify" means to eliminate as many monomials as possible in the Taylor expansion of the vector field. In this subsection, we will explain a standard technique for finding such normal forms up to a given order.

Throughout this subsection, we use the differential operator notation of a vector field v. Namely, with the local coordinates $x=(x^1,\cdots,x^n)$, a vector field v with the i-th component $v^{(i)}$, or equivalently an ordinary differential equation $\dot{x}^i=v^{(i)}(x^1\cdots,x^n)$ $(i=1,\cdots,n)$, is identified with a differential operator

$$v = \sum_{i=1}^{n} v^{(i)}(x^1 \cdots, x^n)\frac{\partial}{\partial x^i}.$$

A homogeneous vector field of degree k refers to a vector field with homogeneous component functions of the same degree k. Let H_k be the vector space consisting of all such homogeneous vector fields of degree k on $\mathbb{R}^n$. The k-jet (see [Bröcker and Lander 1975] for the definition) of a vector field v at an equilibrium point O is represented by an element

$$(v_l)_{l=1}^k \in \bigoplus_{l=1}^{k} H_l.$$

In particular, we identify the linear part v_1 with its matrix representation A, namely $A = (a_{ij})$ where $a_{ij} = \dfrac{\partial v^{(i)}}{\partial x^j}$. Let B_k denote the image of the linear map

$$L_A : H_k \to H_k;\ L_A(h_k) = [h_k, v_1]$$

defined from A, and let G_k denote a subspace of H_k complementary to B_k. Here the bracket stands for the Lie bracket for vector fields, namely,

$$[v, w] = \sum_{i,j=1}^{n} \left(v^{(j)}\frac{\partial w^{(i)}}{\partial x^j} - w^{(j)}\frac{\partial v^{(i)}}{\partial x^j} \right) \frac{\partial}{\partial x^i},$$

where $v = \sum_{i=1}^{n} v^{(i)} \frac{\partial}{\partial x^i}$ and $w = \sum_{i=1}^{n} w^{(i)} \frac{\partial}{\partial x^i}$.

Theorem 3.3.5. *For the homogeneous part $v_k \in H_k$ of degree k of a vector field v at its equilibrium point O, terms in v_k belonging to B_k $(k \geq 2)$ can be eliminated by a smooth coordinate transformation. Therefore the k-jet of v at O can be put into the following form:*

$$v_1 + g_2 + \cdots + g_k, \quad g_i \in G_i\ (2 \leq i \leq k). \tag{3.3.6}$$

Proof. This theorem is proven by induction. Suppose the k-jet of the vector field v can be put into

$$v_1 + g_2 + \cdots + g_{k-1} + v_k.$$

From $H_k = B_k \oplus G_k$, v_k can be decomposed as $v_k = b_k + g_k$ where $b_k = [h_k, v_1]$ for some h_k from the definition of B_k. Take the transformation

$$\varphi_k(x) = x + h_k(x).$$

Then the transformed vector field becomes

$$\begin{aligned} D\varphi_k(\varphi_k(x))^{-1} \cdot v(\varphi_k(x)) \\ &= (I - Dh_k(x)) \cdot (v(x) + Dv_1(x)h_k(x)) + h.o.t. \\ &= v(x) - Dh_k(x)v_1(x) + Dv_1(x)h_k(x) + h.o.t. \\ &= v(x) - [h_k, v_1](x) + h.o.t. \\ &= v_1(x) + g_2(x) + \cdots + g_{k-1}(x) + g_k(x) + h.o.t. \end{aligned}$$

Here *h.o.t.* stands for terms of order higher than k. Therefore we obtain the desired conclusion. □

For more detail as well as further information, see [Arnold 1983], [Takens 1974a], and [Vanderbauwhede 1989].

Example 3.3.6.

(1) Let a vector field v on $\mathbb{R}^2$ with the local coordinates (x, y) have an equilibrium point at the origin O where the linearization matrix A takes $\begin{pmatrix} 0 & -1 \\ 1 & 0 \end{pmatrix}$. Then we have $G_k = \{0\}$ if k is even, and

$$G_k = \mathrm{span}_{\mathbb{R}} \left\{ (x^2+y^2)^l \left(x\frac{\partial}{\partial x} + y\frac{\partial}{\partial y} \right),\ (x^2+y^2)^l \left(-y\frac{\partial}{\partial x} + x\frac{\partial}{\partial y} \right) \right\}$$

if $k = 2l + 1$ is odd, and hence the normal form of order $(2l + 1)$ becomes

$$\begin{aligned} \dot{x} &= -y + (a_1x - b_1y)(x^2+y^2) + \cdots + (a_lx - b_ly)(x^2+y^2)^l + h.o.t. \\ \dot{y} &= x + (b_1x + a_1y)(x^2+y^2) + \cdots + (b_lx + a_ly)(x^2+y^2)^l + h.o.t., \end{aligned}$$

where *h.o.t.* stands for higher order terms. The expression becomes simpler in the polar coordinates

$$x = r\cos\theta, \quad y = r\sin\theta,$$

namely,

$$\begin{aligned} \dot{r} &= a_1r^3 + \cdots + a_lr^{2l+1} + h.o.t. \\ \dot{\theta} &= 1 + b_1r^2 + \cdots + b_lr^{2l} + h.o.t. \end{aligned}$$

The unfolding of this example gives the general case of the Hopf bifurcation treated in the next subsection (see also Example 3.2.32).

(2) Let a vector field v on $\mathbb{R}^2$ have an equilibrium point O with the linearization matrix $A = \begin{pmatrix} 0 & 1 \\ 0 & 0 \end{pmatrix}$. Then in general, we have

$$\left[x^k y^l \frac{\partial}{\partial x}, y\frac{\partial}{\partial x}\right] = -kx^{k-1}y^{l+1}\frac{\partial}{\partial x},$$

$$\left[x^k y^l \frac{\partial}{\partial y}, y\frac{\partial}{\partial x}\right] = x^k y^l \frac{\partial}{\partial x} - kx^{k-1}y^{l+1}\frac{\partial}{\partial y},$$

and hence we can take

$$G_k = \mathrm{span}_{\mathbb{R}}\left\{x^k\frac{\partial}{\partial y},\ x^{k-1}y\frac{\partial}{\partial y}\right\},$$

and therefore its normal form of order k becomes

$$\begin{aligned}\dot{x} &= y & +h.o.t.\\ \dot{y} &= a_1x^2 + b_1xy + \cdots + a_{k-1}x^k + b_{k-1}x^{k-1}y & +h.o.t.\end{aligned}$$

The unfolding of this example gives the general case of the Bogdanov-Takens bifurcation treated in Subsection 3.3.4.

Remark 3.3.7.

(1) There can be a choice of the complementary subspace G_k. For instance, one may also take

$$G_k = \mathrm{span}_{\mathbb{R}}\left\{x^k\frac{\partial}{\partial x},\ x^k\frac{\partial}{\partial y}\right\}$$

in Example 3.3.6(2), which results in another normal form

$$\begin{aligned}\dot{x} &= y + c_1x^2 + c_2x^3 + \cdots + c_{k-1}x^k + h.o.t.\\ \dot{y} &= d_1x^2 + d_2x^3 + \cdots + d_{k-1}x^k + h.o.t.\end{aligned}$$

Therefore, the normal form given in Theorem 3.3.5 is not generally unique.

(2) Two methods have been implemented for obtaining normal forms in a systematic manner up to any degree: one is to take G_k as the orthogonal complement to B_k by introducing the inner product structure in H_k ([Elphick, Tirapegui, Brachet, Coullet, and Iooss 1987]), and the other is to use the representation of the Lie algebra $sl(2,\mathbb{R})$ ([Cushman and Sanders 1986]). See also [Vanderbauwhede 1989].

(3) It is usually possible to eliminate some more terms remaining in the normal forms given in Theorem 3.3.5. This has been carried out in [Takens 1973a], [Ushiki 1984], [Chua and Kokubu 1988], and [Baider and Sanders 1992], and is still under further investigation.

Example 3.3.8. We again take the Rössler equation and make a normal form calculation of it. In Example 3.3.4, we have computed the second order approximation of the center manifold corresponding to the double zero eigenvalue of the equation (3.3.3), which is equivalent to the Rössler equation. Since the linear part contains the Jordan normal form $\begin{pmatrix} 0 & 1 \\ 0 & 0 \end{pmatrix}$, the normal form in Example 3.3.6(2) should be applied to this case. Here we shall only compute the second order normal form of the equation on the center manifold, namely we shall compute the explicit coefficients a_1 and b_1 in Example 3.3.6(2) in terms of the parameters a, b, c of the Rössler equation.

Recall that the equation on the center manifold takes the form

$$\begin{aligned} \dot{u} &= v - \tfrac{1}{\alpha^2}\varphi(u, v, h(u,v)) \\ \dot{v} &= -\tfrac{1}{\alpha}\varphi(u, v, h(u,v)), \end{aligned}$$

where $\alpha = a - c$ and

$$\begin{aligned} \varphi(u,v,w) &= (au - v - cw)(u - av + (\alpha c - 1)w), \\ h(u,v) &= pu^2 + quv + rv^2 + O(3), \end{aligned}$$

and the coefficients p, q, r are given in Example 3.3.4. Here we do not need the explicit form of $h(u,v)$ for the computation of the coefficients a_1 and b_1.

We first eliminate the nonlinear terms in the right-hand side of the $\dot{u}$-equation by changing the variables as follows:

$$X = u, \quad Y = v - \frac{1}{\alpha^2}\varphi(u, v, h(u,v)),$$

which yields the equation

$$\begin{aligned} \dot{X} &= Y \\ \dot{Y} &= -\tfrac{1}{\alpha}(aX - Y)(X - aY) - \tfrac{1}{\alpha^2}(2aX - (a^2+1)Y)Y + O(3) \\ &= AX^2 + BXY + CY^2 + O(3), \end{aligned}$$

where $O(3)$ stands for terms of order higher than or equal to three in X and Y, and the coefficients A, B, C are given by

$$A = -\frac{a}{\alpha}, \quad B = \frac{(\alpha+1)(a^2+1) - 2a\alpha}{\alpha^2}, \quad C = -\frac{a}{\alpha}.$$

In this computation, we have used

$$u = X, \quad v = Y + O(2), \quad h(u,v) = O(2),$$

which is enough for our purpose, since we are only interested in second order terms of the normal form.

Now we need to eliminate the term $CY^2\frac{\partial}{\partial Y}$ by making a further change of variables. Recall from Theorem 3.3.5 that terms in the image of the linear map $L_A(h_k) = [h_k, v_1]$ can be eliminated by the transformation of the form $\varphi_k(x) = x + h_k(x)$. As for $v_1 = Y\frac{\partial}{\partial X}$, we have

$$\left[-\frac{C}{2}X^2\frac{\partial}{\partial X}-CXY\frac{\partial}{\partial Y}, Y\frac{\partial}{\partial X}\right]=XY^2\frac{\partial}{\partial Y},$$

and hence we shall take the transformation

$$x=X-\frac{C}{2}X^2, \quad y=Y-CXY.$$

This yields

$$\begin{aligned}\dot{x} &= \dot{X}-CX\dot{X}=Y-CXY=y\\ \dot{y} &= \dot{Y}-C\dot{X}Y-CX\dot{Y}\\ &= AX^2+BXY+CY^2-CY^2+O(3)\\ &= Ax^2+Bxy+O(3),\end{aligned}$$

and thus we get the desired normal form, whose coefficients are

$$a_1=A=-\frac{a}{\alpha}, \quad b_1=B=\frac{(\alpha+1)(a^2+1)-2a\alpha}{\alpha^2}.$$

The computation of the higher order normal form is, in principle, the same. However, it requires us to take account of enough approximations of the center manifold and becomes more and more tedious as the order goes up. Again, see [Rand and Armbruster 1987] and [Sanders 1991] for normal form calculations using symbolic manipulation systems.

3.3.3 Codimension One Bifurcations

As we have explained, we can reduce the dynamics near a non-hyperbolic equilibrium point to that on its center manifold, for which the linearization has only eigenvalues with zero real parts (central eigenvalues). Then the normal form theorem can be applied to obtain a simplified expression of vector fields on the center manifold depending on the type of central eigenvalues. Here we take the two simplest cases of such non-hyperbolic equilibrium points and study their bifurcations, namely, the dynamics of their unfoldings. These two cases are: the case of simple zero as a central eigenvalue (saddle-node), and that of a simple pair of pure imaginary eigenvalues (Hopf). In these cases, one can even give their versal families.

A Saddle-Node Bifurcation Let v be a C^∞ vector field on $\mathbb{R}^1$ with O being an equilibrium point at which the linearization $v_1=0$. In this case its second order part v_2 is generally non-zero. The singularity given by such a vector field is called a *generic saddle-node singularity*.

Theorem 3.3.9. *The family*

$$v_\mu(x)=(\pm x^2+\mu)\frac{\partial}{\partial x} \quad (\mu\in\mathbb{R}),$$

gives a versal family for the generic saddle-node singularity.

Proof. The generic saddle-node singularity v can be put into the form $v = cx^2(1+h(x))\frac{\partial}{\partial x}$ where $c \neq 0$ and $h(0) = 0$. By changing the time variable t into $|c|(1+h(x))t$, v is C^∞ equivalent to $\pm x^2 \frac{\partial}{\partial x}$. Similarly, we may assume that any unfolding v_λ of v satisfies

$$v_\lambda = f(x,\lambda)\frac{\partial}{\partial x} \quad \text{with} \quad f(x,0) = \pm x^2.$$

Then $f(x,\lambda)$ takes the form

$$f(x,\lambda) = G(x,\lambda)(x - \varphi(\lambda))^2 + a(\lambda),$$

with $G(0,0) = \pm 1$. In fact, since $\frac{\partial^2}{\partial x^2} f(0,0) = \pm 2 \neq 0$, the implicit function theorem shows the existence of a function $x = \varphi(\lambda)$ satisfying

$$\frac{\partial}{\partial x} f(\varphi(\lambda),\lambda) \equiv 0 \quad \text{and} \quad \varphi(0) = 0.$$

Hence by Taylor's theorem, we have

$$\begin{aligned} f(x,\lambda) &= f(\varphi(\lambda),\lambda) + \frac{\partial}{\partial x} f(\varphi(\lambda),\lambda)(x-\varphi(\lambda)) + G(x,\lambda)(x-\varphi(\lambda))^2 \\ &= a(\lambda) + G(x,\lambda)(x-\varphi(\lambda))^2. \end{aligned}$$

It is easy to see $G(0,0) = \pm 1$.

Let us put $\alpha = a(\lambda)$ and denote $\bar\lambda = (\alpha,\lambda)$, considering α to be a new parameter. This extends the unfolding family $f(x,\lambda)$ to

$$\bar f(x,\bar\lambda) = \alpha + G(x,\lambda)(x-\varphi(\lambda))^2 \quad (\bar\lambda = (\alpha,\lambda)).$$

If we can show that $\dot x = \bar f(x,\bar\lambda)$ is locally fiber C^0 equivalent to some $\dot x = v_{\mu(\bar\lambda)}(x) = \pm x^2 + \mu(\bar\lambda)$, which is induced from the versal family v_μ through an appropriate change of parameters $\mu = \mu(\bar\lambda)$, then the same conclusion holds as well for the subfamily $\dot x = f(x,\lambda) = \bar f(x, a(\lambda),\lambda)$, hence proving the desired assertion. Therefore, in what follows, we only consider this extended family $\dot x = \bar f(x,\bar\lambda)$.

Then we immediately see that the zero set $\{(x,\alpha) | \bar f(x,\alpha,\lambda) = 0\}$ for each fixed λ forms a parabola-shaped curve given by

$$\alpha = -G(x,\lambda)(x-\varphi(\lambda))^2$$

in the (x,α)-space, which is tangent to the x-axis at $(x,\alpha) = (0,0)$.

Let us now consider the case $G(0,0) < 0$ and make a fiber C^0 equivalence between the family

$$\dot x = \bar f(x,\bar\lambda) = \alpha + G(x,\lambda)(x-\varphi(\lambda))^2$$

and the family

$$\dot x = -x^2 + \alpha.$$

When $\alpha < 0$, both families take strictly negative right-hand sides and hence, in a neighborhood in $\mathbb{R}$ of the origin, there exists only one orbit in each vector field.

Therefore, for each α and λ, we can obviously take a homeomorphism between appropriate neighborhoods of the origin sending the unique orbit of one family to the other. If $\alpha = 0$, both families have only one equilibrium point at the origin and the unique orbit with negative initial point leaves from the origin while the one with a positive initial point converges to the origin. Hence it is also easy to take a homeomorphism which gives the local topological equivalence between these two vector fields. Finally, for $\alpha > 0$, there are two equilibrium points in each of these two families, one of which is stable and the other is unstable. The orbits other than these equilibrium points are going toward or away from the equilibrium points, and again, one can immediately construct a topological equivalence homeomorphism for these vector fields. Since the fiber C^0 equivalence does not ask the continuity of the equivalence homeomorphisms on the parameters, this completes the proof of Theorem 3.3.9. □

Remark 3.3.10. More precise argument in fact shows even the continuous dependence of equivalence homeomorphisms on the parameters. See [Newhouse, Palis and Takens 1983; Theorem 3.5] for its proof.

Furthermore the Malgrange preparation theorem (see for instance [Bröcker and Lander 1975]) tells us that the mapping $f(x, \lambda)$ with $\frac{\partial^2}{\partial x^2} f(0,0) \neq 0$ can be put into the form

$$f(x, \lambda) = L(x, \lambda) \left\{ b(\lambda) + (x - \psi(\lambda))^2 \right\},$$

with some smooth functions L, b, ψ satisfying $L(0,0) \neq 0$. This apparently shows the (C^∞, C^∞) versality of the family $v_\mu = (\pm x^2 + \mu)\frac{\partial}{\partial x}$, since the smooth transformation

$$y = x - \psi(\lambda), \quad \mu = \mathrm{sgn}(L(0,0)) \cdot b(\lambda), \quad \tau = |L(x, \lambda)| t$$

brings the family $\frac{dx}{dt} = f(x, \lambda)$ into

$$\frac{dy}{d\tau} = \pm y^2 + \mu.$$

Since the Malgrange preparation theorem is a highly non-trivial theorem, we have given here a more elementary but lengthy proof. See the discussion of [Arrowsmith and Place 1990]. We also note that the idea of constructing the fiber C^0 equivalence given in the proof of Theorem 3.3.9 works as well for other bifurcations as we shall see later.

Figure 3.3.2 illustrates the bifurcation in the versal family for a generic saddle-node bifurcation with the minus sign in the second order part, namely $v_\mu(x) = (-x^2 + \mu)\frac{\partial}{\partial x}$. When $\mu > 0$, there is a pair of equilibrium points $\pm\sqrt{\mu}$, one of which is stable and the other is unstable. These equilibrium points collide at $\mu = 0$ and disappear when $\mu < 0$. Such a picture which indicates the various dynamics of corresponding bifurcation parameters is referred to in general as a *bifurcation diagram.*

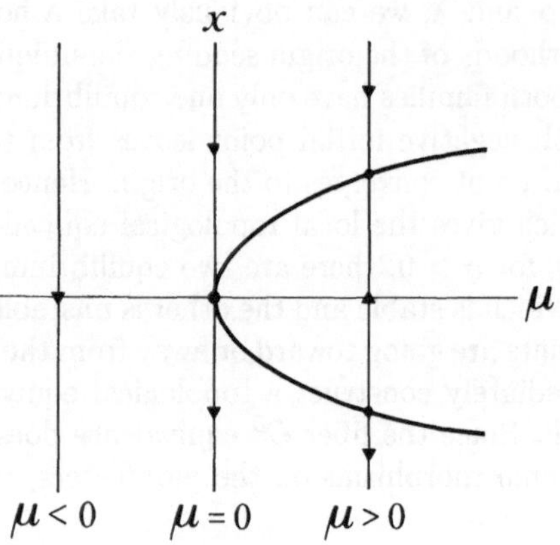

Fig. 3.3.2. Bifurcation diagram for the saddle-node bifurcation.

B Hopf Bifurcation Let v be a C^∞ vector field in $\mathbb{R}^2$ with O being an equilibrium point at which the linearization matrix is given by $A = \begin{pmatrix} 0 & -1 \\ 1 & 0 \end{pmatrix}$. Then it can be transformed to the form in Example 3.3.6(1) due to the normal form theorem (Theorem 3.3.5). In this case the coefficient a_1 in the normal form is generally non-zero, and the singularity given by such a vector field is called a *generic Hopf singularity.*

Theorem 3.3.11. *The family*

$$\begin{aligned} v_\mu(x,y) &= \{(-y + (a_1 x - b_1 y)(x^2+y^2))\tfrac{\partial}{\partial x} \\ &\quad +(x + (b_1 x + a_1 y)(x^2+y^2))\tfrac{\partial}{\partial y}\} \\ &\quad +\mu\left(x\tfrac{\partial}{\partial x} + y\tfrac{\partial}{\partial y}\right), \qquad (\mu \in \mathbb{R}), \end{aligned}$$

or equivalently in polar coordinates,

$$v_\mu(r,\theta) = (\mu r + a_1 r^3)\frac{\partial}{\partial r} + (1 + b_1 r^2)\frac{\partial}{\partial \theta}$$

with $a_1 \neq 0$ gives a versal family of the generic Hopf singularity. In particular, if $a_1 < 0$ (respectively $a_1 > 0$), then a stable (respectively unstable) periodic orbit appears when $\mu > 0$ (respectively $\mu < 0$).

The Hopf bifurcation is called *supercritical*, if a stable equilibrium point loses the stability at the bifurcation point and gives rise to a stable periodic orbit together with the destabilized equilibrium point. On the other hand, it is called *subcritical* if a stable equilibrium point and an unstable periodic orbit collapse at the bifurcation point and become an unstable equilibrium point. In the above theorem, a supercritical (respectively subcritical) Hopf bifurcation occurs when $a_1 < 0$ (respectively $a_1 > 0$).

Proof. Let w_λ be an unfolding of a generic Hopf singularity at $\lambda = 0$, namely, let w_λ be a family of vector fields on $\mathbb{R}^2$ where, for $\lambda = 0$, it satisfies the condition of the generic Hopf singularity at an equilibrium point, say O. We shall show that for each fixed small λ, the phase portrait of w_λ in a neighborhood of O is (locally) topologically equivalent to that of the versal family $v_{\varphi(\lambda)}$ for a smooth function of parameters $\mu = \varphi(\lambda)$.

First of all, we may assume that the equilibrium point of w_λ is the origin O for all λ: $w_\lambda(O) \equiv 0$, since the linearization matrix of a generic Hopf singularity is non-singular and hence the implicit function theorem shows that the equilibrium point of w_λ is parameterized by λ, which thus can be shifted to the origin by an appropriate smooth change of coordinates.

Secondly, from the assumption, the linearization $Dw_\lambda(O)$ has a pair of complex conjugate eigenvalues $\alpha(\lambda) \pm i\omega(\lambda)$ for sufficiently small λ, which satisfies

$$\alpha(0) = 0, \quad \omega(0) = 1.$$

Here we may also assume

$$\alpha'(0) \neq 0$$

without loss of generality, or we may instead introduce an additional parameter α so that the original family w_λ is a subfamily of $\bar{w}_{(\lambda,\alpha)}$ with $\alpha = \alpha(\lambda)$. Then, if we can show that each $\bar{w}_{(\lambda,\alpha)}$ is locally topologically equivalent to some $v_{\varphi(\lambda,\alpha)}$, the same conclusion holds as well for the subfamily $w_\lambda = \bar{w}_{(\lambda,\alpha(\lambda))}$, thereby proving the assertion. In what follows, we adopt such an extension of the family and include α as an independent parameter.

Thirdly, we can put the Hopf singularity into its normal form up to third order terms as in (3.3.7). After this, we take the polar coordinate and rewrite the whole family into the following form:

$$\begin{aligned} \dot{r} &= f(r,\theta;\bar{\lambda}) \\ \dot{\theta} &= g(r,\theta;\bar{\lambda}), \end{aligned} \qquad \bar{\lambda} = (\lambda, \alpha).$$

From the above assumptions, we get

$$\begin{aligned} &f(0,\theta;\bar{\lambda}) = 0, \quad \frac{\partial f}{\partial r}(0,0;\bar{\lambda}) = \alpha, \\ &\frac{\partial^2 f}{\partial r^2}(0,0;0) = 0, \quad \frac{\partial^3 f}{\partial r^3}(0,0;0) = a_1 \neq 0, \\ &g(0,\theta;\bar{\lambda}) = \omega(\bar{\lambda}). \end{aligned}$$

Since $\omega(0) = 1$, we have $g(r,\theta;\bar{\lambda}) > 0$ for sufficiently small $(r,\bar{\lambda})$, and hence it makes sense to consider the differential equation

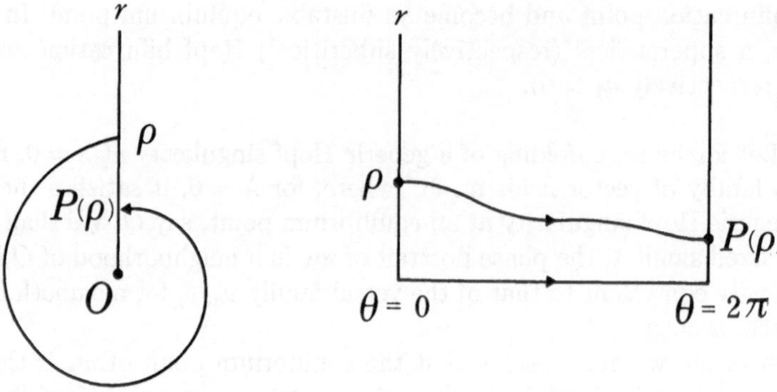

Fig. 3.3.3. The Poincaré map around the origin.

$$\frac{dr}{d\theta} = \frac{f(r,\theta;\bar{\lambda})}{g(r,\theta;\bar{\lambda})} \qquad (r \geq 0,\ 0 \leq \theta \leq 2\pi). \tag{3.3.7}$$

Let $r = r(\rho,\theta;\bar{\lambda})$ be the solution of (3.3.7) with the initial condition $r(\rho,0;\bar{\lambda}) = \rho$, and let $P(\rho;\bar{\lambda}) = r(\rho,2\pi;\bar{\lambda})$. This map P is the mapping assigning an initial point with the point after the angle 2π turn, and is called the Poincaré map. See Fig. 3.3.3. From the condition $f(0,\theta;\bar{\lambda}) \equiv 0$, we get $P(0;\bar{\lambda}) \equiv 0$, and hence $P(\rho;\bar{\lambda})$ takes the form:

$$P(\rho,\bar{\lambda}) \equiv \rho R(\rho;\bar{\lambda}).$$

We shall study the function $R(\rho;\bar{\lambda})$ in detail. The fundamental observation is that if $R(\rho;\bar{\lambda}) = 1$ has a non-zero solution $\rho = \rho_*(\bar{\lambda})$, then it corresponds to a fixed point of P, and hence, to a periodic orbit of the original vector field $\bar{w}_{\bar{\lambda}}$.

A simple but tedious computation shows the following:

Lemma 3.3.12.

(1) $R(0;\lambda,\alpha) = e^{2\pi\alpha}$;

(2) $\dfrac{\partial R}{\partial \rho}(0;\lambda,0) = 0$;

(3) $\dfrac{\partial^2 R}{\partial \rho^2}(0;\lambda,0) = 4\pi a_1$.

From Lemma 3.3.12(1), we have

$$R(0;\lambda,0) = 1 \quad \text{and} \quad \frac{\partial R}{\partial \alpha}(0;\lambda,0) = 2\pi \neq 0,$$

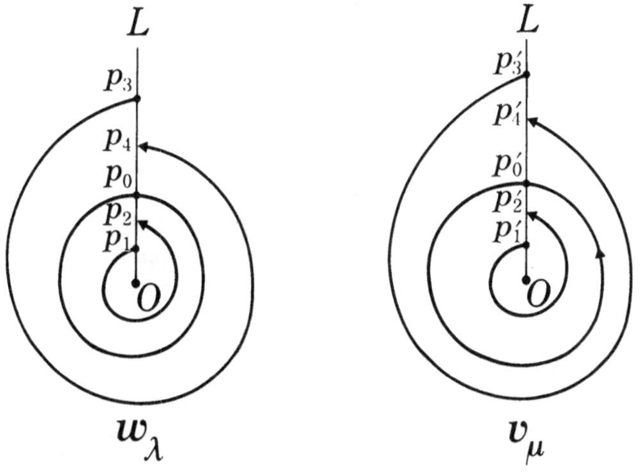

Fig. 3.3.4. Construction of an equivalence homeomorphism for the Hopf bifurcation.

which implies, from the implicit function theorem, the existence of $\alpha = \alpha_*(\rho;\lambda)$ satisfying

$$R(\rho;\lambda,\alpha_*(\rho;\lambda)) \equiv 1, \quad \alpha_*(0;\lambda) = 0.$$

Then

$$\frac{\partial \alpha_*}{\partial \rho}(0;0) = 0 \quad \text{and} \quad \frac{\partial^2 \alpha_*}{\partial \rho^2}(0;0) = -2a_1$$

follow from Lemma 3.3.12(2) and (3), and therefore we have obtained the unique branch

$$\alpha = \alpha_*(\rho;\lambda) = -a_1\rho^2 + O(\rho^3) + O(\rho\lambda)$$

of periodic orbits of $\bar{w}_{\bar{\lambda}}$ bifurcating from the origin.

Finally, we shall show the local topological equivalence of $\bar{w}_{\bar{\lambda}}$ with the versal family v_μ. The vector field v_μ takes the following form in the polar coordinates:

$$\dot{r} = \mu r + a_1 r^3$$
$$\dot{\theta} = 1 + b_1 r^2.$$

Put simply, $\mu = \varphi(\lambda,\alpha) = \alpha$. Assuming $a_1 < 0$, we first consider the case $\mu = \alpha > 0$, when the bifurcated periodic orbit exists. Take the half line L emanating from the origin O in the plane $\mathbb{R}^2$ corresponding to the angle coordinate $\theta = 0$, and let p_0 and p'_0 be the intersection points of the unique periodic orbit of $\bar{w}_{\bar{\lambda}}$ and v_μ with L, respectively. Then take a point p_1 (respectively p'_1) between O and p_0 (respectively O and p'_0) on L. Iterate it once by the Poincaré map P_w of $\bar{w}_{\bar{\lambda}}$ or P_v of v_μ with L, respectively, and denote the image point on L by p_2 (respectively p'_2). Similarly, take a point p_3 (respectively p'_3) between p_0 and $+\infty$ (respectively

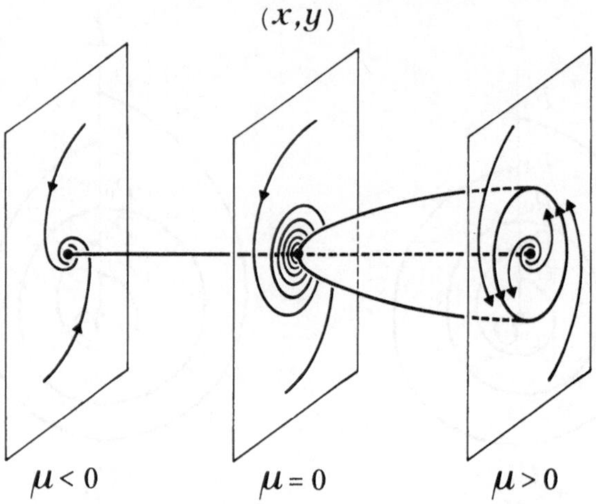

Fig. 3.3.5. Bifurcation diagram for the Hopf bifurcation.

p'_0 and $+\infty$) on L and denote its image after one return by p_4 (respectively p'_4). See Fig. 3.3.4. Now we take any homeomorphism h on L satisfying

$$h(O) = O, \quad h(p_i) = p'_i \quad (i = 0, 1, 2, 3, 4),$$

and conjugating the maps P_w and P_v, namely,

$$h \circ P_w = P_v \circ h.$$

Such a homeomorphism can be easily found first by taking an arbitrary homeomorphism from $[p_1, p_2]$ onto $[p'_1, p'_2]$ and from $[p_3, p_4]$ onto $[p'_3, p'_4]$, and then extending it by the action of P_w and P_v in order to satisfy the conjugacy relation. Clearly this extension is well-defined and gives a homeomorphism on at least a one-sided neighborhood of O in the half line L. Finally, we furthermore extend it to a homeomorphism H on a neighborhood of O in $\mathbb{R}^2$ which gives the local topological equivalence between $\bar{w}_{\bar{\lambda}}$ and v_μ. This can be achieved by parameterizing orbits in terms of the angle coordinate. More precisely, we define H by

$$H(p_w(\theta, x)) = p_v(\theta, h(x)),$$

where $p_w(\theta, x)$ is the point with the angle θ from $x \in L$ along its orbit for $\bar{w}_{\bar{\lambda}}$, and $p_v(\theta, x)$ is similarly defined for v_μ. This obviously gives the topological equivalence.

The other cases can similarly be treated and the proof of Theorem 3.3.11 is thus completed. □

Remark 3.3.13. More careful argument shows that the homeomorphism for the topological equivalence can be taken continuously depending on the parameters.

To the authors' knowledge, there is no written proof for this fact. See [Annabi 1991] for a related result in more degenerate cases.

The bifurcation diagram for the versal family are depicted in Fig. 3.3.5 in the case of $a_1 < 0$, namely, the supercritical Hopf bifurcation. The equilibrium point O is stable when $\mu < 0$. It then loses its stability at $\mu = 0$ and becomes unstable when $\mu > 0$, from which a stable periodic orbit is born instead. The radius of the bifurcating periodic orbit is given by $\sqrt{-\frac{\mu}{a_1}}$.

See [Marsden and McCracken 1976], [Guckenheimer and Holmes 1983], [Vanderbauwhede 1989] for more information as well as application of the Hopf bifurcation theorem.

3.3.4 Bogdanov-Takens Bifurcation

Let v be a C^∞ vector field in $\mathbb{R}^2$ with O being an equilibrium point at which the linearization matrix is given by $A = \begin{pmatrix} 0 & 1 \\ 0 & 0 \end{pmatrix}$. Then it can be transformed to the form in Example 3.3.6(2). In this case the coefficients a_1 and b_1 in the normal form are generically non-zero, and the singularity given by such a vector field is called the *Bogdanov-Takens singularity.* Without loss of generality, we can make $a_1 = b_1 = 1$ by a similarity transformation together with inversion of the time variable, if necessary.

Theorem 3.3.14. *The Bogdanov-Takens singularity v is of codimension two and its versal family is given by*

$$v_{\mu,\nu}(x,y) = \left\{ y\frac{\partial}{\partial x} + (x^2 + xy)\frac{\partial}{\partial y} \right\} + (\mu + \nu y)\frac{\partial}{\partial y}, \quad (\mu, \nu \in \mathbb{R}). \tag{3.3.8}$$

Proof. The proof of this theorem consists of the following seven steps:

(Step 1.) Show that any unfolding w_λ of the Bogdanov-Takens singularity can be put into the form

$$\begin{aligned} \dot{x} &= y \\ \dot{y} &= \mu(\lambda) + \nu(\lambda)y + x^2 + xy + h.o.t., \end{aligned} \tag{3.3.9}$$

by a C^∞ change of coordinates and parameters. Without loss of generality, we may assume the mapping

$$\lambda \mapsto (\mu(\lambda), \nu(\lambda))$$

is of maximal rank two, or we can instead extend the family by introducing the new parameters (μ, ν) as has been done in the proof of the Hopf bifurcation theorem (Theorem 3.3.11). In what follows these two parameters (μ, ν) only play a role.

(Step 2.) Show that a generic saddle-node bifurcation occurs at the line $\mu = 0$, and hence there is no equilibrium point when $\mu > 0$, whereas there are two when $\mu < 0$, one of which is always of saddle type.

(Step 3.) Show that a generic Hopf bifurcation occurs for the equilibrium point other than the saddle at the curve

$$\mu = h_1(\nu) \approx -\nu^2 \quad \text{with} \quad \nu > 0,$$

in the (μ, ν)-parameter space, and, near this curve, the equilibrium point is an unstable focus when $\mu > h_1(\nu)$, while it changes to a stable focus when $\mu < h_1(\nu)$, together with an unstable limit cycle bifurcating off from the equilibrium point. This shows that the Hopf bifurcation is subcritical.

(Step 4.) Prove that there exists a homoclinic orbit based at the saddle point when

$$\mu = h_2(\nu) \approx -\frac{49}{25}\nu^2 \quad \text{with} \quad \nu > 0.$$

(Step 5.) Prove that there exists a unique periodic orbit when

$$h_2(\nu) < \mu < h_1(\nu), \quad \nu > 0.$$

(Step 6.) Verify that there is no other bifurcation than those described above.

(Step 7.) Construct a homeomorphism which gives the fiber C^0 equivalence between (3.3.8) and (3.3.9).

Steps 1-3 can be carried out by elementary calculations as described in, for instance, [Carr 1981], [Guckenheimer and Holmes 1983], and [Broer, Dumortier, van Strien and Takens 1991; Chapter 7], and hence we omit them here. In what follows, we shall give an outline of the proof of Steps 4 and 5, which are the crucial parts of the whole proof. Step 6 is also less trivial and it is verified by using what is called the rotational property. See again [Broer, Dumortier, van Strien and Takens 1991; Chapter 7] for details. Once all these steps are performed, the last assertion in Step 7 follows almost immediately because we have given the complete description of the bifurcation for an unfolding.

Make a change of variables and parameters in (3.3.9) as

$$x = \varepsilon^2\bar{x},\ y = \varepsilon^3\bar{y},\ \mu = -\varepsilon^4,\ \nu = \varepsilon^2\bar{\nu},\ t = \tau/\varepsilon,$$

then we obtain

$$\begin{aligned} \bar{x}' &= \bar{y} \\ \bar{y}' &= \bar{x}^2 - 1 + \varepsilon(\bar{\nu}\bar{y} + \bar{x}\bar{y}) + O(\varepsilon^2), \end{aligned} \qquad \left(' = \frac{d}{d\tau} \right). \tag{3.3.10}$$

When $\varepsilon = 0$, this equation gives the Hamiltonian system in Example 3.2.33, which possesses a homoclinic orbit $(\bar{x}_*(\tau), \bar{y}_*(\tau))$ based at an equilibrium point $(1, 0)$. In an unfolding (3.3.10) of the equation, we take $\bar{x}_{\pm}(\varepsilon,\bar{\nu})$ as shown in Fig. 3.3.6, then the condition for the existence of a homoclinic orbit in (3.3.10) is given by

$$H(\bar{x}_{-}(\varepsilon,\bar{\nu})) - H(\bar{x}_{+}(\varepsilon,\bar{\nu})) = 0,$$

with the Hamiltonian function given in Example 3.2.33.

By a simple computation which is detailed in [Carr 1981], the above condition is equivalent to

$$\bar{\nu} = -\frac{\displaystyle\int_{-\infty}^{\infty} \bar{x}_*(\tau)\bar{y}_*(\tau)^2 d\tau}{\displaystyle\int_{-\infty}^{\infty} \bar{y}_*(\tau)^2 d\tau} + O(\varepsilon),$$

and the integrals involved can be explicitly evaluated by using the exact homoclinic solution

$$(\bar{x}_*(\tau), \bar{y}_*(\tau)) = \left(1 - 3\mathrm{sech}^2\left(\frac{\tau}{\sqrt{2}}\right), 3\sqrt{2}\mathrm{sech}^2\left(\frac{\tau}{\sqrt{2}}\right)\tanh\left(\frac{\tau}{\sqrt{2}}\right)\right),$$

yielding

$$\bar{\nu} = \frac{5}{7} + O(\varepsilon).$$

Going back to the original parameters (μ,ν), this gives a curve of the form

$$\mu = h_2(\nu) = -\left(\frac{5}{7}\right)^2 \nu^2 + o(\nu^2),$$

with $\nu > 0$ as desired, and the proof of Step. 4 is thus completed.

The existence of a periodic orbit is given by the equation

$$\bar{\nu} = -P(h) + O(\varepsilon)$$

using similar reasoning, where

$$P(h) = \frac{\displaystyle\oint \bar{x}_h(\tau)\bar{y}_h(\tau)^2 d\tau}{\displaystyle\oint \bar{y}_h(\tau)^2 d\tau} = \frac{\displaystyle\int_{\Gamma_h} xy\,dx}{\displaystyle\int_{\Gamma_h} y\,dx} \quad \left(-\frac{2}{3} \le h \le \frac{2}{3}\right).$$

This means that, for $\bar{\nu}$ given above, the corresponding vector field (3.3.10) has a periodic orbit which persists from the periodic orbit $\Gamma_h : (\bar{x}_h(\tau), \bar{y}_h(\tau))$ of the unperturbed Hamiltonian system corresponding to the energy level $H(x,y) = h$.

The integrals in the numerator and the denominator of $P(h)$ are Abelian integrals since the level curve $H(x,y) = h$ defines an elliptic curve. This observation helps to study the dependence of the function $P(h)$ on h and one can show that

$$P'(h) > 0 \quad \text{for} \quad -\frac{2}{3} \le h \le \frac{2}{3}, \tag{3.3.11}$$

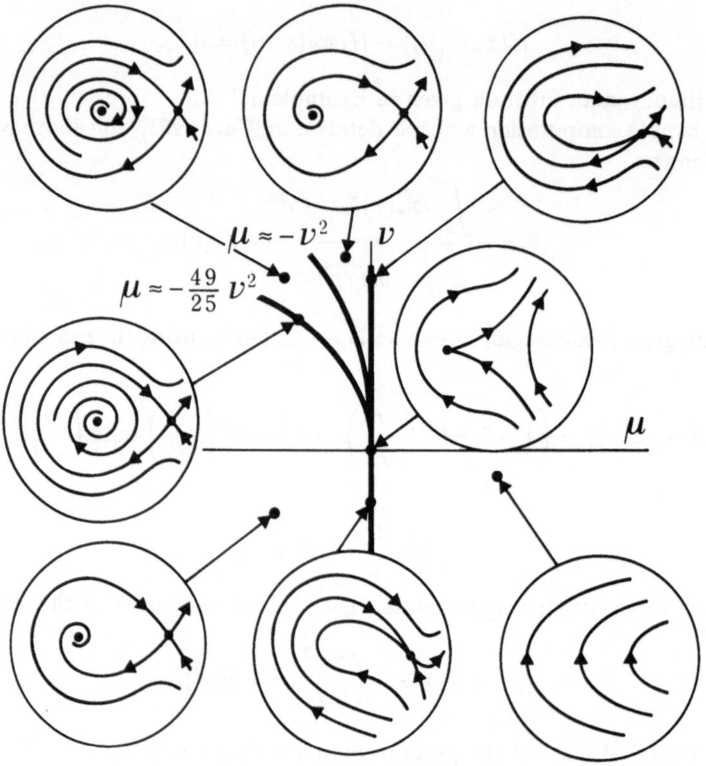

Fig. 3.3.6. Bifurcation diagram for the Bogdanov-Takens bifurcation.

by careful computation using the fact that $P(h)$ satisfies a Ricatti differential equation

$$P'(h) = \frac{1}{4-9h^2}\{7P(h)^2 - 3hP(h) - 5\}.$$

See [Carr 1981] and [Cushman and Sanders 1985] for details.

From (3.3.11), the function $P(h)$ is monotone in h, and hence the periodic orbit corresponding to $\bar{\nu}$ is unique. This completes the proof of Step 5.

We remark that the integrals appearing above are those now known as the Melnikov integrals. Melnikov integrals for homoclinic orbits will be studied in Section 3.5 in detail and Melnikov integrals for periodic orbits are explained in [Guckenheimer and Holmes 1983]. □

Figure 3.3.6 illustrates changes of the phase portraits of $v_{\mu,\nu}$ as the parameter (μ, ν) varies. Starting from some parameter (μ, ν) in the first quadrant where the corresponding vector field has no equilibrium point, we follow the change of phase portraits along a circle around the origin in the parameter space counterclockwise. First we observe the appearance of a saddle-node singularity as the parameter passes the positive ν-axis, from which a pair of one saddle and one source bifurcate as the parameter moves into the second quadrant. Then the source undergoes the subcritical Hopf bifurcation for (μ, ν) satisfying $\mu = h_1(\nu) \approx -\nu^2$ and an unstable periodic orbit appears from the source, which then grows its radius as the parameter moves further and finally touches the saddle when $\mu = h_2(\nu) \approx -\frac{49}{25}\nu^2$ holds. At this moment, the periodic orbit and the saddle equilibrium turn to a homoclinic orbit which disappears immediately after further movement of the parameter. The saddle and a sink still remain until the parameter crosses the negative ν-axis until which no further bifurcation occurs. In the fourth quadrant, these two equilibrium points disappear through the saddle-node bifurcation, and we finally come back to the initial simple phase portrait. These bifurcation curves for saddle-node, Hopf and homoclinic bifurcations emanate from the codimension two point in the parameter space corresponding to the Bogdanov-Takens singularity.

Remark 3.3.15. The Bogdanov-Takens bifurcation has been repeatedly studied, given alternative proofs, improved and simplified by many people since [Bogdanov 1976a,b] and [Takens 1974b], some of which are referred to in the sketch of the proof presented above. Our proof basically follows the idea of [Carr 1981] together with [Broer, Dumortier, van Strien and Takens 1991; Chapter 7].

Note that the equivalence homeomorphism bringing an unfolding into the versal family (3.3.8) can be in fact taken continuously depending on the parameters. This has already been claimed by [Bogdanov 1976a,b] but the proof recently turned out to be incorrect and has been corrected by [Annabi, Annabi and Dumortier 1992] and [Dumortier and Roussarie 1990].

Note also that the uniqueness of the periodic orbit between the Hopf bifurcation and the homoclinic bifurcation provides an example of the finite cyclicity result for planar vector fields. The problem showing the finiteness of limit cycles in planar vector fields is now known as a part of the Hilbert 16th problem, which attempts to give an estimate of the number of limit cycles in polynomial vector fields on the plane. Here we only refer to [Ye 1986] and [Il'yashenko 1991] for a general reference of this topic.

3.3.5 Symmetry and Bifurcations

Symmetry is an additional structure of dynamical systems which often provides us with a better understanding of their dynamical properties.

Definition 3.3.16.

(1) A vector field v admits *symmetry* if the equation

$$\frac{dx}{dt} = v(x)$$

is invariant under a transformation γ of (x, t).

(2) A family of vector fields v_μ admits *symmetry* if the equation

$$\frac{dx}{dt} = v_\mu(x)$$

is invariant under a transformation γ of (x, t, μ).

The transformation γ itself is also called a symmetry of the vector field or the family of vector fields.

Example 3.3.17.

(1) The Lorenz equation

$$\begin{aligned} \dot{x} &= \sigma(y - x) \\ \dot{y} &= rx - y - xz \\ \dot{z} &= -bz + xy \end{aligned}$$

is invariant under

$$\gamma : (x, y, z) \mapsto (-x, -y, z),$$

hence admitting symmetry.

(2) The equation

$$\begin{aligned} \dot{x} &= y \\ \dot{y} &= x^2 - 1 \end{aligned}$$

from Example 3.2.7 is invariant under

$$\gamma : (x, y, t) \mapsto (x, -y, -t),$$

and hence it is a symmetry of the equation. Since the symmetry reverses the time-axis, such a system is called a reversible system.

(3) The family

$$\begin{aligned} \dot{x} &= \mu x - y - x(x^2 + y^2) \\ \dot{y} &= x + \mu y - y(x^2 + y^2) \end{aligned}$$

from Example 3.2.32 has a symmetry of the family given by

$$\gamma : (x, y, t, \mu) \mapsto (x \cos\theta - y \sin\theta, x \sin\theta + y \cos\theta, t, \mu),$$

for any θ, since it is rotationally invariant. This symmetry is in fact a symmetry of each individual vector field of the family, because it fixes μ.

(4) The equation

$$\dot{x} = y$$
$$\dot{y} = x^2 - 1 + \mu y$$

from Example 3.2.33 has a symmetry of the family given by

$$\gamma : (x, y, t, \mu) \mapsto (x, -y, -t, -\mu),$$

which cannot be reduced to a symmetry of individual vector fields of the family (except at $\mu = 0$).

The Double Scroll circuit also admits a symmetry given by reflection with respect to the origin as explained in the previous chapters. This symmetry of the Double Scroll circuit originally comes from the symmetrical nature of the electrical circuit, namely the odd-symmetry of the nonlinear resistor. Likewise, a symmetry of a concrete physical system implies that in its model equation, if the equation is derived by taking account of the symmetry. For instance, the simple pendulum has a reflectional symmetry with respect to the direction of gravity, and hence its model equation

$$\ddot{x} = -\sin x$$

has a symmetry $\gamma_1 : x \mapsto -x$. Furthermore it has another symmetry $\gamma_2 : t \mapsto -t$ which comes from the conservative nature of the system. In fact it has no longer the symmetry γ_2 if we add the friction:

$$\ddot{x} = -\sin x - c\dot{x} \qquad (c \neq 0).$$

If a system has a symmetry, a property of a particular orbit is inherited by its symmetric image under the symmetry transformation. For instance, if a system with a symmetry γ has a periodic orbit which is not invariant under γ, there exists another periodic orbit obtained as the image by γ of the original periodic orbit. Such an effect of symmetry makes dynamics and bifurcations special, far from generic, and far from stable under general non-symmetric perturbations. Nevertheless it has its own importance since we often encounter systems with certain symmetries. In this subsection we do not give a general description of dynamical systems with symmetry but only refer to [Golubitsky and Schaeffer 1985], [Golubitsky, Stewart and Schaeffer 1988], and [King and Stewart 1992] as well as references therein for general literature of the topic. However, we give the simplest result of bifurcations with symmetry, which is called the *pitchfork bifurcation*.

Theorem 3.3.18. *Let $\dot{x} = v(x, \mu)$ be a one-parameter family of vector fields on the real line $\mathbb{R}$ with the symmetry $\gamma : x \mapsto -x$, namely it satisfies*

$$v(-x, \mu) = -v(x, \mu).$$

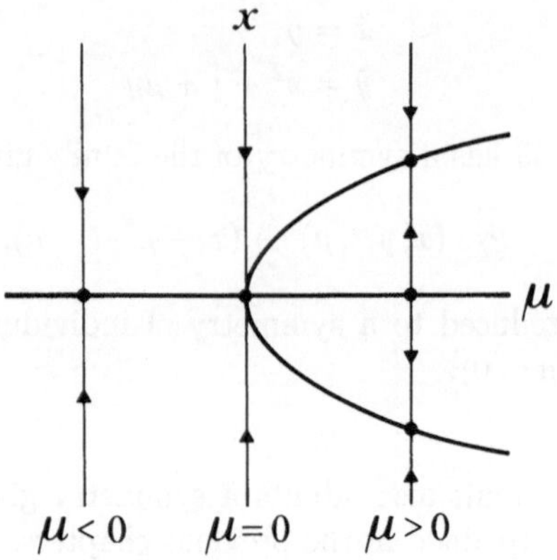

Fig. 3.3.7. Bifurcation diagram for the pitchfork bifurcation.

This in particular implies $v(0, \mu) \equiv 0$, *and hence it has an equilibrium point* $x = 0$ *for any* μ.

Suppose, furthermore, it satisfies the following conditions at $\mu = 0$:

$$\frac{\partial v}{\partial x}(0,0) = 0, \quad \frac{\partial^3 v}{\partial x^3}(0,0) \neq 0, \quad \frac{\partial^2 v}{\partial x \partial \mu}(0,0) \neq 0.$$

Then there exists a branch of equilibrium points bifurcating off from $x = 0$ *at* $\mu = 0$, *which is given by*

$$\mu = \varphi(x^2), \quad \varphi'(0) \neq 0$$

for a smooth function φ. *This is a unique bifurcating branch except the trivial branch* $x \equiv 0$. *See Fig. 3.3.7 for the bifurcation diagram.*

The proof of this theorem proceeds in almost the same way as for the saddle-node bifurcation theorem (Theorem 3.3.9).

Proof. From the symmetry, the vector fields assume the form

$$v(x, \mu) = x \cdot V(x^2, \mu)$$

for a smooth function $V(y, \mu)$, for which the assumption of the theorem implies

$$V(0,0) = 0, \quad \frac{\partial V}{\partial y}(0,0) \neq 0, \quad \frac{\partial V}{\partial \mu}(0,0) \neq 0.$$

Therefore, the implicit function theorem shows the existence of a solution $\mu = \varphi(y)$ satisfying

$$V(y, \varphi(y)) \equiv 0, \quad \varphi(0) = 0, \quad \varphi'(0) \neq 0,$$

which concludes

$$v(x, \varphi(x^2)) = xV(x^2, \varphi(x^2)) \equiv 0.$$

The proof is hence completed. □

Remark 3.3.19. As noted above, dynamical systems with symmetry is far from being generic. This observation also applies to their bifurcations. Theorem 3.3.18 is basically for vector fields with a simple zero eigenvalue in the linearization at a bifurcation point for which generically (without any symmetry constraint) the saddle-node bifurcation occurs as described in Theorem 3.3.9. However, the symmetry $\gamma : x \mapsto -x$ forces the second order derivative $\frac{\partial^2 v}{\partial x^2}(0,0)$ to automatically vanish, and any even order derivatives as well. Therefore, the bifurcation aspect in this situation becomes quite different from that of the saddle-node bifurcation as shown in Fig. 3.3.2. If the family of vector fields is perturbed in a non-symmetric way, it suddenly loses its symmetric character, for instance the existence of the trivial branch $x \equiv 0$, and recovers a saddle-node bifurcation at a nearby point. Such a bifurcation, sometimes called the imperfect bifurcation, is an interesting subject, but we do not pursue it here. Interested readers can find the detail in [Golubitsky and Schaeffer 1985].

Example 3.3.20. Let us take the Lorenz equation with fixed $\sigma \neq 0$ and $b > 0$, and consider r as a bifurcation parameter. Then the equilibrium point O undergoes the pitchfork bifurcation at $r = 1$. In fact, at $r = 1$, the linearization matrix at O possesses a simple zero eigenvalue and the reduced center manifold equation verifies all the assumptions of Theorem 3.3.18 together with the symmetry condition which comes from the original symmetry of the Lorenz equation. We leave the explicit calculation to readers and instead, we directly calculate the branch of equilibrium points of the Lorenz equation.

The origin O is always an equilibrium point for any σ, r, b. Besides it, the Lorenz equation can have an equilibrium point if the following equation has a solution:

$$\begin{aligned} y - x &= 0 \\ rx - y - xz &= 0 \qquad (x,y,z) \neq (0,0,0), \\ -bz + xy &= 0, \end{aligned}$$

which is equivalent to

$$x^2 = b(r-1), \quad y = x, \quad z = r - 1.$$

This has a non-trivial solution if and only if $r > 1$, and hence it fits with the bifurcation diagram in Fig. 3.3.7. Note that the equation of the branch contains the relation

$$r = \frac{1}{b}x^2 + 1,$$

which corresponds to the equation of the bifurcating branch $\mu = \varphi(x^2)$ of Theorem 3.3.18.

For a single vector field or a family of vector fields, the set of all symmetries clearly forms a group. For instance, the group of symmetry of the Lorenz equation is $\mathbb{Z}_2$, which is a finite group with only two elements, whereas the symmetry group of Example 3.3.17(3) is $SO(2) \cong S^1$, the group of rotations in the plane. Therefore, the group theory helps very much the study of dynamical systems with symmetries. See [Golubitsky, Stewart, and Schaeffer 1985, 1988] again for details.

Finally, we remark that dynamical systems with symmetries sometimes appear even when we study general systems without symmetry a priori. For instance, as we have seen in Example 3.3.6(1), the (finite order truncation of) normal form of the Hopf singularity admits rotational symmetry, and therefore we can make use of this symmetry for the study of the Hopf bifurcation in order to describe the dynamics of the normal form, and then we add the effect of non-symmetric flat terms which may or may not change the conclusion obtained from the normal forms with symmetry. In fact, we do not need to proceed in this way in the case of the generic Hopf bifurcation, but this point of view turns out to be very useful for the study of other degenerate singularities. See, for instance [Guckenheimer and Holmes 1983] and references therein. Sometimes this point of view clarifies a very subtle property of dynamical systems caused by highly symmetric systems together with very small but non-symmetric flat perturbations. Such a subtlety can be found in [Takens 1973b,1974a] and [Broer and Vegter 1984].

3.3.6 Other Degenerate Singularities

Equilibrium points with eigenvalues having zero real parts are sometimes called *degenerate singularities*. We have seen several such degenerate singularities and their unfoldings in the previous subsections. One may also see that normal forms provide us with good representatives for the classification of such degenerate singularities. The study of singularities for vector fields received a strong influence from the theory of singularities of smooth mappings ([Bröcker and Lander 1975]), or the catastrophe theory ([Thom 1972], [Arnold 1984]). In fact, the catastrophe theory may be said as the theory of degenerate singularities for gradient vector fields. In this sense, the study of degenerate singularities and their unfoldings for (general) vector fields is a generalization of the (so-called elementary) catastrophe theory. Compared with the theory of singularities for smooth mappings, the study of degenerate singularities for vector fields is still far from complete: singularities in two dimensions are nicely classified by [Dumortier 1977], but we soon encounter a pathological situation when we go to higher dimensions. For instance, a non-stabilizable jet arises in singularities of low codimensions. Namely, there exists a singularity with the property that, even if one specifies any order jet of the singularity, higher order terms added to the jet can change its topological character. This has first been found by [Takens 1973b]. See [Broer, Dumortier, van Strien and Takens 1991; Chapter 7] for more information on the non-stabilizable jets and related topics. As for the study of unfoldings of degenerate singularities, much less is known even for two-dimensional vector fields. Nevertheless, the study of the singularities and their unfoldings are important, not only for the mathematical interest but also for practical reasons. Indeed we encounter many kinds of degenerate singularities in concrete model equations of physical, chemical, biological, and mechanical systems.

Degenerate singularities of codimension two or more at equilibrium points other than the Bogdanov-Takens singularity are listed as follows:

(1) degenerate saddle-node: a simple zero eigenvalue in the linearization with additional degeneracy in its higher order terms ([Broer, Dumortier, van Strien and Takens 1991; Chapter 7]);

(2) degenerate Hopf: a pair of pure imaginary eigenvalues in the linearization with additional degeneracy in its higher order terms ([Takens 1973c], [Annabi 1991]);

(3) the interaction of simple zero and pure imaginary eigenvalues in the linearization with non-degenerate higher order terms ([Langford 1979], [Langford and Iooss 1980], [Guckenheimer 1981, 1984], [Scheurle and Marsden 1984], [Broer and Vegter 1984], [Carr, Chow and Hale 1985], [van Gils 1985]);

(4) the interaction of two pairs of pure imaginary eigenvalues (which are rationally independent) in the linearization with non-degenerate higher order terms ([Guckenheimer 1984], [Żoładek 1983]).

Moreover, in addition to the above,

(5) degenerate Bogdanov-Takens: a zero eigenvalue of multiplicity two with additional degeneracy in higher order terms ([Dumortier, Roussarie and Sotomayor 1987, 1991]. See also [Mardešić 1990, 1992]);

(6) the case of zero eigenvalue of multiplicity three with non-degenerate higher order terms ([Medveď 1984], [Arneodo Coullet, Spiegel, and Tresser 1985], [Yu and Huseyin 1988])

have been investigated as degenerate singularities of codimension three. Among these singularities, the complete versal families are only obtained for (1), (2) and (5), which are all one or two-dimensional vector fields. For cases (3) and (4), certain "reduced" planar vector fields are completely understood, but the entire vector fields on whole $\mathbb{R}^3$ or $\mathbb{R}^4$ have more complicated dynamics which are not yet clarified (see the last paragraph of Subsection 3.3.5). In general, if we go to more than two-dimensional vector fields, we have to face a lot of difficulties. As we saw in the study of the Bogdanov-Takens bifurcation, bifurcations of degenerate singularities are often closely related to those of global orbits in the vector field, such as homoclinic orbits, and in particular, one cannot avoid the existence of chaos in higher dimensional vector fields. Therefore, we find it impossible to reach a complete understanding of these bifurcations without studying such global and complicated dynamics. Section 3.5 will give some information of those bifurcations of global orbits. We shall come back again to the connection between local and global bifurcations at the end of Section 3.5.

3.4 Dynamics and Bifurcations for Discrete Dynamical Systems

In this section, we shall summarize some results on discrete dynamical systems and their bifurcations. Since we are concerned with continuous dynamical systems, we only give some very fundamental notions and ideas for discrete dynamical systems which are sometimes similar to their continuous counterparts. We do not give proofs for most results explained in this section, and instead we mainly work with several basic examples exhibiting typical and important dynamical behaviors.

3.4.1 Discrete Dynamical Systems

As we have seen in Section 3.2, a continuous dynamical system is a continuous one-parameter family of mappings $\{\varphi^t\}_{t\in\mathbb{R}}$ satisfying the group property (G1) and (G2) with respect to the continuous time parameter t. In contrast, a discrete dynamical system is a discrete one-parameter family as defined below.

Definition 3.4.1. A (smooth) *discrete dynamical system* on a manifold M is given by a smooth diffeomorphism $\varphi : M \to M$. Such a diffeomorphism φ generates the family $\{\varphi^n\}_{n\in\mathbb{Z}}$ of mappings given by its n-times iterate parameterized by $n \in \mathbb{Z}$, and hence it is called a discrete dynamical system. We sometimes consider a homeomorphism φ, or a (non-invertible) mapping $\varphi : M \to M$ as a discrete dynamical system as well. The latter in particular is called a discrete semi-dynamical system.

The notion of orbits, periodic orbits, limit sets and invariant sets are defined in a similar way as those for continuous dynamical systems. In particular, a point x fixed by a discrete dynamical system φ, that is $\varphi(x) = x$, is called a *fixed point* instead of an equilibrium point in the continuous case. We also define the ω-limit set $\omega(x)$ (respectively α-limit set $\alpha(x)$) of a point x as the set of all accumulation points of the positive orbit $\{\varphi^n(x)\}_{n\geq 0}$ (respectively the negative orbit $\{\varphi^n(x)\}_{n\leq 0}$). The limit set $L(\varphi)$ for a discrete dynamical system given by φ is the closure of the union of all $\alpha(x)$ and $\omega(x)$ for $x \in M$.

Example 3.4.2.

(1) A linear map $\varphi(x) = Ax$ on $\mathbb{R}^n$ gives a discrete dynamical system. If $\det A = 0$, it is a semi-dynamical system. If any eigenvalue of A is of modulus less than 1, then the origin O is stable, since, for any x, $\varphi^n(x) \to 0$ as $n \to \infty$. On the contrary, if the matrix A has an eigenvalue with modulus bigger than 1, then we have $|\varphi^n(x)| \to \infty$ as $n \to \infty$ for x being an eigenvector associated with the eigenvalue, and hence O is unstable. A linear invertible map $\varphi(x) = Ax$ is called hyperbolic if A has no eigenvalues with modulus 1.

(2) The map

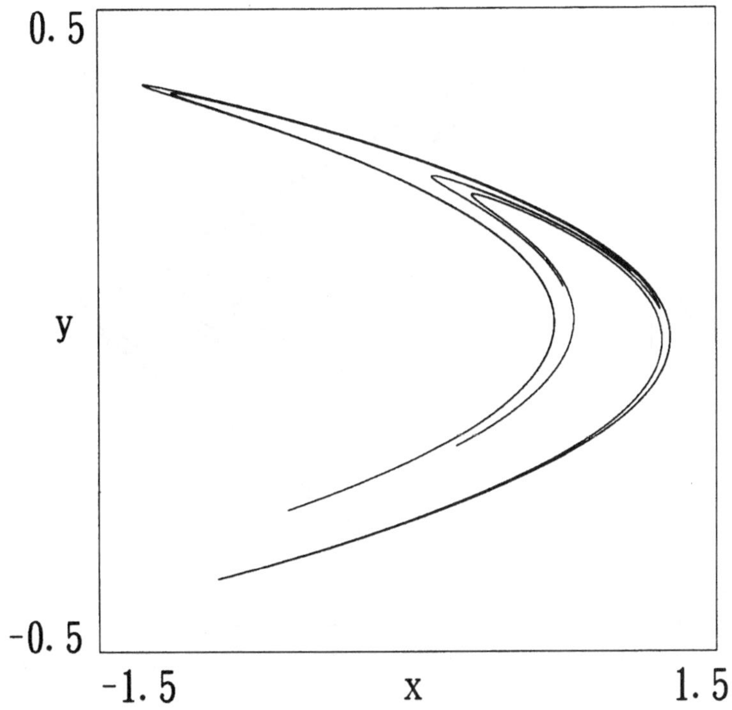

Fig. 3.4.1. The Hénon attractor.

$$H_{a,b} : \begin{pmatrix} x \\ y \end{pmatrix} \mapsto \begin{pmatrix} y + 1 - ax^2 \\ bx \end{pmatrix} \tag{3.4.1}$$

on $\mathbb{R}^2$ is invertible with a polynomial inverse

$$H^{-1}_{a,b} \begin{pmatrix} X \\ Y \end{pmatrix} = \begin{pmatrix} \frac{Y}{b} \\ X - 1 + \frac{aY^2}{b^2} \end{pmatrix}$$

for any a and non-zero b, and hence is a diffeomorphism. This is called the *Hénon mapping*, since it was originally introduced by [Hénon 1976], and according to the computer simulation given in that paper, the map exhibits a seemingly very irregular chaotic behavior for $a = 1.4$ and $b = 0.3$, as in Fig. 3.4.1.

There exists a piecewise-linear analogue to the Hénon mapping: the *Lozi mapping* ([Lozi 1978]), which takes the form

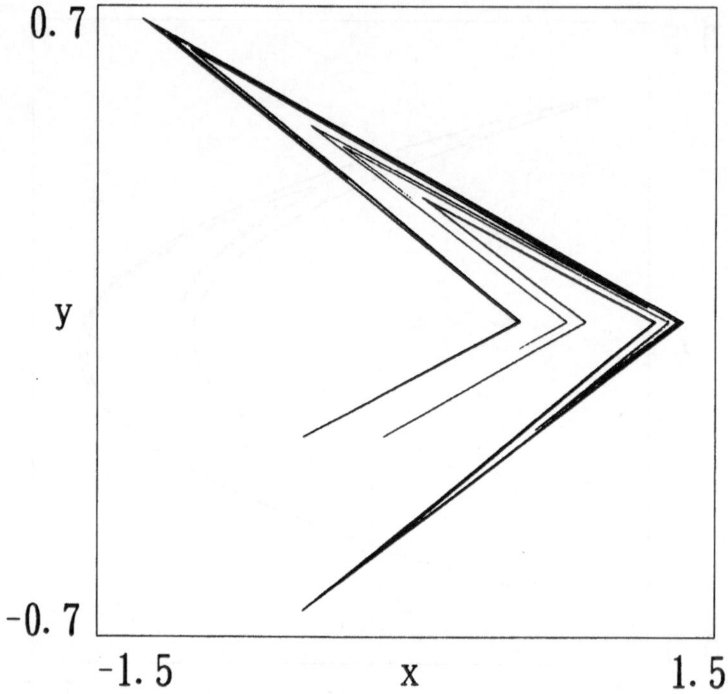

Fig. 3.4.2. The Lozi attractor.

$$L_{a,b} : \begin{pmatrix} x \\ y \end{pmatrix} \mapsto \begin{pmatrix} y + 1 - a|x| \\ bx \end{pmatrix}.$$

Computer simulation ([Lozi 1978]) with $a = 1.7$ and $b = 0.5$ shows a similar kind of chaotic behavior as in Fig. 3.4.2, and unlike the attractor in the Hénon mapping, this attractor has been well-understood (see B of Subsection 3.4.7).

(3) The map $\varphi(x) = 4x(1-x)$ is a (non-invertible) mapping defined on the unit interval $[0,1]$ into itself, and hence it gives a semi-dynamical system. This map is a special case of the so-called logistic family $\varphi_a(x) = ax(1-x)$ with $a = 4$. The dynamics of this mapping is easily seen after changing the variable $x = \sin^2 \pi\theta$, by which we have the piecewise-linear mapping $\psi(x) = 1 - |2x - 1|$. Figure 3.4.3 depicts the graphs of φ and ψ. From its expression, we can show that ψ has a periodic point of an arbitrary period, and hence so does φ. See C of Subsection 3.4.4 for details.

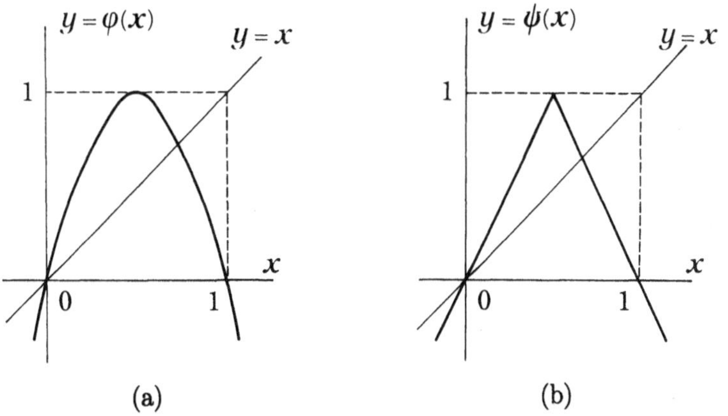

Fig. 3.4.3. Graphs of the one-dimensional mappings. (a) The quadratic map. (b) The piecewise-linear map.

(4) Consider an ordinary differential equation

$$\dot{x} = f(x,t) \qquad (x \in \mathbb{R}^n) \tag{3.4.2}$$

depending explicitly on time t, and assume that

$$f(x, t+T) = f(x,t), \qquad (T > 0).$$

In other words, this is a time-periodic ordinary differential equation. Such an equation frequently appears in the study of systems with time periodic forcing. For instance, a periodically forced pendulum is described by the equation

$$\ddot{x} = -\omega^2 \sin x - c\dot{x} + \varepsilon \cos t,$$

and the so-called Duffing equation ([Duffing 1918], [Moon 1987])

$$\ddot{x} + \delta\dot{x} - x + x^3 = \gamma \cos \omega t \tag{3.4.3}$$

describes the dynamics of a buckled beam with time periodic external force. For the study of such time-periodic ordinary differential equations, it is convenient to take the period map or the Poincaré map defined as follows. Let $x(t, x_0)$ be a solution to the equation (3.4.2) with the initial condition x_o at $t = 0$. Then the period map Π is defined by $\Pi(x_0) = x(T, x_0)$, that is, the period map assigns the point after one period from the initial condition. This map is visualized in Fig. 3.4.4 by extending the equation as

$$\begin{aligned} \dot{x} &= f(x,t) \\ \dot{t} &= 1. \end{aligned} \tag{3.4.4}$$

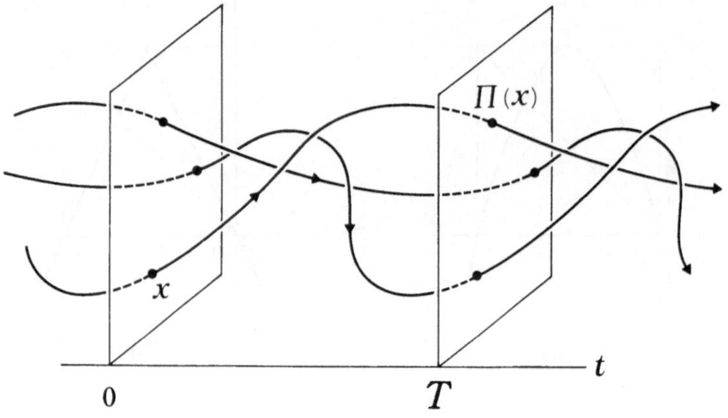

Fig. 3.4.4. The period map for Equation (3.4.2).

Therefore, it is enough to study the period map Π for the investigation of the dynamics of (3.4.2). In particular, a T-periodic orbit in (3.4.2) corresponds to a fixed point of the map Π, while an nT-periodic orbit for (3.4.2) corresponds to an n-periodic point for Π.

The *Poincaré map* can be defined in more general situations. Namely, a continuous dynamical system with a periodic orbit induces a flow-defined map on a (sufficiently small) disc which is transverse to the periodic orbit. Moreover, it is usually possible in the same way that even in a neighborhood of a homoclinic orbit or of a cycle consisting of several heteroclinic orbits, a discrete map induced from the flow is defined at least on a part of the transverse disc to the homoclinic or heteroclinic orbit. See Section 3.5 for more precise definition. Any of such maps are called the Poincaré map.

On the other hand, if we have a diffeomorphism f on a manifold M, we can construct a flow on the fiber space $\tilde{M} = M \times [0,1]/\sim$, where the quotient is taken by identifying the points $(x, 1)$ with $(f(x), 0)$ in $M \times [0,1]$. The flow φ^t on the space $\tilde{M}$ is given by

$$\varphi^t(x, s) = (f^k(x), t + s - k) \qquad (0 \leq s \leq 1),$$

where k is the greatest integer less than $t + s$. We can even embed the flow on $\tilde{M}$ into a bigger space. In particular, the ordinary differential equation (3.4.4) can be thought of as the flow constructed in this way from the Poincaré map or the period map Π. This procedure of constructing a flow from a diffeomorphism is called the *suspension*. See [Palis and de Melo 1982] for more details.

In this way, continuous and discrete dynamical systems are closely related, and hence the results given in this section apply to continuous dynamical systems as well under appropriate interpretation. We will see some such applications in Section 3.5 in the context of homoclinic and heteroclinic bifurcations.

3.4.2 Basic Theorems and Structural Stability

Analogous statements such as the Hartman-Grobman theorem and the stable manifold theorem also hold for discrete dynamical systems.

Definition 3.4.3. (Topological Conjugacy) Two discrete dynamical systems on M given by diffeomorphisms φ_1 and φ_2 are said to be *topologically conjugate* if there exists a homeomorphism $h : M \to M$ such that

$$\varphi_1 \circ h = h \circ \varphi_2$$

holds.

Definition 3.4.4. (Hyperbolic Periodic Points) A periodic point p of a diffeomorphism φ is called *hyperbolic* if $D\varphi^n(p)$ has no eigenvalues with modulus 1, where n is the period of p. If $n = 1$, then it is called a hyperbolic fixed point. A hyperbolic periodic point p of a diffeomorphism φ with period n is a hyperbolic fixed point of φ^n by definition.

Theorem 3.4.5. (Hartman-Grobman) *A diffeomorphism φ in a neighborhood of its hyperbolic fixed point is locally topologically conjugate to its linearization $D\varphi(p)$.*

Theorem 3.4.6. (Stable Manifold Theorem) *For a hyperbolic fixed point O of a C^r diffeomorphism φ on M, the set*

$$W^s(O) = \{x \in M \mid \lim_{n\to+\infty} \varphi^n(x) = O\}$$

is a C^r injectively immersed submanifold in M with $T_O W^s(O) = E^s$ and is called the stable manifold of O, where E^s is the stable subspace of the linearization $D\varphi(O)$, that is, the sum of the generalized eigenspaces corresponding to eigenvalues with a modulus less than 1.

The same holds for the unstable manifold of O:

$$W^u(O) = \{x \in M \mid \lim_{n\to-\infty} \varphi^n(x) = O\}$$

as well. Analogous to the case of continuous dynamical systems, this theorem is implied by its local version:

Theorem 3.4.7. *Let φ be a C^r diffeomorphism on M and O be its hyperbolic fixed point. Let ε be a positive number and define*

$$W^s_{loc}(O) = \{x \in M \mid \mathrm{dist}(\varphi^n(x), O) < \varepsilon \ \textit{ for all } n \geq 0\}.$$

If ε is sufficiently small, then $W^s_{loc}(O)$ is a C^r embedded disc and is tangent to E^s at O. Furthermore, it has the following characterization:

$$W^s_{loc}(O) = \{x \in M \mid \mathrm{dist}(\varphi^n(x), O) < Ce^{-\alpha n} \ \textit{ for all } n \geq 0\}$$

for some positive constants, α and C.

The proof of this last theorem can be done along the same lines as the proof of the local stable manifold theorem for flows indicated in Subsection 3.2.4 by just taking the discrete analogue of the differential equation and the integral equation. See for instance, [Palmer 1988]. However, a more conventional proof uses a slightly different idea called the graph transformation method. This type of proof can be found in [Hirsch and Pugh 1970].

3.4.3 Elementary Bifurcations

In this subsection, we shall give three kinds of bifurcations near a fixed point of a mapping as a discrete dynamical system. These are very fundamental bifurcations and can be seen in one-parameter families of mappings.

A Saddle-Node Bifurcation Consider a family of smooth mappings $\varphi_\mu : \mathbb{R} \to \mathbb{R}$. Assume that, when $\mu = 0$, the mapping φ_0 possesses a fixed point, say $x = 0$, and the linearization of φ_0 at 0 vanishes, that is

$$D\varphi_0(0) = 1, \ \text{and } D^2\varphi_0(0) \neq 0.$$

We call such a mapping φ_0 a *generic saddle-node singularity* for mappings.

Theorem 3.4.8. *The mapping*

$$x \mapsto x \pm x^2 + \mu$$

gives a versal family of the generic saddle-node singularity.

Proof. The proof of this theorem relies on Theorem 3.3.9. We see that a mapping ψ_λ unfolding a generic saddle-node singularity is a local one-dimensional orientation preserving diffeomorphism on a neighborhood of $x = 0$ for sufficiently small λ from the assumption $D\psi_0(0) = 1$, and hence only the position of fixed points is important. Therefore, we consider the equation

$$f(x, \lambda) = \psi_\lambda(x) - x.$$

From the assumptions, we have

$$f(0,0) = 0, \quad \frac{\partial f}{\partial x}(0,0) = 0, \quad \frac{\partial^2 f}{\partial x^2}(0,0) \neq 0.$$

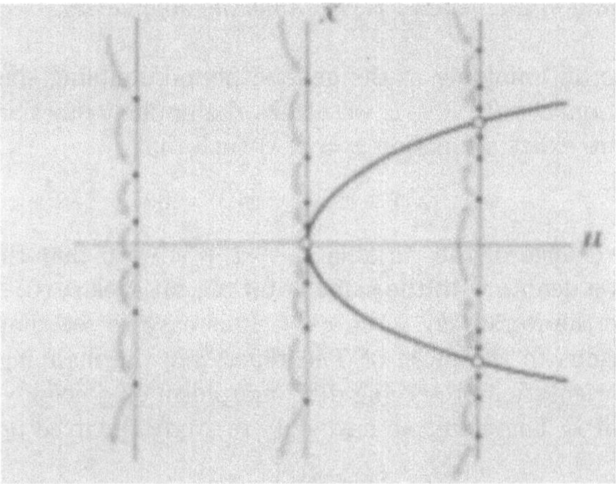

$\mu < 0$ $\mu = 0$ $\mu > 0$

Fig. 3.4.5. Bifurcation diagram for the saddle-node bifurcation for mappings.

Furthermore, we may assume

$$\frac{\partial f}{\partial \lambda}(0,0) \neq 0,$$

by the reasoning we have used in Section 3.3, from which we can make use of Theorem 3.3.9 to conclude that the fixed point of the map ψ_λ bifurcates in exactly the same manner as the equilibrium points of the saddle-node bifurcation for flows.

The construction of the fiber C^0 equivalence is given similarly. □

However, unlike the continuous case, it is not easy to prove the continuous dependence of the equivalence map on the parameters. An interested reader should refer to [Newhouse, Palis and Takens 1983].

The bifurcation diagram of the versal family is given in Fig. 3.4.5 in the case of the minus sign, where the fixed points $\pm\sqrt{\mu}$ appear when $\mu > 0$, one of which is stable and the other of which is unstable.

B Period-Doubling Bifurcation Consider a family of smooth mappings $\varphi_\mu : \mathbb{R} \to \mathbb{R}$. Assume that, when $\mu = 0$, the mapping φ_0 possesses a fixed point $x = 0$ at which

$$D\varphi_0(0) = -1 \quad \text{and} \quad D^3(\varphi_0)^2(0) \neq 0,$$

where $(\varphi_0)^2 = \varphi_0 \circ \varphi_0$. We call it a *generic period-doubling singularity* for mappings.

Theorem 3.4.9. *The mapping*

$$x \mapsto -x \pm x^3 + \mu x$$

gives a versal family of the generic period-doubling singularity.

Proof. Let ψ_λ be an unfolding of the generic period-doubling singularity. Then from the assumption $D\psi_0(0) = -1$, we utilize the implicit function theorem and conclude that there exists a function $x = x(\lambda)$ such that

$$\psi_\lambda(x(\lambda)) \equiv x(\lambda), \quad x(0) = 0.$$

We then make a change of the variable $x \mapsto x + x(\lambda)$ so that the transformed mapping, which we denote with the same notation, satisfies $\psi_\lambda(0) \equiv 0$.

Now we take the iterate $\Psi_\lambda = \psi_\lambda \circ \psi_\lambda$. It is easy to see that $D\Psi_0(0) = 1$, and hence, analogous to the proof of Theorem 3.4.8, the mapping Ψ_λ is a local one-dimensional orientation preserving diffeomorphism on a neighborhood of 0 for a sufficiently small λ. Therefore, we have only to study the fixed point equation

$$f(x,\lambda) = \Psi_\lambda(x) - x.$$

The property $\psi_\lambda(0) \equiv 0$ implies $f(0,\lambda) \equiv 0$, and thus the function $f(x,\lambda)$ takes the form

$$f(x,\lambda) = x \cdot F(x,\lambda)$$

for some smooth function $F(x,\lambda)$. Then the assumptions of the generic period-doubling singularity imply

$$F(0,0) = 0, \quad \frac{\partial F}{\partial x}(0,0) = 0, \quad \frac{\partial^2 F}{\partial x^2}(0,0) \neq 0,$$

and

$$\frac{\partial F}{\partial \lambda}(0,0) \neq 0.$$

The last condition again follows from the standard technique for bifurcation parameters, which we have used in Section 3.3.

Therefore, we conclude that there exists a unique branch of fixed points of the mapping Ψ_λ given by the form

$$\lambda = \lambda(x) \quad \text{with} \quad \lambda(0) = 0, \ \lambda'(0) = 0, \ \lambda''(0) \neq 0.$$

Notice that the fixed point of Ψ_λ other than $x = 0$ is nothing but the periodic point of period two of the mapping ψ_λ, hence describing the complete bifurcation of the original mapping ψ_λ.

The construction of the equivalence homeomorphism is again similar to the previous cases and hence omitted. Moreover, in this case, this obvious construction of the equivalence homeomorphism shows that it is in fact continuous on the parameters, which follows from the existence of the fixed point for all small λ. □

The bifurcation diagram of the versal family is given in Fig. 3.4.6 in the case of the minus sign for the coefficient of x^3. The fixed point $x = 0$ is stable when $\mu < 0$ and loses its stability at $\mu = 0$. When $\mu > 0$, the fixed point becomes unstable and instead a pair of attractive periodic points $\pm\sqrt{\mu}$ with period two appears from the fixed point.

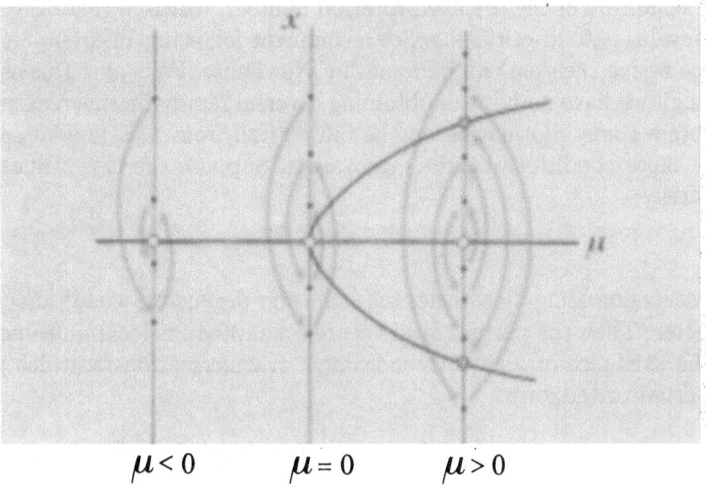

Fig. 3.4.6. Bifurcation diagram for the period-doubling bifurcation for mappings.

C Hopf Bifurcation Consider a family of smooth mappings $\varphi_\mu : \mathbb{R}^2 \to \mathbb{R}^2$ ($\mu \in \mathbb{R}$) on the plane, and assume that, when $\mu = 0$, the mapping φ_0 possesses a fixed point O at which the linearization $D\varphi_0(O)$ has complex conjugate eigenvalues with modulus one, $e^{\pm i\alpha}$, with $\alpha \neq 0, \pi$. We call such a mapping φ_0 a *Hopf singularity* for mappings.

In this case we cannot expect to have a versal family as seen in the following theorem.

Theorem 3.4.10. *Let φ_μ be a family given above such that the eigenvalue $\lambda(\mu)$, $\overline{\lambda(\mu)}$ of $D\varphi_\mu(O_\mu)$ satisfies*

$$\left.\frac{d}{d\mu}\right|_{\mu=0} |\lambda(\mu)| \neq 0, \tag{3.4.5}$$

where O_μ is the unique fixed point of φ_μ with $O_0 = O$. Then there exists an arbitrarily nearby family $\bar{\varphi}_\mu$ in the C^∞ topology which is not fiber C^0 equivalent to φ_μ.

Suppose there exists a versal family, it would be natural to ask the condition (3.4.5) for it, since this means that the critical eigenvalue crosses the unit circle with non-zero speed as parameter varies. In fact, if a family does not satisfy this condition, we can make a change of the parameter in such a way that the condition holds for the new family, as we have been done repeatedly. Then the above theorem tells us that such a family never be stable under perturbation which contradicts to the definition of the versal family. This difficulty essentially comes from the fact that the dynamics of a homeomorphism on the circle strongly depends on an arithmetic condition of the so-called rotation number, which in this case is related to the eigenvalue $\lambda(0)$ at criticality. See a comment following Theorem 3.4.11. The proof of the above theorem can be found in [Newhouse, Palis and Takens 1983].

Although we have no hope of obtaining a versal family by this reason, we can at least obtain some information on the bifurcation from the Hopf singularity, if we assume more conditions on the eigenvalues. Suppose the eigenvalues $e^{\pm i\alpha}$ of $D\varphi_0(O)$ satisfy

$$\alpha \neq 2\pi \frac{q}{p} \quad \text{with} \quad p, q \in \mathbb{Z}, \ |p| \leq 4, \tag{3.4.6}$$

and suppose its unfolding φ_μ satisfies the same non-degeneracy condition (3.4.5) on the parameter. Then the normal form theorem adapted for local diffeomorphisms leads to the existence of a smooth coordinate transformation by which the map φ_μ is converted to the form

$$\tilde{\varphi}_\mu(r, \theta) = (\delta(\mu)r + a(\mu)r^3 + h.o.t., \ \theta + \alpha(\mu) + b(\mu)r^2 + h.o.t.)$$

using the polar coordinates. The above assumptions give

$$\delta_0 = \left.\frac{d}{d\mu}\right|_{\mu=0} |\delta(\mu)| \neq 0,$$
$$\alpha(0) \neq 2\pi \frac{q}{p} \quad \text{with} \quad p, q \in \mathbb{Z}, \ |p| \leq 4,$$

and we moreover assume

$$a_0 = a(0) \neq 0.$$

We call such a singularity the *generic Hopf singularity without strong resonance.* There are two types of bifurcations from this singularity depending on the sign of δ_0 and a_0. We call it *supercritical* (respectively *subcritical*) if $\delta_0 a_0 < 0$ (respectively $\delta_0 a_0 > 0$).

Theorem 3.4.11. *Let φ_μ be as above. Then there exists an invariant circle of φ_μ for $\mu > 0$ (respectively $\mu < 0$) if $a_0 < 0$ (respectively $a_0 > 0$), and the invariant circle is attracting (respectively repelling) if it is supercritical (respectively subcritical).*

See [Ruelle and Takens 1971] for the proof. Note that, in [Ruelle and Takens 1971], the strong resonance condition (3.4.6) is formulated as with $|p| \leq 5$, which was later weakened by [Iooss 1979] as above. See also [Marsden and McCracken 1976].

The bifurcation diagram of a supercritical family is given in Fig. 3.4.7. The stable fixed point for $\mu < 0$ loses its stability at $\mu = 0$ and after that an invariant circle appears around the fixed point. The dynamics on the invariant circle depends sensitively on the eigenvalues at the singularity and on the whole family itself. In some cases it is topologically conjugate to the rigid rotation with an irrational rotation number, but it is also likely that there are periodic orbits on the circle.

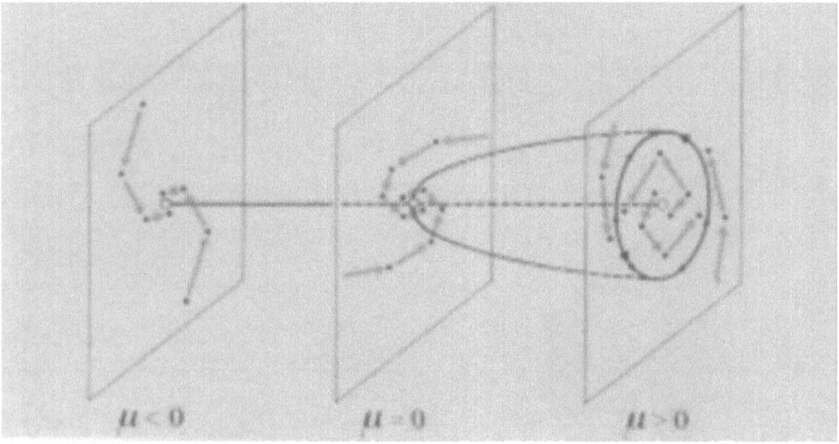

Fig. 3.4.7. Bifurcation diagram for the Hopf bifurcation for mappings.

See [Arnold 1983] and [de Melo and van Strien 1993] for more information on mappings on the circle.

This theorem explains an obstruction for the existence of versal families. Indeed, a family unfolding a generic Hopf singularity can be approximated by families which generate an invariant circle as described in the above theorem. However the dynamics on an invariant circle sensitively depends on its rotation number, and hence a small perturbation can change the dynamics drastically. See [Newhouse, Palis and Takens 1983] for details.

For more information about these elementary bifurcations for mappings, see [Iooss 1979], [Arnold 1983], [Newhouse, Palis and Takens 1983], [Whitley 1983], [Devaney 1986], and [Arrowsmith and Place 1990]. We remark that all such bifurcations can occur as well for periodic orbits in continuous dynamical systems through the Poincaré map as noted before.

3.4.4 One-Dimensional Mapping (1)

In this and the next subsections we shall deal with two families of one-dimensional mappings, which are of fundamental importance in the study of chaotic dynamical systems. This subsection is devoted to the study of the family of mappings

$$\varphi_\mu(x) = x^2 + \mu \qquad (x \in \mathbb{R},\ \mu \in \mathbb{R}), \tag{3.4.7}$$

which is sometimes called the quadratic family. This family is equivalent, by an affine coordinate transformation, to

$$x \mapsto 1 - ax^2$$

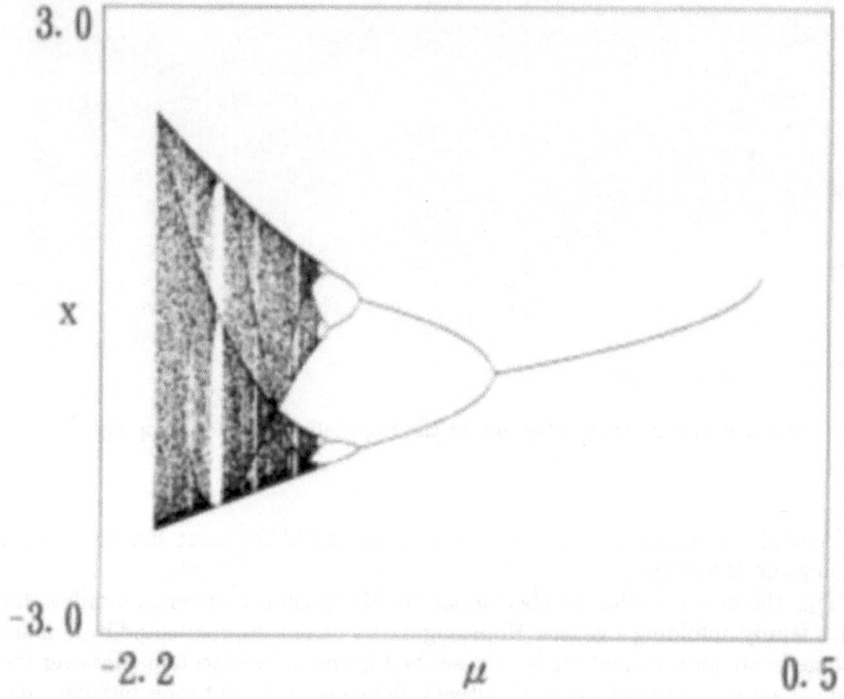

Fig. 3.4.8. The period-doubling cascade for the quadratic family.

or to the logistic family ([May 1976])

$$x \mapsto ax(1-x).$$

A Elementary Bifurcations for Quadratic Family The quadratic family (3.4.7) has a fixed point if $x^2+\mu=x$ has a real solution. This fixed point equation has a unique solution $x=\frac{1}{2}$ when $\mu=\frac{1}{4}$, and if $\mu>\frac{1}{4}$, it has no solution, while if $\mu<\frac{1}{4}$, it has two solutions given by

$$x_{\pm}=\frac{1\pm\sqrt{1-4\mu}}{2}.$$

It is thus easy to verify that the quadratic family undergoes a generic saddle-node bifurcation when $\mu=\frac{1}{4}$.

For $\mu<\frac{1}{4}$, the fixed point x_+ is always unstable since

$$D\varphi_\mu(x_+) = 2x_+ = 1 + \sqrt{1-4\mu} > 1,$$

while x_- is stable for μ satisfying $|1 - \sqrt{1-4\mu}| < 1$, that is,

$$-\frac{3}{4} < \mu < \frac{1}{4}.$$

When $\mu = -\frac{3}{4}$, $D\varphi_\mu(x_-) = -1$, and it undergoes a generic period-doubling bifurcation, hence yielding 2-periodic points. Such a 2-periodic point is a solution of the equation $\varphi_\mu^2(x) = x$, that is,

$$x^4 + 2\mu x^2 - x + \mu^2 + \mu = 0.$$

We know that the fixed points $x_\pm$ must also satisfy the equation, and hence the equation factors through

$$x^4 + 2\mu x^2 - x + \mu^2 + \mu = (x^2 - x + \mu)(x^2 + x + \mu + 1) = 0,$$

which gives the 2-periodic points

$$x_\pm^{(2)} = \frac{-1 \pm \sqrt{-3-4\mu}}{2} \quad \left(\mu < -\frac{3}{4}\right).$$

The linearization at $x_\pm^{(2)}$ is given by

$$D\varphi_\mu^2(x_\pm^{(2)}) = 4(\mu + 1),$$

and therefore the periodic point is stable for $-\frac{5}{4} < \mu < -\frac{3}{4}$. When $\mu = -\frac{5}{4}$, $D\varphi_\mu^2(x_\pm^{(2)}) = -1$ and hence it again undergoes a period-doubling bifurcation producing 4-periodic points.

It is difficult to keep going on such explicit computation further. However, a computer simulation shows that a sequence of period-doubling bifurcations successively occurs and accumulates to some μ_∞, as observed in Fig. 3.4.8. The computer picture is drawn by taking an initial point, iterating it under the map φ_μ many times, and plotting these points after ignoring some transient part of the orbit, with changing the parameter gradually from $\mu = 0.5$ to $\mu = -2.2$. Thus the figure shows a long term behavior of the orbit for each parameter value. In Fig. 3.4.8, we can clearly observe the first and second period-doubling bifurcations which are explicitly computed above, and a few more period-doubling bifurcations as well as a variety of asymptotic behaviors as the change of parameter. For this family, accumulation of such successive period-doubling bifurcations actually occurs, and moreover it satisfies a beautiful universal scaling property. To be a little more precise, let μ_n be the value of the parameter at which the n-th period-doubling bifurcation occurs and produces 2^n-periodic points. Then the following limit exists

$$\lim_{n\to\infty} \frac{\mu_{n+1} - \mu_n}{\mu_n - \mu_{n-1}} = \delta \approx 4.6692\cdots,$$

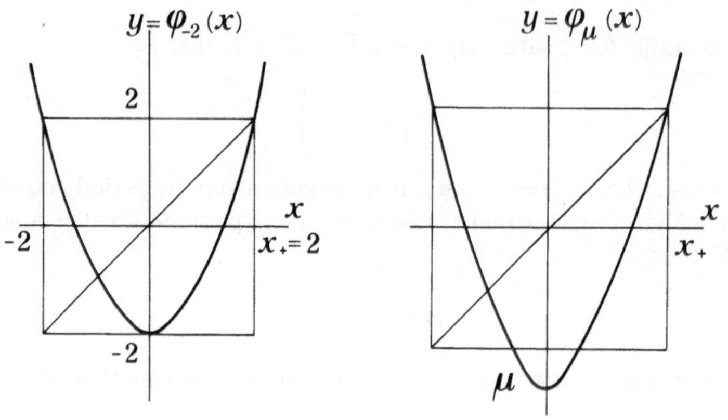

Fig. 3.4.9. Graphs of the mapping φ_μ. (a) $\mu = -2$. (b) $\mu < -2$.

and the constant δ is a universal constant in the sense that it does not depend on a particular form of families of one-dimensional mappings. This universality was first observed by [Feigenbaum 1978, 1979] and [Coullet and Tresser 1978]. Moreover, [Feigenbaum 1978, 1979] suggests a mechanism of this fascinating phenomenon in terms of a nonlinear mapping (called the renormalization operator or the doubling operator) acting on the infinite dimensional space of unimodal one-dimensional mappings. This was the origin of a number of subsequent theoretical and/or computer-assisted works attempting to justify the beautiful picture presented by Feigenbaum. We do not give any result along this line but only refer to [Collet and Eckmann 1980], [Lanford 1982] and [de Melo and van Strien 1993] for the detail.

B The Case of $\mu < -2$ Let us turn to consider the case of $\mu < -2$. When $\mu = -2$, we have $x_+ = -\mu$ and hence the map $\varphi_\mu(x) = x^2 + \mu$ sends the interval $[-2, 2]$ to itself as depicted in Fig. 3.4.9(a). Therefore, when $\mu < -2$, the map φ_μ has no interval which is mapped onto itself. Consider the interval $I = [-x_+, x_+]$ and define the set

$$C = \{x \in I \mid \varphi_\mu^n(x) \in I \text{ for } \forall n \geq 0\}.$$

By definition, for any $x \notin C$, $\varphi_\mu^n(x)$ eventually leaves the interval I and goes to $+\infty$ as $n \to +\infty$. The dynamics of φ_μ for $\mu < -2$ is thus completely understood once we understand the structure and the dynamics in the set C.

It is easy to see that C is a Cantor set. Indeed, letting J be the maximal subinterval in I satisfying $\varphi_\mu(J) < -x_+$, the set C is described by

$$C = I \setminus \bigcup_{n=0}^{\infty} \varphi_\mu^{-n}(J).$$

Since φ_μ is monotone decreasing on $\{x < 0\}$ and increasing on $\{x > 0\}$, the intervals in $\varphi_\mu^{-n}(J)$ are all disjoint, and hence the complement C is a Cantor set.

The dynamics of φ_μ on the Cantor set C is well-described by introducing symbolic sequences as follows. Let us take $+1$ and -1 as two symbols. There are two ways of assigning a symbolic sequence consisting of the symbols $\{+1, -1\}$ to a point in C. The first one is dynamic: define, for each $x \in C$,

$$c_n(x) = \begin{cases} +1 & \text{if } \varphi_\mu^n(x) > 0 \\ -1 & \text{if } \varphi_\mu^n(x) < 0, \end{cases}$$

which gives a sequence $c_0(x), c_1(x), c_2(x), \cdots$ for $x \in C$. The second one is static: first look at the two intervals given by

$$I \setminus J = I_1^+ \cup I_1^-$$

and assign ± 1 to x if $x \in I_1^\pm$, respectively. In the next stage of constructing the Cantor set C, the middle part corresponding to $\varphi_\mu^{-1}(J)$ in $I_1^\pm$ is taken out and we obtain four intervals. We then assign

$-1, -1$ to points in the left subinterval in I_1^-;

$-1, +1$ to points in the right subinterval in I_1^-;

$+1, -1$ to points in the left subinterval in I_1^+;

$+1, +1$ to points in the right subinterval in I_1^+.

We continue this procedure infinitely many times and finally we obtain a sequence $\varepsilon_0(x), \varepsilon_1(x), \varepsilon_2(x), \cdots$ of $+1$ and -1 corresponding to any point x in C. These two kinds of symbolic representations are related as follows:

$$\varepsilon_n(x) = \prod_{k=0}^{n} c_k(x). \tag{3.4.8}$$

By this correspondence, we can easily obtain the periodic points of φ_μ. For instance, dynamic sequences

$$+1, +1, +1, +1, \cdots \quad \text{and} \quad -1, -1, -1, -1, \cdots$$

clearly give fixed points of φ_μ and hence the position of these points in C is given by corresponding static sequences

$$+1, +1, +1, +1, \cdots \quad \text{and} \quad -1, +1, -1, +1, \cdots.$$

Similarly, the position of a 2-periodic point

$$+1, -1, +1, -1, +1, -1, +1, -1, \cdots$$

in dynamic representation is given by the static representation

$$+1, -1, -1, +1, +1, -1, -1, +1, \cdots .$$

Since (3.4.8) gives bijective correspondence between dynamic and static sequences, we can readily see that there are infinitely many periodic points in C. Furthermore, we can show that periodic points are dense in C and that there is an orbit which itself is dense in C. Indeed, this easily follows if one notes that C is homeomorphic to the space of all such sequences $\{\varepsilon_n\}$ endowed with the metric

$$d(\{\varepsilon_n\}, \{\varepsilon'_n\}) = \sum_{n=0}^{\infty} \frac{|\varepsilon_n - \varepsilon'_n|}{2^n},$$

where $|\varepsilon_n - \varepsilon'_n|$ stands for the difference of ε_n and ε'_n as numbers. We denote the space by $\{+1, -1\}^{\mathbb{N}}$. This is a complete metric space with the above metric.

C The Case of $\mu = -2$ When $\mu = -2$, the map $\varphi_\mu(x) = \varphi_{-2}(x) = x^2 - 2$ sends the interval $[-2, 2]$ onto itself, and hence any point from the interval keeps staying in it under the iteration of φ_{-2}. Furthermore, this mapping φ_{-2} has the following very special property: putting $x = 2\cos\theta$,

$$\varphi_{-2}(x) = 4\cos^2\theta - 2 = 2\cos 2\theta.$$

In other words, by the mapping

$$x = h(\theta) = 2\cos\left(\theta - \frac{\pi}{2}\right) \qquad \left(-\frac{\pi}{2} \le \theta \le \frac{\pi}{2}\right),$$

φ_{-2} is topologically conjugate to the piecewise-linear mapping

$$\psi(\theta) = |2\theta| - \frac{\pi}{2} \qquad \left(-\frac{\pi}{2} \le \theta \le \frac{\pi}{2}\right).$$

The assignment of dynamic sequences described in B of this subsection also applies to almost all points in $\left[-\frac{\pi}{2}, \frac{\pi}{2}\right]$ for the mapping ψ. More precisely, unless $\theta \in \left[-\frac{\pi}{2}, \frac{\pi}{2}\right]$ eventually falls into 0 under the iteration ψ, we can define

$$c_n(\theta) = \begin{cases} +1 & \text{if} \quad \psi^n(\theta) \in \left[-\frac{\pi}{2} 0\right), \\ -1 & \text{if} \quad \psi^n(\theta) \in \left(0, \frac{\pi}{2}\right]. \end{cases}$$

Conversely, for any sequence in $\{+1, -1\}^{\mathbb{N}}$, there exists one and only one point in $\left[-\frac{\pi}{2}, \frac{\pi}{2}\right]$ whose corresponding dynamic sequence is the prescribed one. Therefore, similar reasoning as before implies that the mapping ψ possesses infinitely many periodic points of arbitrary period together with an orbit which itself is dense in the interval $\left[-\frac{\pi}{2}, \frac{\pi}{2}\right]$. Note that there are only countably many points which eventually fall to O. Since the conjugacy $x = h(\theta)$ is a homeomorphism, the map φ_{-2} has the same property as well.

In addition to such a topological property, we can say more about the statistical properties of the dynamics of the mappings φ_{-2} and ψ. Since ψ has constant expansion by 2 everywhere except 0, that is, $|\psi'(\theta)| = 2$ for $\theta \neq 0$, the asymptotic distribution of orbits is also constant $\frac{1}{\pi}$. To be more precise, take any point $\xi \in \left[-\frac{\pi}{2}, \frac{\pi}{2}\right]$ and consider the asymptotic distribution of orbits of ξ. Namely, for any fixed subinterval $A \subset \left[-\frac{\pi}{2}, \frac{\pi}{2}\right]$, we count the average rate of $\psi^n(\xi)$ falling into the set A:

$$\lim_{n\to\infty} \frac{1}{n} \#\{i \mid \psi^i(\xi) \in A \ (i = 0, 1, \cdots, n-1)\},$$

where # indicates the number of elements of the set. This average rate can be more sophisticatedly written as

$$\lim_{n\to\infty} \frac{1}{n} \sum_{1=0}^{n-1} \chi_A(\psi^i(\xi)),$$

where χ_A is the characteristic function of the subinterval A, that is,

$$\chi_A(\xi) = \begin{cases} 1 & \text{if } \xi \in A \\ 0 & \text{otherwise.} \end{cases}$$

The celebrated Birkhoff individual ergodic theorem (see [Mañé 1983] for the precise statement) tells us that this time average in fact equals the space average:

$$\lim_{n\to\infty} \frac{1}{n} \sum_{1=0}^{n-1} \chi_A(\psi^i(\xi)) = \frac{1}{\pi} \int \chi_A(\theta) d\theta = \int \frac{1}{\pi} d\theta$$

for almost all points $\xi \in \left[-\frac{\pi}{2}, \frac{\pi}{2}\right]$, since the Lebesgue measure m is clearly invariant under ψ from its definition, namely,

$$m(\psi^{-1}(A)) = m(A)$$

for any subinterval $A \subset \left[-\frac{\pi}{2}, \frac{\pi}{2}\right]$, and since the map ψ is ergodic with respect to the Lebesgue measure, meaning that, if there exists an invariant set A for ψ: $\psi^{-1}(A) = A$, then either A has the measure 0 or else it has the full measure of the total space, $\left[-\frac{\pi}{2}, \frac{\pi}{2}\right]$ in this case. This assertion again follows from the piecewise-linearity of the map ψ. Under these conditions, the Birkhoff ergodic theorem can be applied to show that the asymptotic distribution of almost all points (with respect to the Lebesgue measure) becomes constant independent of the initial points.

This asymptotic distribution, or the invariant probability measure, is transformed to that for the quadratic mapping φ_{-2} through the conjugacy homeomorphism $h(\theta) = 2\cos\left(\theta - \frac{\pi}{2}\right)$. Indeed, the resulting invariant probability measure for φ_{-2} is given by

$$P(X) = \int_X \frac{dx}{\pi\sqrt{4 - x^2}} \qquad (X \subset [-2, 2]),$$

since $dx = -2\sin\left(\theta - \frac{\pi}{2}\right) d\theta = \sqrt{4 - x^2} d\theta$. This measure P is not the (constant multiple of) Lebesgue measure on $[-2, 2]$ because its density function $\frac{1}{\pi\sqrt{4-x^2}}$ is not constant, but it is absolutely continuous with respect to the Lebesgue measure, namely, if $m(X) = 0$ then $P(X) = 0$. Moreover, the converse is also valid: $P(X) = 0$ implies $m(X) = 0$, and hence they are said to be equivalent as measures.

This invariant measure P for φ_{-2} is important, because it gives the asymptotic distribution of almost all points $y \in [-2,2]$ under iteration of φ_{-2},

$$\lim_{n\to\infty} \frac{1}{n} \#\{i \mid \varphi_{-2}^i(y) \in X \ (i = 0, 1, \cdots, n-1)\} = \int_X \frac{dx}{\pi\sqrt{4-x^2}},$$

which says that orbits of almost all points distribute not constantly but uniformly over the whole interval $[-2,2]$ with the density $\frac{1}{\pi\sqrt{4-x^2}}$. Therefore, this measure clearly describes the quite random nature of the dynamics of φ_{-2}. In general, from the construction of the measure P, the ergodicity of the map ψ implies the ergodicity of φ_{-2} with respect to this measure P, and hence, for any integrable function $f(x)$, we have

$$\lim_{n\to\infty} \frac{1}{n} \sum_{i=0}^{n-1} f(\varphi_{-2}^i(y)) = \int_{[-2,2]} \frac{f(x)}{\pi\sqrt{4-x^2}} dx$$

for almost all $y \in [-2,2]$ with respect to the Lebesgue measure.

To summarize, the case of $\mu = -2$ is an extreme case where the dynamics of φ_μ is very well described although it exhibits fully random (or chaotic) behaviors.

Remark 3.4.12. The exposition given here on the family $\varphi_\mu(x) = x^2 + \mu$, or on its equivalent families, is very limited, and a number of remarkable achievements are left unexplained. Indeed, this family has been studied extensively by many authors, since it is considered as one of typical and the simplest families which exhibit chaotic dynamics and bifurcations toward it. The simple one-dimensional nature of the quadratic family helps us a lot to understand its dynamical behavior in detail and, up to now, very interesting and deep results have been obtained for this and similar but more general one-dimensional mappings, which are certainly beyond our scope of this book. See A of Subsection 3.4.7 for a more explanation on the quadratic family.

3.4.5 One-Dimensional Mapping (2)

In this subsection, we shall consider another important one-parameter family of one-dimensional mappings, namely

$$\psi_\mu(x) = \mathrm{sgn}(x)(|x|^\gamma - \mu) \quad (x \in \mathbb{R},\ \mu \in \mathbb{R}),$$

where $0 < \gamma < 1$ is a fixed constant. This type of mapping is observed in a numerical simulation of the Lorenz equation (3.2.3), and hence here we call ψ_μ the family of Lorenz maps. See, for instance, [Sparrow 1982] for a relation between the Lorenz attractor and corresponding one-dimensional mappings.

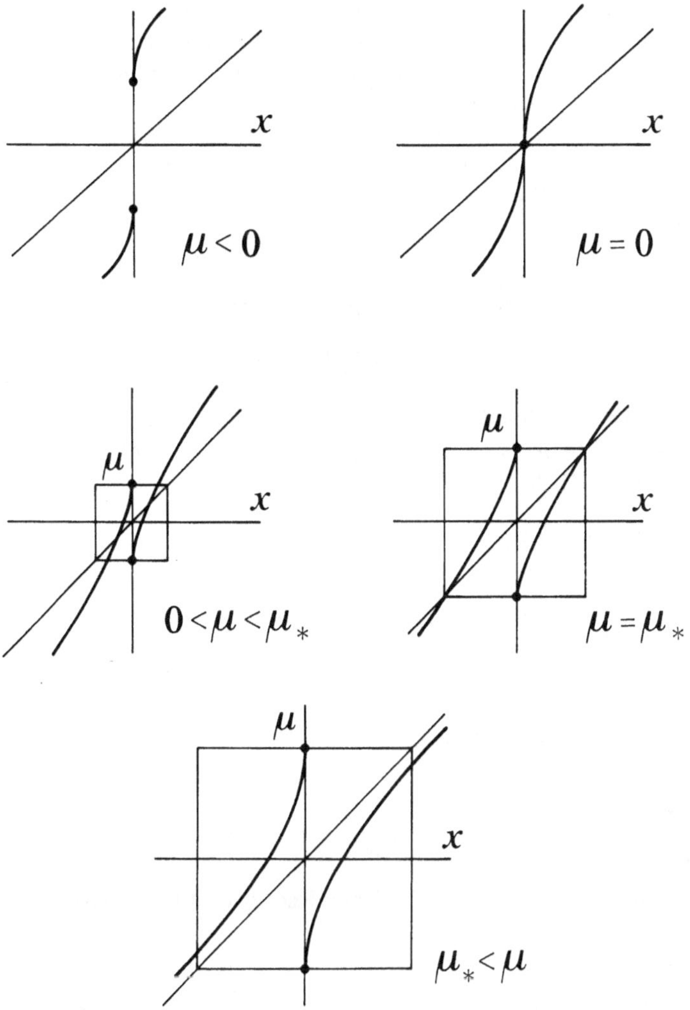

Fig. 3.4.10. The change of the family of Lorenz maps for various μ.

Figure 3.4.10 illustrates the change of the graph of ψ_μ for various parameter values where μ_* is the solution of the equation

$$\mu_*^\gamma - \mu_* = \mu_*,$$

that is,

$$\mu_* = 2^{\frac{1}{\gamma-1}}.$$

It is easy to see from Fig. 3.4.10 that, when $\mu < 0$, there is no bounded invariant set of ψ_μ; and when $\mu = 0$, $x = 0$ is the unstable fixed point and any other orbits diverge to infinity. If we assume $2\gamma > 1$, then we can show that, when $0 < \mu < \mu_*$, there exists an invariant set of ψ_μ in the interval $[-\mu, \mu]$ which is homeomorphic to the Cantor set.

The dynamics of ψ_μ becomes the most interesting when $\mu > \mu_* (> 0)$. In this case, the whole interval $[-\mu, \mu]$ is invariant under ψ_μ. If we assume that γ is taken sufficiently close to 1, then the derivative of ψ_μ is computed as

$$\psi'_\mu(x) = \gamma|x|^{\gamma-1} \geq \gamma|\mu|^{\gamma-1} > \sqrt{2} \quad (\forall x \in [-\mu, \mu])$$

for a range of the parameter, say $\mu_* < \mu < 2\mu_*$. Also we have

$$\psi_\mu(\mu) = -(|\mu|^\gamma - \mu) > 0$$

in the same parameter range, since $2\mu_* < 1$.

Therefore the map $\psi_\mu : [-\mu, \mu] \to [-\mu, \mu]$ is one of those maps $f : [-a, a] \to [-a, a]$ $(a > 0)$ satisfying the following conditions:

(LM1) f is discontinuous only at $x = 0$;

(LM2)

$$\lim_{x\to 0-} f(x) = a, \quad \lim_{x \to 0+} f(x) = -a \quad \text{and} \quad f(-a) < 0 < f(a);$$

(LM3) $f \in C^1$ on $[-a, a] \setminus \{0\}$ and

$$f'(x) > \sqrt{2} \quad \text{for} \quad \forall x \in [-a, a];$$

(LM4)

$$\lim_{x \to 0\pm} f'(x) = +\infty.$$

The dynamics of such a map is nicely described in the following way.

Definition 3.4.13. Let f be a mapping on an interval I into itself.

(1) The map f is *locally eventually onto* if, for any open subinterval J of I, there exists a positive integer k such that $f^k(J) = I$.

(2) The map f is *topologically transitive* if there exists a point in I whose orbit is dense in it.

Remark 3.4.14. It is easy to show that a locally eventually onto map is topologically transitive.

Theorem 3.4.15. *The Lorenz map ψ_μ satisfying (LM1)-(LM4) is locally eventually onto, and hence topologically transitive. Furthermore, the periodic points of ψ_μ form a dense subset of $[-\mu, \mu]$.*

See [Williams 1979] for the proof.

In order to give a more precise description of the dynamics of the maps with (LM1)-(LM4), we introduce symbolic sequences as follows:

Let

$$k_0(x) = \begin{cases} +1 & (x > 0) \\ 0 & (x = 0) \\ -1 & (x < 0) \end{cases}$$

for $x \in [-a, a]$, and define the sequence $\{k_n(x)\}$ by

$$k_n(x) = k_0(f^n(x))$$

for a map satisfying (LM1)-(LM4). We introduce the distance between two such sequences, as we have done in B of the previous subsection: for two such sequences $\{k_n\}$ and $\{k'_n\}$, define

$$d(\{k_n\}, \{k'_n\}) = \sum_{n=0}^{\infty} \frac{|k'_n - k_n|}{2^n}.$$

This gives the metric on the set of all such sequences, and hence it permits us to take the limit

$$\{k_n^{\pm}\} = \{k_n^{\pm}(f)\} = \lim_{x \to 0\pm} \{k_n(x)\}.$$

It is easy to see that the limit exists, and we call the pair of limit sequences the *kneading sequence* of the map f.

Theorem 3.4.16. ([Rand 1978], [Williams 1979]) *Two maps f and g satisfying (LM1)-(LM4) are topologically conjugate, provided that the corresponding kneading sequences are identical, namely, $\{k_n^+(f)\} = \{k_n^+(g)\}$ and $\{k_n^-(f)\} = \{k_n^-(g)\}$.*

Furthermore, one can show that the kneading sequences generate the complete set of invariants for the topological conjugacy among those maps, and, in particular, this implies that, if $\mu_* < \mu_1, \mu_2 < 2\mu_*$ with $\mu_1 \neq \mu_2$, then the corresponding Lorenz maps ψ_{μ_1} and ψ_{μ_2} are not topologically conjugate. In other words, the parameter μ of the Lorenz maps gives a modulus of stability, namely the topological conjugacy classes form a continuum so that different parameter values correspond to different points in it.

Since the Lorenz map has the uniform expanding property due to (LM3), a theorem in [Keller 1985] shows that it admits an invariant measure which is absolutely continuous with respect to the Lebesgue measure. Moreover, the Lorenz map is ergodic and is even mixing with this invariant measure.

All the above properties show that the dynamics of the Lorenz map is very chaotic yet well-understood in a mathematically rigorous way.

3.4.6 Horseshoe

The horseshoe is a diffeomorphism formulated by [Smale 1963] as a non-trivial structurally stable dynamical system. In this subsection, we shall give a very brief idea of the horseshoe and explain it as simply as possible. For details, see [Palis and de Melo 1982] or [Moser 1973].

A Topological Horseshoe Let $f : \mathbb{R}^2 \to \mathbb{R}^2$ be a diffeomorphism which maps the rectangle $R = [0,1] \times [0,1]$ to a horseshoe-shaped region as shown in Plate 3.1. More precisely, the rectangle is first contracted horizontally and stretched vertically, and then folded and put back to the original place as depicted, so that the image of the rectangle which is colored in red in Plate 3.1 intersects with the original rectangle R in two vertical subrectangles.

On the contrary, if we map back the horseshoe-shaped region together with the rectangle in the opposite way, then the red region returns to the white original rectangle, whereas the white rectangle in the right picture of Plate 3.1 changes again to a horseshoe-shaped region, which is colored in green in Plate 3.2.

Plate 3.3 shows several forward iterates of the rectangle R under the map f^n with $n = 1, 2, 3, 4$, and one can imagine that similar pictures are obtained by the backward iterates of the rectangle by f^{-n}.

From these pictures, we see that once a point goes out from the rectangle under iteration of f, it never returns to the rectangle R. Since we do not have any further information on the dynamics outside R, we are only interested in those points which keep staying in R under all iteration of f. Such a set becomes somewhat complicated. For instance, Plate 3.4 shows the set of points which stay within R up to four forward or backward iterates of f as the intersection of red and green figures depicts. Namely, the intersection which consists of $32^2 = 1024$ number of small subrectangles indicates the set

$$\bigcap_{-4 \le n \le 4} f^n(R).$$

In general, we call the set given by

$$\Lambda(R) = \bigcap_{n \in \mathbb{Z}} f^n(R)$$

the maximal invariant subset $\Lambda(R)$ of f in R. Our purpose is to understand the structure of the set and the dynamics of f on $\Lambda(R)$. Note that, since we have made no assumption of the rate of contraction and expansion of the rectangle in the process of forming the horseshoe-shaped region up to now, we do not a priori know if connected components in the set $\bigcap_{-N \le n \le N} f^n(R)$ become smaller as N tends to $+\infty$, although the pictures might indicate so. However, this is in fact the case after imposing further assumptions on the uniform expansion and contraction rates of the mapping f, which will be described later. For the moment, we vaguely assume that somehow the rectangle is mapped to a horseshoe-shaped region intersecting with the original rectangle in two vertical subrectangles as depicted in Plate 3.1. In this case, the map f is called a topological horseshoe.

In order to describe the structure of the maximal invariant subset $\Lambda(R)$ in R, we first introduce its symbolic coding. We name the two vertical subrectangles $V(+1)$ and $V(-1)$. Then for any point x in $\Lambda(R)$, we can associate the sequence $\{c_n(x)\}_{n\in\mathbb{Z}}$ consisting of $+1$ and -1 in the following way:

$$c_n(x) = \pm 1 \quad \text{if} \quad f^n(x) \in V(\pm 1).$$

Clearly from the above figures, this assignment of symbols is well-defined and, moreover, any symbolic sequences of $+1$ and -1 are possible for this type of map. It is also clear that, the symbolic sequence of the point $f(x)$ is the same as the sequence of x except that the sequence of $f(x)$ is shifted one step left from the sequence of x. In other words,

$$c_n(f(x)) = c_{n+1}(x). \tag{3.4.9}$$

We have seen several types of symbolic coding in the preceding subsections. Such symbolic coding in general define what is called the shift dynamics.

Definition 3.4.17. Take m numbers of symbols ($m \geq 2$), say, $1, \cdots, m$, and let S_m be the set of two-sided infinite sequences consisting of these m symbols, namely, an element in S_m is a sequence $c = \{c_n\}_{n\in\mathbb{Z}}$, where $c_n \in \{1, \cdots, m\}$. We also denote S_m by $\{1, \cdots, m\}^{\mathbb{Z}}$. If we only consider one-sided infinite sequences $c = \{c_n\}_{n\in\mathbb{N}}$ with m symbols, we denote the set of all such sequences by $S_m^+ = \{1, \cdots, m\}^{\mathbb{N}}$.

The set S_m or S_m^+ is endowed with the metric

$$d(\{c_n\}, \{c_n'\}) = \sum_{n\in I} \frac{|c_n - c_n'|}{2^{|n|}} \qquad (I = \mathbb{Z} \text{ or } \mathbb{N}),$$

where $|c_n - c_n'|$ stands for the difference of c_n and c_n' as numbers, and by this metric, S_m or S_m^+ becomes a complete metric space.

Define the shift map σ on S_m or S_m^+ by

$$\sigma(c)_n = c_{n+1} \qquad (c = \{c_n\}),$$

namely, the n-th entry of $\sigma(c)$ is the $(n+1)$-th entry of c. The map σ is a homeomorphism on S_m and a continuous endomorphism on S_m^+. We call the pair (σ, S_m) or (σ, S_m^+) the *shift dynamics* with m symbols.

Proposition 3.4.18. *The shift dynamics (σ, S_m) has the following properties:*

(1) There are countably many periodic points with different periods in S_m;

(2) The set of all periodic points is dense in S_m;

(3) *σ is topologically transitive. Namely, there exists a point $c \in S_m$ whose orbit is dense in S_m;*

(4) *σ is expansive, that is, there exists a positive number ε such that any two points which stay within the distance ε under all forward and backward iterates of σ must be identical;*

(5) *S_m is homeomorphic to a Cantor set.*

Proof. The proof is not hard if one notices the following facts:

- The definition of the metric on S_m means that two sequences are within the distance $\frac{m}{2^{N-1}}$ if the entries of these sequences coincide from $(-N)$-th site to N-th site;
- Periodic points of the shift dynamics are nothing but periodic symbolic sequences.

□

The symbolic representation of the points in the maximal invariant set $\Lambda(R)$ of the diffeomorphism f given by (3.4.9) shows that

$$c : \Lambda(R) \to S_2 = \{+1, -1\}^{\mathbb{Z}}$$

is a *semi-conjugacy* between the diffeomorphism f acting on $\Lambda(R)$ and the shift dynamics (σ, S_2) with two symbols. Namely, the map c is a continuous surjective (but not necessarily injective) map satisfying the conjugacy relation

$$c \circ f = \sigma \circ c.$$

If, furthermore, c is injective, then it is a homeomorphism and hence it gives a topological conjugacy between $(f, \Lambda(R))$ and (σ, S_2), from which the dynamics of f on $\Lambda(R)$ is completely described by the symbolic representations. This will be shown below under an additional hyperbolicity condition. The semi-conjugacy to the shift dynamics shows the sufficiently complicated structure and dynamics of the topological horseshoe $(f, \Lambda(R))$.

B Hyperbolicity In order for the semi-conjugacy to be the true conjugacy, we need to show that subrectangles appearing in the forward and backward iterations of the diffeomorphism f become thinner and thinner so that they shrink to just lines in the limit, which implies the injectivity of the symbolic representation. This uniform shrinking property is formulated as the hyperbolicity, which is one of the most important notion in the theory of dynamical systems. If the diffeomorphism f satisfies the hyperbolicity condition explained below, f is called a horseshoe.

Definition 3.4.19. The map f is called *hyperbolic* on its invariant set Λ if, for each $x \in \Lambda$, there exists a direct sum decomposition of the tangent space $T_x\mathbb{R}^2$ at x into

$$T_x\mathbb{R}^2 = E_x^u \oplus E_x^s,$$

which depends continuously on $x \in \Lambda$, such that the differential of the map f at x

$$Df(x) : E_x^u \oplus E_x^s \to E_{f(x)}^u \oplus E_{f(x)}^s$$

preserves the decomposition and acts on each summand in the following way:

$$\begin{aligned} &\exists C > 0,\ \exists \lambda \in (0,1), \forall n \geq 0, \\ &\qquad \| D(f^n)^{-1}(x)v_x \| \leq C\lambda^n, \quad \forall v_x \in E_x^u, \\ &\qquad \| D(f^n)(x)w_x \| \leq C\lambda^n, \quad \forall w_x \in E_x^s. \end{aligned}$$

Roughly speaking, hyperbolicity tells us that the dynamics of f on Λ is decomposed in a continuous way into uniform expanding and contracting directions. It is clear that a diffeomorphism is hyperbolic on a hyperbolic periodic orbit in the above sense since the invariant set Λ consists of the periodic orbit itself and the unstable and stable eigenspaces give the desired decomposition. In general, a diffeomorphism is called *hyperbolic* if it is hyperbolic on its limit set.

Definition 3.4.20. The diffeomorphism $f : \mathbb{R}^2 \to \mathbb{R}^2$ which maps the rectangle, $R = [0,1] \times [0,1]$, into the horseshoe-shaped region as in Plate 3.1, is called a *horseshoe* if it is hyperbolic on the maximal invariant subset $\Lambda(R)$.

In general it is not an easy task to find the precise direct sum decomposition of hyperbolicity for a given diffeomorphism. Therefore, we need a more convenient criterion to verify the hyperbolicity. The most popular way is using what is called the cone condition, which we shall explain below.

Definition 3.4.21.

(1) A curve in the rectangle R is called a *vertical curve* (respectively a *horizontal curve*) if it takes the form

$$\{(v(y), y) \in R \mid y \in [0,1]\}$$
$$(\text{respectively} \quad \{(x, h(x)) \in R \mid x \in [0,1]\})$$

by a smooth function v (respectively h) : $[0,1] \to [0,1]$ with the Lipschitz constant $\rho < 1$.

(2) A region in the rectangle R given by two disjoint vertical (respectively horizontal) curves is called a *vertical* (respectively *horizontal*) *strip*. The *width* of a vertical strip V given by v_1 and v_2 is defined by

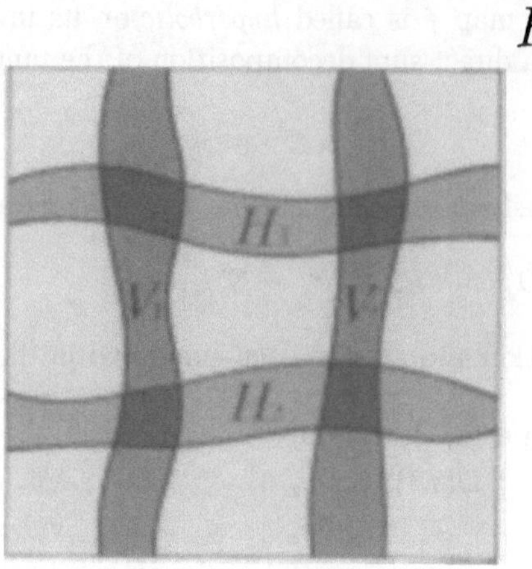

Fig. 3.4.11. Horizontal and vertical strips in R.

$$\text{width}(V) = \sup_{y\in[0,1]} |v_1(y) - v_2(y)|.$$

The definition is similar for a horizontal strip H.

The above diffeomorphism f is assumed to send two horizontal strips H_1 and H_2 in the rectangle R onto two vertical strips V_1 and V_2, respectively, as in Fig. 3.4.11. Here we consider the following condition.

(Cone condition) There exist fields of unstable and stable cones C^u on $V_1 \cup V_2$ and C^s on $H_1 \cup H_2$, namely, for each point $x \in V_1 \cup V_2$ and $y \in H_1 \cup H_2$, there exist cones

$$\begin{aligned} C_x^u &= \{(\xi,\eta) \in T_x\mathbb{R}^2 \mid |\xi| < \rho|\eta|\} \\ C_y^s &= \{(\xi,\eta) \in T_y\mathbb{R}^2 \mid |\eta| < \rho|\xi|\}, \end{aligned}$$

with $0 < \rho < 1$ such that

$$Df(x)C_x^u \subset C_{f(x)}^u \quad \text{and} \quad Df^{-1}(y)C_y^s \subset C_{f^{-1}(y)}^s.$$

Furthermore, in the unstable cone C_x^u, the tangent vectors are expanded uniformly by the differential map as:

$$(\xi,\eta) \in C_x^u \text{ and } (\xi_+,\eta_+) = Df(x)(\xi,\eta) \in C_{f(x)}^u, \text{ then } |\eta_+| \geq \rho^{-1}|\eta|,$$

and in the stable cone C_y^s, they are contracted as:

$$(\xi,\eta) \in C_y^s \text{ and } (\xi_-,\eta_-) = Df^{-1}(y)(\xi,\eta) \in C_{f^{-1}(y)}^s, \text{ then } |\xi_-| \geq \rho^{-1}|\xi|.$$

The number ρ is called the constant of hyperbolicity for the cones.

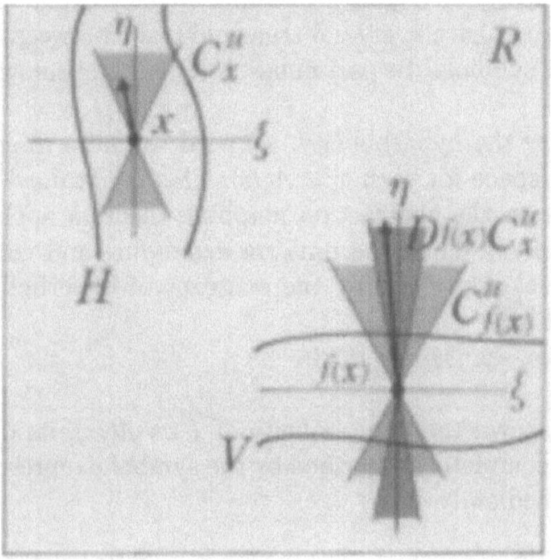

Fig. 3.4.12. The cone condition.

See Fig. 3.4.12.

Theorem 3.4.22. *Let $f : \mathbb{R}^2 \to \mathbb{R}^2$ be a diffeomorphism for which there exist two disjoint horizontal strips H_1, H_2 and two disjoint vertical strips V_1, V_2, all in the rectangle $R = [0,1] \times [0,1]$ such that*

$$f(H_i) = V_i \quad (i = 1, 2).$$

Suppose f satisfies the cone condition for these strips and suppose the Lipschitz constants of the boundary horizontal and vertical curves of these strips, together with the constant of hyperbolicity of the cones, are smaller than a constant $\rho < \frac{1}{2}$. Then the maximal invariant subset $\Lambda(R)$ of f in R is homeomorphic to the Cantor set, and $(f, \Lambda(R))$ is topologically conjugate to the shift dynamics on the set $\{+1, -1\}^{\mathbb{Z}}$ of all bi-infinite sequences with two symbols.

If, moreover,

$$\sup_{x \in \Lambda(R)} \max\{|\det Df(x)|, \ |\det Df^{-1}(x)|\} \leq \frac{1}{2}\rho^{-2},$$

then f is hyperbolic on $\Lambda(R)$.

Proof. The proof essentially needs to show that the limit of the width of nested infinite horizontal or vertical strips corresponding to a given symbolic sequence goes to zero, which determines a horizontal or vertical curve in R, and that any given pair of horizontal and vertical curves intersect at exactly one point. This last statement implies the injectivity of the semi-conjugacy explained in the beginning of this subsection, and hence it gives a true conjugacy between f on $\Lambda(R)$ and the shift map with two symbols. In particular $\Lambda(R)$ is homeomorphic to the Cantor set.

In order to show the hyperbolicity, we need to find a direct sum decomposition of the tangent space for each $x \in \Lambda(R)$. This is obtained by using the fixed point argument due to the contraction mapping theorem applied to the fields of trial directions on $\Lambda(R)$. Then the uniform expanding and contracting rate λ in Definition 3.4.19 is clearly given by the constant of hyperbolicity for the cones, which is dominated by $\rho < \frac{1}{2}$.

For more details, see [Moser 1973]. □

Once we have shown the hyperbolicity of f on $\Lambda(R)$, its dynamical property for the horseshoe is completely described by the symbolic representation. Therefore the following is immediately clear:

- f has countably many hyperbolic periodic orbits with different periods in $\Lambda(R)$;
- The set of all periodic points is dense in $\Lambda(R)$;
- f is topologically transitive, namely, there exists a point $x \in \Lambda(R)$ whose orbit is dense in $\Lambda(R)$;
- f is expansive, that is, there exists a positive number ε such that any two points which stay within the distance ε under all forward and backward iterations of f must be identical;
- f on $\Lambda(R)$ is structurally stable, namely, a diffeomorphism g close to f on R has the maximal invariant set $\Lambda_g(R)$ on which it is topologically conjugate to f on $\Lambda(R)$.

The Hénon map $H_{a,b}$ given in Example 3.4.2(2) is proved by [Devaney and Nitecki 1979] to possess a horseshoe for a sufficiently large a. Indeed the Hénon map changes the dynamics from very simple to the one with a horseshoe as the parameter varies. Therefore, the Hénon map provides a very good example for studying bifurcation in the creation of a horseshoe, which is one of the most important problems in the study of dynamical systems at present. See also Subsection 3.4.7.

C Transverse Homoclinic Points and Horseshoes Now we shall give a remarkable theorem due to Poincaré, Birkhoff and Smale, which claims the existence of a horseshoe near a transverse homoclinic point to a hyperbolic fixed point.

Definition 3.4.23. Let $f : \mathbb{R}^2 \to \mathbb{R}^2$ be a diffeomorphism which admits a hyperbolic fixed point O. A *transverse homoclinic point* x to the fixed point is defined

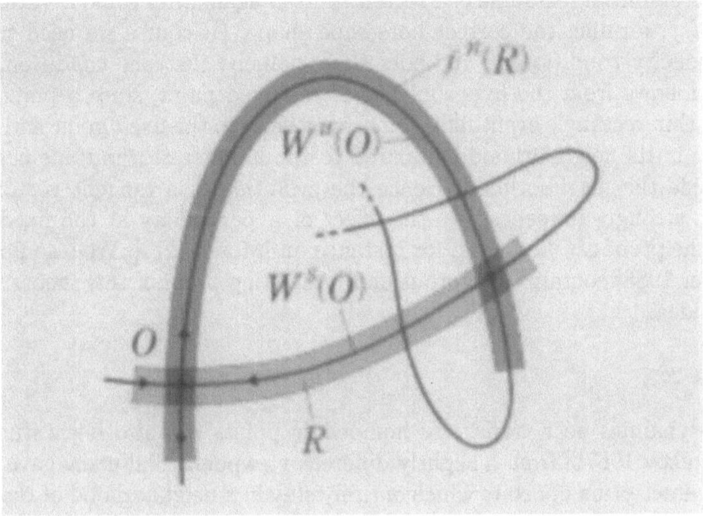

Fig. 3.4.13. A transverse homoclinic point and a horseshoe.

as a point in the transverse intersection of the stable and unstable manifolds of O, namely

$$x \in W^u(O) \cap W^s(O) \setminus \{O\}$$

where $W^u(O)$ and $W^s(O)$ intersect transversely at x. See Fig. 3.4.13.

Remark 3.4.24. The stable and unstable manifolds of an identical hyperbolic fixed point can intersect transversely for diffeomorphisms, whereas for vector fields, the intersection cannot be transverse since the intersection has to contain at least one orbit which is of dimension one. Therefore, a homoclinic orbit is not (locally) structurally stable for vector fields while it can be so for diffeomorphisms if it is transverse.

Theorem 3.4.25. (Poincaré-Birkhoff-Smale) *Let $f : \mathbb{R}^2 \to \mathbb{R}^2$ be a diffeomorphism which admits a hyperbolic fixed point O. Suppose its stable and unstable manifolds intersect transversely at x, then there exists an invariant subset Λ of f near the orbit of x on which some iterates of f is hyperbolic, and moreover there exists a positive integer N such that (f^N, Λ) is topologically conjugate to the shift dynamics with two symbols.*

The idea of the proof of this theorem is clear from Fig. 3.4.13: we first take a very thin rectangle along the stable manifold of O and then iterate it by f. Then the rectangle shrinks along the stable manifold and stretches along the unstable manifold because of the hyperbolic nature of the fixed point. Therefore, an appropriately chosen initial rectangle can return to itself again after some iterations, say N-times, of f, forming the correct horseshoe shape. Of course we need to check the hyperbolicity condition on the horseshoe, namely, the cone condition, which essentially follows from the hyperbolicity of the fixed point, since a point in the initial very thin rectangle eventually comes very close to the fixed point and spends a long time in its neighborhood compared to the number of iterations needed to travel outside the neighborhood. Hence the evolution of a tangent vector along the orbit is strongly influenced by the effect of hyperbolicity at the fixed point. Details of the proof can be found, for instance, in [Moser 1973]. We also point out that [Palmer 1988] contains a rigorous and interesting proof of this theorem using a different idea.

Remark 3.4.26.

(1) The dynamics near transverse homoclinic points has also been studied by [Shil'nikov 1967b] from a slightly different viewpoint. Shil'nikov gave a complete description of orbits which entirely stay in a neighborhood of the homoclinic orbit by using the symbolic dynamics with countably many symbols. Recent works done by [Hale and Lin 1986b], [Deng 1989c] and [Deng 1990c] are inspired by the Shil'nikov's idea.

(2) Almost all results presented in this subsection remain valid for general hyperbolic diffeomorphisms, and furthermore, many other interesting and important properties are known as well. Indeed the structure theory for the limit set of hyperbolic diffeomorphisms has been widely developed, and as a result, we have now a nice characterization of structurally stable diffeomorphisms ([Mañé 1987]). Here we explain only a few basic ideas of the entire theory. Interested readers can find further details in e.g. [Smale 1967], [Moser 1973], [Bowen 1970], and [Palis and de Melo 1982].

We end this subsection by giving a brief discussion on the definition of the term "chaos". Chaotic dynamics literally refers to a complicated, seemingly random behavior of orbits in a dynamical system. As we have seen in this and the previous subsections, the quadratic map, the Lorenz map, and the horseshoe diffeomorphism exhibit such complicated behaviors, which can be described by means of the symbolic representation of orbits. Therefore we are tempted to call these examples "chaotic".

No unanimously accepted mathematical definition has been, however, given to the notion "chaos" up to now, simply because its mathematical theory has not yet been fully developed. In fact, there are number of examples of dynamical systems, such as the Rössler equation (3.2.4), which seem to exhibit chaotic dynamics but are not proved to be so in any reasonable sense.

Throughout this book, by chaos we mean a dynamical system which contains a shift dynamics. To be more precise, a discrete dynamical system φ on M is said to be chaotic if there exists an invariant subset on which some iterates of φ is topologically semi-conjugate to the shift dynamics (σ, S_m) with a finite, say m symbols. A continuous dynamical system is chaotic if so is the Poincaré map on a suitable cross section. Therefore the existence of a transverse homoclinic point provides a simple and useful criterion for the existence of chaotic dynamics. It also works for continuous dynamical systems by taking account of an appropriate Poincaré mapping along a hyperbolic periodic orbit. Some examples of how this criterion works can be found in [Guckenheimer and Holmes 1983], and in the next section, we shall see more examples of chaos in continuous dynamical systems. We also refer to a strange attractor or a chaotic attractor as an attractor on which the dynamical system is chaotic in the above sense.

The definition of chaos given above only formalizes a way of detecting some kind of irregular behavior appearing in a dynamical system by relating it to symbolic sequences, as we have done before. Therefore we do not claim that this definition is a suitable definition for general chaotic dynamics, since that it might miss an important feature of true chaotic behaviors in dynamical systems beyond the symbolic representation, although we believe that this definition captures at least some of the most typical chaotic behaviors. For different aspects of "chaos" as well as other definitions in literature, we refer to [Li and Yorke 1975], [Guckenheimer and Holmes 1983; Section 5.4], [Devaney 1986; Section 1.8] (see also [Takens 1988] and [Banks, Brooks, Cairns, Davis and Stacey 1992]), [Bergé, Pomeau and Vidal 1984], [Palis and Takens 1993], and [Broer, Dumortier, van Strien and Takens 1991; Chapter 5].

3.4.7 Further Developments

We close this section by giving some indications of important achievements for discrete dynamical systems with emphasis on strange attractors.

A One-Dimensional Quadratic Family There has been a lot of works for one-dimensional mappings, in particular for the quadratic family given in Subsection 3.4.4 or its equivalent families. Such one-dimensional mappings are studied from combinatorial, topological and metrical points of view (see [de Melo and van Strien 1993]). The combinatorial study concerns the symbolic representation of orbits and investigates the dynamical behavior of the mapping from the corresponding symbolic sequences, especially the so-called kneading sequence, a special symbolic sequence corresponding to critical points of the map. This gives a lot of information on the dynamical properties of the mapping, for instance, on the number of periodic points and their combinatorial types. One of the most remarkable result belonging to this level of the study is the Šarkovskii's theorem [Šarkovskii 1964], which gives an order relation among periods of periodic points appearing in general one-dimensional continuous mappings. See [Collet and Eckmann 1980] and [Milnor and Thurston 1988] for more information on the kneading sequence.

The investigation of qualitative nature of more general invariant sets for one-dimensional mappings is the task of the topological study. Topological properties of attractors are of particular interest here. This type of the theory also includes the conjugacy problem which asks when two mappings are topologically conjugate with each other. This type of results are summarized in [de Melo and van Strien 1993].

One-dimensional mappings including the quadratic family are studied as well from the metrical or measure theoretical point of view, which studies typical behavior of orbits by means of a measure associated to the mapping. In this theory, the existence of absolutely continuous invariant measures with respect to the Lebesgue measure becomes important. As we have seen in C of Subsection 3.4.4, such an absolutely continuous invariant measure sometimes gives the asymptotic distribution of orbits and is related to the random and chaotic behavior of the dynamics. One of the most remarkable results along this line is the one due to [Jakobson 1981]. We state it here in a somewhat stronger form given by [Benedicks and Carleson 1985].

Theorem 3.4.27. (Jakobson-Benedicks-Carleson) *For the quadratic family* $Q_a(x) = 1 - ax^2$, *there exists a positive Lebesgue measure set* A *of parameter values* a *in* $[0, 2]$ *such that, for any* $a \in A$, *the map* $Q_a : [-1, 1] \to [-1, 1]$ *satisfies the following two conditions:*

(1) $\liminf_{n\to\infty} \frac{1}{n} \log |DQ_a^n(1)| > 0$;

(2) Q_a *admits an invariant probability measure which is absolutely continuous with respect to the Lebesgue measure.*

In fact, the second condition follows from the first one. The first condition means that there exists a positive constant c (which can indeed be chosen close to $\log 2$) such that $|DQ_a^n(1)| \geq (\text{const}) \cdot e^{cn}$ for all $n \geq 0$ as long as the parameter a belongs to the positive measure set A, which shows that the tangent vector expands in an exponential rate under the iteration.

Due to the importance of this result, a number of alternative proofs have been given to this theorem since [Jakobson 1981]. To name a few, [Guckenheimer 1987], [Rychlik 1988], [Benedicks and Carleson 1985], [Tsujii 1992, 1993] provide alternatives proofs and its generalizations.

These are a brief overlook for the study of dynamics given by the quadratic family and more general one-dimensional mappings on an interval in the real line. Yet another way to look at the quadratic family is to regard it as a family of holomorphic mappings from the complex plane to itself. The complexification provides us many deep results together with beautiful computer pictures of invariant sets in the complex plane. Some of these results have implications to the corresponding results for real one-dimensional mappings, most of which cannot be obtained without using such complex analytic argument. See [Devaney 1986], [Milnor 1990], and references therein. All these aspects of the quadratic family and more general one-dimensional mappings are explained in detail in [de Melo and van Strien 1993].

B Lozi Map The piecewise-linearity of the Lozi map

$$L_{a,b} : \begin{pmatrix} x \\ y \end{pmatrix} \mapsto \begin{pmatrix} y + 1 - a|x| \\ bx \end{pmatrix}$$

helps to understand its dynamics well. In fact, [Misiurewicz 1980] already showed the following remarkable results.

Theorem 3.4.28. (Misiurewicz) *There exists an explicitly given positive measure set of parameters such that the Lozi map $L_{a,b}$ with parameters from this set satisfies the following conditions:*

(1) There exists a compact invariant subset Λ and its open neighborhood U such that any point from U approaches to Λ under positive iteration of $L_{a,b}$;

(2) The invariant set Λ is topologically transitive;

(3) The set Λ is the closure of the unstable manifold of a hyperbolic fixed point of $L_{a,b}$;

(4) The map $L_{a,b}$ is hyperbolic on Λ except for countably many points.

Following this work, [Rychlik 1983] and [Collet and Levy 1984] independently proved the existence of a nice invariant measure on the attractor Λ, which satisfies the so-called Bowen-Ruelle-Sinai property for the Lozi map. Moreover, it was also shown that the Lozi map is ergodic, and furthermore it is isomorphic to the Bernoulli shift with respect to this invariant measure (see [Mañé 1983] for the definition). Therefore we may well call the invariant set Λ a strange attractor, since it is an attractor on which the dynamics is completely random. Later this result has been generalized for more general almost hyperbolic homeomorphisms by [Young 1985] and [Pesin 1992].

C Hénon Map Unlike the Lozi map, it is very hard to show the existence of a strange attractor for the Hénon map

$$H_{a,b} : \begin{pmatrix} x \\ y \end{pmatrix} \mapsto \begin{pmatrix} y + 1 - ax^2 \\ bx \end{pmatrix},$$

despite the similarity of the forms of these two mappings. The main difficulty comes from its smooth nature, since the Lozi map consists of two linear maps which are glued together continuously at a line in the plane at which the map loses its differentiability. Thanks to this piecewise-linear property, we are able to show the hyperbolicity of the Lozi map for points whose orbits never fall onto the exceptional line. The Hénon map has, on the other hand, a smooth folding nature which prevents us from characterizing the set of points at which orbits lose the hyperbolicity. However, [Benedicks and Carleson 1991] have succeeded to show the following remarkable results on the dynamics of the Hénon mapping under the strong dissipative situation:

Theorem 3.4.29. (Benedicks-Carleson) *For the Hénon map $H_{a,b}$ with sufficiently small b, there exists a positive measure set $A(b)$ of parameter values a such that, for any $a \in A(b)$, the Hénon map satisfies the following conditions:*

(1) There exists a compact invariant subset Λ and an open set U in the (x, y)-plane, such that any point from U converges to Λ;

(2) There exists a point $z \in \Lambda$ satisfying that:

(i) the orbit of z is dense in Λ;

(ii) $\liminf_{n\to\infty} \frac{1}{n} \log \| DH^n_{a,b}(z)v \| > c$ where $c > 0$ is a constant close to $\log 2$ and v is the vertical unit vector $v = \frac{\partial}{\partial x}$.

This means that the Hénon map has an invariant set which attracts nearby points from a positive measure set and this invariant set has a dense orbit on which the mapping is expanding with an exponential growth rate in some direction. Readers should notice the resemblance of this statement to Jakobson's theorem. In fact, these two theorems are proved in a similar way, although the latter is much more involved because of its two-dimensionality and the lack of specific orbit corresponding to the critical point for one-dimensional mappings.

Ergodic theoretical study for the attractor has been done by [Benedicks and Young 1992] following the above result showing the existence of a unique Bowen-Ruelle-Sinai measure for the Hénon map with parameters from the set described in the above theorem. Moreover the dynamics of the Hénon map on the attractor Λ with respect to the invariant measure is isomorphic to the Bernoulli shift. In this sense, the dynamics of the Hénon map can also be said chaotic, if the parameters belong to the positive measure set as described, and hence we may call it a strange attractor. These results have been extended to a more general situation of a homoclinic point of a diffeomorphism losing its transversality, which will be briefly explained below.

As noted in the previous subsection, the Hénon map provides us with a good explicit example of a family of diffeomorphisms which creates a horseshoe. Namely, for the parameter a of the Hénon map small enough, it exhibits only a simple dynamics, whereas for sufficiently large a, it has a horseshoe as proved in [Devaney and Nitecki 1979]. Since it is important to understand the creation of horseshoes for the study of dynamical systems and their bifurcations, much attention has been paid to the Hénon map as well as the Lozi map which can be thought of as a piecewise linear variant of the Hénon map.

D Homoclinic Tangency In Subsection 3.4.6, we have given the Poincaré-Birkhoff-Smale theorem on the existence of horseshoes near transverse homoclinic

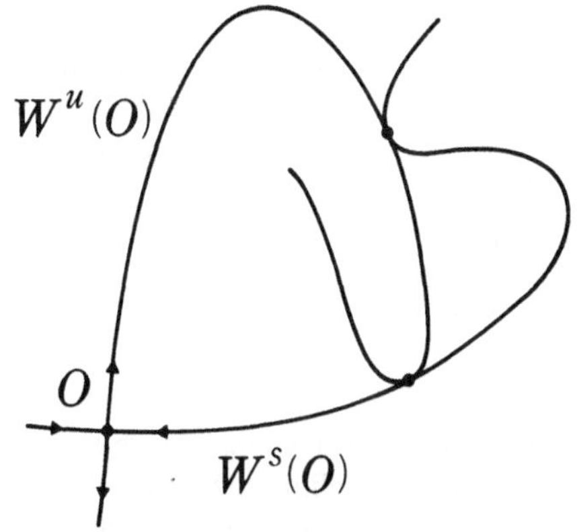

Fig. 3.4.14. A homoclinic tangency.

points, which is of fundamental importance in the theory of hyperbolic diffeomorphisms, and we now well understand the dynamics near such transverse homoclinic points. Then the next natural task is to study what happens when a transverse homoclinic point is created, namely, the study of bifurcations of a family of diffeomorphisms which has no homoclinic points at a parameter value and has a transverse homoclinic point at another parameter value. Since the transverse homoclinic point is given by a transverse intersection of the stable and unstable manifolds, such a family must have a tangency of these manifolds in-between these two parameter values as depicted in Fig. 3.4.14, which is now called a homoclinic tangency. The homoclinic tangency has been studied in the last two decades mainly by Newhouse, Palis and Takens. A forthcoming book [Palis and Takens 1993] provides us with a good exposition of interesting dynamics and bifurcations near the homoclinic tangency.

The homoclinic tangency is an important subject not only because it is a natural objective following the study of hyperbolic diffeomorphisms, but also because it is closely related to the understanding of some complicated chaotic dynamics, in particular "strange attractors" as observed in the Hénon map (3.4.1). In fact, inspired by the theorem of Benedicks and Carleson given above, [Mora and Viana 1991] extended this result to more general diffeomorphisms with homoclinic tangencies. Namely, for a generic one-parameter family which unfolds a homoclinic tangency associated to a dissipative hyperbolic periodic point, they have shown that, if the parameter belongs to a prescribed positive measure set, a Hénon-like attractor having the properties described in Theorem 3.4.29 appears for the diffeomorphism corresponding to the parameter.

We do not explain here any details of these remarkable achievements for discrete dynamical systems. Readers should remember that these results also give a lot of insight to the understanding of the dynamics and bifurcations in continuous dynamical systems, since discrete dynamical systems naturally appear from continuous ones by taking the Poincaré map. In the next section, We will make use of the results explained in this section in order to understand the dynamics near periodic and homoclinic/heteroclinic orbits in vector fields.

3.5 Bifurcations of Homoclinic and Heteroclinic Orbits in Vector Fields

Going back to continuous dynamical systems, in this section we mainly study bifurcations of homoclinic and heteroclinic orbits in vector fields connecting hyperbolic equilibrium points. We also discuss how such a bifurcation is related to chaotic dynamics from the point of view of the bifurcation theory.

3.5.1 Persistence of Homoclinic/Heteroclinic Orbits and the Melnikov Integral

Consider a vector field $\dot{x} = v(x)$ on $\mathbb{R}^n$ having two hyperbolic equilibrium points $O_\pm$. If there exists an orbit Γ given by a solution $h(t)$ satisfying $\lim_{t\to\pm\infty} h(t) = O_\pm$, or equivalently, if $W^u(O_-) \cap W^s(O_+) \neq \emptyset$, then the orbit is called an (O_-, O_+)-*heteroclinic orbit*. In particular, it is called a *homoclinic orbit to* O if, moreover, $O_- = O_+ = O$. Throughout this section, we identify an orbit with a solution giving the orbit. As we have seen in Section 3.2, although hyperbolic equilibrium points are structurally stable, homoclinic and some heteroclinic orbits are not, and hence these orbits do not persist under perturbation in general. Therefore, we begin with our study of homoclinic and heteroclinic orbits by taking a (generic) unfolding of them and by describing the set of parameters in which the homoclinic or heteroclinic orbits persist. This set of parameters is called the *bifurcation set* for the homoclinic/heteroclinic orbits.

Suppose a family of vector fields

$$\dot{x} = v(x, \mu) \qquad (\mu \in \mathbb{R}^k)$$

possesses two hyperbolic equilibrium points $O_\pm(\mu)$ and admits an (O_-, O_+)-heteroclinic orbit $h(t)$ at $\mu = \mu_0$. The index of the heteroclinic orbit which we denote by $ind(h(t))$ is defined by

$$ind(h(t)) = \dim W^u(O_+(\mu)) - \dim W^u(O_-(\mu)) + 1.$$

We assume the following two generic conditions.

(Ind) Two submanifolds $W^u(O_-(\mu_0))$ and $W^s(O_+(\mu_0))$ are in a general position along the heteroclinic orbit $h(t)$. Namely, it holds that

$$\dim\left(T_{h(t)}W^u(O_-(\mu_0)) + T_{h(t)}W^s(O_+(\mu_0))\right) = n - d,$$

where $d = \max\{0, ind(h(t))\}$.

This condition is equivalent to the existence of d solutions to the adjoint of the variational equation

$$\dot{\hat{z}} = -\hat{z} \cdot Dv(h(t), \mu_0) \qquad (3.5.1)$$

along $h(t)$, which are bounded over $\mathbb{R}$ and are linearly independent. Denote these solutions by $\hat{q}_1(t),\ \hat{q}_2(t), \ldots, \hat{q}_d(t)$. We may assume $k \geq d$ without loss of generality, where k is the dimension of the parameter space, by embedding the original family into one with more parameters.

(Unf) Assume that the family $\dot{x} = v(x, \mu)$ ($\mu \in \mathbb{R}^k$) generically unfolds the heteroclinic orbit $h(t)$. This assumption is formulated as that the d vectors given by

$$M_i = \int_{-\infty}^{\infty} \hat{q}_i(t) \cdot \frac{\partial}{\partial \mu} v(h(t), \mu) dt \in \mathbb{R}^k, \qquad (i = 1, \ldots, d) \qquad (3.5.2)$$

determined from $\hat{q}_1(t),\ \hat{q}_2(t), \ldots, \hat{q}_d(t)$ are linearly independent.

Under these conditions, we can show the following.

Theorem 3.5.1. *Suppose $ind(h(t)) \geq 1$. Then the bifurcation set*

$$B = \{\mu \mid W^u(O_-(\mu)) \cap W^s(O_+(\mu)) \neq \emptyset\}$$

for the $(O_-(\mu), O_+(\mu))$-heteroclinic orbit forms a submanifold of codimension $d = ind(h(t))$ in a neighborhood of μ_0, and the vectors $(M_1, \cdots, M_d)$ are perpendicular to B at μ_0.

Proof. The geometric idea of the proof of this theorem is very simple. We first take a transverse cross section Σ to the heteroclinic orbit $h(t)$ at $x = h(0)$, and look at the intersection σ^u and σ^s of $W^u(O_-(\mu))$ and $W^s(O_+(\mu))$ with the cross section Σ, respectively. These submanifolds σ^u and σ^s in Σ are nicely described as

$$\sigma^u = \text{graph}\left[(\xi^{(u)}, \mu) \mapsto (\xi_-^{(s)}(\xi^{(u)}, \mu), \xi_-^{(c)}(\xi^{(u)}, \mu))\right]$$

$$\sigma^s = \text{graph}\left[(\xi^{(s)}, \mu) \mapsto (\xi_+^{(u)}(\xi^{(s)}, \mu), \xi_+^{(c)}(\xi^{(s)}, \mu))\right],$$

by choosing suitable basis vectors on Σ. Then we take a displacement function $\Xi(\mu)$ which measures the "distance" of the stable and unstable manifolds along the $\xi^{(c)}$-direction defined by

$$\Xi(\mu) = \xi_+^{(c)}(\xi_*^{(s)}(\mu), \mu) - \xi_-^{(c)}(\xi_*^{(u)}(\mu), \mu),$$

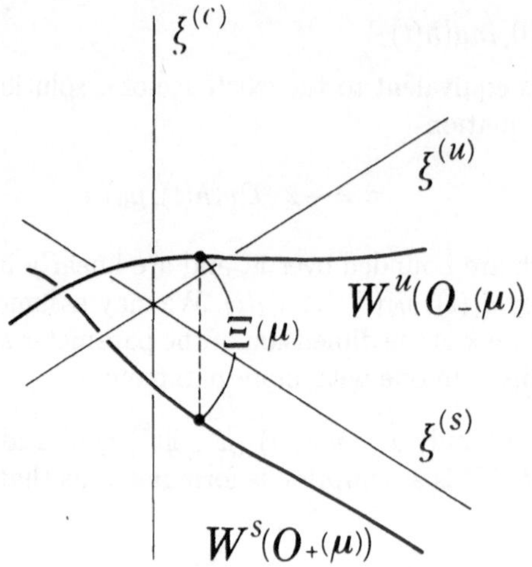

Fig. 3.5.1. Geometric meaning of the function $\Xi(\mu)$ $(d=1)$.

where $\xi_*^{(s)}(\mu)$ and $\xi_*^{(u)}(\mu)$ are the unique solutions to the equation

$$\xi_-^{(s)}(\xi^{(u)},\mu)=\xi^{(s)},\quad \xi_+^{(u)}(\xi^{(s)},\mu)=\xi^{(u)},$$

respectively. See Fig. 3.5.1. This displacement function $\Xi(\mu)$ takes values in $\mathbb{R}^d$ from (Ind) and the bifurcation set B is nothing but the zeros of the function $\Xi(\mu)$; $B=\Xi^{-1}(0)$. Thus, B forms a submanifold of codimension d if the Jacobian matrix $\frac{\partial}{\partial\mu}\Xi(\mu_0)$ is of maximal rank.

Finally, it turns out that $\frac{\partial}{\partial\mu}\Xi(\mu_0)$ is given by the integrals $M=(M_1,\cdots,M_d)$ since the basis vectors giving the coordinates $\xi^{(c)}\in\mathbb{R}^d$ generate fundamental solutions whose adjoints are bounded solutions to (3.5.1). □

If, on the other hand, $ind(h(t))\le 0$ under the condition (Ind), the stable and unstable manifolds intersect transversely along the heteroclinic orbit $h(t)$, and hence the bifurcation set B forms a neighborhood of $\mu=0$. In other words, such a heteroclinic orbit is locally structurally stable.

The integrals in the condition (Unf) are called *Melnikov integrals*, the idea of which was originally introduced by [Poincaré 1899], and then extensively used by [Melnikov 1963] for the study of periodic solutions in a periodically forced Hamiltonian system. This was was then reformulated and generalized by [Holmes and Marsden 1982], [Gruendler 1985], and others. Similar results have been obtained by [Chow, Hale and Mallet-Paret 1980] and [Palmer 1984] using a different idea for the derivation of the Melnikov integrals. The general formula valid for any autonomous ordinary differential equations with a homoclinic/heteroclinic orbit is due to [Hale and Lin 1986a]. See also [Guckenheimer and Holmes 1983], [Sanders 1982], [Robinson 1988], [Kokubu 1988], [Wiggins 1988], [Battelli and Lazzari 1990], and [Gruendler 1992] for related works.

Example 3.5.2. Consider the following ordinary differential equation with three parameters $\mu = (\mu_1, \mu_2, \mu_3)$:

$$\begin{aligned}
\dot{x} &= y \\
\dot{y} &= 2x(x-1)(x-2) \quad +\mu_1 z + \mu_2(x-2)^2 \quad +\mu_3(x-2) \\
\dot{z} &= \left(\tfrac{3}{2}x - 1\right) z \qquad\quad +\mu_1 \qquad\qquad\qquad +\mu_3 y.
\end{aligned}$$

When $\mu = (\mu_1, \mu_2, \mu_3) = (0,0,0)$, the equation has two hyperbolic equilibrium points

$$O_1 = (0,0,0) \quad \text{and} \quad O_2 = (2,0,0)$$

and a heteroclinic orbit $h(t)$ from O_1 to O_2 given by

$$h(t) = (h_1(t), h_2(t), 0) = \left(\frac{2e^{2t}}{e^{2t}+1}, \frac{4e^{2t}}{(e^{2t}+1)^2}, 0 \right).$$

Furthermore, the index of this (O_1, O_2)-heteroclinic orbit is two.

In order to compute the Melnikov integrals, we first get two independent bounded solutions to the adjoint equation (3.5.1), which are in fact given by

$$\begin{aligned}
\hat{q}_1(t) &= (-\dot{h}_2(t), \dot{h}_1(t), 0) \\
\hat{q}_2(t) &= (0, 0, \delta(t)),
\end{aligned}$$

where

$$\delta(t) = \exp\left[-\int_0^t \left(\frac{3}{2}h_1(s) - 1 \right) ds \right].$$

Then the Melnikov integrals take the forms

$$M_1 = \int_{-\infty}^{\infty} \left(0, \dot{h}_1(t)(h_1(t)-2)^2, \dot{h}_1(t)(h_1(t)-2)\right) dt$$

and

$$M_2 = \int_{-\infty}^{\infty} (\delta(t), 0, \delta(t)h_2(t))\, dt.$$

It is easy to evaluate these integrals. In fact, for M_1, we change the variable $x = h_1(t)$ and get

$$M_1 = \int_0^2 (0, (x-2)^2, x-2) dx = (0, \tfrac{2}{3}, 2).$$

In order to compute M_2, we first note that

$$\int_0^t \left(\frac{3}{2}h_1(s) - 1\right) ds = \frac{3}{2}\log\left(\frac{e^{2t}+1}{2}\right) - t$$

and hence

$$\delta(t) = e^t \cdot \left(\frac{e^{2t}+1}{2}\right)^{-\frac{3}{2}}.$$

From this we easily obtain

$$M_2 = \left(2\sqrt{2}, 0, \frac{16}{15}\sqrt{2}\right).$$

Therefore, we conclude from Theorem 3.5.1 that the (O_1, O_2)-heteroclinic orbit persists along a one-dimensional curve in the parameter space of $\mu = (\mu_1, \mu_2, \mu_3)$, whose tangent vector at $\mu = 0$ is given by $(\frac{8}{15}, 3, -1)$, which is perpendicular to the vectors M_1 and M_2. This example is taken from [Kokubu 1993] with a slight modification, where other bifurcation aspects of this example can also be found.

The Melnikov integrals as well as analogous integrals obtained by a similar idea can be used to study the global nature of various kinds of dynamics which are reduced to the existence of homoclinic and heteroclinic orbits. For instance, the non-integrability of a Hamiltonian system ([Holmes and Marsden 1982], [Guckenheimer and Holmes 1983]), chaotic dynamics for a periodically forced system such as the Duffing equation ([Holmes 1979], [Guckenheimer and Holmes 1983], [Robinson 1988]), the existence of layer solutions to a singularly perturbed ordinary differential equations ([Palmer 1986], [Lin 1989, 1990b], [Sakamoto 1990], [Szmolyan 1991], [Kaper and Wiggins 1991], [Oka 1992]), among others. Various kinds of Melnikov integrals are extensively obtained in [Wiggins 1988], which are suitable mainly for studying some kinds of Hamiltonian systems with several degrees of freedom. We have also seen the Melnikov integral in the study of the Bogdanov-Takens bifurcation in Subsection 3.3.4.

Example 3.5.3. Recall that the vector field (3.3.10):

$$\begin{aligned}\dot{x} &= y\\ \dot{y} &= x^2 - 1 + \varepsilon(\nu y + xy) + O(\varepsilon^2),\end{aligned}$$

has a homoclinic orbit $h(t) = (x(t), y(t))$ when $\varepsilon = 0$. The bounded solution to the adjoint of the variation equation along the homoclinic orbit is simply given by

$$\dot{h}^{\perp}(t) = (-\dot{y}(t), \dot{x}(t)).$$

Therefore, the Melnikov integral takes the form

$$\int_{-\infty}^{\infty} \dot{h}^{\perp}(t) \cdot \begin{pmatrix} 0 \\ \nu y(t) + x(t)y(t) \end{pmatrix} dt$$
$$= \int_{-\infty}^{\infty} \dot{x}(t)\{\nu y(t) + x(t)y(t)\}dt$$
$$= \int_{-\infty}^{\infty} \{\nu y(t)^2 + x(t)y(t)^2\}dt.$$

If this integral is non-zero, then the homoclinic orbit is generically unfolded by the parameter ε and hence it disappears. Thus, the first approximation of the bifurcation set for the homoclinic orbit is given by

$$\int_{-\infty}^{\infty} \{\nu y(t)^2 + x(t)y(t)^2\}dt = 0,$$

which implies

$$\nu = -\frac{\int_{-\infty}^{\infty} x(t)y(t)^2 dt}{\int_{-\infty}^{\infty} y(t)^2 dt} = \frac{5}{7}$$

from the explicit form of the homoclinic orbit given in Subsection 3.3.4.

For such local bifurcation problems, the Melnikov integrals often take the form of Abelian integrals involving parameters, and the study of the parameter dependence of such Abelian integrals by using, for instance, a complex analytic method often provides us with a detailed information for the bifurcation ([Arnold 1983], [Il'yashenko 1978], [Carr 1981], [Cushman and Sanders 1985], [Carr, Chow and Hale 1985], [van Gils 1985], [Khorozov 1979], and [Żoładek 1991]).

Theorem 3.5.1 gives only the condition of parameters for which the homoclinic/heteroclinic orbit persists in an unfolding. We do not know in general whether any transverse family to the bifurcation set gives a versal family for the bifurcation of such a homoclinic/heteroclinic orbit. However it is the case if the vector field is two-dimensional. Namely, we have the following theorem.

Theorem 3.5.4. *Let $\dot{x} = v(x, \mu)$ be a family of vector fields on the plane $\mathbb{R}^2$ which possesses a homoclinic orbit $h(t)$ to a hyperbolic saddle point O for $\mu = 0$. Assume that the eigenvalues of the linearization $Dv(O, 0)$ do not have an equal modulus, that is,* $\mathrm{tr}Dv(O, 0) \neq 0$, *and that*

$$\int_{-\infty}^{\infty} \hat{q}(t) \cdot \frac{\partial v}{\partial \mu}(h(t), 0)dt \neq 0,$$

where $\hat{q}(t)$ is a non-zero bounded solution to the adjoint of the variational equation along the homoclinic orbit $h(t)$. Then the family is a versal family for the homoclinic orbit.

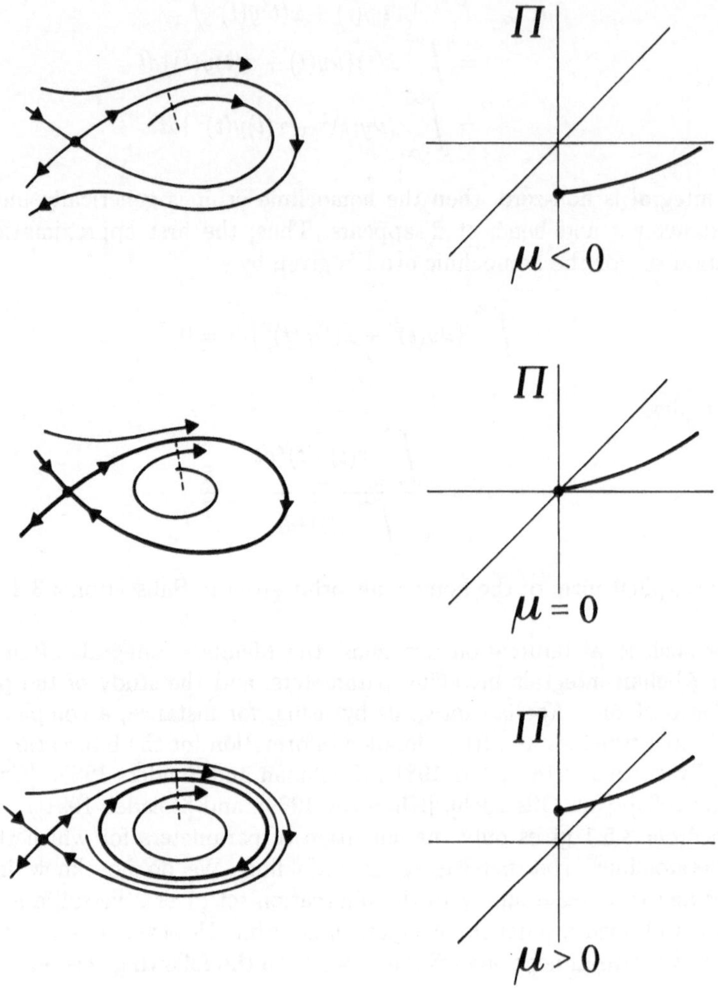

Fig. 3.5.2. The bifurcation of a planar homoclinic orbit.

The bifurcation diagram is depicted in Fig. 3.5.2. Readers should notice that this bifurcation is a part of the generic Bogdanov-Takens bifurcation given in Subsection 3.3.4.

Proof. We first take a good local coordinates (x^1, x^2) in a small neighborhood of O in $\mathbb{R}^2$ such that the vector field when $\mu = 0$ is C^1 linearized in the coordinates. This is due to the remark following Belitskii's theorem (Theorem 3.2.20). In particular we can take the unstable and stable manifolds to O as being the coordinate axes in the neighborhood:

$$W^u(O) = \{x^1 = 0\} \quad \text{and} \quad W^s(O) = \{x^2 = 0\}.$$

Then we define the cross sections to the homoclinic orbit as follows:

$$\Sigma^s = \{x^1 = \delta\} \quad \text{and} \quad \Sigma^u = \{x^2 = \delta\}$$

for a sufficiently small $\delta > 0$.

The main idea of the proof is to reduce the bifurcation problem for the planar homoclinic orbit to one for a one-dimensional flow-defined mapping along the homoclinic orbit. For that purpose, we take the maps

$$\Pi_{loc} : \Sigma^s \to \Sigma^u, \quad \text{and} \quad \Pi_{far} : \Sigma^u \to \Sigma^s,$$

and compose them to get the entire Poincaré return map

$$\Pi = \Pi_{far} \circ \Pi_{loc}.$$

To be more precise, first we consider the local map Π_{loc}. Since the vector field is already C^1 linearized in a neighborhood of the equilibrium point O, it is easy to obtain the form

$$\Pi_{loc}(u) = u^\nu$$

where $\nu = \frac{|\lambda^s|}{\lambda^u}$. We may assume that $\nu > 1$ without losing generality, otherwise we take the time reversed flow. Then the global map Π_{far} is simply an orientation preserving diffeomorphism and hence can be expanded as

$$\Pi_{far}(v) = av + O(v^2),$$

with $a > 0$. Therefore, the entire return map Π is given by

$$\Pi(u) = au^\nu + o(u^\nu) \qquad (a > 0,\ \nu > 1).$$

This is the return map along the homoclinic orbit for $\mu = 0$ and the graph of this map is indicated in Fig. 3.5.2(b).

Now we perturb the vector field. Then the corresponding one-dimensional map depends on μ, denoted by $\Pi(u;\mu)$. The assumption of the theorem implies that

$$\frac{\partial \Pi}{\partial \mu}(0;0) \neq 0,$$

which means the end point of the graph moves away from zero with non-zero speed when the parameter μ passes through 0. Therefore, we obtain the two different possibilities of the perturbed mapping as indicated in Fig. 3.5.2(a) and (c). From these pictures, we can immediately see that, in the case of (a), every orbit eventually leaves the neighborhood of the origin under the iteration of the mapping, whereas there exists a unique fixed point of the one-dimensional map in the case of (c), which corresponds to a periodic orbit for the flow. It is not difficult to show that this analysis is enough to conclude the versality. □

This theorem was proved by [Shil'nikov 1968] besides the versality result and in a more general context. The versality proof has been given by [Dumortier and Roussarie 1990] in a stronger form. Namely, it is shown that one can find an equivalence homeomorphism bringing an unfolding of the planar homoclinic orbit to its versal family as being continuous on the unfolding parameter.

Theorem 3.5.4 shows that the dynamics near a single homoclinic orbit in the plane is generically rather simple. This is not the case in general for higher dimensional vector fields. In the next subsection, we will see that the dynamics in a neighborhood of a homoclinic orbit associated to a saddle-focus singularity can possess a chaotic dynamics, which is now known as the Shil'nikov theorem. Homburg has recently proved that, for homoclinic orbits to a real saddle in $\mathbb{R}^3$ ($n \geq 3$), the nearby dynamics is also chaotic under some extra open condition on what he calls the generalized homoclinic orbit. See [Homburg 1992] for the detail.

More degenerate homoclinic and heteroclinic orbits are also studied by many authors. A homoclinic orbit to a nonhyperbolic equilibrium point is one of such degenerate homoclinic orbits. Schecter studied a homoclinic orbit to a saddle-node singularity ([Schecter 1987]) and a heteroclinic orbit to a pitchfork singularity ([Schecter 1990]), and Roussarie ([Roussarie 1992]) studied a homoclinic orbit to the Bogdanov-Takens singularity, all in the plane. For the first two cases, even versal families are given. Analogous studies have also been done for higher dimensional vector fields by [Chow and Lin 1990] (homoclinic orbits to saddle-node singularities), [Deng 1990a] (homoclinic orbits to nonhyperbolic equilibrium points), [Deng and Sakamoto 1989] and [Lin 1993] (homoclinic orbits to Hopf singularities). From Subsection 3.5.3 and later, we will consider different types of degenerate homoclinic/heteroclinic orbits, which are also of interest and of importance.

There are a lot of works for homoclinic and heteroclinic orbits which we cannot explain here in detail. See [Glendinning 1988] and [Wiggins 1988] as general references for this topic.

3.5.2 Shil'nikov Theorem

A homoclinic orbit persists along a submanifold of codimension one in the parameter space because $d = 1$ in Theorem 3.5.1. The dynamics around the homoclinic orbit can, however, be very rich in some cases, since the flow near a homoclinic orbit is recurrent and hence various orbits other than the persistent homoclinic orbit can also appear under perturbation. Therefore, the bifurcation set for the entire dynamics is in general more complicated. Among such homoclinic orbits, a homoclinic orbit of Shil'nikov type is of particular interest. Restricting it to three-dimensional vector fields for simplicity, it is a homoclinic orbit based at a saddle-focus equilibrium point at which the linearization matrix has a positive real eigenvalue $\lambda^u > 0$ and a pair of complex conjugate eigenvalues λ^s, $\overline{\lambda^s}$ (with

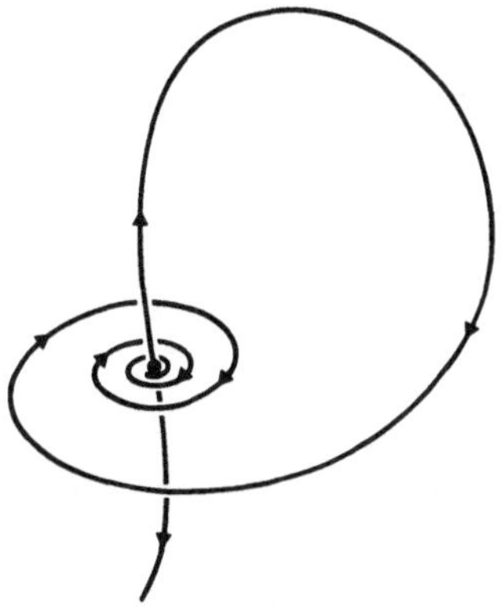

Fig. 3.5.3. The Shil'nikov-type homoclinic orbit.

$Im\lambda^s \neq 0$) satisfying

$$\lambda^u > |Re\lambda^s|.$$

The last condition is crucial in what follows, and is referred to as the *Shil'nikov condition.* See Fig. 3.5.3.

For such a homoclinic orbit Γ in $\mathbb{R}^3$, one can in general take a flow-defined map along Γ, which is defined on an appropriate domain of definition D in a cross section Σ taking values in Σ itself, where Σ is a two-dimensional disc transverse to Γ at a point in its tubular neighborhood. This map is called the Poincaré map along Γ. The Poincaré map $\Pi : D(\subset S) \to S$ along a homoclinic orbit of Shil'nikov type admits the following remarkable property.

Theorem 3.5.5. (Shil'nikov) *For any positive integer $m \geq 2$ there exists a Π-invariant subset $V_m \subset D$ such that $\Pi|_{V_m}$ is topologically conjugate to the shift dynamics*

$$\sigma : S_m = \{1, \ldots, m\}^{\mathbb{Z}} \to S_m \; ; \; (s_i)_{i=-\infty}^{\infty} \mapsto (s_{i+1})_{i=-\infty}^{\infty}.$$

In other words, it admits a homeomorphism $h : S_m \to V_m$ satisfying

$$\Pi|_{V_m} \circ h = h \circ \sigma.$$

Proof. This theorem, which was shown by [Shil'nikov 1965, 1967a] for the first time, is proved by describing the Poincaré map Π along the Shil'nikov-type homoclinic orbit as a composition of the local map Π_{loc} near the equilibrium point and the global map Π_{far} outside, and by showing that the map $\Pi = \Pi_{far} \circ \Pi_{loc}$ involves arbitrarily many horseshoes (Subsection 3.4.6). Since a number of proofs

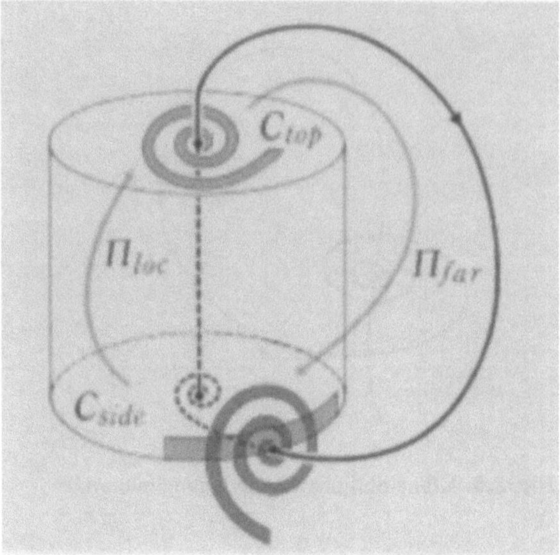

Fig. 3.5.4. The Poincaré map Π along a Shil'nikov-type homoclinic orbit.

already appear in the literature ([Tresser 1984], [Guckenheimer and Holmes 1983], [Wiggins 1988]), we shall only give an outline of the proof for the existence of a single horseshoe, namely, the shift dynamics with two symbols.

First, we note that the Belitskii's theorem (Theorem 3.2.20) and a remark following it show that a vector field $\dot{x} = v(x)$ on $\mathbb{R}^3$ is locally C^1 linearizable near a saddle-focus equilibrium point. Therefore, we may assume the vector field with a Shil'nikov-type homoclinic orbit is already made linear in a neighborhood U of the saddle-focus equilibrium point, say O. Inside this neighborhood with the coordinates $x = (x^1, x^2, x^3)$, we can take the (x^1, x^2)-plane as the stable manifold and the x^3-axis as the unstable manifold, and take a sufficiently small cylinder given by

$$C = \{|x^1|^2 + |x^2|^2 \leq \delta^2\} \times \{0 \leq x^3 \leq \delta\}$$

as in Fig. 3.5.4. Since the vector field is linear in U, we can easily obtain the local flow-defined map Π_{loc} from the side-boundary

$$C_{side} = \{|x^1|^2 + |x^2|^2 = \delta^2\} \times \{0 < x^3 \leq \delta\}$$

to the top boundary

$$C_{top} = \{|x^1|^2 + |x^2|^2 \leq \delta^2\} \times \{x^3 = \delta\}$$

as

$$\Pi_{loc}(\delta \cos\theta, \delta \sin\theta, \xi) = (\delta^{1+\nu}\xi^{-\nu} \cos(\theta + a(\xi)), \delta^{1+\nu}\xi^{-\nu} \sin(\theta + a(\xi)), \delta)\,,$$

where

$$\nu = \frac{Re\lambda^s}{\lambda^u} \quad \text{and} \quad a(\xi) = \frac{Im\lambda^s}{\lambda^u} \log\left(\frac{\delta}{\xi}\right).$$

On the other hand, the global map $\Pi_{far} : C_{top} \to C_{side}$ is a diffeomorphism and the Poincaré map along the homoclinic orbit is then given by the composition $\Pi = \Pi_{far} \circ \Pi_{loc}$.

Because the local map Π_{loc} is given explicitly and the global map has an almost constant non-singular Jacobian matrix in a sufficiently small neighborhood of the homoclinic orbit, we can show that a short vertical segment

$$\{\theta = \text{constant}\} \times \{0 < x^3 < \xi_0\}$$

in C_{side} is mapped by Π to an almost logarithmic spiral, as in Fig. 3.5.4, whose maximum height is approximately of order $\xi_0^{-\nu} \gg \xi_0$ because of the Shil'nikov condition

$$0 < -\nu = \frac{|Re\lambda^s|}{\lambda^u} < 1.$$

This implies that we can find an appropriate rectangle R on C_{side} which is mapped by Π into C_{side} with the shape of a horseshoe. See Fig. 3.5.4.

We then need to verify the cone condition (Subsection 3.4.6) in order to conclude that this is a true horseshoe. This is essentially shown by computing the derivative of the local map Π_{loc} which is again obtained explicitly, since the orbits stay in the neighborhood U of the equilibrium point much longer than outside. Hence the essential dynamical feature such as hyperbolicity is determined by the information from the local dynamics. The details can be found in the references cited in the beginning of the proof. □

Remark 3.5.6.

(1) Theorem 3.5.5 can be generalized to vector fields of arbitrary dimension ([Shil'nikov 1970]). For this case, the Belitskii's theorem is not in general applicable in order to derive the explicit form of the local map, since we need certain condition on the eigenvalues for the C^1 linearization. However, we can instead make use of a technique given by [Deng 1989b] using what he calls the exponential expansion (see Subsection 3.5.3) adapted to the saddle-focus equilibrium points. This has been carried out in [Deng 1990b], where a stronger result than Theorem 3.5.5 is proved.

(2) We have seen, in Subsection 3.4.6, that a transverse homoclinic point to a hyperbolic periodic orbit in a continuous dynamical system implies the existence of chaotic dynamics. This criterion especially works for time periodic differential equation such as the Duffing equation (3.4.3), since it is relatively easy to analyze the global Poincaré map. See [Guckenheimer and Holmes 1983] for the explicit computation for the Duffing equation. Theorem 3.5.5 gives another useful criterion for the existence of chaos in vector fields, which is more suitable for autonomous ordinary differential equations. Since

the proof of the theorem is valid for piecewise-linear vector fields as well, it can also be used to verify the chaotic dynamics of the Double Scroll circuit as in Section 1.4.

Shil'nikov-type homoclinic orbits are observed in many ways, as listed below:

- in some concrete examples of ordinary differential equations, for instance, the Lorenz equation ([Glendinning and Sparrow 1986]), the Rössler equation ([Arneodo, Coullet and Tresser 1981, 1982], [Gaspard, Kapral and Nicolis 1983, 1984], [Glendinning and Sparrow 1984]), an equation of nerve impulse dynamics ([Evans, Fenichel and Feroe 1982], [Hastings 1982]), and the Double Scroll circuit (Section 1.4);
- in an unfolding of degenerate singularities ([Guckenheimer and Holmes 1983], [Broer and Vegter 1984], [Arneodo, Coullet, Spiegel and Tresser 1985]);
- in an unfolding of degenerate homoclinic/heteroclinic orbits ([Rodriguez 1986], [Kokubu 1991, 1993]).

The study of Shil'nikov type homoclinic orbits is, however, far from being complete, in spite of many studies of it. One-parameter unfolding of the Shil'nikov-type homoclinic orbit is studied by [Gaspard 1984], [Gaspard, Kapral and Nicolis 1984], and [Glendinning and Sparrow 1984], where complicated subsidiary bifurcations are observed as shown in the following theorem.

Theorem 3.5.7. *Consider a one-parameter family of three-dimensional smooth vector fields with a parameter $\mu \in \mathbb{R}$. Suppose this system with $\mu = 0$ possesses a Shil'nikov-type homoclinic orbit. Then, under a suitable additional genericity condition, the following holds:*

(1) For $\mu \neq 0$, the system has an unboundedly growing number of periodic orbits of saddle type as $\mu \to 0$.

(2) On the two sides of $\mu = 0$, there exists a countable set $\{\mu_n^{SN}\}$ of saddle-node bifurcations of periodic orbits accumulating to $\mu = 0$ at the rate

$$\lim_{n\to\infty} \frac{\mu_{n+1}^{SN} - \mu_n^{SN}}{\mu_n^{SN} - \mu_{n-1}^{SN}} = \exp\left(\frac{2\pi\rho}{\omega}\right),$$

where $\rho = |Re\lambda^s|$ and $\omega = |Im\lambda^s|$. Each saddle-node bifurcation is accompanied by a period-doubling bifurcation occurring at μ_n^{PD}. Moreover, putting $\nu = \lambda^u$,

(i) if $0 < \frac{\rho}{\nu} < \frac{1}{2}$, then the saddle-node bifurcation occurs subcritically, and

$$\mu_n^{PD} = \mu_n^{SN} + O(\exp\left(\frac{4\pi n\rho}{\omega}\right));$$

(ii) if $\frac{1}{2} < \frac{\varrho}{\nu} < 1$, then the saddle-node bifurcation occurs supercritically, and

$$\mu_n^{PD} = \mu_n^{SN} + O(\exp(\frac{2\pi n\nu}{\omega})).$$

(3) There also exists a twofold countable set μ_n^i with $i = 1, 2$ of parameters at which the system possesses a homoclinic orbit (called a doubled homoclinic orbit) which crosses twice any hypersurface transverse to the original homoclinic orbit at $\mu = 0$. The parameters μ_n^i accumulate to $\mu = 0$ for each $i = 1, 2$, and satisfy

$$\lim_{n\to\infty} \frac{\mu_{n+1}^i - \mu_n^i}{\mu_n^i - \mu_{n-1}^i} = \exp(\frac{2\pi\nu}{\omega}), \qquad i = 1, 2.$$

The information given in the above theorem is, however, apparently incomplete. According to the recent results due to [Palis and Takens 1993] and [Mora and Viana 1991], among others, one may expect to have a very rich and complicated dynamics and bifurcations in the creation of horseshoes, for instance, infinitely many periodic attractors and Hénon-like attractors. See Subsection 3.4.7. Such phenomena should be observed in the unfolding of Shil'nikov-type homoclinic orbit. To the authors' knowledge, very little is known along this line up to now.

A homoclinic orbit to a saddle-focus equilibrium point without satisfying the Shil'nikov condition does not in general give rise to chaotic dynamics and complicated bifurcations. See [Tresser 1984] and [Glendinning 1988]. Nevertheless, the dynamics can exhibit some kind of irregular behavior near the homoclinic orbit. Indeed, it has been proved that, if a vector field has a symmetry and admits a symmetric pair of homoclinic orbits to a saddle-focus which does not satisfy the Shil'nikov condition, then the limit set in a neighborhood of the homoclinic orbits is simply given by only these orbits, but nearby orbits show a complicated transient behavior which is described by symbolic sequences in some way. See [Holmes 1980] for the detail.

3.5.3 Gluing Bifurcations for Heteroclinic Orbits and Exponential Expansion

It sometimes happens in an unfolding of a vector field with more than one homoclinic/heteroclinic orbit that not only a single homoclinic/heteroclinic orbit persists but also a homoclinic/heteroclinic orbit of different type appears under perturbation as illustrated in Fig. 3.5.5. For instance, there may be a heteroclinic orbit from O_1 to O_3 passing near O_2 in a perturbation of the case (a) of Fig. 3.5.5, and a homoclinic orbit to O_1 passing near O_2 in the case (b). In the case (c), we may have a new type of homoclinic orbit to O in an appropriate perturbation which follows both of the two homoclinic orbits in the unperturbed vector field several times before coming back to O. Such a bifurcation was first studied

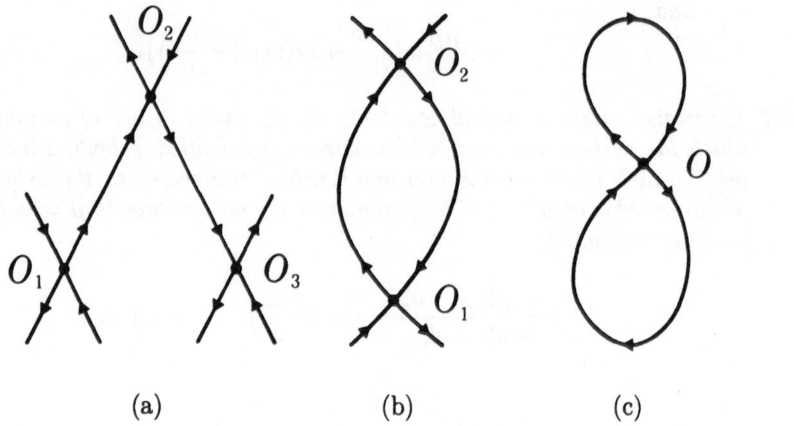

Fig. 3.5.5. Three kinds of codimension two heteroclinic and homoclinic orbits.

by [Gambaudo, Glendinning and Tresser 1984, 1988] in one case with the name "gluing bifurcation". Other cases were then investigated by [Chow, Deng and Terman 1990], [Kokubu 1988], and [Lin 1990a]. We shall explain here the case of Fig. 3.5.5(a) below as the simplest case, since cases (b) and (c) are those obtained by identifying some of the equilibrium points in (a), and hence they exhibits an additional dynamical behavior as we will see later.

Assume the following for an equilibrium point O_i $(i = 1, 2, 3)$:

(Ev) The positive and negative principal eigenvalues λ_i^u, λ_i^s at O_i are unique, real, and simple.

Here a positive (respectively negative) *principal eigenvalue* is an eigenvalue with positive and the smallest (respectively negative and the largest) real part.

We furthermore assume that, at $\mu = \mu_0$, the vector field has two heteroclinic orbits, one from O_1 to O_2 and the other from O_2 to O_3, and that, for the (O_i, O_{i+1})-heteroclinic orbit $h_i(t)$ $(i = 1, 2)$, the conditions (Ind) with $ind(h_i(t)) = 1$ and (Unf) in Subsection 3.5.1 are satisfied together with the following generic condition.

(Asy) The heteroclinic orbit $h_i(t)$ is tangent as $t \to -\infty$ (respectively $+\infty$) to the eigenspace associated with the positive (respectively negative) principal eigenvalue at O_i (respectively O_{i+1}). In other words, the following limits exist and are non-zero:

$$\lim_{t \to -\infty} e^{-\lambda_i^u t} |h_i(t) - O_i|, \quad \lim_{t \to \infty} e^{\lambda_{i+1}^s t} |h_i(t) - O_{i+1}|.$$

Under these conditions, Theorem 3.5.1 implies that the bifurcation sets $B_{i,i+1}$ $(i = 1, 2)$ for the (O_i, O_{i+1})-heteroclinic orbit $h_i(t)$ form submanifolds of

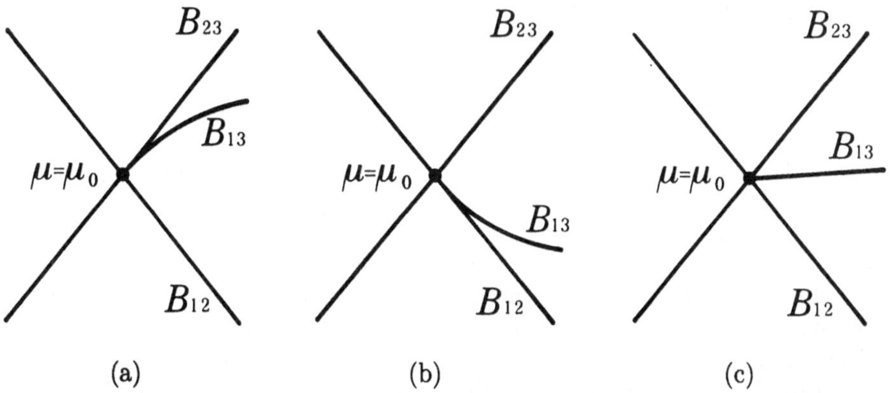

Fig. 3.5.6. Bifurcation diagram for the codimension two heteroclinic orbits.

codimension one, respectively. If, moreover, the corresponding two Melnikov integrals $M_{i,i+1}$ $(i = 1, 2)$ are linearly independent, then B_{12} and B_{23} intersect transversely at μ_0.

On the other hand, the bifurcation set B_{13} for the (O_1, O_3)-heteroclinic orbit becomes as follows:

Theorem 3.5.8. *Under the conditions (Ev), (Ind), (Unf) and (Asy) as above, the following holds:*

(1) If $\lambda_2^u \neq |\lambda_2^s|$, then B_{13} is a submanifold of codimension one with the boundary $B_{12} \cap B_{23}$, and moreover,

(i) if $\lambda_2^u < |\lambda_2^s|$ then B_{13} is tangent to B_{23} at μ_0;

(ii) if $\lambda_2^u > |\lambda_2^s|$ then B_{13} is tangent to B_{12} at μ_0.

(2) In the case of $\lambda_2^u = |\lambda_2^s|$, if we further assume the linear independence of the vector

$$M_{13} = \left.\frac{d}{d\mu}\right|_{\mu=\mu_0} \{\lambda_2^u(\mu) - \lambda_2^s(\mu)\}$$

together with the Melnikov integrals $M_{i,i+1}$ $(i = 1, 2)$, then the bifurcation set B_{13} is again a submanifold of codimension one with the boundary $B_{12} \cap B_{23}$, and in this case, B_{13} is not tangent to B_{12} or B_{23} at μ_0.

See Fig. 3.5.6.

The result of this form is given by [Kokubu 1988]. Similar results are obtained in [Gambaudo, Glendinning and Tresser 1988] for the case of Fig. 3.5.5(b), and in [Chow, Deng and Terman 1990], [Deng 1991a] for the case of Fig. 3.5.5(c). In these cases, bifurcation sets are more complicated and even infinitely many different homoclinic/heteroclinic bifurcations may occur besides the bifurcation sets above. Here we only refer to [Gambaudo, Glendinning and Tresser 1988] and [Deng 1991a] for details. In particular, [Gambaudo, Glendinning and Tresser 1988] nicely described these complicated bifurcations by using so-called Farey sequences. The case of complex principal eigenvalues can be dealt with in a similar way though it does not appear in the literature.

The proof of Theorem 3.5.8 uses the local expressions of $W^u(O_1(\mu_o))$ and $W^s(O_3(\mu_o))$ given in the proof of Theorem 3.5.1 and compares them in a neighborhood of the middle equilibrium point O_2. For that we need some detailed information on orbits around hyperbolic equilibrium points. This is given by the analysis using the smooth linearization or the method of exponential expansion attributed to Shil'nikov and Deng. For the former, we make a differentiable change of variables in a neighborhood of the equilibrium point O_2 so that the vector field becomes linear in the neighborhood, which has been already used in the previous subsections. The existence of such a smooth coordinate transformation is, however, implied by Belitskii's theorem (Theorem 3.2.20) under an additional condition among eigenvalues at O_2, and hence, we have to assume, in general, this extra condition in order to make a smooth linearization. On the contrary, the latter method does not require any condition of this kind, but gives less information compared with the former. Here we shall briefly explain the latter method based on [Deng 1989a,b]. It should be noted that [Lin 1990a] provided us with yet another method to study this type of the problem, which has also been used by [Vanderbauwhede and Fiedler 1992].

We first take local coordinates $(x, y) \in E^s \oplus E^u$ in which the local stable and unstable manifolds are identical with the stable and unstable eigenspaces E^s and E^u, respectively, and consider a problem of finding an orbit $(x(t), y(t))$ satisfying $x(0) = x_0$, $y(\tau) = y_1$ for a given data $(x_0, y_1, \tau) \in E^s \times E^u \times \mathbb{R}_+$. See Fig. 3.5.7. This problem is called the *Shil'nikov problem.* The existence and uniqueness of solutions to the Shil'nikov problem is proved analogously to the fundamental theorem of ordinary differential equations (Theorem 3.2.1). Indeed, we first rewrite the vector field around the hyperbolic equilibrium point, say O, as

$$\begin{aligned} \dot{x} &= Ax + f(x, y) \\ \dot{y} &= By + g(x, y), \end{aligned} \tag{3.5.3}$$

where A and B are the linearization matrices on E^s and E^u, respectively, and f, g are smooth functions satisfying

$$\begin{aligned} f(0, y) &\equiv 0, \qquad Df(0,0) = 0, \\ g(x, 0) &\equiv 0, \qquad Dg(0,0) = 0. \end{aligned} \tag{3.5.4}$$

The last condition follows from the assumption that the local stable and unstable manifolds coincide with the stable and unstable eigenspaces in the local coordinates. We then consider the integral equation

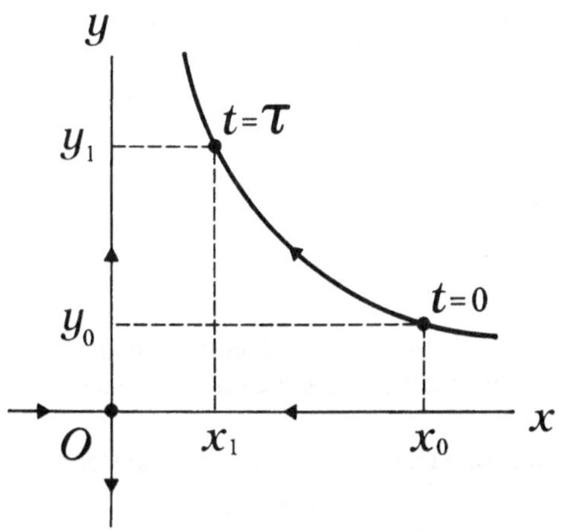

Fig. 3.5.7. The Shil'nikov problem.

$$
\begin{aligned}
x(t) &= e^{At}x_0 + \int_0^t e^{A(t-s)} f(x(s), y(s))ds \\
y(t) &= e^{B(t-\tau)}y_1 + \int_\tau^t e^{B(t-s)} g(x(s), y(s))ds,
\end{aligned}
$$

which is equivalent to the Shil'nikov problem for (3.5.3). The contraction mapping theorem applied to this integral equation yields a unique solution to the Shil'nikov problem for sufficiently small (x_0, y_1) and any $\tau > 0$. We denote this solution by

$$(x(t; \tau, x_0, y_1), y(t; \tau, x_0, y_1)).$$

This argument also shows that the solution to the Shil'nikov problem depends smoothly on the data (τ, x_0, y_1).

In order to study the dynamics near the hyperbolic equilibrium point O, we need more precise information of the solution $(x(t; \tau, x_0, y_1), y(t; \tau, x_0, y_1))$ with respect to (τ, x_0, y_1). The exponential expansion given below works for such a purpose. For the exposition, we first need some notations. We decompose the linear parts A and B into

$$A = \begin{pmatrix} A_1 & 0 \\ 0 & A_2 \end{pmatrix} \quad \text{and} \quad B = \begin{pmatrix} B_1 & 0 \\ 0 & B_2 \end{pmatrix},$$

where

$$
\begin{aligned}
\sup Re(spec(A_2)) &< \lambda^s = Re(spec(A_1)) < 0, \\
\sup Re(spec(B_2)) &> \lambda^u = Re(spec(B_1)) > 0,
\end{aligned}
$$

and $Re(spec(L))$ denotes the set of the real parts of all eigenvalues of the matrix L. We also define modified matrices A_* and B_* as follows:

$$A_* = \begin{pmatrix} A_1 & 0 \\ 0 & (\lambda^s - \varepsilon)I_{A_2} \end{pmatrix}, \quad B_* = \begin{pmatrix} B_1 & 0 \\ 0 & (\lambda^u + \varepsilon)I_{B_2} \end{pmatrix},$$

where $\varepsilon > 0$ is taken so that

$$\sup Re(spec(A_2)) < \lambda^s - \varepsilon < 0 < \lambda^u + \varepsilon < \sup Re(spec(B_2)),$$

and I_L stands for the identity matrix of the same size as L. Notice that λ^s and λ^u are the principal stable and unstable eigenvalues, respectively.

Theorem 3.5.9. (Shil'nikov-Deng) *There exists a suitable change of coordinates to (3.5.3) preserving the condition (3.5.4) so that the solution to the Shil'nikov problem for the transformed system admits the expansion:*

$$\begin{aligned} x(t;\tau,x_0,y_1) &= e^{A_* t}\varphi(\tau - t, x_0, y_1) + Q(\tau - t; \tau, x_0, y_1) \\ y(t;\tau,x_0,y_1) &= e^{B_*(t-\tau)}\psi(t, x_0, y_1) + R(t;\tau,x_0,y_1), \end{aligned}$$

where $\varphi(t,x_0,y_1)$, $\psi(t,x_0,y_1)$, $Q(t;\tau,x_0,y_1)$, $R(t;\tau,x_0,y_1)$ are smooth functions in (t,τ,x_0,y_1). Furthermore, φ and ψ satisfy the following properties:

$$\begin{aligned} \frac{\partial\varphi}{\partial x_0}(t,0,0) &= \begin{pmatrix} I_{A_1} & 0 \\ 0 & 0 \end{pmatrix}, \\ \frac{\partial\varphi}{\partial y_1}(t,0,0) &= \begin{pmatrix} 0 & 0 \\ 0 & 0 \end{pmatrix}, \\ \frac{\partial\psi}{\partial x_0}(t,0,0) &= \begin{pmatrix} 0 & 0 \\ 0 & 0 \end{pmatrix}, \\ \frac{\partial\psi}{\partial y_1}(t,0,0) &= \begin{pmatrix} I_{B_1} & 0 \\ 0 & 0 \end{pmatrix}, \end{aligned}$$

while Q and R are the remainder terms in the expansion. Namely, there exist constants $K > 0$ and $\sigma > 0$ independent of (t,τ,x_0,y_1) such that, for all $k \geq 0$,

$$\begin{aligned} |\partial^k Q(t;\tau,x_0,y_1)| &\leq Ke^{(\lambda^s-\sigma)t}, \\ |\partial^k R(t;\tau,x_0,y_1)| &\leq Ke^{(\lambda^u+\sigma)(t-\tau)}, \end{aligned}$$

for $0 \leq \forall t \leq \tau$, where ∂^k stands for derivatives of order k with respect to (t,τ,x_0,y_1) containing derivatives with τ of order at most one.

The expansion introduced above is called the *exponential expansion*, which shows that the orbit structure in a hyperbolic equilibrium point is quite similar to that in the linearized flow, and by that we can study the orbits near equilibrium points without using linearization. The proof of Theorem 3.5.9 requires quite elaborate estimates, hence it is omitted here. See [Deng 1989a,b] for details. It should also be noted that the exponential expansion has been generalized for nonhyperbolic equilibrium points to which the smooth linearization technique never be applied, and hence it can be used for the study of bifurcations of orbits homoclinic to nonhyperbolic equilibrium points. This generalization as well as its application to such type of homoclinic bifurcations was done by [Deng 1990a] and [Deng and Sakamoto 1989]. See also [Schecter 1987, 1990], [Chow and Lin 1990] and [Lin 1990a] for related works using different approach.

3.5.4 T-points and Gluing Bifurcations with Different Saddle-Indices

In the previous subsection, we dealt with the gluing bifurcation of two heteroclinic orbits whose indices are identically one. This subsection is, then, devoted to a partial study of heteroclinic orbits with indices different from one. Here we only focus on those orbits $h_i(t)$ $(i = 1, 2)$ with

(Ind) $\quad ind(h_1(t)) = 2 \quad \text{and} \quad ind(h_2(t)) = 0.$

We remark that our motivation comes not only from the purely mathematical interest of pursuing a generalization of known results in an abstract manner but also from a concrete example exhibiting interesting dynamical behaviors such as in the Lorenz equations (3.2.3). We shall briefly discuss the relation of our results to these examples later.

Let us consider a smooth family of vector fields with a k-dimensional parameter μ:

$$\dot{x} = v(x, \mu), \qquad x \in \mathbb{R}^n \ (n \geq 3), \ \mu \in \mathbb{R}^k \ (k \geq 2), \tag{3.5.5}$$

and suppose that the vector field $\dot{x} = v(x, \mu_0)$ has three hyperbolic equilibrium points O_i $(i = 1, 2, 3)$ and two (O_i, O_{i+1})-heteroclinic orbits $h_i(t, \mu_0)$ $(i = 1, 2)$ satisfying thc above condition (Ind). Then the bifurcation set B is again divided into three parts

$$B = B_{12} \cup B_{23} \cup B_{13}$$

where

$$B_{ij} = \{\mu \in \mathbb{R}^k \mid \exists (O_i, O_j)\text{-heteroclinic orbit for } \mu\}.$$

Theorem 3.5.1 particularly shows that the set B_{12} forms a hypersurface of codimension two whereas B_{23} forms a neighborhood of μ_0. In order to examine the set B_{13}, we assume, besides (Ind), the following three hypotheses:

(Ev) The principal eigenvalues λ_i^u and λ_i^s at O_i $(i = 1, 2, 3)$ are simple. Moreover, λ_2^u and λ_2^s satisfy

$$|Re\lambda_2^s| > Re\lambda_2^u;$$

(Asy) The (O_i, O_{i+1})-heteroclinic orbit $h_i(t, \mu_0)$ $(i = 1, 2)$ converges to O_i (respectively O_{i+1}) as $t \to -\infty$ (respectively $t \to \infty$) along the eigendirection with respect to the principal eigenvalue λ_i^u (respectively λ_{i+1}^s);

(Unf) The linear ordinary differential equation

$$\dot{\hat{z}}_1 = -\hat{z}_1 \cdot \frac{\partial}{\partial x} v(h_1(t, \mu_0), \mu_0)$$

has two linearly independent bounded solutions $\hat{q}_{1j}(t, \mu_0)$ $(j = 1, 2)$ for which the vectors given by the integrals

$$M_{1j} = \int_{-\infty}^{\infty} \hat{q}_{1j}(t, \mu_0) \frac{\partial}{\partial \mu} v(h_1(t, \mu_0), \mu_0) dt, \quad (j = 1, 2)$$

are linearly independent, whereas

$$\dot{\hat{z}}_2 = -\hat{z}_2 \cdot \frac{\partial}{\partial x} v(h_2(t, \mu_0), \mu_0)$$

has no nontrivial bounded solution.

Under these hypotheses (Ev), (Asy) and (Unf), together with (Ind), we can show that:

Theorem 3.5.10. *The bifurcation set B_{13} locally forms*

(1) a hypersurface of codimension one with the boundary B_{12}, if λ_2^u is real.

(2) a logarithmic scroll with the center B_{12}, if λ_2^u is not real, where a logarithmic scroll with the center C is the image of

$$\{(r, \theta) \mid r = e^{-\theta}\} \times (-1, 1)^{k-2}$$

by a local diffeomorphism $\varphi : (-1, 1)^k \to \mathbb{R}^k$ with

$$C = \varphi(\{0\} \times (-1, 1)^{k-2}).$$

The idea of the proof of the above theorem is the same as the proof of Theorem 3.5.8. See [Kokubu 1991] for details. Similar results are obtained by [Bykov 1991].

Theorem 3.5.10 holds for any dimension of the phase space ≥ 3 and the number of parameters ≥ 2, and can be considered as a generalization of a result of [Glendinning and Sparrow 1986], in which they showed a bifurcation result very similar to Theorem 3.5.10 using a heuristic reduction of the number of variables in the bifurcation equation. Though their argument is intuitive, it deals with an interesting bifurcation phenomenon occurring at what they call a *T-point* for the Lorenz equation. We shall briefly explain this phenomenon and discuss it as an application of our results.

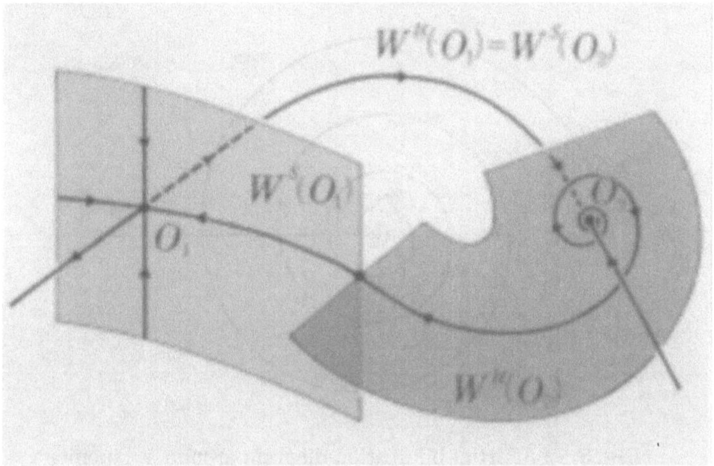

Fig. 3.5.8. A heteroclinic loop at a T-point.

In what follows, we consider only three-dimensional vector fields for simplicity.

First, we investigate the case that O_1 and O_3 coincide. In this case, the heteroclinic orbits make a loop consisting of two equilibrium points O_1 and O_2 and two heteroclinic orbits, from O_1 to O_2 and vise versa, as in Fig. 3.5.8. The condition

$$ind(h_1(t)) = 2 \quad \text{and} \quad ind(h_2(t)) = 0$$

implies

$$\dim W^u(O_1) = 1 \quad \text{and} \quad \dim W^u(O_2) = 2$$

since the vector field is of three dimension.

Let λ_i^s and λ_i^u be the principal stable and unstable eigenvalues at O_i, respectively, and suppose

$$Re\lambda_1^u < |Re\lambda_1^s|, \qquad Re\lambda_2^u > |Re\lambda_2^s|,$$

then we can apply Theorem 3.5.10 to this case. Here we focus on the case $\lambda_1^u \in \mathbb{R}$ and $\lambda_2^s \in \mathbb{C}\backslash\mathbb{R}$. Then we obtain the bifurcation diagram as given in Fig. 3.5.9, where the curve $\mathcal{A}$ indicates a bifurcation set for a homoclinic orbit based at O_2 while the curve $\mathcal{B}$ indicates that based at O_1. Note that the latter can be obtained by applying Theorem 3.5.10 to the time reversed system of the original one.

In this case, the equilibrium point O_2 is of saddle-focus type satisfying the Shil'nikov condition on the eigenvalues, and hence we can conclude from Theorem 3.5.7 that, along any generic arc crossing the curve $\mathcal{A}$ transversely in Fig. 3.5.9, we should observe very complicated bifurcation phenomena including the shift dynamics, accumulation of saddle-node bifurcations and period-doubling bifurcations for certain types of periodic orbits, and bifurcations to doubled homoclinic orbits.

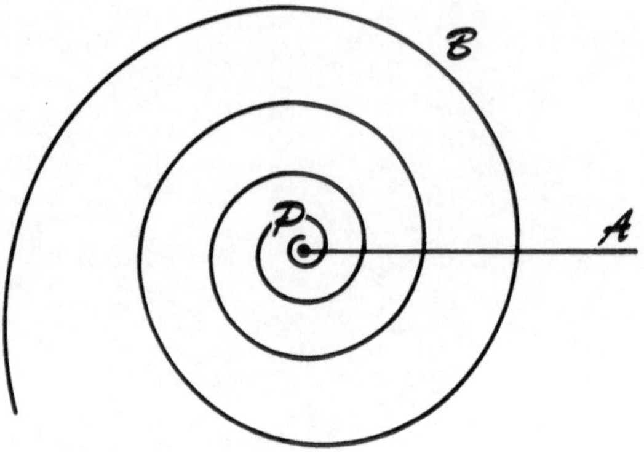

Fig. 3.5.9. Partial bifurcation diagram around a T-point.

Since these curves will appear in a neighborhood of any point on the curve $\mathcal{A}$, these infinitely many curves accumulate to the T-point at $\mu = 0$. Up to now, we know very little about how such complicated bifurcation structures come together around the codimension two singularity. This situation, however, looks similar to a bifurcation aspect numerically observed by [Gaspard, Kapral and Nicolis 1984] in the study of a simple quadratic mapping which is derived heuristically from the Poincaré map along a Shil'nikov-type homoclinic orbit.

According to the result in [Gaspard, Kapral and Nicolis 1984], the specific feature of the bifurcation structure near such singularity lies in the accumulation of cusp points on the saddle-node bifurcation curves, around which these curves as well as the period-doubling bifurcation curves form a unique shape which is called a "fishhook" or a "swallow-tail". See Fig. 3.5.10. Moreover, such a shape of bifurcation curves as in the fishhook seems rather universal in the sense that it is observed not only in bifurcation diagrams for three-dimensional vector fields including the Double Scroll circuit ([Gaspard, Kapral and Nicolis 1984] and Section 1.4) but also in bifurcation diagrams for cubic one-dimensional mappings ([MacKay and Tresser 1986, 1987]), for complex dynamical systems ([Ushiki 1989]), and for two-dimensional diffeomorphisms such as the Hénon map ([Hitzle and Zele 1985], [El-Hamouly and Mira 1981, 1982], [Sannami 1987, 1989], [Shibayama 1989]), all of which are related to chaotic dynamics or strange attractors.

[Alfsen and Frøyland 1985] numerically showed that the Lorenz equation possesses a homoclinic loop at $\sigma = 10, r \approx 30.475, b \approx 2.623$, which has similar properties to those of the heteroclinic loop discussed above, except that the heteroclinic loop is symmetric because of the $\mathbb{Z}_2$-symmetry in the Lorenz equation. Consequently, the bifurcation diagram becomes slightly more complicated which is studied by [Glendinning and Sparrow 1986] using somewhat intuitive argument. Furthermore, this type of heteroclinic loop may provide a clue for understand-

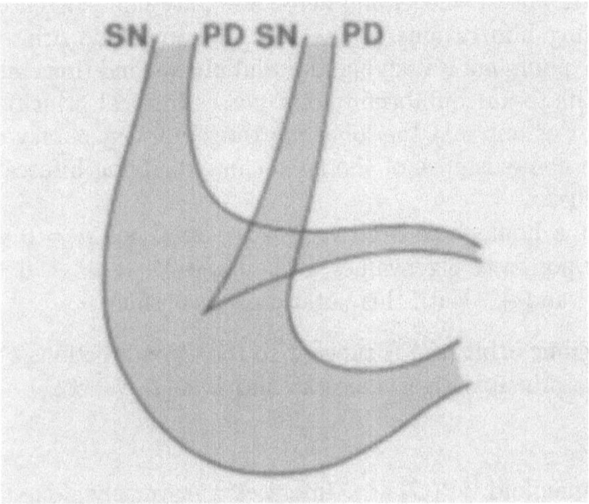

Fig. 3.5.10. A fishhook configuration.

ing some kind of complicated dynamics and bifurcations observed in the Lorenz equation as described in [Sparrow 1982]. It seems, however, difficult to prove the existence of such a heteroclinic loop in the Lorenz equation itself, and hence we are still far from mathematically rigorous understanding of the Lorenz equation or similar explicitly given ordinary differential equations.

3.5.5 Homoclinic Doubling Bifurcation

Here and in the next subsection, we restrict ourselves to three-dimensional vector fields for the sake of simplicity of explanation. Most of the results given are valid in general under slight modification.

A Motivation Consider a family of vector fields v_μ on $\mathbb{R}^3$ with a hyperbolic equilibrium point O and suppose that it admits a homoclinic orbit to the equilibrium point for $\mu = 0$. Then for sufficiently small μ, v_μ may possess a homoclinic orbit rounding twice in a small tubular neighborhood of the original homoclinic orbit. Such a bifurcation is referred to as *homoclinic doubling bifurcation* and the bifurcating homoclinic orbit is called a twice rounding homoclinic orbit or a doubled homoclinic orbit with respect to the original one which is called the primary homoclinic orbit.

The homoclinic doubling bifurcation was first studied in [Evans, Fenichel and Feroe 1982] in the case of the Shil'nikov-type homoclinic orbit. In other words, this corresponds to the assertion (3) of Theorem 3.5.7 given independently by [Gaspard 1984], [Gaspard, Kapral and Nicolis 1984].

Furthermore, Evans, Fenichel and Feroe also gave non-existence theorems for homoclinic doubling bifurcations in the case of homoclinic orbits to a saddle-focus equilibrium point not satisfying the Shil'nikov condition, and in the case of homoclinic orbits to an equilibrium point possessing real principal eigenvalues. In particular, it turns out that the following three non-degeneracy conditions are important for the non-existence of the homoclinic doubling bifurcation with real principal eigenvalues.

Let v_μ admit a homoclinic orbit $h(t)$ based at O for $\mu = 0$ where the linearization matrix possesses eigenvalues $\lambda^u > 0$ and $\lambda^{ss} < \lambda^s < 0$. The principal eigenvalues are λ^u and λ^s. With this notation, we consider:

(Asy) The homoclinic orbit $h(t)$ is tangent to the eigendirections e^u and e^s associated to the principal eigenvalues λ^u and λ^s as $t \to \pm\infty$;

(Ev) $\lambda^u \neq |\lambda^s|$;

(Inc) The stable manifold $W^s(O)$ at O intersect transversely along the homoclinic orbit with the two-dimensional extended unstable manifold $W^{eu}(O)$ which is tangent at O to the linear space spanned by e^u and e^s.

Remark 3.5.11. The proof for the center manifold theorem (Theorem 3.3.1) works as well for the existence of the extended unstable manifold $W^{eu}(O)$. See [Hirsch, Pugh and Shub 1977] for the proof. This invariant manifold is not unique, but has the unique tangent space along the homoclinic orbit, and hence the condition (Inc), which is sometimes referred to as the strong inclination property ([Deng 1989a]), is independent of the choice of the extended unstable manifold.

The non-existence of the homoclinic doubling bifurcation is shown in [Evans, Fenichel and Feroe 1982] under these generic conditions. Then the question arises: does homoclinic doubling bifurcation occur in the case of real principal eigenvalues if one of these conditions breaks? This has been answered affirmatively by [Yanagida 1987], namely he claimed that there are three possibilities of homoclinic orbits with real principal eigenvalues which can generate a doubled homoclinic orbit under perturbation. These three are listed as follows:

(1) a homoclinic orbit with resonant eigenvalues;

(2) an inclination-flip homoclinic orbit;

(3) an orbit-flip homoclinic orbit.

Each of them is defined by breaking one of the above genericity conditions in the following way:

(1) a homoclinic orbit with resonant eigenvalues satisfies (Asy) and (Inc) but has $\lambda^u = |\lambda^s|$;

(2) an inclination-flip homoclinic orbit satisfies (Asy) and (Ev) but $W^{cu}(O)$ is tangent to $W^s(O)$ along the homoclinic orbit;

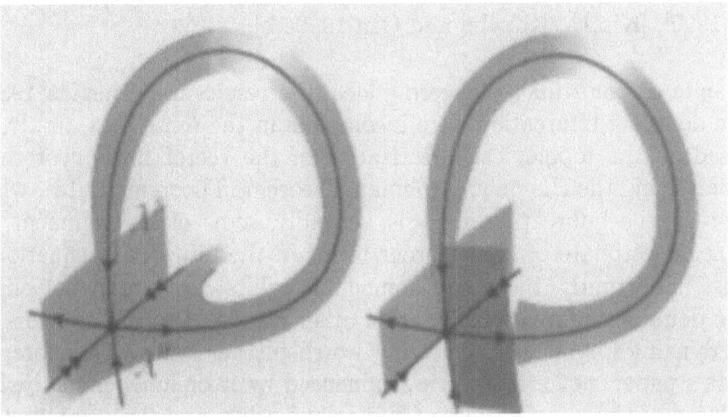

(a) (b)

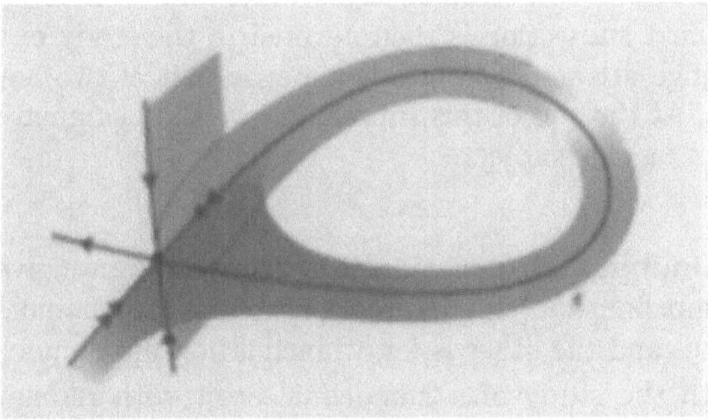

(c)

Fig. 3.5.11. Three types of degenerate homoclinic orbits.

(3) an orbit-flip homoclinic orbit satisfies (Ev) and (Inc) but it is tangent to the eigendirection associated to λ^{ss}.

See Fig. 3.5.11.

Remark 3.5.12. The studies of homoclinic doubling bifurcations done by [Evans, Fenichel and Feroe 1982] and [Yanagida 1987] were both motivated by the study of nerve impulses. Because of this, they dealt only with the case in which the unstable manifold is of one dimension while the stable manifold can be of any dimension, which is a consequence of the model equation describing the dynamics of nerve impulses. The above genericity conditions can be, however, formulated without any restrictions on dimension. See, for instance, [Kokubu 1988], [Chow, Deng and Fiedler 1990], [Kisaka, Kokubu and Oka 1992a,b].

In spite of Yanagida's pioneering idea, the results in [Yanagida 1987] for homoclinic doubling bifurcation were incomplete in the sense that, firstly, his proof was based on the topological linearization of the vector fields near the equilibrium point using the Hartman-Grobman Theorem (Theorem 3.2.15), which is not sufficient for the bifurcation analysis; secondly, some of his argument contained an erroneous estimate of higher order terms in the bifurcation equation; thirdly, and most importantly, the above homoclinic orbits can exhibit not only the homoclinic doubling bifurcations but also other rich dynamical behaviors including chaotic dynamics, which are certainly worth further study. Therefore, since the Yanagida's paper, not a few works, influenced by it or independently, have been done related to these three types of homoclinic orbits and the study is still rapidly progressing.

In this subsection, we have focused only on the homoclinic doubling bifurcation for homoclinic orbits with resonant eigenvalues and for inclination-flip homoclinic orbits. The next subsection is then devoted to the study of the bifurcation of Lorenz-like strange attractors from a symmetric pair of two homoclinic orbits of these two types. At the end of this subsection, we shall summarize more recent developments given on this subject.

B Homoclinic Doubling Bifurcation Theorems Here we give two theorems concerning the homoclinic doubling bifurcations; one is for a homoclinic orbit with resonant eigenvalues and the other is for an inclination-flip homoclinic orbit.

We begin with the study of a homoclinic orbit with resonant eigenvalues, namely, a homoclinic orbit based at a hyperbolic equilibrium point O satisfying the conditions (Asy), (Inc) and the resonance condition $\lambda^u = |\lambda^s|$ for the principal eigenvalues when $\mu = 0$. From the condition (Asy), we have two kinds of such homoclinic orbits. In order to see these two kinds, we reverse the time-axis of the flow generated by the vector field v_0 for $\mu = 0$ which admits the homoclinic orbit with resonant eigenvalues. Then the stable manifold $W^s(O)$ comes back along the homoclinic orbit to the equilibrium point O, and hence, (Asy) implies that the stable manifold is expanded toward the eigendirection associated with the eigenvalue λ^{ss} by the time-reversed flow. This assertion is known as the strong inclination property ([Deng 1989a]), and as its consequence, the stable manifold forms a two-dimensional surface in a neighborhood of the homoclinic orbit, which is either a cylinder or a Möbius band. If it is a Möbius band, the homoclinic orbit is called *twisted*, otherwise it is called *non-twisted*. See Fig. 3.5.12.

Theorem 3.5.13. *Let v_μ be a generic two-parameter family of vector fields which admits a homoclinic orbit Γ with resonant eigenvalues at $\mu = 0$.*

(1) If Γ is non-twisted, then the homoclinic doubling bifurcation does not occur;

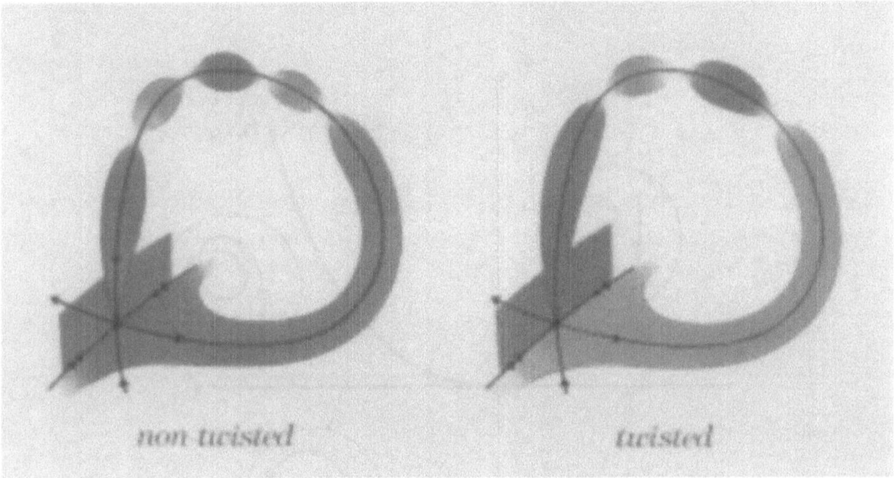

Fig. 3.5.12. Twisted and non-twisted homoclinic orbits.

(2) *If Γ is twisted, then the homoclinic doubling bifurcation occurs. More precisely, there exists a local change of parameters at $\mu = 0$*

$$\varepsilon = (\varepsilon_1, \varepsilon_2) = \varepsilon(\mu)$$

and a function

$$\varepsilon_2 = \kappa_{2H}(\varepsilon_1) \quad (\varepsilon_1 > 0)$$

in the parameter space such that a primary homoclinic orbit persists along $\varepsilon_2 = 0$ whereas a doubled homoclinic orbit bifurcates along the curve $\varepsilon_2 = \kappa_{2H}(\varepsilon_1)$. Moreover,

$$\kappa_{2H}(\varepsilon_1) \approx (\text{const})^{-1/\varepsilon_1}.$$

The bifurcation diagram is depicted in Fig. 3.5.13, which shows that the bifurcation curve for the doubled homoclinic orbit emanates from the origin of the parameter space corresponding to the homoclinic orbit with resonant eigenvalues, and the curve has exponentially flat contact with the curve for the persistence of the original primary homoclinic orbit, which is in fact the ε_1-axis in Fig. 3.5.13. We should remark that this is not a complete bifurcation diagram for the bifurcation from the homoclinic orbit with resonant eigenvalues. See D of this subsection and [Chow, Deng and Fiedler 1990] for additional information on the bifurcation.

Next we turn to the inclination-flip homoclinic orbit, that is the homoclinic orbit Γ based at an equilibrium point O satisfying (Ev), (Asy) and that the stable manifold $W^s(O)$ is tangent to the extended unstable manifold $W^{eu}(O)$ along Γ. Here we need some more notations for stating the theorem. Consider the family of vector fields v_μ with a hyperbolic equilibrium point O. Define

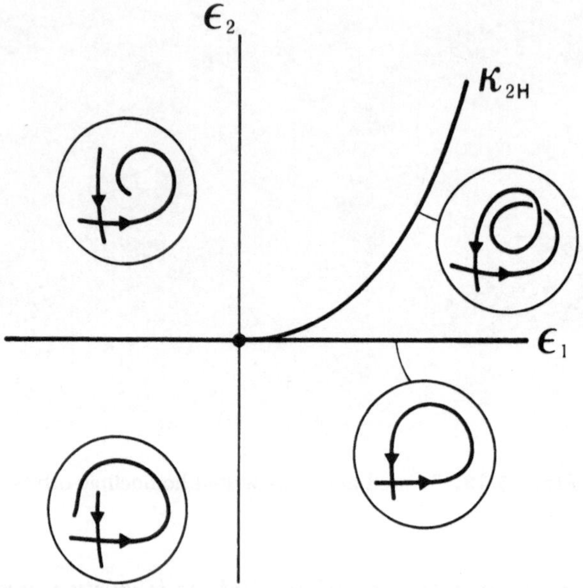

Fig. 3.5.13. Partial bifurcation diagram for the homoclinic doubling bifurcation.

$$\nu(\mu) = \frac{|\lambda^s(\mu)|}{\lambda^u(\mu)}$$

and

$$gap = |\lambda^{ss}(0) - \lambda^s(0)|, \tag{3.5.6}$$

where $\lambda^s(\mu)$ and $\lambda^u(\mu)$ are principal stable and unstable eigenvalues for the linearization $Dv_\mu(O)$ and $\lambda^{ss}(\mu)$ is the other (non-principal) eigenvalue.

Theorem 3.5.14. *Let v_μ be a generic two-parameter family of vector fields which has an inclination-flip homoclinic orbit Γ at $\mu = 0$. Then, under the gap condition,*

$$|\lambda^u(0) + \lambda^s(0)| < gap \tag{3.5.7}$$

the following holds:

(1) If $1 < \nu(0) < 2$, then the homoclinic doubling bifurcation does not occur.

(2) If $\frac{1}{2} < \nu(0) < 1$, then the homoclinic doubling bifurcation occurs. More precisely, there exists a local change of parameters at $\mu = 0$

$$\varepsilon = (\varepsilon_1, \varepsilon_2) = \varepsilon(\mu)$$

and a function

$$\varepsilon_2 = \kappa_{2H}(\varepsilon_1) \quad (\varepsilon_1 > 0)$$

in the parameter space such that a primary homoclinic orbit persists along $\varepsilon_2 = 0$ *whereas a doubled homoclinic orbit bifurcates along the curve* $\varepsilon_2 = \kappa_{2H}(\varepsilon_1)$. *Moreover,*

$$\kappa_{2H}(\varepsilon_1) \approx \varepsilon_1^{\frac{1}{1-\nu(0)}}.$$

The bifurcation diagram is the same (Fig. 3.5.13) as that for homoclinic orbits with resonant eigenvalues except the order of tangency of the bifurcation curve for doubled homoclinic orbits with that for primary homoclinic orbits. Namely, for the resonant case, κ_{2H} is of exponentially flat contact with the ε_1-axis, whereas in the inclination-flip case, they are of a power order contact. Again this is a partial bifurcation diagram. See also D of this subsection and [Kisaka, Kokubu and Oka 1992a,b] for more information.

C Proof of the Homoclinic Doubling Bifurcation Theorems We give an outline of the proof of these two homoclinic doubling bifurcation theorems using the Lyapunov-Schmidt reduction. This proof mainly follows an idea of [Chow, Deng and Fiedler 1990] and the details can be found in [Chow, Deng and Fiedler 1990] for the resonant case and in [Kisaka, Kokubu and Oka 1992b] for the inclination-flip case. An alternative proof using the idea of invariant foliation has been given in [Kisaka, Kokubu and Oka 1992a]

Proof. We begin by constructing the Poincaré map

$$\Pi = \Pi_{far} \circ \Pi_{loc}$$

as composition of the local map Π_{loc} and the global map Π_{far}, where Π_{loc} is a flow-defined map near the equilibrium point O and Π_{far} is a flow-defined diffeomorphism along the homoclinic orbit Γ outside. For that purpose, we first make a suitable coordinate change so that in the new coordinates $(x^1, x^2, y) \in \mathbb{R}^3$, we can assume

$$v_\mu(O) = O, \quad Dv_\mu(O) = \begin{pmatrix} \lambda^s(\mu) & 0 & 0 \\ 0 & \lambda^{ss}(\mu) & 0 \\ 0 & 0 & \lambda^u(\mu) \end{pmatrix},$$

and

$$|\lambda^{ss}(\mu) - \lambda^s(\mu)| > \frac{1}{2} gap > 0$$

for a sufficiently small range of the parameter μ around $\mu = 0$. Then we take the cross sections

$$\begin{aligned} \Sigma^s &= \{(x^1, x^2, y) \mid x^1 = \delta,\ |x^2|, |y| \le \delta\} \\ \Sigma^u &= \{(x^1, x^2, y) \mid y = \delta,\ |x^1|, |x^2| \le \delta\} \end{aligned}$$

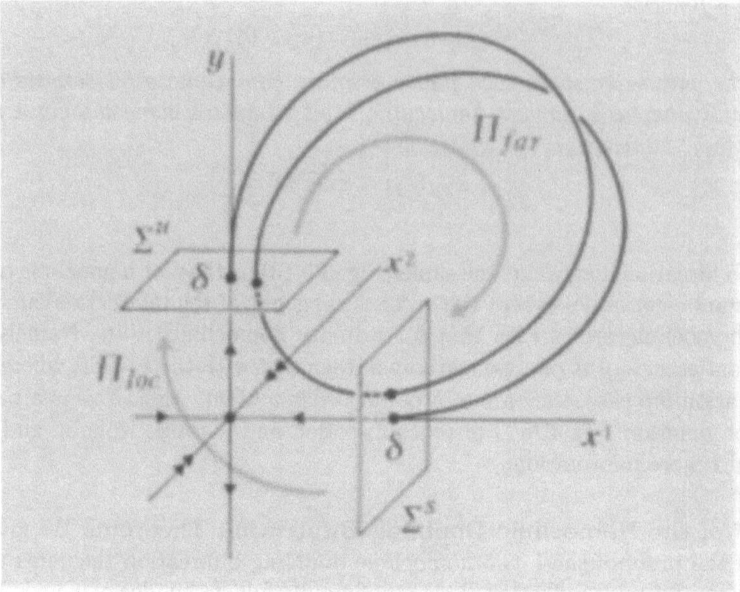

Fig. 3.5.14. The Poincaré map along the homoclinic orbit Γ.

for small fixed $\delta > 0$. For these cross sections, we define

$$\Pi_{loc} : \Sigma^s \to \Sigma^u, \quad \Pi_{far} : \Sigma^u \to \Sigma^s$$

as flow-defined maps. Note that the local map Π_{loc} is not everywhere defined on Σ^s.

In order to prove the existence of a doubled homoclinic orbit, we have to solve the following set of equations:

$$\begin{aligned} \Pi_{far}(0,0,\delta) &= (\delta, x_1^2, y_1) \\ \Pi_{loc}(\delta, x_1^2, y_1) &= (x_2^1, x_2^2, \delta) \\ \Pi_{far}(x_2^1, x_2^2, \delta) &= (\delta, x_3^2, 0). \end{aligned} \tag{3.5.8}$$

See Fig. 3.5.14. By using the solutions to the Shil'nikov problem introduced in Subsection 3.5.3, the points (δ, x_1^2, y_1) and (x_2^1, x_2^2, δ) are related as

$$\begin{aligned} \begin{pmatrix} x_2^1 \\ x_2^2 \end{pmatrix} &= x(\tau; \tau, x_1^2; \mu) \\ &= r^{\nu(\mu)}\{\varphi(x_1^2; \mu) + Q(\tau, x_1^2; \mu)\} \\ y_1 &= y(0; \tau, x_1^2; \mu) \\ &= r\{\psi(x_1^2; \mu) + R(\tau, x_1^2; \mu)\}, \end{aligned}$$

where $r = e^{-\lambda^u(\mu)\tau}$, which reduces (3.5.8) to

$$\Pi_{far}(0,0,\delta) = (\delta, x_1^2, r\{\psi(x_1^2;\mu) + R(\tau, x_1^2;\mu)\})$$
$$\Pi_{far}(r^{\nu(\mu)}\{\varphi(x_1^2;\mu) + Q(\tau, x_1^2;\mu)\}, x_2^2, \delta) = (\delta, x_3^2, 0).$$

For simplicity, we write the above equation as

$$\Psi(r, x_1^2, x_3^2;\mu) = \begin{pmatrix} \Psi_1 \\ \Psi_2 \end{pmatrix} = 0, \tag{3.5.9}$$

since τ is equivalent to r by definition. Note that Equation (3.5.9) has a trivial solution $(r, x_1^2, x_3^2;\mu) = 0$ corresponding to the original homoclinic orbit Γ for $\mu = 0$.

Next we reduce the number of equations and the variables x_1^2, x_3^2 by applying the so-called Lyapunov-Schmidt reduction ([Chow and Hale 1982]) as follows. Take the unit vector $p(\mu)$ in Σ^s parallel to the principal unstable eigendirection e^u, and define the projection P_μ by

$$P_\mu \Psi = \begin{pmatrix} \langle p(\mu), \Psi_1\rangle \\ \langle p(\mu), \Psi_2\rangle \end{pmatrix}$$

for $\Psi = \begin{pmatrix} \Psi_1 \\ \Psi_2 \end{pmatrix}$. Then the equation

$$(Id - P_\mu)\Psi(r, x_1^2, x_3^2;\mu) = 0$$

can be solved in a neighborhood of the trivial solution $(r, x_1^2, x_3^2;\mu) = 0$ as

$$x_1^2 = x_1^2(r,\mu) \quad \text{and} \quad x_3^2 = x_3^2(r,\mu),$$

and hence Equation (3.5.9) is reduced to

$$P_\mu \Psi(r, x_1^2(r,\mu), x_3^2(r,\mu);\mu) = 0.$$

Now we need to distinguish two cases in order to obtain the precise expression of the reduced equations using the information from exponential expansion (Theorem 3.5.9).

(Case: Resonant eigenvalues.) If the homoclinic orbit has resonant eigenvalues for $\mu = 0$, then the corresponding reduced equation takes the form

$$\begin{aligned} -c_0(\mu) + c_1(\mu)r \qquad &+O(|r|^{1+\omega}) = 0 \\ -c_0(\mu) - c_2(\mu)r^{\nu(\mu)} \quad &+O(|r|^{1+\omega}) = 0, \end{aligned} \tag{3.5.10}$$

where $\nu(0) = 1$ from the resonance condition and $c_0(\mu) = 0$ for the persistence of the primary homoclinic orbit. We can also make $c_1(0) \geq 0$. Moreover, it holds that $\frac{c_2(0)}{c_1(0)} > 0$ (respectively < 0) if the original primary homoclinic orbit is non-twisted (respectively twisted). Therefore, under the genericity conditions,

- $\dfrac{\partial \nu}{\partial \mu}(0)$ and $\dfrac{\partial c_0}{\partial \mu}(0)$ are linearly independent;
- $c_1(0) > 0$;
- $\left|\dfrac{c_2(0)}{c_1(0)}\right| \neq 0, 1$;

we can make the change of parameters

$$\varepsilon_1 = \nu(\mu) - 1, \qquad \varepsilon_2 = \frac{c_0(\mu)}{c_1(\mu)},$$

so that Equation (3.5.10) becomes

$$\begin{aligned} r &= \varepsilon_2 && +O(|r|^{1+\omega}) \\ 0 &= \varepsilon_2 + a(\varepsilon) r^{1+\varepsilon_1} && +O(|r|^{1+\omega}), \end{aligned}$$

where $a(\varepsilon) = \dfrac{c_2(\mu)}{c_1(\mu)}$.

If the original primary homoclinic orbit is non-twisted, namely, $a(0) > 0$, then so are $a(\varepsilon)$ and hence there is no non-zero solution r which concludes the non-existence of doubled homoclinic orbits. If, on the other hand, $a(0) < 0$, then we may assume $-1 < a(0) < 0$ without loss of generality by taking another choice of $(\varepsilon_1, \varepsilon_2)$ if necessary (see [Chow, Deng and Fiedler 1990] for the detail), and hence we have a solution $r = r(\varepsilon_1)$ defined for $\varepsilon_1 > 0$ satisfying

$$1 = -a(\varepsilon_1, 0) r^{\varepsilon_1} + O(r^{\omega}),$$

and hence, the corresponding bifurcation curve $\varepsilon_2 = \kappa_{2H}(\varepsilon_1)$ satisfies

$$\varepsilon_2 \approx r(\varepsilon_1) \approx \left(-\frac{1}{a(0)}\right)^{-1/\varepsilon_1}.$$

(Case:Inclination-flip.) In this case, letting $\gamma = \min\{1, \inf_\mu \nu(\mu)\}$ where the infimum is taken over sufficiently small range of the parameter μ around $\mu = 0$, we get similar reduced equations

$$\begin{aligned} -c_0(\mu) + c_1(\mu) r + O(|r|^{\gamma+\omega}) &= 0 \\ -c_0(\mu) - c_2(\mu) r^{\nu(\mu)} + O(|r|^{\gamma+\omega}) &= 0, \end{aligned}$$

where $c_0(\mu) = 0$ for the persistence of the primary homoclinic orbits and $c_2(0) = 0$ from the inclination-flip condition for $\mu = 0$. In fact this can be done under some restriction on the eigenvalues. Namely, if $\nu(0) > 1$ (respectively $\nu(0) < 1$), we need to assume $1 < \nu(0) < 2$ (respectively $\frac{1}{2} < \nu(0) < 1$) in order to find a constant ω for which the above reduced equations are valid. Here we can even choose ω satisfying $\gamma + \omega - 1 > 0$.

Therefore, if $\dfrac{\partial c_0}{\partial \mu}(0)$ and $\dfrac{\partial c_2}{\partial \mu}(0)$ are linearly independent, then the equation becomes

$$\begin{aligned} r &= \varepsilon_2 && +O(|r|^{\gamma+\omega}) \\ 0 &= \varepsilon_2 - \varepsilon_1 r^{\nu(\varepsilon)} && +O(|r|^{\gamma+\omega}). \end{aligned} \tag{3.5.11}$$

By taking the difference of the two equations and dividing it by r, we have

$$1 = \varepsilon_1 r^{\nu(\varepsilon)-1} + O(|r|^{\gamma+\omega-1}),$$

where $\gamma + \omega - 1 > 0$.

If $1 < \nu(0) < 2$, then this equation has no solution $(\varepsilon_1, \varepsilon_2, r)$ bifurcating from $(0,0,0)$ since, for such a solution, the right hand side of the equation would be much smaller than 1 and hence it is impossible to have equality. If, on the other hand, $\frac{1}{2} < \nu(0) < 1$, then Equation (3.5.11) is solved as $\varepsilon_2 = \kappa_{2H}(\varepsilon_1)$ by eliminating r using the implicit function theorem, and it is easy to see that

$$\kappa_{2H}(\varepsilon_1) \approx \varepsilon_1^{\frac{1}{1-\nu(0)}}.$$

This completes the proofs of Theorems 3.5.13 and 3.5.14. □

D Further Development Not only the homoclinic doubling bifurcations but also other bifurcations are already known for homoclinic orbits of the types considered in this subsection.

It was shown in [Chow, Deng and Fiedler 1990] that the period-doubling bifurcation or the saddle-node bifurcation for a periodic orbit occurs in a generic two-parameter unfolding of a homoclinic orbit with resonant eigenvalues, depending whether the homoclinic orbit is twisted or non-twisted. Similar bifurcations are also shown for inclination-flip homoclinic orbits ([Kisaka, Kokubu and Oka 1992a,b], where the term "critically twisted" was used instead of the "inclination-flip") with the ratio ν of the principal eigenvalues satisfying $\frac{1}{2} < \nu < 1$. On the other hand, if the ratio ν is smaller than $\frac{1}{2}$, then more complicated dynamics such as the shift dynamics accompanied by rich bifurcation phenomena in their creation possibly appear ([Deng 1991c], [Homburg, Kokubu and Krupa 1993]).

As for the orbit-flip homoclinic orbit, recently B. Sandstede has announced that homoclinic doubling and homoclinic N-tupling bifurcation ($N \geq 3$) as well as the shift dynamics occur in its unfolding. There is also a numerical simulation done by [Iori, Yanagida and Matsumoto 1993] for piecewise-linear vector fields involving an orbit-flip homoclinic orbit which shows interesting bifurcation curves for N-homoclinic orbits with $2 \leq N \leq 11$. For this numerical work, a method developed in Chapter 2 played an important role.

In spite of these works, we are still far from the complete understanding of bifurcations from these three types of homoclinic orbits studied in this subsection. The next subsection will be devoted to a brief explanation of the remarkable result by [Rychlik 1990] which asserts that a strange attractor resembling the Lorenz attractor bifurcates from a symmetric pair of inclination-flip homoclinic orbits.

3.5.6 Bifurcation Generating Geometric Lorenz Attractors from Homoclinic Orbits

The Lorenz attractor (Example 3.2.5) is an attractor numerically observed by [Lorenz 1963] for the study of a three-dimensional ordinary differential equation (3.2.3) which is reduced from a partial differential equation describing a thermal convection of fluid. In spite of the simple form of the equation, there is no rigorous mathematical proof that this equation really exhibits such a (strange) attractor as observed by the computer simulation. However, there exists a well-defined mathematical model which seems to describe the dynamics nicely on the attractor in Equation (3.2.3), which is the geometric Lorenz attractor formulated and studied by Guckenheimer and Williams ([Guckenheimer 1976], [Guckenheimer and Williams 1979], [Williams 1979]). The purpose of this subsection is to show that this geometric Lorenz attractor can be generated through bifurcations of certain homoclinic orbits, namely, a symmetric pair of inclination-flip homoclinic orbits or those with resonant eigenvalues. This remarkable result was obtained by [Rychlik 1990] for the inclination-flip case and by [Robinson 1989, 1992] for the resonant case.

We begin with the definition of the geometric Lorenz attractor. Throughout this subsection, vector fields are assumed to be of three dimension and have the symmetry given by the linear map $\gamma = \begin{pmatrix} -1 & 0 & 0 \\ 0 & -1 & 0 \\ 0 & 0 & 1 \end{pmatrix}$. Namely, a vector field v on $\mathbb{R}^3$ is assumed to satisfy

$$v(\gamma x) = \gamma v(x), \qquad \forall x = (x^1, x^2, x^3) \in \mathbb{R}^3.$$

We call such a vector field a *γ-equivariant vector field.*

Consider a γ-equivariant vector field v on $\mathbb{R}^3$ satisfying the following conditions:

(GL1) $v(O) = 0$ and $Dv(O) = \begin{pmatrix} \lambda^u & 0 & 0 \\ 0 & \lambda^{ss} & 0 \\ 0 & 0 & \lambda^s \end{pmatrix}$ where

$$0 < -\lambda^s < \lambda^u < -\lambda^{ss}.$$

In particular, the x^3-axis is the eigendirection to the principal stable eigenvalue λ^s and at the same time it is invariant under the symmetry γ.

(GL2) There exists a rectangular cross section R transverse to the flow generated by v which intersects with the x^3-axis, such that the rectangle R is mapped back by the flow-defined Poincaré map Π into itself as in Fig. 3.5.15. The image of R consists of two cusp-shaped regions as indicated.

(GL3) There exist local coordinates (ξ, η) on R such that the Poincaré map Π : $R \to R$ takes the form

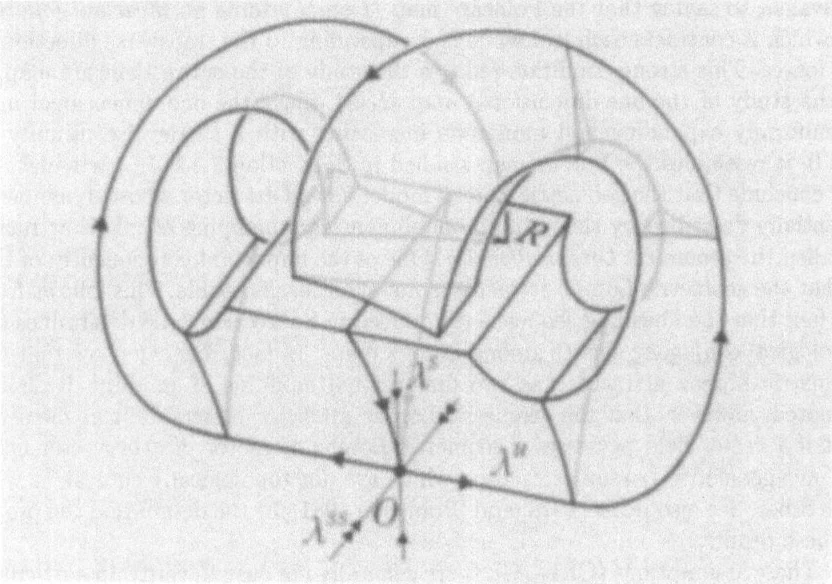

Fig. 3.5.15. The geometric Lorenz attractor.

$$\Pi(\xi,\eta) = (\Pi_1(\xi), \Pi_2(\xi,\eta)) \tag{3.5.12}$$

for $\xi \neq 0$ with smooth functions Π_1, Π_2 satisfying

- $\Pi(-\xi,-\eta) = -\Pi(\xi,\eta)$;
- the line $\{\xi = 0\}$ corresponding to the stable manifold $W^s(O) \cap R$;
- $\Pi_1'(\xi) > \sqrt{2}$ and $\lim_{\xi \to 0} \Pi_1'(\xi) = +\infty$;
- the existence of a constant c with $0 < c < 1$ such that

$$0 < \frac{\partial \Pi_2}{\partial \eta}(\xi,\eta) < c < 1 \quad \text{for} \quad \xi \neq 0$$

and

$$\lim_{\xi \to 0} \frac{\partial \Pi_2}{\partial \eta}(\xi,\eta) = 0.$$

A γ-equivariant vector field v satisfying these conditions is called a *geometric Lorenz model.* Intuitively, these conditions seem to fit nicely with the computer picture from the original Lorenz equation. However, two strong conditions are imposed: one is that the Poincaré map is assumed to preserve lines given by ξ = constant due to the form of the map (3.5.12), and the other is the hyperbolicity; the expansion in ξ-direction and the contraction in η-direction. This is equivalent to saying that the Poincaré map Π on R admits an *invariant foliation* for which it contracts each leaf while it is expanding to the transverse direction of the leaves. This strong condition reduces the study of the entire Poincaré map Π to the study of the one-dimensional map $\Pi_1(\xi)$. Since the one-dimensional map is uniformly expanding and monotone increasing with a single discontinuity at $\xi = 0$, it resembles the Lorenz map studied in Subsection 3.4.5. In particular, we can conclude that the geometric Lorenz model has an attractor whose dynamics is essentially described by the reduced one-dimensional mapping Π_1. This attractor is called the *geometric Lorenz attractor*. One of the important consequences of this is that the geometric Lorenz attractor is not structurally stable. This follows from the fact that the kneading sequence introduced in Subsection 3.4.5 determines the topological conjugacy classes among Lorenz maps. In fact, one can show that the geometric Lorenz attractor has two dimensional modulus of stability. It should be noted, however, that the geometric Lorenz attractor is persistent in the sense that if a vector field possesses a geometric Lorenz attractor, nearby vector fields all have geometric Lorenz attractors, which are not topologically equivalent with each other. See also [Rand 1978] and [Robinson 1981] for the details and the proofs of these results.

These assumptions (GL1)-(GL3) are generally not easy to verify. In particular, there has been given no proof for the existence of a geometric Lorenz attractor in the original Lorenz equation (3.2.3). However, it is possible to show the existence of the geometric Lorenz attractor in the context of homoclinic bifurcations. We shall give below a brief explanation of such an idea given by [Rychlik 1990]. Note that, in a general context, a topological study of the dynamics of Lorenz-like systems was done in [Afraĭmovich, Bykov and Shil'nikov 1982] and [Afraĭmovich and Pesin 1987], whose results are independent but closely related to [Guckenheimer and Williams 1979] explained above. This work also provides us with a sufficient condition for the existence of continuous invariant foliation for the Poincaré map.

Let v be a γ-equivariant vector field on $\mathbb{R}^3$ satisfying (GL1) and, instead of (GL2,3), we assume that it admits an inclination-flip homoclinic orbit Γ_+ to the origin O. Then the symmetry implies that there exists another inclination-flip homoclinic orbit $\Gamma_- = \gamma(\Gamma_+)$ as depicted in Fig. 3.5.16. We call such a pair $\Gamma_\pm$ a *γ-symmetric pair of inclination-flip homoclinic orbits.*

Theorem 3.5.15. (Rychlik) *Let v be a γ-equivariant vector field satisfying (GL1) with a symmetric pair of inclination-flip homoclinic orbits to a hyperbolic equilibrium point whose eigenvalues $\lambda^{ss} < \lambda^s < 0 < \lambda^u$ satisfy*

$$\frac{1}{2} < \frac{|\lambda^s|}{\lambda^u} < 1 < \frac{|\lambda^{ss}|}{\lambda^u}, \tag{3.5.13}$$

and let v_μ be its generic unfolding with $v_0 = v$. Then there exists an arbitrarily small μ such that v_μ possesses a geometric Lorenz attractor.

Here we can take μ to be a two-dimensional parameter under the symmetry. We remark that the condition

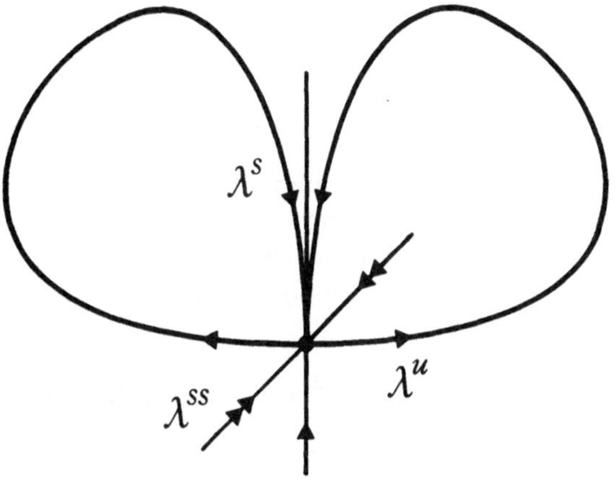

Fig. 3.5.16. A symmetric pair of inclination-flip homoclinic orbits.

$$\lambda^u < -\lambda^{ss}$$

is equivalent to the gap condition (3.5.7) in this context.

Proof. The outline of the proof of Theorem 3.5.15 is the following: take a cross section transverse to $\Gamma_\pm$ and construct a one-dimensional $C^{1+\delta}$ foliation which is invariant under the Poincaré map along $\Gamma_\pm$. This induces a one-dimensional interval mapping from the Poincaré map for which we show that it admits an invariant measure which is absolutely continuous with respect to the Lebesgue measure. The construction of the invariant foliation is due to a method given in [Robinson 1981], and the existence of an invariant measure is implied by a result of [Keller 1985] as explained in Subsection 3.4.5. The fundamental idea underlying the proof is again that an orbit close to the homoclinic orbit stays a very long time in a neighborhood of the equilibrium point O and hence the eigenvalues of the linearization matrix essentially govern the dynamics of the Poincaré map.

We take the local coordinates $(x^1, x^2, x^3) \in \mathbb{R}^3$ near O satisfying (GL1) such that the unstable and stable manifolds coincide with the unstable and stable eigenspaces, respectively, near the equilibrium point O in the coordinates. Then we take the cross sections $\Sigma^u_\pm$ and Σ^s to the homoclinic orbits $\Gamma_\pm$ given by

$$\Sigma^u_\pm = \{x^1 = \pm\delta\} \quad \text{and} \quad \Sigma^s = \{x^3 = \delta\},$$

and describe the Poincaré map $\Pi = \Pi_{far} \circ \Pi_{loc}$ along $\Gamma_\pm$ as the composition of the local map $\Pi_{loc} : \Sigma^s \to \Sigma^u_\pm$ and the global map $\Pi_{far} : \Sigma^u_\pm \to \Sigma^s$. Since the vector field is γ-equivariant and since the cross sections chosen are γ-symmetric, the resulting Poincaré map $\Pi = \Pi(x^1, x^2)$ respects the symmetry as well, namely, it satisfies

$$\Pi(-x^1, -x^2) = -\Pi(x^1, x^2).$$

Therefore the form of Π is determined only from the form for $x^1 > 0$ and Π for $x^1 < 0$ is given by

$$\Pi(x^1, x^2) = \text{sgn}(x^1)\Pi(|x^1|, \text{sgn}(x^1)x^2).$$

Now we employ the technique repeatedly used before in order to obtain the form of the local and global maps Π_{loc} and Π_{far}. For the local map, we use smooth linearization ([Rychlik 1990]), or the exponential expansion ([Kisaka, Kokubu and Oka 1992a]), and the global map is simply a local diffeomorphism which is expanded as

$$\Pi_{far}(x^2, x^3) = \mathbf{p}(\mu) + \mathbf{q}(\mu)x^3 + \mathbf{r}(\mu)x^2 + O(|x^2|^2 + |x^3|^2).$$

Therefore, the resulting Poincaré map turns out to be of the form

$$\Pi(x^1, x^2) = \mathbf{p}(\mu) + \mathbf{r}(\mu)h(x^2, \mu)(x^1)^{\nu} + \mathbf{k}(x^1, x^2, \mu)(x^1)^{\eta} \tag{3.5.14}$$

for $x^1 > 0$, where $\nu = \frac{|\lambda^s|}{\lambda^u}$ and η are positive numbers satisfying

$$\frac{1}{2} < \nu < 1 < \eta$$

for $\mu = 0$, and $h, \mathbf{k}$ are smooth functions with $h(0, \mu) = 1$. We remark that this condition follows from the assumption (3.5.13) on the eigenvalues. From the assumption for the inclination-flip homoclinic orbit, we have $\mathbf{p}(0) = \binom{0}{0}$ since the homoclinic orbit passes through $(x^1, x^2, x^3) = (\delta, 0, 0)$, and $\mathbf{r}(0) = \binom{0}{*}$ since the homoclinic orbit is of inclination-flip at $\mu = 0$, and hence Π_{far} maps the unit tangent vector $\frac{\partial}{\partial x^1}$ to a vector parallel to $\frac{\partial}{\partial x^2}$.

Recall that the persistency condition of the primary homoclinic orbit is given by $p_1(\mu) = 0$, where $p_1(\mu)$ is the first component of the vector $\mathbf{p}(\mu)$ and that the inclination-flip condition for the homoclinic orbits is given by $r_1(\mu) = 0$ as shown above. Our two-parameter family of vector fields is assumed in such a way that it generically unfolds the inclination-flip homoclinic orbit at $\mu = 0$. Therefore, we can take the new parameters

$$\varepsilon_2 = p_1(\mu) \quad \text{and} \quad \varepsilon_1 = r_1(\mu) \tag{3.5.15}$$

as unfolding parameters for the symmetric pair of inclination-flip homoclinic orbits instead of the original parameter μ. In other words, we may assume that (3.5.15) generically defines a diffeomorphic change of parameters, and hence the Poincaré map Π given by (3.5.14) takes the form of

$$\Pi(x^1, x^2) = \begin{pmatrix} \varepsilon_2 \\ p_2(\varepsilon) \end{pmatrix} + \begin{pmatrix} \varepsilon_1 \\ r_2(\varepsilon) \end{pmatrix} h(x^2, \varepsilon)(x^1)^{\nu(\varepsilon)} + \mathbf{k}(x^1, x^2, \varepsilon)(x^1)^{\eta(\varepsilon)}$$

for $x^1 > 0$.

We then define the scaling set $M_\sigma(c)$ by

$$M_\sigma(c) = \left\{ (\varepsilon_1, \varepsilon_2) \;\middle|\; \varepsilon_1 = -\sigma \frac{c}{h(0,\varepsilon)} |\varepsilon_2|^{1-\nu(\varepsilon)}, \quad \sigma = \mathrm{sgn}(\varepsilon_2) \right\},$$

which gives decomposition of a punctured neighborhood of $(\varepsilon_1, \varepsilon_2) = (0,0)$ with $(\varepsilon_1, \varepsilon_2) = (0,0)$ itself removed into infinitely many curves. For each $(\varepsilon_1, \varepsilon_2) \in M_\sigma(c)$ with $\varepsilon_1 \neq 0$, we take the linear transformation

$$\varphi : \begin{pmatrix} u \\ v \end{pmatrix} \longmapsto \begin{pmatrix} \varepsilon_2 & 0 \\ \frac{\varepsilon_2}{\varepsilon_1} r_2 & |\varepsilon_2|^\rho \end{pmatrix} \begin{pmatrix} u \\ v \end{pmatrix} = \begin{pmatrix} \varepsilon_2 u \\ \frac{\varepsilon_2}{\varepsilon_1} r_2 u + |\varepsilon_2|^\rho v \end{pmatrix},$$

where $\rho < \min(\nu, 1-\nu)$.

Computing the composition $\varphi^{-1} \circ \Pi \circ \varphi$ and denoting it by the same notation, we obtain the transformed Poincaré map Π of the form

$$\Pi(u,v) = \Pi_0(u,v) + H(u,v) \quad \text{on} \quad (u,v) \in (-1,1) \times (-1,1)$$

with

$$\begin{aligned} \Pi_0(u,v) &= \mathrm{sgn}(\varepsilon_2 u) \left\{ \begin{pmatrix} 1 \\ p_2(\varepsilon) \end{pmatrix} - \begin{pmatrix} c \\ 0 \end{pmatrix} h(\mathrm{sgn}(u)v, \varepsilon)|u|^\nu \right\} \\ H(u,v) &= \mathrm{sgn}(\varepsilon_2 u)\mathbf{k}(|u|, \mathrm{sgn}(u)v, \varepsilon)|u|^\eta \end{aligned} \tag{3.5.16}$$

having the following properties:

$$\begin{aligned} p_2(\varepsilon) &= O(|\varepsilon_2|^\zeta), \\ h(0,\varepsilon) &= 1, \\ \frac{\partial h}{\partial v}(v,\varepsilon) &= O(|\varepsilon_2|^\zeta), \end{aligned}$$

and

$$\left\| \frac{\partial^{k+l} H}{\partial u^k \partial v^l} \right\| \leq (\mathrm{const})|\varepsilon_2|^\zeta |u|^{\eta-k},$$

for some $\zeta > 0$.

From this form of the Poincaré map, we intend to find an invariant stable direction field on Σ^s, namely, to each $(u,v) \in \Sigma^s$ is associated a direction vector

$$\xi(u,v)\frac{\partial}{\partial u} + \frac{\partial}{\partial v}$$

which is invariant under the derivative of Π. This invariance is formulated as

$$\begin{pmatrix} a(u,v) & c(u,v) \\ b(u,v) & d(u,v) \end{pmatrix} \begin{pmatrix} \xi(u,v) \\ 1 \end{pmatrix} // \begin{pmatrix} \xi(\Gamma(u,v)) \\ 1 \end{pmatrix},$$

where $D\Pi = \begin{pmatrix} a & c \\ b & d \end{pmatrix}$, which implies the equation

$$\xi = \frac{d \cdot \xi \circ \Pi - c}{a - b \cdot \xi \circ \Pi}.$$

Define

$$\mathcal{F}(\xi) = \frac{d \cdot \xi \circ \Pi - c}{a - b \cdot \xi \circ \Pi}$$

and regard it as the fixed point problem for the map $\mathcal{F}$. After careful estimates of the terms involved, we can show that the map $\mathcal{F}$ is a contraction on a suitable function space, hence possessing a unique fixed point ξ_*, which is in fact an almost vertical invariant direction field on Σ^s. Then the fundamental theorem of ordinary differential equations implies that this invariant direction field determines an invariant foliation on Σ^s, that is, a family of curves preserved under the Poincaré map Π. This invariant foliation has sufficient smoothness to conclude that there exists a change of coordinates sending this foliation into the family of straight lines such that it satisfies (GL2) and (GL3) of the assumptions of the geometric Lorenz model.

From the form of the Poincaré map (3.5.16), the reduced one-dimensional map in (GL3) takes approximately the form

$$\Pi_1(u) \approx \mathrm{sgn}(u)(1 - c|u|^{\nu}).$$

Readers should notice that this form of the one-dimensional map is equivalent to the Lorenz map studied in Subsection 3.4.5.

The details, especially the estimates required in the arguments, are somewhat elaborate and we refer to [Rychlik 1990] for the detail. □

Remark 3.5.16. A similar result on the existence of geometric Lorenz attractors has been obtained by [Robinson 1989, 1992] for a γ-symmetric pair of homoclinic orbits with resonant eigenvalues. Robinson claims that the construction of the invariant foliation is different from and is easier than that in the case of inclination-flip homoclinic orbits. In fact, Robinson's earlier work ([Robinson 1981]) indicates two methods of constructing the foliations, one of which is adopted in [Robinson 1989, 1992] and the other of which is adopted in [Rychlik 1990], since the former method cannot be applied to inclination-flip homoclinic orbits. Here we do not give the details of Robinson's results since they need more preliminary work. Interested readers should consult the papers cited above.

At the end of this subsection, we shall give two examples of ordinary differential equations which exhibit geometric Lorenz attractors due to the Rychlik's theorem.

Example 3.5.17. We first start from the original Lorenz equation:

$$\begin{aligned}
\dot{x} &= \sigma(y - x)\\
\dot{y} &= rx - y - xz\\
\dot{z} &= -bz + xy,
\end{aligned}$$

and make a change of coordinates as follows ([Ushiki, Oka and Kokubu 1984]):

$$X = \frac{x}{\sqrt{2}}, \quad Y = \frac{\sigma}{\sqrt{2}}(y - x), \quad Z = \frac{\sigma}{2\sigma - b}\left(z - \frac{x^2}{2\sigma}\right),$$

which embeds the Lorenz equation into

$$\begin{aligned} \dot{X} &= Y \\ \dot{Y} &= aX - X^3 + AY + BXZ \\ \dot{Z} &= -bZ + X^2. \end{aligned} \tag{3.5.17}$$

In fact, when

$$a = \sigma(r-1), \quad A = -(\sigma+1), \quad B = -(2\sigma - b),$$

then the system (3.5.17) is equivalent to the original Lorenz equation. Notice that the transformed equation still preserves the symmetry γ.

Then [Rychlik 1990] observes that, if the parameter A and B are put to zero while keeping

$$a > 0, \quad 0 < b < \sqrt{a},$$

then the system

$$\begin{aligned} \dot{X} &= Y \\ \dot{Y} &= aX - X^3 \\ \dot{Z} &= -bZ + X^2 \end{aligned} \tag{3.5.18}$$

admits a γ-symmetric pair of inclination-flip homoclinic orbits. Indeed, the first two equations are independent on the Z-variable and are integrable due to the Hamiltonian structure. Thus, we can obtain the explicit homoclinic solution $(X(t), Y(t))$ to the hyperbolic equilibrium point $(0,0)$, where the linearization has the eigenvalues $\pm\sqrt{a}$. Then, we substitute the known function $X(t)$ into the third equation of (3.5.18) and solve it using the variation of constants formula yielding

$$Z(t) = e^{-bt}Z(0) + \int_0^t e^{-b(t-s)}X(s)^2 ds.$$

Take $Z(0) = -\int_0^{-\infty} e^{bs}X(s)^2 ds$, namely, take

$$Z(t) = \int_{-\infty}^t e^{-b(t-s)}X(s)^2 ds.$$

We claim that this solution $h(t) = (X(t), Y(t), Z(t))$ is an inclination-flip homoclinic orbit to the equilibrium point $O = (0,0,0)$. In fact, since the eigenvalues at O are given by $\lambda^u = \sqrt{a}, \lambda^s = -b, \lambda^{ss} = -\sqrt{a}$, the condition (Ev) is satisfied from the assumption $0 < b < \sqrt{a}$. Then the orbit $h(t)$ is indeed homoclinic to O, since $X(t)$ has the following asymptotic behavior:

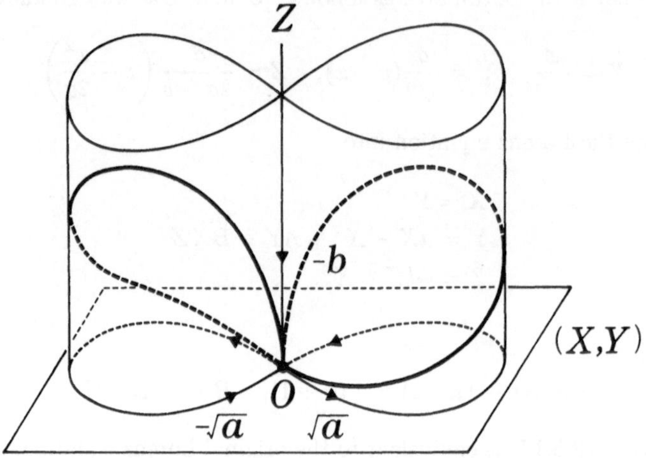

Fig. 3.5.17. The inclination-flip homoclinic orbits in Example 3.5.17.

$$|X(t)| = O(e^{\sqrt{a}t}) \text{ as } t \to -\infty \quad \text{and} \quad |X(t)| = O(e^{-\sqrt{a}t}) \text{ as } t \to +\infty,$$

from which it is easy to show that $Z(t) \to 0$ as $t \to \pm\infty$. The same estimate in fact shows that $|Z(t)| = O(e^{-bt})$ as $t \to +\infty$, which verifies the condition (Asy). Finally, in order to check the inclination-flip condition, it is sufficient to see that the surface given by the direct product of the homoclinic orbit $(X(t), Y(t))$ in the (X, Y)-plane with the Z-axis is invariant under the flow. This implies that the surface is nothing but the stable manifold and at the same time the extended unstable manifold $W^{eu}(O)$. See Fig. 3.5.17.

However, this system (3.5.18) does not satisfy Condition (3.5.13) in Rychlik's theorem, since $\lambda^u = |\lambda^{ss}| = \sqrt{a}$, and therefore we cannot directly apply the theorem to (3.5.18), but instead we need to add more parameters to perturb the equation so that it recovers the right eigenvalue condition $\lambda^u < |\lambda^{ss}|$ without breaking the symmetric pair of inclination-flip homoclinic orbits.

Such a perturbation is not easy in general, since homoclinic orbits are not structurally stable and hence they are easy to disappear under perturbation. In [Rychlik 1990], the following equation is presented:

$$\begin{aligned}\dot{x} &= y \\ \dot{y} &= x - 2x^3 + \alpha y + \beta xy + \gamma x^2 y \\ \dot{z} &= -\kappa z + x^2.\end{aligned}$$

Apparently, although some coefficients are different, this is essentially an extension of (3.5.17). It has been shown that, for $\frac{1}{2} < \kappa < 1$ with appropriate values of α and β while $\gamma = 0$, the equation satisfies the assumption of Theorem 3.5.15, and hence certain perturbations of it, satisfying $\gamma \neq 0$, generate geometric Lorenz attractors.

The main idea here is to use results of [Khorozov 1979] and [Carr 1981], which provide us with a symmetric version of the Bogdanov-Takens bifurcation on the plane. This enables us to perturb (3.5.18) so as to satisfy the right eigenvalue condition while keeping the inclination-flip homoclinic orbits, since, when $\gamma = 0$, the existence of the inclination-flip homoclinic orbit in the entire three-dimensional space is equivalent to that of a homoclinic orbit in the plane as explained above.

Instead of using a planar result, one may be able to employ a Melnikov-like technique adapted to this situation in order to keep the inclination-flip homoclinic orbits under perturbation. This idea was suggested in [Rychlik 1990] and is carried out in [Dumortier, Kokubu and Oka 1992]. As a result, we can show that the equation

$$\begin{aligned} \dot{x} &= y \\ \dot{y} &= ax - x^3 + Ay + Bxz + Cyz \\ \dot{z} &= -bz + x^2 \end{aligned} \tag{3.5.19}$$

generates geometric Lorenz attractors for certain parameter values satisfying $0 < b < \sqrt{a}$.

Any one of these equations needs to include an extra term to (3.5.18), and hence the existence of geometric Lorenz attractors in these equations does not imply an analogous result for the original Lorenz equation.

Finally, we remark that two other sets of ordinary differential equations have been given in [Robinson 1989, 1992] as those generating geometric Lorenz attractors for a symmetric pair of homoclinic orbits with resonant eigenvalues.

3.5.7 Local Bifurcations and Global Bifurcations

Local bifurcations near equilibrium points and those for more global orbits such as homoclinic or heteroclinic bifurcations are intimately related to each other. For instance, the Bogdanov-Takens bifurcation studied in Subsection 3.3.4 involves the simplest bifurcation for the single planar homoclinic orbit given in Theorem 3.5.4. The bifurcation of equilibrium points with a simple zero eigenvalue and a pair of pure imaginary eigenvalues at linearization is expected to involve a Shil'nikov-type homoclinic orbit in its unfolding (see [Broer and Vegter 1984]), and hence it may be accompanied by complicated dynamics due to the Shil'nikov's theorem (Theorem 3.5.5). In this way, one can say that a global bifurcation is compressed into a degenerate singularity, and the higher is the codimension of a singularity, the richer becomes the dynamics generated from it. This is the main idea of the catastrophe theory, and in this sense [Thom 1972] called a degenerate singularity an organizing center of dynamics coming out from the singularity. The singularity theory and the bifurcation theory can therefore be thought of as a mathematical justification of (part of) the catastrophe theory.

Among such degenerate singularities, particular interest is given to an organizing center of a chaotic attractor. There are numerically suggested results indicating that a degenerate singularity with zero eigenvalues of multiplicity three possibly exhibits a chaotic attractor resembling the Rössler attractor ([Rössler 1979b]). This observation was first given by [Arneodo, Coullet, Spiegel and Tresser 1985]. See also [Gaspard, Kapral and Nicolis 1984], [Glendinning and Sparrow 1984] and [Chua and Kokubu 1989].

Similar observations were given to the Lorenz equation as well ([Ushiki, Oka and Kokubu 1984]). Take the transformed equation (3.5.17) from the Lorenz equation and make a rescaling of the system as follows:

$$\bar{X} = \varepsilon X, \quad \bar{Y} = \varepsilon^2 Y, \quad \bar{Z} = \varepsilon Z, \quad \tau = \frac{t}{\varepsilon}.$$

Then it brings (3.5.17) into

$$\begin{aligned} \bar{X}' &= \bar{Y} \\ \bar{Y}' &= \varepsilon^2 a\bar{X} - \bar{X}^3 + \varepsilon A\bar{Y} + \varepsilon B\bar{X}\bar{Z} \qquad \left(' = \frac{d}{d\tau}\right), \\ \bar{Z}' &= -\varepsilon b\bar{Z} + \bar{X}^2 \end{aligned}$$

and therefore, taking the limit $\varepsilon \to 0$, we obtain the degenerate system

$$\begin{aligned} \bar{X}' &= \bar{Y} \\ \bar{Y}' &= -\bar{X}^3 \\ \bar{Z}' &= \bar{X}^2. \end{aligned} \tag{3.5.20}$$

From this, one can observe that any dynamics appearing in the original Lorenz equation including a numerically generated Lorenz attractor for the standard parameters $\sigma = 10, r = 28, b = \frac{3}{8}$, can be put into an arbitrarily small perturbation of the degenerate system (3.5.20). Of course this does not show the existence of chaotic attractors in an unfolding of the degenerate singularity (3.5.20), because there has been no rigorous proof of the existence of chaotic attractors in the original Lorenz equation up to now. However, recently, [Dumortier, Kokubu and Oka 1992] show the following result:

Theorem 3.5.18. *A degenerate singularity whose 3-jet is given by*

$$y\frac{\partial}{\partial x} - x^3\frac{\partial}{\partial y} + x^2\frac{\partial}{\partial z}$$

has an unfolding which generates geometric Lorenz attractors.

The main idea of the proof is to reduce an unfolding of the singularity to the equation (3.5.19), which then shows the existence of geometric Lorenz attractors as explained in Example 3.5.17. In particular, a connection between the local bifurcation around a degenerate singularity and the global bifurcation for certain homoclinic orbits plays a role here as well. This singularity is fairly degenerate, and it may be possible to obtain similar results with less degenerate singularities. The details of the proof and more information on the singularity can be found in [Dumortier, Kokubu and Oka 1992].

An analogous problem are proposed for diffeomorphisms. In this context, the results of [Mora and Viana 1991] may be significant. They assert that a Hénon-like strange attractor bifurcates from a homoclinic tangency of the stable and unstable manifolds of a dissipative saddle periodic point. This result may be important for the search of degenerate vector field singularities generating a Rössler-like attractor, because computer simulation shows that the cross section of the Rössler attractor looks very much like the Hénon attractor.

References

Afraĭmovich, V.S., Bykov, V.V., and Shil'nikov, L.P. [1982]. On structurally unstable attracting limit sets of Lorenz attractor type. *Trans. Moscow Math. Soc.* **44**(1983), 153–216. Russian original: *Trudy Moskov. Mat. Obshch.* **44**, 150–212.

Afraĭmovich, V.S. and Pesin, Ya.B. [1987]. Dimension of Lorenz type attractors. *Sov. Sci. Rev. C Math./Phys.* **6**, 169–241.

Alfsen, K.H. and Frøyland, J. [1985]. Systematics of the Lorenz model at $\sigma = 10$. *Physica Scripta* **31**, 15–20.

Annabi, M.L. [1991]. Dépendence continue par rapport aux paramètres dans la bifurcation de Hopf-Takens de codimension 3. *Annal. Faculté Sci. Toulouse* **XII**, 295–327.

Annabi, H., Annabi, M.L., and Dumortier, F. [1992]. Continuous dependence on parameters in the Bogdanov-Takens bifurcation. In *Geometry and Analysis in Nonlinear Dynamics* (Eds. H.W. Broer and F. Takens), Pitman Research Notes, Vol. 222, 1–21. Longman Scientific & Technical.

Andronov, A.A., Gordon, I.I., Leontvich, E.A., and Maier, A.G. [1973a]. *Qualitative Theory of Second-Order Dynamic Systems.* John Wiley & Sons: New York.

Andronov, A.A., Gordon, I.I., Leontvich, E.A., and Maier, A.G. [1973b]. *Theory of Bifurcations of Dynamic Systems on a Plane.* John Wiley & Sons: New York.

Arneodo, A., Coullet, P., and Tresser, C. [1981]. Possible new strange attractors with spiral structure. *Comm. Math. Phys.* **79**, 573–579.

Arneodo, A., Coullet, P., and Tresser, C. [1982]. Oscillators with chaotic behavior: An illustration of a theorem by Shil'nikov. *J. Stat. Phys.* **27**, 171–182.

Arneodo, A., Coullet, P., Spiegel, E.A., and Tresser, C. [1985]. Asymptotic chaos. *Physica D* **14**, 327–347.

Arnold, V.I. [1961]. Small denominators. I: Mapping of the circumference onto itself. *Amer. Math. Soc. Transl., Ser.2.* **46**(1965), 213–284. Russian original: *Izv. Akad. Nauk SSSR* **25**, 21–86.

Arnold, V.I. [1973]. *Ordinary Differential Equations.* MIT Press: Cambridge.

Arnold, V.I. [1983]. *Geometrical Methods in the Theory of Ordinary Differential Equations.* Springer-Verlag: New York.

Arnold, V.I. [1984]. *Catastrophe Theory.* Third Edition 1992. Springer-Verlag: New York.

Arrowsmith, D.K. and Place, C.M. [1990]. *An Introduction to Dynamical Systems.* Cambridge University Press.

Baider, A. and Sanders, J. [1992]. Further reduction of the Takens-Bogdanov normal form. *J. Diff. Eq.* **99**, 205–244.

Bamón, R., Labarca, R., Mañé, R., and Pacífico, M.J. [1992]. Explosion of singular cycles. Preprint.

Banks, J., Brooks, J., Cairns, G., Davis, G., and Stacey, P. [1992]. On Devaney's definition of chaos. *Amer. Math. Monthly* **99**, 332–334.

Battelli, F. and Lazzari, C. [1990]. Exponential dichotomies, heteroclinic orbits, and Melnikov functions. *J. Diff. Eq.* **86**, 342–366.

Belitskii, G.R. [1977]. Equivalence and normal forms of germs of smooth mappings. *Russian Math. Surveys* **33**(1978), 107–177. Russian original: *Uspekhi Mat. Nauk* **33**, 95–155.

Beloqui, J.A. [1986]. Modulus of stability for vector fields on 3-manifolds. *J. Diff. Eq.* **65**, 374–395.

Benedicks, M. and Carleson, L. [1985]. On iterates of $x \mapsto 1 - ax^2$ on $(-1, 1)$. *Ann. Math.* **122**, 1–25.

Benedicks, M. and Carleson, L. [1991]. The dynamics of the Hénon map. *Ann. Math.* **133**, 73–169.

Benedicks, M. and Young, L.-S. [1992]. Sinai-Bowen-Ruelle measures for certain Hénon maps. Preprint.

Bergé, P., Pomeau, Y., and Vidal, C. [1984]. *L'ordre dans le Chaos.* Hermann: Paris.

Berger, M.S. [1977]. *Nonlinearity and Functional Analysis.* Academic Press: New York.

Bogdanov, R.I. [1976a]. Bifurcation of the limit cycle of a family of plane vector fields. *Selecta Math. Sovietica* **1**(1984), 373–387. Russian original: *Trudy Sem. Petrovsk.* **2**, 23–35.

Bogdanov, R.I. [1976b]. Versal deformations of a singularity of a vector field on the plane. *Selecta Math. Sovietica* **1**(1984), 389–421. Russian original: *Trudy Sem. Petrovsk.* **2**, 37–65.

Bowen, R. [1970]. *Equilibrium States and the Ergodic Theory of Anosov Diffeomorphisms.* Lecture Notes in Math., Vol. 470, Springer-Verlag: New York.

Bröcker, Th. and Lander, L. [1975]. *Differentiable Germs and Catastrophes.* Cambridge University Press.

Broer, H.W., Dumortier, F., van Strien, S.J., and Takens, F. [1991]. *Structures in Dynamics.* North-Holland: Amsterdam.

Broer, H.W. and Vegter, G. [1984]. Subordinate Šil'nikov bifurcations near some singularities of vector fields having low codimension. *Ergod. Th. and Dynam. Sys.* **4**, 509–525.

Bronson, S.D., Dewey, D., and Linsay, P.S. [1983]. Selfreplicating attractor of a driven semiconductor oscillator. *Phys. Rev. A* **28**, 1201–1203.

Burchard, A., Deng, B., and Lu, K. [1992]. Smooth conjugacy of centre manifolds. *Proc. Royal Soc. Edinburgh* **120A**, 61–77.

Bykov, V.V. [1991]. Bifurcations of separatrix contours and chaos (in Russian). In *Methods of Qualitative Theory and Theory of Bifurcations* (Ed. L.P. Shil'nikov), 84–104. Gor'kov. Gos. Univ.

Carr, J. [1981]. *Applications of Centre Manifold Theory.* Springer-Verlag: New York.

Carr, J., Chow, S.-N., and Hale, J.K. [1985]. Abelian integrals and bifurcation theory. *J. Diff. Eq.* **59**, 413–436.

Cascais, J., Dilao, R., and Norondacosta, A. [1983]. Chaos and reverse bifurcation in a RCL circuit. *Phys. Lett.* **93A**, 213–216.

Chow, S.-N., Deng, B., and Fiedler, B. [1990]. Homoclinic bifurcation at resonant eigenvalues. *J. Dynam. Diff. Eq.* **2**, 177–244.

Chow, S.-N., Deng, B., and Terman, D. [1990]. The bifurcation of homoclinic and heteroclinic orbits from two heteroclinic orbits. *SIAM J. Math. Anal.* **21**, 179–204.

Chow, S.-N., Deng, B., and Terman, D. [1991]. The bifurcation of homoclinic orbits from two heteroclinic orbits – a topological approach. *Applicable Analysis* **42**, 275–299.

Chow, S.-N. and Hale, J.K. [1982]. *Methods of Bifurcation Theory.* Springer-Verlag: New York.

Chow, S.-N., Hale, J.K., and Mallet-Paret, J. [1980]. An example of bifurcation to homoclinic orbits. *J. Diff. Eq.* **37**, 351–373.

Chow, S.-N. and Lin, X.-B. [1990]. Bifurcation of a homoclinic orbit with a saddle-node equilibrium. *Differential Integral Equations* **3**, 435–466.

Chua, L.O and Ying, R.L.P. [1983]. Canonical piecewise-linear analysis. *IEEE Trans. Circuits and Systems* **CAS30**, 125–140.

Chua, L.O and Deng, A.C. [1988]. Canonical piecewise-linear representation. *IEEE Trans. Circuits and Systems* **CAS35**, 101–111.

Chua, L.O. and Kokubu, H. [1988]. Normal forms for nonlinear vector fields, Part I. *IEEE Trans. Circuits and Systems* **CAS35**, 863–880.

Chua, L.O. and Kokubu, H. [1989]. Normal forms for nonlinear vector fields, Part II. *IEEE Trans. Circuits and Systems* **CAS36**, 51–70.

Chua, L.O., Komuro, M., and Matsumoto, T. [1986]. The Double Scroll family. *IEEE Trans. Circuits and Systems* **CAS33**, 1073–1118.

Coddington, E.A. and Levinson, N. [1955]. *Theory of Ordinary Differential Equations.* McGraw Hill: New York.

Collet, P. and Eckmann, J.-P. [1980]. *Iterated Maps of the Interval as Dynamical Systems.* Birkhäuser: Boston.

Collet, P. and Levy, Y. [1984]. Ergodic properties of the Lozi mappings. *Comm. Math. Phys.* **93**, 461–481.

Coomes, B.A. [1989]. The Lorenz system does not have a polynomial flow. *J. Diff. Eq.* **82**, 386–407.

Coullet, P. and Tresser, C. [1978]. Itération d'endomorphismes et groupe de renormalisation. *C. R. Acad. Sci. Paris* **287**, 577–580.

Cushman, R. and Sanders, J. [1985]. A codimension two bifurcation with a third order Picard-Fuchs equation. *J. Diff. Eq.* **59**, 243–256.

Cushman, R. and Sanders, J. [1986]. Nilpotent normal forms and representation theory of $sl(2,\mathbb{R})$. In *Multiparameter Bifurcation Theory* (Eds. J. Guckenheimer and M. Golubitsky), 31–51. AMS: Providence.

de Melo, W. and van Strien, S.J. [1993]. *One-Dimensional Dynamics.* To appear from Springer-Verlag: New York.

Deng, B. [1989a]. The Sil'nikov problem, exponential expansion, strong λ-lemma, C^1-linearization, and homoclinic bifurcation. *J. Diff. Eq.* **79**, 189–231.

Deng, B. [1989b]. Exponential expansion with Šil'nikov's saddle-focus. *J. Diff. Eq.* **82**, 156–173.

Deng, B. [1989c]. The transverse homoclinic dynamics and their bifurcations at nonhyperbolic fixed points. Preprint.

Deng, B. [1988]. Exponential expansion with principal eigenvalues. Preprint.

Deng, B. [1990a]. Homoclinic bifurcations with nonhyperbolic equilibria. *SIAM J. Math. Anal.* **21**, 693–720.

Deng, B. [1990b]. On Šilnikov's homoclinic-saddle-focus theorem. Preprint.

Deng, B. [1990c]. Symbolic dynamics for chaotic systems. IMA Preprint #701.

Deng, B. [1991a]. The existence of countable connections from a twisted heteroclinic loop. *SIAM J. Math. Anal.* **22**, 653–679.

Deng, B. [1991b]. The existence of infinitely many traveling front and back waves in the FitzHugh-Nagumo equations. *SIAM J. Math. Anal.* **22**, 1631–1650.

Deng, B. [1991c]. Homoclinic twisting bifurcations and cusp horseshoe maps. Preprint.

Deng, B. and Sakamoto, K. [1989]. The shift dynamics for the attractor near an orbit homoclinic to a Hopf equilibrium. Preprint.

Denjoy, A. [1932]. Sur les courbes définies par les équations differentielles à la surface du tore. *J. Math.* **17**, 333–375.

Devaney, R. [1986]. *An Introduction to Chaotic Dynamical Systems.* Benjamin Commings: Menlo Park.

Devaney, R. and Nitecki, Z. [1979]. Shift automorphisms in the Hénon mapping. *Comm. Math. Phys.* **67**, 137–146.

Duffing, G. [1918]. *Erzwungene Schwingungen bei Veränderlicher Eigenfrequenz.* Braunschweig.

Dumortier, F. [1977]. Singularities of vector fields on the plane. *J. Diff. Eq.* **23**, 53–106.

Dumortier, F. [1978]. *Singularities of Vector Fields.* Monografias de Matemática, Vol. 32, IMPA: Rio de Janeiro.

Dumortier, F., Kokubu, H., and Oka, H. [1992]. A degenerate singularity generating geometric Lorenz attractors. Preprint.

Dumortier, F. and Roussarie, R. [1990]. On the saddle loop bifurcation. In *Bifurcations of planar vector fields* (Eds. P. Françoise and R. Roussarie), Lecture Notes in Math., Vol. 1455, 44–73. Springer-Verlag: New York.

Dumortier, F., Roussarie, R., and Sotomayor, J. [1987]. Generic 3-parameter families of vector fields on the plane, unfolding a singularity with nilpotent linear part: the cusp case of codimension 3. *Ergod. Th. and Dynam. Sys.* **7**, 375–413.

Dumortier, F., Roussarie, R., and Sotomayor, J. [1991]. Generic 3-parameter families of vector fields on the plane, unfoldings of saddle, focus and elliptic singularities with nilpotent linear parts. In *Bifurcations of Planar Vector Fields.* Lecture Notes in Math., Vol. 1480, 1–164. Springer-Verlag: New York.

El-Hamouly, H. and Mira, C. [1981]. Lien entre les propriétés d'un endomorphisme de dimension un et celles d'un difféomorphisme de dimension deux. *C. R. Acad. Sci. Paris* **293**, 525–528.

El-Hamouly, H. and Mira, C. [1982]. Singularités dues au feuilletage du plan des bifurcations d'un difféomorphisme bi-dimensionnel. *C. R. Acad. Sci. Paris* **294**, 387–390.

Elphick, E., Tirapegui, E., Brachet, M.E., Coullet, P., and Iooss, G. [1987]. A simple global characterization for normal forms of vector fields. *Physica D* **29**, 95–127.

Endo, T., and Saito, T. [1990] Chaos in electrical and electronic circuits and systems. *Trans. IEICE.* **E73**, 763–771.

Evans, J.W., Fenichel, N., and Feroe, J.A. [1982]. Double impulse solutions in nerve axon equations. *SIAM J. Appl. Math.* **42**, 219–234.

Feigenbaum, M. [1978, 1979]. Quantitative universality for a class of nonlinear transformations. *J. Stat. Phys.* **19**, 25–52; *ibid.* **21**, 669–709.

Feroe, J.A. [1982]. Existence and stability of multiple impulse solutions of a nerve equations. *SIAM J. Appl. Math.* **42**, 235–247.

Fleming, W. [1965]. *Functions of Several Variables.* Addison-Wesley: Reading.

Fujimoto, R., Komuro, M., Tokunaga, R., and Matsumoto, M. [1990]. Bifurcation analysis of the Shil'nikov's chaos. In *Bifurcation Phenomena in Nonlinear Systems and Theory of Dynamical Systems* (Ed. H. Kawakami), Advanced Series in Dynamical Systems, Vol. 8, 125–142. World Scientific: Singapore.

Gambaudo, J.M., Glendinning, P., and Tresser, C. [1984]. Collage de cycles et suites de Farey. *C. R. Acad. Sci. Paris* **299**, 711-714.

Gambaudo, J.M., Glendinning, P., and Tresser, C. [1988]. The gluing bifurcation: I Symbolic dynamics of the closed curves. *Nonlinearity* **1**, 203–214.

Gantmacher, F.R. [1959]. *The Theory of Matrices.* Chelsea: New York.

Gaspard, P. [1984]. Generation of a countable set of homoclinic flows through bifurcation in multidimensional systems. *Bull. Class. Sci. Acad. Roy. Belg.* **LXX**, 61–83.

Gaspard, P. and Nicolis, G. [1983]. What can we learn from homoclinic orbits in chaotic dynamics? *J. Stat. Phys.* **31**, 499–518.

Gaspard, P., Kapral, R., and Nicolis, G. [1984]. Bifurcation phenomena near homoclinic systems: A two-parameter analysis. *J. Stat. Phys.* **35**, 697–727.

George, D.P. [1986]. Bifurcations in a piecewise linear system. *Phys. Lett. A* **118**, 17–21.

Glazier, J.A. and Libchaber, A. [1988]. Quasi-periodicity and dynamical systems: An experimentalist's view. *IEEE Trans. Circuits and Systems* **CAS35**, 790–809.

Glendinning, P. [1984]. Bifurcations near homoclinic orbits with symmetry. *Phys. Lett. A* **103**, 163–166.

Glendinning, P. [1988]. Global bifurcations in flows. In *New Directions in Dynamical Systems* (Eds. T. Bedford and J. Swift), 120–149. Cambridge University Press.

Glendinning, P. and Sparrow, C. [1984]. Local and global behavior near homoclinic orbits. *J. Stat. Phys.* **35**, 645–697.

Glendinning, P. and Sparrow, C. [1986]. T-points: a codimension two heteroclinic bifurcation. *J. Stat. Phys.* **43**, 479–486.

Golubitsky, M. and Schaeffer, D.G. [1985]. *Singularities and Groups in Bifurcation Theory, Part I.* Springer-Verlag: New York.

Golubitsky, M., Stewart, I., and Schaeffer, D.G. [1988]. *Singularities and Groups in Bifurcation Theory, Part II.* Springer-Verlag: New York.

Grebogi, C., Ott, E., and Yorke, J. [1982]. Chaotic attractor in crisis. *Phys. Rev. Lett* **48**, 1507–1510.

Grobman, D.M. [1959]. Homeomorphisms of systems of differential equations (in Russian). *Dokl. Akad. Nauk. SSSR* **128**, 880–881.

Gruendler, J. [1985]. The existence of homoclinic orbits and the method of Melnikov for systems in $\mathbb{R}^n$. *SIAM J. Math. Anal.* **16**, 907–931.

Gruendler, J. [1992]. Homoclinic solutions for autonomous dynamical systems in arbitrary dimension. *SIAM J. Math. Anal.* **23**, 702–721.

Guckenheimer, J. [1976]. A strange strange attractor. In *The Hopf Bifurcation and Its Applications* (Eds. J. E. Marsden and M. McCracken), 368–381. Springer-Verlag: New York.

Guckenheimer, J. [1981]. On a codimension two bifurcation. In *Dynamical Systems and Turbulence* (Eds. D. Rand and L.-S. Young), Lecture Notes in Math., Vol. 898, 99–142. Springer-Verlag: New York

Guckenheimer, J. [1984]. Multiple bifurcation problems of codimension two. *SIAM J. Math. Anal.* **15**, 1-49.

Guckenheimer, J. [1987]. Renormalization of one dimensional mappings and strange attractors. In *The Lefschetz Centennial Conference, Part III* (Ed. A. Verjovsky), Contemporary Math., Vol. 58.III, 143–160. AMS: Providence.

Guckenheimer, J. and Holmes, P. [1983]. *Nonlinear Oscillations, Dynamical Systems and Bifurcations of Vector Fields.* Third Printing 1989. Springer-Verlag: New York.

Guckenheimer, J. and Williams, R.F. [1979]. Structural stability of Lorenz attractors. *Publ. Math. IHES* **50**, 59-72.

Hale, J.K. [1969]. *Ordinary Differential Equations.* Second Edition 1980. Krieger: Malabar.

Hale, J.K. [1988]. *Asymptotic Behavior of Dissipative Systems.* AMS: Providence.

Hale, J.K. and Lin, X.-B. [1986a]. Heteroclinic orbits for retarded functional differential equations. *J. Diff. Eq.* **65**, 175–202.

Hale, J.K. and Lin, X.-B. [1986b]. Symbolic dynamics and nonlinear semiflows. *Ann. Mat. Pura Appl.* **144**, 229–260.

Hartman, P. [1960]. A lemma in the theory of structural stability of differential equations. *Proc. Amer. Math. Soc.* **11**, 610–620.

Hartman, P. [1963]. On the local linearization of differential equations. *Proc. Amer. Math. Soc.* **14**, 568–573.

Hastings, S.P. [1982]. Single and multiple pulse waves for the FitzHugh-Nagumo equations. *SIAM J. Appl. Math.* **42**, 247–260.

Hénon, M.A. [1976]. A two-dimensional mapping with a strange attractor. *Comm. Math. Phys.* **50**, 69–77.

Herman, M.R. [1979]. Sur la conjugaison différentiable des difféomorphismes du cercle à des rotations. *Publ. Math. IHES* **49**, 5–234.

Hirsch, M.W. and Pugh, C. [1970]. Stable manifolds and hyperbolic sets. In *Global Analysis* (Eds. S.S. Chern and S. Smale) Proc. Sympos. Pure Math., Vol. 14, 133–163. AMS: Providence.

Hirsch, M.W., Pugh, C., and Shub. M. [1977]. *Invariant Manifolds.* Lecture Notes in Math., Vol. 583. Springer-Verlag: New York.

Hirsch, M.W. and Smale, S. [1974]. *Differential Equations, Dynamical Systems and Linear Algebra.* Academic Press: New York.

Hitzle, D.L. and Zele, F. [1985]. An exploration of the Hénon quadratic map. *Physica D* **14**, 305–326.

Holmes, P.J. [1979]. A nonlinear oscillator with a strange attractor. *Phil. Trans. Royal Soc. A* **292**, 419–448.

Holmes, P.J. [1980]. A strange family of three-dimensional vector fields near a degenerate singularity. *J. Diff. Eq.* **37**, 382–403.

Holmes, P.J. and Marsden, J.E. [1982]. Melnikov's method and Arnold diffusion for perturbations of integrable Hamiltonian systems. *J. Math. Phys.* **23**, 669–675.

Homburg, A.J. [1992]. Some global aspects of homoclinic bifurcations of vector fields. Preprint.

Homburg, A.J., Kokubu, H., and Krupa, M. [1993]. The cusp horseshoe and its bifurcations from inclination-flip homoclinic orbits. Preprint.

Il'yashenko, Yu.S. [1978]. The multiplicity of limit cycles arising from perturbations of the form $w' = p_2/q_1$ of a Hamiltonian equation in the real and complex domain. *Amer. Math. Soc. Transl.* **118**(1982), 191–202. Russian original: *Trudy Sem. Petrovsk.* **3**, 49–60.

Il'yashenko, Yu.S. [1991]. *Finiteness Theorems for Limit Cycles.* AMS: Providence.

Il'yashenko, Yu.S. and Yakovenko, S.Yu. [1991]. Finitely-smooth normal forms of local families of diffeomorphisms and vector fields. *Russian Math. Surveys* **46**(1991), 1–43. Russian original: *Uspekhi Mat. Nauk* **46**, 3–39.

Iooss, G. [1979]. *Bifurcation of Maps and Applications.* North-Holland: Amsterdam.

Iori, K., Yanagida, E., and Matsumoto, T. [1993]. N-homoclinic bifurcations of piecewise-linear vector fields, In *Structure and Bifurcations of Dynamical Systems* (Ed. S. Ushiki), Advanced Series in Dynamical Systems, Vol. 11, 82–97. World Scientific: Singapore.

Irwin, M.C. [1980]. *Smooth Dynamical Systems.* Academic Press: New York.

Jakobson, M.V. [1981]. Absolutely continuous invariant measures for one-parameter families of one-dimensional maps. *Comm. Math. Phys.* **81**, 39–88.

Kaper, T. and Wiggins, S. [1991]. Lobe area in adiabatic Hamiltonian systems. *Physica D* **51**, 205–212.

Keller, G. [1985]. Generalized bounded variation and applications to piecewise monotonic transformations. *Z. Wahrscheinlichkeitstheorie verw. Gebiete* **69**, 461–478.

Khorozov, E. [1979]. Versal deformations of equivariant vector fields for the case of symmetries of order 2 and 3. In *Topics in Modern Mathematics* (Ed. O.A. Oleinik), (1985), 207–243. Consultants Bureau: New York. Russian original: *Trudy Sem. Petrovsk.* **5**, 164–192.

King, G.P. and Stewart, I.N. [1992]. Symmetric chaos. In *Nonlinear Equations in the Applied Sciences* (Eds. W.F. Ames and C. Rogers), 257–315. Academic Press: New York.

Kirchgraber, U. and Palmer, K.J. [1990]. *Geometry in the Neighborhood of Invariant Manifolds of Maps and Flows and Linearization.* Pitman Research Notes in Math., Vol. 233. Longman Scientific & Technical.

Kisaka, M., Kokubu, H., and Oka, H. [1992a]. Supplement to homoclinic doubling bifurcation in vector fields. To appear in *Dynamical Systems* (Eds. R. Bamón, R. Labarca, J. Lewowicz and J. Palis, Jr), Pitman Research Notes in Math. Longman Scientific & Technical.

Kisaka, M., Kokubu, H., and Oka, H. [1992b]. Bifurcations to N-homoclinic orbits and N-periodic orbits in vector fields. To appear in *J. Dynam. Diff. Eq.*

Kokubu, H. [1988]. Homoclinic and heteroclinic bifurcations in vector fields. *Japan J. Appl. Math.* **5**, 455–501.

Kokubu, H. [1991]. Heteroclinic bifurcations associated with different saddle indices. In *Dynamical Systems and Related Topics, Proceedings of International Conference on Dynamical Systems, Nagoya, 1990* (Ed. K. Shiraiwa), Advanced Series in Dynamical Systems, Vol. 9, 236–260. World Scientific: Singapore.

Kokubu, H. [1993]. A construction of three-dimensional vector fields which have a codimension two heteroclinic loop at Glendinning-Sparrow T-point. To appear in *Z. Angeb. Math. Phys.* **44**.

Kokubu, H., Nishiura, Y., and Oka, H. [1990]. Heteroclinic and homoclinic bifurcations in bistable reaction diffusion systems. *J. Diff. Eq.* **86**, 260–341.

Komuro, M. [1988a]. Normal forms of continuous piecewise-linear vector fields and chaotic attractors: Part I. *Japan J. Appl. Math.* **5**, 257–304.

Komuro, M. [1988b]. Normal forms of continuous piecewise-linear vector fields and chaotic attractors: Part II. *Japan J. Appl. Math.* **5**, 503–549.

Komuro, M. [1992]. Bifurcation equations of continuous piecewise-linear vector fields. *Japan J. Ind. Appl. Math.* **9**, 269–312.

Komuro, M., Tokunaga, R., Matsumoto, T., Chua, L.O., and Hotta, A. [1991] A global bifurcation analysis of the Double Scroll circuit. *Int. J. Bifurcation and Chaos*, **1**, 139–182.

Lanford, O.E. [1982]. A computer-assisted proof of the Feigenbaum conjectures. *Bull. Amer. Math. Soc.* **6**, 427–434.

Langford, W. [1979]. Periodic and steady state mode interactions lead to tori. *SIAM J. Appl. Math.* **37**, 22–48.

Langford, W. and Iooss, G. [1980]. Interactions of Hopf and pitchfork bifurcations. In *Bifurcation Problems and Their Numerical Solution* (Eds. H.D. Mittelmann and H. Weber), 103–134. Birkhäuser: Boston.

Li, T.-Y. and Yorke, J. [1975]. Period three implies chaos. *Amer. Math. Monthly* **82**, 985–992.

Lin, X.-B. [1989] Shadowing lemma and singularly perturbed boundary value problems. *SIAM J. Appl. Math.* **49**, 26–54.

Lin, X.-B. [1990a] Using Melnikov's method to solve Silnikov's problems. *Proc. Royal Soc. Edinburgh* **116A**, 295–325.

Lin, X.-B. [1990b] Heteroclinic bifurcation and singularly perturbed boundary value problems. *J. Diff. Eq.* **84**, 319–382.

Lin, X.-B. [1993] . Preprint.

Linsay, P.S. [1981] Period doubling and chaotic behavior in a driven anharmonic oscillator. *Phys. Rev. Lett.* **47**, 1349–1352.

Lorenz, E. [1963]. Deterministic nonperiodic flow. *J. Atmos. Sci.* **20**, 130–141.

Lozi, R. [1978]. Un attracteur étrange? du type attracteur de Hénon. *J. Phys.* **39**, 9–10.

MacKay, R.S. and Tresser, C. [1986, 1987]. *Physica D* **19**, 206–237; *ibid.* **27**, 412–422.

Mañé, R. [1983]. *Ergodic Theory and Differentiable Dynamics.* Springer-Verlag: New York.

Mañé, R. [1987]. A proof of the C^1 stability conjecture. *Publ. Math. IHES* **66**, 161-210.

Mardešić, P. [1990]. The number of limit cycles of polynomial deformations of a Hamiltonian vector field. *Ergod. Th. and Dynam. Sys.* **10**, 523–529.

Mardešić, P. [1992]. Le déploiement versel du cusp d'ordre n. *C. R. Acad. Sci. Paris* **315**, 1235–1239.

Marsden, J. E. and McCracken, M. [1976]. *The Hopf Bifurcation and Its Applications.* Springer-Verlag: New York.

Matsumoto, T. [1976]. On the dynamics of electrical networks. *J. Diff. Eq.* **32**, 179–196.

Matsumoto, T. [1987]. Chaos in electronic circuits. *Proc. IEEE* **75**, 1033–1055.

Matsumoto, T., Chua, L.O., and Ayaki, K. [1988] Reality of chaos in the Double Scroll circuit: A computer assisted proof. *IEEE Trans. Circuits and Systems* **CAS35**, 909–925.

Matsumoto, T., Chua, L.O., and Kobayashi, K. [1986]. Hyperchaos: Laboratory experiment and numerical confirmation. *IEEE Trans. Circuits and Systems* **CAS33**, 1143–1147.

Matsumoto, T., Chua, L.O., and Komuro, M. [1985]. The Double Scroll. *IEEE Trans. Circuits and Systems* **CAS32**, 797–818.

Matsumoto, T., Chua, L.O., and Komuro, M. [1986]. The Double Scroll bifurcations. *Int. J. Circuit Theory Appl.* **14**, 117–146.

Matsumoto, T., Chua, L.O., and Komuro, M. [1987]. Birth and death of the Double Scroll. *Physica D* **24**, 97–124.

Matsumoto, T., Chua, L.O., and Tanaka, S. [1984]. Simplest chaotic nonautonomous circuit. *Phys. Rev. A* **30**, 1155–1157.

Matsumoto, T., Chua, L.O., and Tokumasu, K. [1986]. Double Scroll via a two-transistor circuit. *IEEE Trans. Circuits and Systems* **CAS33**, 828–835.

Matsumoto, T., Chua, L.O., and Tokunaga, R. [1987]. Chaos via torus breakdown. *IEEE Trans. Circuits and Systems* **CAS34**, 240–253.

Matsumoto, T. and Salam, F. [1988]. Special Issue on Chaotic Circuits and Systems. *IEEE Trans. Circuits and Systems* **CAS35**.

May, R.M. [1976]. Simple mathematical models with very complicated dynamics. *Nature* **261**, 459–466.

Medveď, M. [1984]. On a codimension three bifurcation. *Časopis. Pěst. Mat.* **109**, 3–26.

Mees, A.I. and Chapman, P.B. [1987]. Homoclinic and heteroclinic orbits in the Double Scroll attractor. *IEEE Trans. Circuits and Systems* **CAS34**, 1115–1120.

Melnikov, V.K. [1963]. On the stability of the center for time periodic perturbations. *Trans. Moscow Math. Soc.* **12**, 1–57.

Milnor, J. [1985]. On the concept of attractor. *Comm. Math. Phys.* **99**, 177–195.

Milnor, J. [1990]. Dynamics in one complex variable: Introductory lectures. SUNY Preprint #1990/5.

Milnor, J. and Thurston, W [1988]. On iterated maps of the interval. In *Dynamical Systems* (Ed. J.C. Alexander), Lecture Notes in Math., Vol. 1342, 465–563. Springer-Verlag: New York.

Misiurewicz, M. [1980]. Strange attractors for the Lozi mappings. In *Nonlinear Dynamics* (Ed. R.H.G. Helleman), Ann. New York Acad. Sci., Vol. 357, 348–358.

Moon, F.C. [1987]. *Chaotic Vibrations.* John Wiley & Sons: New York.

Moore, R.E. [1979]. *Methods and Applications of Interval Analysis.* SIAM:Philadelphia.

Mora, L. and Viana, M. [1991]. Abundance of strange attractors. To appear in *Acta Math.*

Moser, J.K. [1973]. *Stable and Random Motion in Dynamical Systems.* Princeton University Press.

Newhouse, S., Palis, J., and Takens, F. [1983]. Bifurcations and stability of families of diffeomorphisms. *Publ. Math. IHES* **57**, 5–71.

Oka, H. [1992]. Singular perturbations and heteroclinic bifurcations for certain types of ordinary differential equations. To appear in *Dynamical Systems* (Eds. R. Bamón, R. Labarca, J. Lewowicz and J. Palis, Jr), Pitman Research Notes in Math. Longman Scientific & Technical.

Oseledec, V.I. [1968]. A multiplicative ergodic theorem: Lyapunov characteristic numbers for dynamical systems. *Trans. Moscow Math. Soc.* **19**, 197–231. Russian original: *Trudy Moskov. Mat. Obshch.* **19**, 179–210.

Palis, J. [1978]. A different invariant of topological conjugacies and moduli of stability. *Astérisque* **51**, 335–346.

Palis, J. and de Melo, W. [1982]. *Geometric Theory of Dynamical Systems.* Springer-Verlag: New York.

Palis, J. and Takens, F. [1977]. Topological equivalence of normally hyperbolic dynamical systems. *Topology* **16**, 335–345.

Palis, J. and Takens, F. [1987]. Hyperbolicity and the creation of homoclinic orbits. *Ann. Math.* **125**, 337–374.

Palis, J. and Takens, F. [1993]. *Hyperbolicity and Sensitive Chaotic Dynamics at Homoclinic Bifurcations.* To appear from Cambridge University Press.

Palmer, K.J. [1984]. Exponential dichotomies and transversal homoclinic points. *J. Diff. Eq.* **55**, 225–256.

Palmer, K.J. [1986]. Transversal heteroclinic points and Cherry's example of nonintegrable Hamiltonian system. *J. Diff. Eq.* **55**, 225–256.

Palmer, K.J. [1988]. Exponential dichotomies, the shadowing lemma and transversal homoclinic points. In *Dynamics Reported* (Eds. U. Kirchgraber and H.O. Walther), Vol. 1, 265–306. John Wiley & Sons: New York.

Peinke, J., Muhlbach, A., Rohricht, B., Wessely, B., Mannhart, J., Parisi, J., and Huebener, R.P. [1986]. Chaos and hyperchaos in the electric avalanche breakdown of p-germanium at 4.2K, *Physica D* **23**, 176–180.

Perko, L. [1991]. *Differential Equations and Dynamical Systems.* Springer-Verlag: New York.

Pesin, Ya. B. [1992]. Dynamical systems with generalized hyperbolic attractors: hyperbolic, ergodic and topological properties. *Ergod. Th. and Dynam. Sys.* **12**, 123–151.

Poincaré, H. [1899]. *Les Méthodes Nouvelles de la Mécaniques Celeste.* Gauthier-Villars: Paris.

Pommeau, Y. and Manneville, P. [1980]. Intermittent transition to turbulence in dissipative dynamical systems. *Comm. Math. Phys.* **74**, 189–197.

Pugh, C. [1969]. On a theorem of P. Hartman. *Amer. J. Math.* **91**, 363–367.

Rand, D. [1978]. The topological classification of Lorenz attractors. *Math. Proc. Camb. Phil. Soc.* **83**, 451–460.

Rand, R.H. and Armbruster, D. [1987]. *Perturbation Methods, Bifurcation Theory and Computer Algebra.* Springer-Verlag: New York.

Robinson, C. [1981]. Differentiability of the stable foliation of the model Lorenz equations. In *Dynamical Systems and Turbulence* (Eds. D. Rand and L.-S. Young), Lecture Notes in Math., Vol. 898, 302–315. Springer-Verlag: New York.

Robinson, C. [1984]. Transitivity and invariant measures for the geometric model of the Lorenz equations. *Ergod. Th. and Dynam. Sys.* **4**, 605–511. Errata: *ibid.* **6**, 323.

Robinson, C. [1988]. Horseshoes for autonomous Hamiltonian systems using the Melnikov integral. In *Charles Conley Memorial Volume* (Eds. J. Moser, M.R. Herman, R. McGehee and E. Zehnder), special issue of *Ergod. Th. and Dynam. Sys.* **8**, 395–409. Cambridge University Press.

Robinson, C. [1989]. Homoclinic bifurcation to a transitive attractor of Lorenz type. *Nonlinearity* **2**, 495–518.

Robinson, C. [1992]. Homoclinic bifurcation to a transitive attractor of Lorenz type, II. *SIAM J. Math. Anal.* **23**, 1255–1268.

Rodriguez, J.A. [1986]. Bifurcations to homoclinic connections of the focus-saddle type. *Arch. Rat. Mech. Anal.* **93**, 81–90.

Rollins, R.W. and Hunt, E.R. [1982]. Exactly solvable model of a physical system exhibiting universal chaotic behavior. *Phys. Rev. Lett.* **49**, 1295–1298.

Rössler, O.E. [1976]. An equation for continuous chaos. *Phys. Lett. A* **57**, 397–398.

Rössler, O.E. [1979a]. Chaotic oscillations–An example of hyperchaos. In *Nonlinear Oscillations in Biology* (Ed. C. Hoppensteadt), Lectures in Applied Mathematics, Vol. 17, 141–156. AMS: Providence.

Rössler, O.E. [1979b]. Continuous chaos–Four prototype equations. In *Bifurcation Theory and Applications in Scientific Disciplines* (Ed. O. Gurel and O.E. Rössler), Ann. New York Acad. Sci., Vol. 316, 376–392.

Roussarie, R. [1989]. Cyclicité finie des lacets et des points cuspidaux. *Nonlinearity* **2**, 73–117.

Roussarie, R. [1992]. Desingularization of unfoldings of cuspidal loops. In *Geometry and Analysis in Nonlinear Dynamics* (Eds. H.W. Broer and F. Takens), Pitman Research Notes in Math., Vol. 222, 41–55. Longman Scientific & Technical.

Routh, E.J. [1905]. *Dynamics of a System of Rigid Bodies.* Macmillan: London

Ruelle, D. and Takens, F. [1971]. On the nature of turbulence. *Comm. Math. Phys.* **20**, 167–192; *ibid.* **23**, 343–344.

Rychlik, M.R. [1983]. Mesures invariantes et principe variationnel pour les applications de Lozi. *C. R. Acad. Sci. Paris* **296**, 19–22.

Rychlik, M.R. [1988]. Another proof of Jakobson's theorem and related results. *Ergod. Th. and Dynam. Sys.* **8**, 93–109.

Rychlik, M.R. [1990]. Lorenz attractors through Šil'nikov-type bifurcation, Part I. *Ergod. Th. and Dynam. Sys.* **10**, 793–821.

Sakamoto, K. [1990]. Invariant manifolds in singular perturbation problems for ordinary differential equations. *Proc. Royal Soc. Edinburgh* **116A**, 45–78.

Sanders, J.A. [1982]. Melnikov's method and averaging. *Celest. Mech.* **28**, 171–181.

Sanders, J.A. [1991]. On the computation of normal forms. In *Computational Aspects of Lie Group Representations and Related Topics* (Ed. A.M. Cohen), CWI Tract, Vol. 84, 129–142. CWI: Amsterdam.

Sannami, A. [1987]. On the structure of the parameter space of the Hénon family. In *Dynamical Systems and Applications* (Ed. N. Aoki), Advanced Series in Dynamical Systems, Vol. 5, 143–157. World Scientific: Singapore.

Sannami, A. [1989]. A topological classification of the periodic orbits of the Hénon family. *Japan J. Appl. Math.* **6**, 291–330.

Sano, M. and Sawada, Y. [1983]. Transition from quasi-periodicity to chaos in a system of coupled nonlinear oscillator. *Phys. Lett. A* **97**, 73–76.

Šarkovskii, A.N. [1964]. Coexistence of cycles of a continuous map of the line into itself (in Russian). *Ukrain. Math. Zh.* **16**, 61–71.

Schecter, S. [1987]. The saddle-node separatrix-loop bifurcation. *SIAM J. Math. Anal.* **18**, 1142–1156.

Schecter, S. [1990]. Simultaneous equilibrium and heteroclinic bifurcation of planar vector fields via the Melnikov integral. *Nonlinearity* **3**, 79–99.

Scheurle and Marsden [1984]. Bifurcation to quasi-periodic tori in the interaction of steady state and Hopf bifurcations. *SIAM J. Math. Anal.* **15**, 1055–1074.

Shibayama, K. [1989]. Connections of periodic orbits in the parameter space of the Lozi family. In *The Study of Dynamical Systems* (Ed. N. Aoki), Advanced Series in Dynamical Systems, Vol. 7, 10–25. World Scientific: Singapore.

Shil'nikov, L.P. [1965]. A case of the existence of a countable set of periodic motions. *Sov. Math. Dokl.* **6** (1965), 163–166. Russian original: *Dokl. Acad. Nauk. SSSR* **160**, 558–561.

Shil'nikov, L.P. [1967a]. The existence of a denumerable set of periodic motions in a four-dimensional space in an extended neighborhood of a saddle-focus. *Sov. Math. Dokl.* **8** (1967), 54–58. Russian original: *Dokl. Acad. Nauk. SSSR* **172**, 54–57.

Shil'nikov, L.P. [1967b]. On a Poincaré-Birkhoff problem. *Math. USSR Sbornik* **3** (1967), 353–371. Russian original: *Mat. Sbornik* **74**, 378–397.

Shil'nikov, L.P. [1968]. On the generation of a periodic motion from trajectories doubly asymptotic to an equilibrium state of saddle type. *Math. USSR Sbornik* **6** (1968), 427–472. Russian original: *Mat. Sbornik* **77**, 461–103.

Shil'nikov, L.P. [1970]. A contribution to the problem of the structure of an extended neighborhood of a rough equilibrium state of saddle-focus type. *Math. USSR Sbornik* **10** (1970), 91–102. Russian original: *Mat. Sbornik* **81**, 92–103.

Shimada, I. and Nagashima, T. [1979] A numerical approach to ergodic problem of dissipative dynamical systems. *Progr. Theor. Phys.* **61**, 1605–1616.

Sijbrand, J. [1985]. Properties of center manifolds. *Trans. Amer. Math. Soc.* **289**, 431–469.

Smale, S. [1963]. Diffeomorphisms with infinitely many periodic points. In *Differential and Combinatorial Topology* (Ed. S.S. Chern), 63–80. Princeton University Press.

Smale, S. [1967]. Differentiable dynamical systems. *Bull. Amer. Math. Soc.* **73**, 747–817.

Sparrow, C.T. [1981]. Chaos in a three-dimensional single loop feedback system with a piecewise-linear feedback function. *J. Math. Anal. Appl.* **83**, 275–291.

Sparrow, C.T. [1982]. *The Lorenz equation: Bifurcations, Chaos and Strange Attractors.* Springer-Verlag: New York.

Sternberg, S. [1958]. On the structure of local homeomorphisms of Euclidean n-space, II. *Amer. J. Math.* **80**, 623–631.

Stowe, D. [1986]. Linearization in two dimension. *J. Diff. Eq.* **63**, 183–226.

Szmolyan, P. [1991]. Transversal heteroclinic and homoclinic orbits in singular perturbation problems. *J. Diff. Eq.* **92**, 252–281.

Takens, F. [1971]. Partially hyperbolic fixed points. *Topology* **10**, 133–147.

Takens, F. [1973a]. Normal forms for certain singularities of vector fields. *Ann. Inst. Fourier* **23**, 163–195.

Takens, F. [1973b]. A nonstabilizable jet of a singularity of a vector field. In *Dynamical Systems* (Ed. M.M. Peixoto), 583–597. Academic Press: New York.

Takens, F. [1973c]. Unfoldings of certain singularities of vector fields: Generalized Hopf bifurcations. *J. Diff. Eq.* **14**, 476–493.

Takens, F. [1974a]. Singularities of vector fields. *Publ. Math. IHES* **43**, 47–100.

Takens, F. [1974b]. Forced oscillations and bifurcations. In *Applications of Global Analysis, I.*, 1–57. Rijksuniversiteit Utrecht.

Takens, F. [1988]. Book review of R.L.Devaney: *An Introduction to Chaotic Dynamical Systems. Acta Appl. Math.* **13**, 221–226.

Tanaka, S., Higuchi, S., and Matsumoto, T. [1993]. A global two-parameter bifurcation analysis of a driven R-L-Diode circuit. In press.

Tanaka, S., Matsumoto, T., and Chua, L.O. [1985]. Bifurcations in a driven R-L-Diode circuit. *Proc. IEEE Int. Symp. on Circuits and Systems*, 851–854.

Tanaka, S., Matsumoto, T., and Chua, L.O. [1987]. Bifurcation scenario in a driven R-L-Diode circuit. *Physica* **28D**, 317–344.

Testa, J., Perez, J., and Jeffries, C. [1982]. Evidence for universal chaotic behavior of a driven nonlinear oscillator. *Phys. Rev. Lett.* **48**, 714–717.

Thom, R. [1972]. *Stabilité Structurelle et Morphogénèse.* 2ème Edition 1977. Benjamin: New York.

Thompson, J.M.T. and Stewart, H.B. [1986] *Nonlinear Dynamics and Chaos.* John Wiley & Sons: New York.

Togawa, Y. [1987]. A modulus of 3-dimensional vector fields. *Ergod. Th. and Dynam. Sys.* **7**, 295–301.

Tokunaga, R., Matsumoto, T., Ida, T., and Miya, K. [1989]. Homoclinic linkage in the Double Scroll circuit and the cusp-constrained circuit. In *The Study of Dynamical Systems* (Ed. N. Aoki), Advanced Series in Dynamical Systems, Vol. 7, 192–209. World Scientific: Singapore.

Tresser, C. [1984]. About some theorems by L.P. Shil'nikov. *Ann. Inst. H. Poincaré: Physique Théorique* **40**, 441–461.

Tsujii, M. [1992]. A proof of Benedicks-Carleson-Jakobson theorem. Preprint.

Tsujii, M. [1993]. Positive Lyapunov exponents in families of one dimensional dynamical systems. *Invent. Math.* **111**, 113–137.

Ueda, Y. [1980]. Steady motions exhibited by Duffing's equation: a picture book of regular and chaotic motions. In *New Approaches to Nonlinear Problems in Dynamics* (Ed. P.J. Holmes), 311–322. SIAM: Philadelphia.

Uehleke, B. [1982]. Chaos in einem stuckweise linearen System: Analytische Resulte. Ph.D. thesis. Univ. Tübingen.

Uehleke, B. and Rössler, O.E. [1984]. Analytical results on a chaotic piecewise-linear O.D.E. *Z. Naturforsch.* **39a**, 342–348.

Ushiki, S. [1984]. Normal forms for singularities of vector fields. *Japan J. Appl. Math.* **1**, 1–37.

Ushiki, S. [1989]. Arnold's tongues and swallow's tails in complex parameter spaces. In *Stability Theory and Related Topics in Dynamical Systems* (Ed. K. Shiraiwa and G. Ikegami), Advanced Series in Dynamical Systems, Vol. 6, 153–178. World Scientific: Singapore.

Ushiki, S., Oka, H., and Kokubu, H. [1984]. Existence d'attracteurs étranges dans le déploiement d'une singularité dégénérée d'un champ de vecteurs invariant par translation. *C. R. Acad. Sci. Paris* **298**, 39–42.

Vanderbauwhede, A. [1989]. Center manifolds, normal forms and elementary bifurcations. In *Dynamics Reported* (Eds. U. Kirchgraber and H.O. Walther), Vol. 2, 89–169. John Wiley & Sons: New York.

Vanderbauwhede, A. and Fiedler, B. [1992]. Homoclinic period blow-up in reversible and conservative systems. *Z. Angeb. Math. Phys.* **44**, 292–318.

Vanderbauwhede, A. and van Gils, S.A. [1987]. Center manifolds and contractions on a scale of Banach spaces. *J. Funct. Anal.* **72**, 209–224.

Van der Pol, B. [1960]. *Selected Scientific Papers.* North-Holland: Amsterdam.

Van der Pol, B. and Van der Mark, J. [1927]. Frequency demultiplication. *Nature* **120**, 363–364.

van Gils, S.A. [1985]. A note on "Abelian Integrals and Bifurcation Theory". *J. Diff. Eq.* **59**, 437–441.

van Strien, S.J. [1979]. Center manifolds are not C^∞. *Math. Z.* **166**, 143–145.

van Strien, S.J. [1982]. One parameter families of vector fields, bifurcations near saddle-connections. Thesis. Rijksuniversiteit Utrecht.

Whitley, D. [1983]. Discrete dynamical systems in dimensions one and two. *Bull. London Math. Soc.* **15**, 177–217.

Wiggins, S. [1988]. *Global Bifurcations and Chaos.* Springer-Verlag: New York.

Williams, R.F. [1979]. The structure of Lorenz attractors. *Publ. Math. IHES* **50**, 321–347.

Yanagida, E. [1987]. Branching of double pulse solutions from single pulse solutions in nerve axon equations. *J. Diff. Eq.* **66**, 243–262.

Yang, E.S. [1987]. *Fundamentals of Semiconductor Devices.* World Scientific: Singapore.

Ye, Y.-Q. [1986]. *Theory of Limit Cycles.* Translations of Mathematical Monographs, Vol. 66. AMS: Providence.

Yoccoz, J.-C. [1991] Polynômes quadratiques et attracteur de Hénon. *Astérisque* **201-202-203**, 143–165.

Yoon, T.H., Song, J.W., Shin, S.Y., and Ra, J.W. [1984]. One-dimensional map and its modification for periodic-chaotic sequence in a driven nonlinear oscillator. *Phys. Rev. A* **30**, 3347–3350.

Young, L.-S. [1983]. Entropy, Lyapunov exponents and Hausdorff dimension in differentiable dynamical systems. *IEEE Trans. Circuits and Systems* **CAS30**, 599–607.

Young, L.-S. [1985]. Bowen-Ruelle measures for certain piecewise hyperbolic maps. *Trans. Amer. Math. Soc.* **287**, 41–48.

Yu, P. and Huseyin, K. [1988]. Bifurcations associated with a three-fold zero eigenvalue. *Quart. Appl. Math.* **XLVI**, 193–216.

Żoładek, H. [1983]. On the versality of a family of symmetric vector fields in the plane. *Math. USSR Sbornik* **48**(1983), 463–492. Russian original: *Math. Sbornik* **120**, 437–499.

Żoładek, H. [1987]. Bifurcations of certain family of planar vector fields tangent to axes. *J. Diff. Eq.* **67**, 1–55.

Żoładek, H. [1991]. Abelian integrals in unfoldings of codimension 3 singular planar vector fields. In *Bifurcations of Planar Vector Fields.* Lecture Notes in Math., Vol. 1480, 165–224. Springer-Verlag: New York.

Index

Credits

(1) Plates 1.1, 1.2, Figs.1.2.4, 1.3.1, 1.4.3: Reprinted with permission from T. Matsumoto, L. O. Chua and M. Komuro, "Birth and death of the Double Scroll", *Physica* **24D**, 97–124. 1986 ©Elsevier Science Publishers.

(2) Plates 1.3-1.5, Figs.1.3.2-1.3.4: Reprinted with permission from T. Matsumoto, L. O. Chua and M. Komuro, "The Double Scroll", *IEEE Trans. Circuits and Systems* **32**, 117–146, 1986. ©1986 IEEE.

(3) Figs. 1.4.1, 1.4.2, 1.4.4-1.4.7: Adapted with permission from L. O. Chua, M. Komuro and T. Matsumoto, "The Double Scroll family", *IEEE Trans. Circuits and Systems* **33**, 1073–1118, 1986. ©1986 IEEE.

(4) Plates 1.11-1.14,Cover Picture, Figs.1.5.1-1.5.8, 1.5.10-1.5.13, 1.5.15-1.5.19, Adapted with permission from M. Komuro, R. Tokunaga, T. Matsumoto, L. O. Chua and A. Hotta, "A global bifurcation analysis of the Double Scroll circuit", *Int. J. Bifurcations and Chaos* **1**, 139–182, 1991. ©1991 World Scientific Publishing Company.

(5) Plates 1.15, 1.16, Figs.1.6.1-1.6.9: Reprinted with permission from T. Matsumoto, L. O. Chua and R. Tokunaga, "Chaos via torus breakdown", *IEEE Trans. Circuits and Systems* **34**, 240–253, 1987. ©1987 IEEE.

(6) Plates 1.11-1.13, Figs.1.7.1, 1.7.2: Reprinted with permission from T. Matsumoto, L. O. Chua and K. Kobayashi, "Hyperchaos: Laboratory experiment and numerical confirmation", *IEEE Trans. Circuits and Systems* **33**, 1143–1147, 1986. ©1986 IEEE.

(7) Plates 1.23, 1.24, Figs.1.9.1-1.9.6: Reprinted with permission from S. Tanaka, T. Matsumoto and L. O. Chua, "Bifurcation scenario in a driven R-L-Diode circuit", *Physica* **28D**, 317–344, 1987. ©Elsevier Science Publishers.

(8) Plates 1.25, 1.26, Figs 1.9.7-1.9.11:Reprinted with permission from S. Tanaka, T. Matsumoto, J. Noguchi and L. O. Chua "Multi-folding: alternative appearance of period-one attractors and chaotic attractors in a driven R-L-diode circuit", *Phys. Lett. A* **157**, 37–43, 1991. ©1991 Elsevier Science Publishers.

(9) Figs.1.8.1: Reprinted with permission from B.Van der Pol and J.Van der Mark, "Frequency demultiplication", *Nature* **120**, 363–364, 1927. ©Macmilan Magazines Ltd.

(10) Figs. 2.4.1-2.4.11: Reprinted with permission from M.Komuro, "Normal Form of Continuos Piecewise Linear Vector Fields and Chaotic Attractors Part II", *Japan J.Appl. Math.*, **5**, 503-549, 1988. ©The JJAM publishing Committee.

(11) Figs. 2.5.1, 2.6.1-2.6.10: Reprinted with permission from M.Komuro, "Normal Form of Continuos Piecewise-Linear Vector Fields", *Japan J.Ind.Appl. Math.*, **9**, 503-549, 1988. ©The JJAM publishing Committee.